Handbook of Energy Conservation for Mechanical Systems in Buildings

About the Editor . . .

Robert W. Roose, P.E., has made many contributions to the industry in which he has devoted his entire professional career. He is a member of the American Society of Heating, Refrigerating and Air-Conditioning Engineers, and has been listed in *Who's Who in Engineering* and *Who's Who in the Midwest*. Mr. Roose was graduated from the University of Illinois, receiving the B.S. and M.S. degrees in mechanical engineering, and is a member of the honorary mechanical engineering fraternity, Pi Tau Sigma. He has served on many industry committees concerned with setting standards of equipment performance and with developing design and installation procedures. Mr. Roose also served as Editor, Editorial Director, and then Publisher of *Heating/Piping/Air Conditioning*, and prior to that as Research Assistant Professor of Mechanical Engineering at the University of Illinois, where he directed research studies sponsored by the industry.

Handbook of Energy Conservation for Mechanical Systems in Buildings

Compiled and Edited by

Robert W. Roose, P.E.

Senior Editor
Heating/Piping/Air Conditioning
Chicago, Illinois

A compilation of articles and information published previously in *Heating/Piping/Air Conditioning*, the Penton/IPC magazine for mechanical systems engineering and energy management.

VNR **VAN NOSTRAND REINHOLD COMPANY**
NEW YORK   CINCINNATI   ATLANTA   DALLAS   SAN FRANCISCO
LONDON   TORONTO   MELBOURNE

Van Nostrand Reinhold Company Regional Offices:
New York   Cincinnati   Atlanta   Dallas   San Francisco

Van Nostrand Reinhold Company International Offices:
London   Toronto   Melbourne

Copyright © 1978 by Litton Educational Publishing, Inc.

Library of Congress Catalog Card Number: 77-11027
ISBN: 0-442-27012-7

Manufactured in the United States of America

Published by Van Nostrand Reinhold Company
450 West 33rd Street, New York, N.Y. 10001

Published simultaneously in Canada by Van Nostrand Reinhold Ltd.

15 14 13 12 11 10 9 8 7 6 5 4 3 2

Library of Congress Cataloging in Publication Data
Main entry under title:

Handbook of energy conservation for mechanical systems
   in buildings.

   Includes index.
   1. Buildings—Energy conservation.   I. Roose,
Robert Welburne
TJ163.5.B84H36   696   77-11027
ISBN 0-442-27012-7

Introduction

The need for energy conservation is a real challenge, worthy of everyone's attention. Utility rates are continuing to climb, natural gas rates are projected to increase, and fuel oil prices will undoubtedly go up. Where these costs will jump to is anyone's guess. Energy costs are not only soaring, causing the waste of energy to become increasingly expensive, but it is also contributing to our national predicament.

In 1971, the United States consumed energy at the rate of 69 quadrillion Btu's per year, as reported by the U.S. Department of Interior. By 1980, this figure is conservatively estimated to increase to 96 quadrillion Btu's, a 39 percent increase, according to the Federal Energy Administration. By 1990, energy consumption is expected to be 140 quadrillion Btu's per year, more than double the energy consumption in 1971. Energy devoted to electrical generation alone is expected to triple by 1990. At the same time, the nation's limited fuel resources are being depleted, and the United States is becoming increasingly dependent upon foreign oil and gas supplies.

Through 1950, the United States was totally self-sufficient regarding its energy needs; it easily met its growing needs for energy with cheap and abundant domestic fuels such as coal, oil, gas, and hydroelectric power. By 1960, imports of crude oil and petroleum products accounted for 15 percent of the total domestic petroleum consumption, and by 1976 imports had jumped to over 40 percent.

The United States now relies on oil for over 45 percent of its energy and on natural gas for over 30 percent, while coal provides only about 15 percent of our consumption; these sources represent between 90 and 95 percent of our total domestic energy resources. Coal is the most abundant fossil fuel in the United States. Proven reserves are estimated at over 660 billion tons, which is equal to more than 800 years of consumption at the 1975 level, according to the Committee on Technology for the U.S. House of Representatives. However, at the current level of consumption, we have only about 10 years of proven oil and gas reserves.

Although conservation of energy must also address the issue of supply, the primary emphasis in this book is on the demand and use side. Therefore, it is with this purpose that this book has been assembled—to aid in the application of the principles and practices of proven energy conservation which are described herein by 44 experienced engineers. Every possible known energy conservation method in the design, installation, operation, and maintenance of the heating, piping, air conditioning, and ventilating systems for all types of buildings is included in the chapters of this book.

Based upon a recent survey conducted among the readers of *Heating/Piping/Air Conditioning* magazine, we know that on the average, a 10 year old building wastes 40 percent of the energy delivered to that building, as verified by the National Bureau of Standards. Also, it is known that over 50 percent of the existing industrial, commercial, institutional, and public buildings are over 10 years old. Of these buildings, over 100,000 are of the industrial type and over 400,000 are of the commercial, institutional, and governmental type representing over 4,000,000 sq ft of space. To further confirm these findings, the Mechanical Contractors Association of America estimates that these buildings, on the average, waste between 40 and 50 percent of the energy consumed in the existing commercial, institutional, industrial, and governmental buildings. Over 73 percent of the engineers and mechanical contractor readers of HPAC have indicated they will be implementing energy conservation programs in the very near future with a goal of saving more than 10 percent of the total fuel consumption for these energy uses.

It is estimated that there are about 100,000 old, inefficient boilers and approximately 200,000 out-dated burners in these commercial, institutional, and governmental buildings and millions of miles of uninsulated piping; all of these were installed when energy was cheap and thought to be plentiful and inexhaustible. Much of this equipment operates at 60 to 70 percent efficiency and consumes millions of gallons of oil and millions of cubic feet of natural gas more than is necessary. By renovating them and by coupling this with other measures, such as the application of insulation, installing solar heating panels and high efficiency air conditioners, the total energy consumption can be reduced by 20 percent or more. Continual care in maintenance and testing, adjusting, and balancing of the mechanical systems will also help in conserving energy as outlined in the four chapters in Section V.

Cooling systems represent 42 percent of the nation's energy needs for building operations during the summer. The National Bureau of Standards research indicates that energy requirements for cooling can be reduced 30 percent with very little sacrifice of comfort.

The application of the American Society of Heating, Refrigerating and Air-Conditioning Engineers' Standard 90-75 in the design of these mechanical systems can reduce energy consumption in new buildings by 27 percent, if adopted in all states, without escalating construction costs. Savings would range from 9 to 15 percent in single family homes and from 30 to 45 percent in commercial structures. These estimates were based upon an objective study of the application of ASHRAE Standard 90-75 conducted by the Arthur D. Little organization commissioned by the Federal Energy Administration. An additional section to this important standard which will evaluate all energy resources as delivered to the building site is expected to be adopted during 1977. This section is indeed necessary if the entire subject of energy conservation is to be effective. The implementation of this standard is, therefore, a measure that will positively aid in achieving conservation of energy by outlining design goals and limitations. As states and communities adopt this standard in their building codes, definite results will be noticed. See Chapters 17, 34, 50, and 63 for direct indications of how ASHRAE Standard 90-75 will reduce energy consumption.

Energy consumption for illumination, according to the National Bureau of Standards, can be effectively reduced by 15 percent in most existing buildings by turning off lights when not needed, paying more attention to matching the amount of lighting used to do jobs being done, and using lights that require less energy.

Each building is different and unique within itself, and therefore energy savings can only be achieved through a combination of various design and operation changes. While it is possible to implement energy conservation measures in new buildings by using correct construction methods at little or no additional cost, most modifications to existing buildings will require an additional investment to achieve the desired results. The cost of such an investment will naturally vary with the type of building, but in most cases, significant results in annual energy savings can be accomplished. Such savings should be considered in terms of their present value to determine their true worth and the payback periods over the life of the buildings.

Some rules of thumb to remember as you read over the contents of this handbook are listed below:

1. A $\frac{1}{8}$ in. diameter steam leak in a 100 psi line will cause the loss of 50,000 lb of steam per month.
2. A $\frac{1}{32}$ in. diameter leak in a water line at 40 psi will cause a waste of about 4800 gal. per month.
3. A leaky hot water faucet could cause a loss of 3280 gal. of heated water per year.
4. A temperature change of 1°F in a duct system represents a loss of 108 Btu's per each 100 cfm of air circulating through the system.
5. A 1 hp motor will consume 2900 kwh per year operating 8 hr per day.
6. An increase of 0.5 in. WG in air system pressure drop above a system pressure drop of 2 in. WG requires an increase of 25 percent in fan horsepower.

ROBERT W. ROOSE, PE

Acknowledgments

The information in this *Handbook* has been obtained from a wide variety of authorities who are specialists in their respective fields of energy conservation in both new and existing buildings. I wish to express my appreciation to each of them for their cooperation in permitting this information to be included. Many of the chapters in this *Handbook* were originally presented at the special series of seminars on the subject "How to Save Energy in Existing Buildings" developed by *Heating/Piping/Air Conditioning*, published by Reinhold Publishing Company, a subsidiary of Penton/IPC, Inc. Also included are many additional chapters from the magazine relating to the subject of energy conservation which makes this valuable reference worthy of being called a *Handbook*.

I would like to express special appreciation to Robert J. Osborn, Publisher of *Heating/Piping/Air Conditioning*, for his cooperation and encouragement in the publication of this *Handbook*.

There is no doubt that there is the need for conserving energy today. It is hoped that this *Handbook*, with its emphasis on the need for energy conservation, will prove to be a valuable tool in the hands of its users.

ROBERT W. ROOSE, P.E.

Contents

Handbook of Energy Conservation for Mechanical Systems in Buildings

SECTION I

The need for energy conservation

The seriousness of the energy shortage problems facing the world as outlined in these 11 chapters includes discussions of the following:

- Detailed definitions of energy.
- How energy is transported.
- The need for understanding energy economics.
- The urgent need for energy conservation.
- How much energy is consumed in the manufacture and construction of building materials for a typical college building.
- The importance of energy conservation in heating, ventilating, and air conditioning systems.
- Detailed discussion of the energy effectiveness factor as related to the application of mechanical systems.

1
The energy scene: a can of worms

The energy problem is a serious one, yet it has a low priority with the public. A nonpartisan approach is required to make headway in evolving overdue solutions to this complex problem.

T. E. PANNKOKE, PE,
Engineering Editor, Heating/Piping/
Air Conditioning, Chicago, Illinois

Of all the problems the United States has faced in its 200 year history, the energy problem is potentially the most serious and disruptive. Virtually every facet of our lives is affected by the supply and cost of energy. These facets include:

- Agriculture
- Commerce
- Environment
- Housing
- Manufacturing
- National defense
- Transportation
- Recreation

A significant aspect of the energy situation is that energy is required to solve other problems. Energy is needed to prevent or alleviate the effects of pollution, rehabilitate urban areas that have fallen into decay, and restore blighted land so that future needs of an expanding population may be met.

Energy is required to obtain and utilize natural resources. No matter how much mineral wealth lies in the ground or in scrap piles, use of the earth's "capital" is denied

Fig. 1-1.

us unless energy is available for extracting, recycling, etc.

The energy problem is therefore part of an overall problem of how to best use, conserve, and preserve our varied resources— land, air, water, and mineral. The earth, large as it seemingly is, is a finite body. Many resources are still available in vast quantities. Rising population, however, means in effect that there is less for everyone; and as the population increases, the reserves of some resources are becoming depleted. The obvious answer is that conservation, recycling, better management of resources, and more careful application of technology with respect to the environment will have to become a way of life. Otherwise, the already difficult and unstable social, economic, and political situations in the world will further deteriorate in the face of competition for many resources.

The United States has weathered many crises. This provides a good basis for optimism, but not for complacency. The so-called prophets of doom still have not been vindicated. But there is growing evidence that the way we have applied our technology has had serious environmental consequences, and we will probably learn of more as time goes on.

Do we see the problem?

In a democratic society, a sufficient portion of the population must agree that a problem exists and that it is of major concern before it can be dealt with effectively.

Humans tend to be crisis motivated creatures. If we are personally affected, it is easy to get our attention. Examples include the 1929 depression and World War II. The public accepted the changes brought about by the New Deal, and they switched from an isolationist mood after the attack on Pearl Harbor. In both cases, people responded because they perceived a threat to their security.

The energy crisis began to receive broad coverage in 1970. At first, it was a remote thing. Lights worked, buildings continued to be heated and cooled (all too often wastefully), and gasoline was plentiful.

The oil embargo of 1973 got nearly everyone's attention. A few states initiated simple gasoline rationing programs, and the federal government set up machinery for nationwide rationing. Prices of oil and gasoline began to rise. A nationwide 55 mph speed limit was established, and we were told to lower and raise winter and summer thermostat settings (this saved energy in some buildings, wasted it in others).

The 55 mph speed limit has been ignored more than enforced. Much nonessential lighting, such as advertising signs and decorative lighting, continues to burn far into the night, sometimes all night. Society is still being conditioned to purchase consumable type products, many of them made of plastic derived from oil.

The Gallup Poll shows that as of early March 1975,[1] only 7 percent of the people polled felt that energy was the number one problem. This represents an improvement over September 1974, when 2 percent said energy was their number one concern. Both times the number one problem was high living costs and inflation—81 percent in September and 60 percent in March. Unemployment was named by 20 percent in March. Crime and dissatisfaction with government also compete for attention with the energy crisis.

Clearly, what is happening to us *now* (rather than what might happen to us in years to come) commands most of our attention. Also, many contradictory views on the energy problem have been widely disseminated, which adds confusion to the complexity. Finally, some people feel that energy companies have either created the shortages or at least are manipulating them for their own gain.

Problems such as the 1929 depression and World War II stayed before us in our day-to-day lives. They commanded most of everyone's attention. The energy situation does not.

Although we have had a taste of what fuel shortages can mean, it may well seem incomprehensible to many that the life style we enjoy could be seriously or permanently affected. Recognizing such a happening, however, may help us to avoid it, or at least lead to an orderly transition to a more modest, yet adequate living standard.

Table 1-1 shows a breakdown of the U.S. population by age groups as of July 1974.[2] Well over half the population—58.5 percent—were born during or after 1940. Approximately 11 percent were born between

[1]Superior numerals indicate references at end of this chapter.

Table 1-1. Age structure of U.S. population shows that more than two-thirds were born during or after 1930.* The life experience of this group has been in an affluent society that has grown accustomed to heavy use of the world's resources.

Age, yr	Population as of 7/1/74, thousands	Percent distribution	Percent born in or after given year	
All	211,909	100		
Under 5	16,304	7.7	1969	7.7
5–13	34,082	16.1	1961	23.8
14–17	16,878	8.0	1957	31.8
18–24	26,908	12.7	1950	44.5
25–34	29,770	14.0	1940	58.5
35–44	22,823	10.8	1930	69.3
45–54	23,821	11.2	1920	80.5
55–64	19,507	9.2	1910	89.7
65 plus	21,815	10.3		

*Adapted from Table A, Reference 2.

1930 and 1939, and for most of these, memories of the depression are vague or nonexistent. Thus, the primary living experience of more than two-thirds of our population has occurred basically in a period of prosperity. They have been exposed to massive advertising, via television and other media, of a bewildering array of products. They have lived in a society that has consumed the world's resources at a staggering rate. They have been conditioned since childhood to a world of plenty in which the continuous purchase, use, and discarding of goods is an accepted way of life. These habits will be difficult to change.

The future becomes now

In 1948, a book was published that clearly foresaw problems we now face.[3] Some of the points that author made then included:

● *We are a wasteful people.* Certainly all we throw away each year attests to that.

● *We waste thousands of tons of coal each year by allowing millions of electric lights to burn needlessly.* Lately we have made a modest effort to turn out some of them.

● *We are probably most wasteful of gasoline.* Today we talk about conserving oil

and gasoline but cannot agree on how to go about it.

● *We might go to war to obtain resources.* The author used tin as an example and was quite disturbed that reclamation of tin instituted during WW II was no longer being practiced. As of 1973, the United States was using about 30 percent of the world's tin production. Neglecting tin stockpiled for strategic purposes, U.S. tin reserves, both hypothetical and speculative, would furnish less than a four-year supply.[4] We import many such materials, some because our reserves are low, others because of economic considerations. Chromium is a material the U.S. has little of, based on current knowledge. In recent years, the USSR has been a major supplier of chromium to this country.

And although we have not heard further talk of going to war for tin, we did hear discussions after the oil embargo of 1973 of taking "extreme measures" if necessary to gain access to oil.

Another book, published in 1972, presented a timetable for an energy crisis.[5] It included the following:

● Curtailed heating and cooling of shopping centers, department stores, theaters, and restaurants in the early 1970s.

● Curtailed heating and cooling of public buildings, schools, houses, and hospitals by the late 1970s.

● Gasoline rationing by the late 1970s, or in the 1980s.

Within a year of publication, as a result of the oil embargo, the federal government requested reduced heating and cooling of buildings, instituted the nationwide 55 mph speed limit, and made plans for gasoline rationing. A future hypothesized in 1972 was becoming reality in 1973.

Some other occurrences suggested in the book were a major depression and world conflict over energy resources, both slotted for the 1980s. And things don't get any better after that, either.

There is widespread agreement among those deeply involved in studying ways to cope with the problem of energy that from now to the end of the century is a critical period. It is critical because of our current dependence on fossil fuels and because of the time lag before alternate energy sources can be expected to make a major contribution. Solar energy today can make a limited contribution; fast breeder reactors and fusion power are down the road. Additional research and development, in a nonconducive economy, are required for all three of these energy sources.

On the other hand, there is marked disagreement among experts involved with various energy technologies. Some examples of the diversified opinions follow:

Early in 1975, 32 prominent scientists, 11 of them Nobel laureates, came out in favor of greatly increased use of both coal and uranium as the only feasible options— in addition to energy conservation—that we have for the remainder of this century.[6] After the views of these men were presented at a press conference in Washington, D.C., Ralph Nader released a copy of a letter to President Ford critical of his decision to speed up the licensing of nuclear plants. The gist of the letter was that greater attention had been given to building these plants quickly than safely. The letter was signed by eight scientists, five of them Nobel laureates.[7]

Others argue that economics may favor the construction of coal generating stations rather than nuclear ones.[8]

Another viewpoint is that even if the projected growth in energy demand to the year 2000 can be cut in half, it would still be impossible to build enough power plants to meet the reduced need. Among the reasons given are problems in obtaining the required capital for financing, problems in obtaining plant sites, and a shortage of skilled labor.[9]

Different views on complex issues are to be expected. The engineer, the scientist, the economist, etc., will each interpret facts and draw conclusions based on his experiences and background. But at some point in time, decisions must be made as to what are the best alternatives to follow. We are five years into the problem (measured from when it began to be discussed on a continuing basis), and there is little time to waste

in coming to grips with what we are going to do about energy and other resources, both in the immediate future and for the long haul. Social, economic, and political problems associated with energy must be solved as well.

Another worm in the can is that it will cost people money to practice energy conservation, even on an individual basis. Fixed costs of utility plants are ongoing regardless of how much or how little energy is distributed. If utilities have excess capacity, resulting from conservation by customers and/or from demand not increasing at an anticipated rate, they will need additional money to pay for facilities built to provide that capacity.

Conflicting interests

Solutions to the energy shortage will require that all resources be managed more effectively because energy use is synonymous with resource use.

Our current mode of living makes it difficult to accept the consequences involved in meaningful energy conservation efforts— consequences having social, economic, and political implications.

The immediate need is to conserve oil and natural gas and to reduce insofar as is practical our dependence on foreign sources of these fuels. Conservation of oil and gas can be achieved in several ways: voluntarily, semi-voluntarily, mandatorily, and by switching to coal where possible.

The voluntary option has already been initiated to some degree by business and industry, particularly in areas where shortages exist or have been threatened. Rationing or allocation of fuels can be implemented in the near future. Use of more coal entails, over the long run, opening of more mines, conversion of existing facilities from oil and gas, and the construction of large scale synthetic natural gas (SNG) plants and more power plants. Therefore, the impact of coal on the energy problem will be felt later than those of the first two options. Long lead times are required for the construction of power plants, both coal and nuclear.

Voluntary conservation entails such simple things as doing less driving—careful planning of shopping trips, for example. Other examples are turning off lights and appliances when not needed, or scheduling operations so that all areas of a plant can be shut down over weekends. Such actions are simply a matter of using one's head. Each one by itself won't make a big dent in the problem, but the sum of many small contributions will help.

Other forms of voluntary conservation require money to implement, but they can yield a considerable return on investment as well as conserve significant amounts of energy. Energy use in existing buildings is a prime example of where this is true. Estimates of possible savings of energy in existing commercial, institutional, and industrial buildings range between 30 and 50 percent. Attendance at the seminars "How To Save Energy in Existing Buildings," initiated by *Heating/Piping/Air Conditioning* (carried on later by ASHRAE), showed that there is considerable interest in the business community in saving energy. This is to be expected because energy is vital to economic activity.

Semi-voluntary conservation would set the price of fuel—gasoline has been the most frequently mentioned—at a level high enough to reduce consumption to a desired level. The user would decide whether he wanted to—or could—pay the higher price.

Involuntary conservation means rationing. Fuel use would be allocated in some manner. One method would permit a user to have a reduced amount based on a previous year's use; another would allocate sufficient fuel to areas deemed vital (agriculture, for example) and ration the rest.

Voluntary conservation has not been effective on an overall basis, and as long as the energy problem is not foremost in everyone's mind, voluntary cooperation will fail to have a major impact.

Raising the price of oil will generate more conservation. Commercial and industrial users can reduce their use more significantly than can the homeowner because more can be done with commercial type HVAC systems and industrial processes than can be accomplished with most residential heating systems. The more affluent could use more of the higher priced fuel than persons in lower income brackets, which in turn would increase the frustrations of people trying to make ends meet. Feelings of resentment and being ripped off would grow.

Rationing would create a host of problems. How do you treat everyone fairly? Some areas of our economy obviously are more important than others. Agriculture would have to receive high priority; so would transportation because food products must be shipped to food processors and retail outlets. Also, raw materials and goods must continuously be transported to keep the economy functioning. Giving "preferential treatment" to certain segments of the economy would certainly create economic problems and political repercussions. Owners, employees, and stockholders of enterprises receiving lower priorities would protest vehemently if their security and/or income were threatened because their economic activity was considered "by someone in Washington" less essential than another.

No simple, overall politically acceptable answer exists to the energy situation. Time is passing, and the unpleasant economic and political factors involved in taking meaningful steps to deal with energy use must be faced and dealt with.

Economic dislocation and some hardship will probably occur in sectors of the economy through efforts to conserve. Yet, intelligent planning can serve to reduce these negative impacts. The alternative is to wait until shortages become even worse and then deal with them using day-to-day improvised solutions. This course of action will result in far greater hardships than if a comprehensive, planned but flexible approach is followed.

A news item of February 5, 1973 stated that Rep. Wright Patman (D. Tex.), outgoing chairman of the Congressional Joint Committee on Defensive Production, expressed the conclusion that the country was moving too slowly in coping with the energy crisis.[10] In the area of oil and gas, the committee's 1972 report concluded that domestic supplies could not meet the additional energy requirements to 1985 and that the United States would be dependent on foreign sources for one-third to one-half of its oil requirements by 1985. Forty-nine of 56 nuclear power plants under construction were reported to have slipped behind schedule an average of 14.3 months each at that time.

A two-year study by the Ford Foundation concluded in 1974 that gasoline rationing should be adopted immediately as part of an overall approach to dealing with the energy situation.[11]

Nonpartisan energy policy

Many persons in and outside of government have been working on the energy problem, and progress is being made. One example is the development of Standard 90-75 by the American Society of Heating, Refrigerating and Air-Conditioning Engineers, Inc. (ASHRAE), which sets energy conserving design criteria for new buildings. This effort involved several thousand people, including everyone who reviewed the various drafts of the document and submitted suggestions.

However, responsibility involving national considerations, ultimately rests with the leadership of the country—the President and the Congress of the United States. Their responsibility is to set overall policy and enact whatever legislation is required to implement that policy.

Setting energy policy is not a simple job in our society. First of all, a large segment of the population appears to give the problem low priority. Consequently, they react negatively to anything causing them inconvenience or hardship. Further, there are strong differences of opinion on what to do and how to do it—the nuclear power plant debate, resorting to strip mining, etc. Also to be considered are the legitimate concerns of the qualified environmentalists and many other factors that apply to the problem. It would be a difficult task to set energy policy even if the President and every Congressman could work at it full time.

A way to skirt some of the difficulties involved would be for the leadership—the President and the Congress—to declare that the energy crisis be placed outside the realm of partisan politics.

The President and the leaders of the two major parties should go before the public to stress the seriousness of the problem and communicate the fact that there are no easy solutions. They should impress on the public that unpopular measures taken now will prevent worse consequences later. Fortunately, this has now been done under the Carter Administration.

The leadership of both parties should then work together in Congress to develop both short term and long term policies to cope with the energy shortage itself and with related economic dislocations that may result from implementation of meaningful energy conservation programs. There should be no Democratic or Republican program to deal with the situation, only a national program, and all involved should share the credit or blame for the results.

With policy goals set, the many government agencies involved with energy would be required to direct their efforts within the framework of these national goals. These agencies presumably would provide valuable input in the formation of the goals since many have been studying various aspects of the problem for a long time.

This program may appear to be a simplistic solution. Admittedly, disagreements will continue on policy matters. It will not be easy to gain consensus on what must be done, but it is necessary that we make a start in dealing with all of the problem's ramifications. Failure to do so now will only cause greater trouble in the future and will ultimately make it more difficult to implement needed measures when the energy crisis *does* become the main problem in the public's view.

References

1. "The Gallup Opinion Index," The American Institute of Public Opinion, April 1975.
2. *Population Estimates and Projections*, U.S. Dept.

of Commerce, Bureau of the Census, Series P-25, No. 529, September 1974.

3. Vogt, Williams, *Road to Survival*, William Sloane Associates, 1948.

4. *United States Mineral Resources*, Geological Survey Professional Paper 820, D. A. Brobst and W. P. Pratt, ed., U.S. Government Printing Office, 1973.

5. Rocks, Lawrence, and Runyon, Richard P., *The Energy Crisis*, Crown Publishers, Inc., New York, 1972.

6. "No Alternative to Nuclear Power," *Bulletin of the Atomic Scientists*, March 1975.

7. *Science*, Jan. 31, 1975, p. 331.

8. Bupp, I. C., Derian, J., and Donsimoni, M., "The Economics of Nuclear Power," *Technology Review*, Feb. 1975, p. 15.

9. "The Crunch To Come," *Technology Review*, Feb. 1975, p. 57.

10. "Patman Report on Energy Crisis Warns U.S. Is Moving too Slowly," *Chicago Sun-Times*, Feb. 5, 1973.

11. "Rationing Urged After Two-Year Energy Study," *Chicago Sun-Times*, Oct. 18, 1974.

2

Education: the key to the energy dilemma

WILLIAM J. COAD, PE,
Vice President, Charles J. R. McClure
& Associates, St. Louis, Missouri

Few will deny that education is a fundamental ingredient in the solution of the vast majority of social problems. Ideally, an educated society is virtually devoid of ignorance. As such, it has an understanding of the interrelationships between the natural laws that dictate its course. Education instills in man's mind a curiosity that hungers for answers to these phenomena that are not readily understood. This curiosity generates explanations for what were previously considered mysteries, and so the cycle continues. An educated society, like an educated person, considers the unknown a challenge to be conquered rather than a threat to be feared.

Historically, totalitarian societies have succeeded temporarily by controlling all education systems and institutions. Such success has inevitably been short-lived, because to maintain the strength of the society, they had to continue to educate. Even though the content of the education programs was controlled, as the population became educated, their curiosity and hunger for knowledge led them to question their society. This curiosity born of education thus becomes the first thrust to topple a totalitarian state's dominance.

Today, the world society is in the midst of an extremely complex and potentially dangerous problem—that of an accelerating consumption of nonreplenishable energy resources. Highly industrialized nations are totally dependent on enormous consumptions of these resources for economic survival. Densely populated nations depend on energy to maintain agricultural economies. Many resources are owned by nations other than those who have a survival need for them. The world is thus close to sitting on a powder keg. The United States government has talked of such wishful solutions as "energy independence," a term coined to give the population peace of mind by thinking that we can get by with what we have. Three years later, however, we are importing more and domestically producing less than in the years preceding "Project Independence."

How do these two observations tie together? The proposition is that if there is one key to the jigsaw puzzle of the so-called energy crisis, it is education!

The United States government currently is spending billions of dollars trying to cope with the energy dilemma, and the lack of success of these efforts is apparent to anyone who has been observing the continuing upward climb of the consumption curves. The reason is, stated bluntly, that when it comes to energy, the populace is ignorant. They can be easily misled by news releases, television commercials, and best-selling books written by self-proclaimed authorities who offer impractical solutions.

If it is accepted that the energy dilemma is a just national issue, then the federal government should channel a large portion of the funds appropriated for solutions into the educational field. We must become an energy educated society. These efforts should not be limited to those in the sciences, but open to all citizens in all areas of interest. If such a thrust were made today, within a very short period of time (say five years), the United States would be the world leader in energy management.

The proposed programs for energy education should address all concepts from the basic definitions, understanding where the energy to brighten the light bulb comes from, and a simple knowledge of how it was transformed from, say, coal into light, to the enormously complex problems of energy economics and socio-political implications.

Energy economics should be taught in elementary schools, as required courses for all students in secondary schools, and as required courses for bachelor degrees in arts and sciences. For those specializing in sciences directly related to energy, such as mechanical engineering, physics, and geology, specific courses in energy economics as re-lated to these disciplines should be required; and economics and political science programs should offer specific courses in energy economics. Further discussions of energy economics as applied to conservation are included in Chapter 5.

Those who have completed their formal education could avail themselves of the required knowledge through adult education programs, educational television, and other such vehicles.

The federal government could serve as a catalyst in this effort through such agencies and organizations as the Department of Health, Education, and Welfare and the National Science Foundation. The programs could (and should) be tailored to fund the development of programs, courses, and textbooks, *not to provide them.*

A population with an understanding of the concepts of energy source and conversion, factual limitations, and impact of continued excesses would not find it necessary to turn to self-proclaimed leaders or to Washington for a solution to their inevitable problem, but would, with understanding naturally develop their own solutions.

3
What is energy?

WILLIAM J. COAD, PE,
Vice President, Charles J. R. McClure
& Associates, St. Louis, Missouri

Since engineering is an applied science, founded to a major degree upon the basic science of physics, engineers tend to think as applied physicists. A natural consequence is the tendency to seek exact definitions when addressing engineering problems.

This chapter will discuss the seemingly parochial problem of defining "energy." Doubtless, there are few engineers who do not know what energy is, just as there are few Americans of any walk of life who must even stop to think of what energy is as they listen to energy related news reports or read endless references to energy shortages, crises, and related problems in the daily newspapers.

Most engineers will probably think back to the fundamental definition in introductory physics: "energy is work or the capacity to do work." Those with a less technical background will form mental images of energy, which takes such diverse forms as gasoline for the family car, warm homes in winter, cool homes in summer, electric bills, gas bills, active children in constant motion, etc.

Whether one relates to the fundamental physics definition or the less structured concept, our society has managed to develop a frightening dependence upon the energy potential stored in the earth's nonreplenishable resources without the benefit of exactly defining the source of our dependence. Partially as a result of this, we find ourselves in the midst of an energy consciousness that has been called everything from a crisis to a mere situation and which has generated the expending of literally tens of millions of manhours and billions of dollars in discussions, studies, research projects, etc., resulting to date in numerous documents drafted and published ranging from policies, standards, and guidelines to federal, state, and municipal laws.

After reading many of these documents, it becomes evident that if the classical definition were applied to the word energy as used therein in the legal context, it would likely be thrown out in the lowest court. Let us then consider a definition of energy as it might relate to the so-called popular movement of energy conservation or the science of energy economics. A brief scattering of some readily available sources reveals the following definitions (or lack thereof) of the word, energy.

- *Webster's Dictionary:* "capacity for performing work."
- High school physics text: "the capacity to do work."
- Engineering thermodynamics text (Stoever): ". . . the something that is transferred to or from a system: a) when work is done on or by the system, and b) when heat is added to or removed from the system."
- Introductory economics text (Samuelson): no definition.
- State of Minnesota Legislation HF No. 2675, an act relating to energy: no definition.

From this brief sampling, it is seen that little has been done to define energy beyond the basic physics definition. Yet, when we start discussing energy in terms of energy economics or energy conservation, most will agree that we are addressing the conservation of tangible things—not simply forces, distances, and relative temperature levels.

Considering the various energy resources available to mankind, starting at the practical source, all energy on earth emanates (or emanated) from the sun in the forms of nuclear and thermal energy transmitted in mass migration with the formation of the solar system, or on an ongoing basis by radiation waves. Following many centuries since its formation, the earth has stored a given amount of the original mass migration energy in relatively fixed or limited quantities. This stored energy is available for conversion to useful forms, generally through chemical conversion processes, then through the first law processes (building heating and industrial thermal needs), or second law processes (transportation, electrical power generation, refrigeration, shaft power). It is these sources that are the rightful target of so-called energy conservation efforts. Few would deny that efforts in energy conservation do not relate to reducing the energy available from the sun, the tides, or the human energies of mankind.

Thus, efforts at energy conservation are actually efforts at reducing the *rate of use* of the limited or finite energy producing potential stored within the earth. All attempts at the control or reduction of energy usage or consumption should therefore address a definition of energy relating to this consumption or depletion of the nonreplenishable resources of the earth regardless of the quantity that is stored. Any finite quantity will eventually be depleted.

Consider then the following definition of energy as it would relate to energy conservation efforts, energy economics, or energy policies: "Energy is the potential for providing useful work or heat stored in the finite resources of the earth."

Although it may appear that the various available forms can be totalized by reducing to a common denominator, such as the potential heat content (say, Btu), the study of energy economics must take into account the available quantity of each form, and the convertible use for the given forms by current state of the art technology.

The resources, within the current context of known technology include three general forms (which could be further subdivided): fossil fuels, nuclear, and geothermal.

4
Energy transportation

WILLIAM J. COAD, PE,
Vice President, Charles J. R. McClure
& Associates, St. Louis, Missouri

A fundamental study of the phenomena of energy transportation is an essential ingredient to any comprehensive development of future energy policies, the optimization of energy related parameters in end use systems, or the sources to serve these systems. Energy transportation has been utilized for centuries, but previous developments have been the result of an immediately available supply to satisfy an immediately available demand. In many cases, the transport phenomena have been confused with the terminal use requirement or the source form. This chapter will simply present an overview of the concept of energy transportation as food for thought.

The first category to be considered is that of transporting the energy from its source through various stages of processing or conversion to the consumer or user. Only the current state of the art will be discussed initially. This category includes:

- Bulk or batch transportation of raw materials, such as coal ore and uranium ore; crude petroleum liquids or gases; processed solid fossil fuels; refined nuclear fuel; various grades of liquid petroleum products; gaseous fuels liquefied for transportation effectiveness or efficiency; gaseous fuels in gaseous phase; dry wastes for processing as fuel; and fuel processed from dry wastes.
- Electromagnetic wave transportation of solar energy to the earth's environment.
- Steady state conduit flow (pipeline) of crude petroleum products to points of refinement or use.
- Pipeline flow of refined petroleum products to either distribution stations for transfer to other transport systems or into distribution piping systems and processed natural gas to distribution system storage and piping systems.
- Distribution piping system transport from storage or primary pipelines to user entities and distribution of various forms of manufactured gas to user entities.
- Transportation of electric energy from generating stations either directly to user entities or to use entities via network transport systems.
- Two phase transport fluid systems, such as steam systems, that convey high level thermal energy from a central conversion plant to user entities or refrigerant systems that convey low level thermal energy from a user entity to a refrigeration plant.
- Single phase transport fluid systems, such as high temperature water or other thermal fluid, that convey high level thermal energy to a user entity or low temperature chilled water or other thermal fluid that conveys low level energy from the user entity to a refrigeration plant.

User entity

The process needs or thermal environmental devices within the user entity (building or plant) can use a variety of energy forms. Also, the transport systems available to the site generally provide a choice of the forms available to satisfy the user needs. Usually, all the various process needs within an entity are not best satisfied by the same form of energy, and the same available form is not necessarily best suited to all the various process needs. Thus, based on the relevant parameters, a system designer or analyst must determine the form best suited to the needs of each individual process, the form(s) of energy available for purchase that could best serve the needs, the proper conversion system(s) to employ, and the method(s) of transport or distribution that is most effective between the point of entrance to conversion and, subsequently, from the point of conversion to the product. These subsystem parameters, although intimately interrelated, should be considered on the basis of individual merit.

The transport systems available within a building include:

- Single phase high level distribution, such as hot water or thermal fluid.
- Two phase high level distribution, such as steam.
- Single phase low grade high level transport systems, such as condenser water systems or energy transfer systems for heat pumps.
- Two phase low grade high level transport systems, such as refrigerant condenser circuits—hot gas and liquid.
- Single phase low level distribution systems, such as chilled water or thermal fluid.
- Two phase low level distribution, such as refrigerant suction vapor.
- Electric energy distribution via electrical conductors.

The systems designer should thoroughly investigate the advantages and disadvantages of each of these distribution forms. Basic considerations should include: parameters of investment cost of both the distribution system and the interface at the process or terminal and at the source, flexibility for changes in source form, distribution system efficiency or energy effectiveness, maintenance requirements (and costs), and adaptability to future system growth or modification.

Although the energy transportation systems *to* the user entity are not of immediate concern to the systems designer, they are of paramount importance to society and its business institutions in the future. The ramifications of energy transportation must be considered in depth by corporate decision makers in both energy industries and the user facility industries and by policy makers at all levels. Energy is a resource; most forms are both being depleted and are highly localized in nature. Thus, the transport concept is the key to future viability of both local and national economies.

Perhaps it is the recognition by many that the universal distribution of solar energy directly to the user entity negates the need to address this complex problem and is thus responsible for the appeal of solar energy. Others have recognized the limitations of solar availability due to limited flux density and, consequently, have turned to the conceptual development of such transport systems as laser beams and extremely high density electromagnetic radiation to address our transportation needs of the future. Until such times as these are developed to the demonstration stage, however, our needs must be satisfied by optimum utilization of current technology in conversion and transport.

5

Energy economics: a needed science

WILLIAM J. COAD, PE,
Vice President, Charles J. R. McClure
& Associates, St. Louis, Missouri

Mechanical engineering has classically been defined as "the applied science of energy conversion." Thus, it is somewhat of a contradiction for practitioners in the HVAC field (a subscience of mechanical engineering) to undertake the goal of "energy conservation." The practice of minimizing energy consumption in building systems, whether in design or operation, might therefore be more accurately described in other terms.

After some reflection and research, we suggest that *energy economics* is an appropriate label to hang on such efforts.

Webster defines *economics* as "a social science concerned chiefly with description and analysis of the production, distribution, and consumption of goods and services." In the past, most efforts involving energy consumption by the building systems design profession were, understandably, motivated by monetary economics. And though it is certain that monetary economics will continue to be a primary design parameter, it is becoming increasingly evident that another parameter will have an even more meaningful effect on both system design and energy source selection—the parameter of pure energy economics. Examples of this include ASHRAE Standard 90-75 and many proposed and adopted national, state and local statutes.

Thus, while we cannot omit monetary economics from our considerations, we must supplement it with a new science of energy economics. At the outset, let us make two basic observations:

- In the definition of *economics*, no limitations are placed on "description and analysis" in monetary terms.
- Energy, as obtained in the world today in its initial or raw material forms, can be considered "goods" in the light of Webster's definition.

After accepting this, we see that the basic hypothesis underlying further development of the independent science of energy economics is:

The forces that motivate and strengthen the monetary economics of a society are diametrically opposed to the conservation of energy resources.

In any socioeconomic system in the world today, the increased use of energy serves to strengthen the so-called economy of the society. Inversely, conserving the energy resources has the effect of retarding economic growth and weakening the economy.

Since none of the commercially available energy sources today are replenishable, however, continued accelerated use of these commodities is tantamount to living on credit, *i.e.*, today's use, and the advantages thereof, will not affect society until tomorrow. This observation leads to the thought that enters the mind of every intelligent and decision-making businessman and homemaker—defining the risk. The risk of using tomorrow's resources today must be taken in a calculated manner; or as the conservative gambler might say, we have to hedge our bet. We must always be able to retain the flexibility to regroup and change the game plan when the need arises.

It is with this concept that the economics of energy must be disassociated from classical monetary economics. In the development of the proposed science, the basic laws of economics can be utilized, keeping in mind the basic difference that money is generated by consumption of resources whereas energy use depletes our finite resources (whether energy or mined commodities). Thus, the basic economic law of supply and demand is the irrefutable footing of the science of energy economics.

Upon this footing can be constructed the science: The supply is our limited resources, and the demand is what a sound monetary economic system must have to survive and grow. This balance between monetary economics and energy economics must be defined and then achieved on national levels (if not worldwide) if mankind and nations are to survive.

The science of energy economics must recognize that technical breakthroughs occur at (hopefully) opportune times in the history of civilization and provide for these eventualities. We should never be in the position of betting all we have on such an eventuality, however. For example, in the late 40s and early 50s, we rested comfortably in the thought that nuclear energy represented a new found infinite source. Now we are beginning to realize that the loop is not closed; we still lack a method of disposing of the wastes.

In summary, a new subscience of energy economics is needed as an evaluation tool for professionals in energy conversion as well as for society in general. This science can use as its footing the basic law of supply and demand and build upon it in much the same manner that monetary economics has developed. In its application, energy economics should interface with monetary economics as well as affording an independent input parameter to the design of all energy converson cycles, systems, and machinery. The science should be established on a worldwide basis, but until current national boundaries are transcended by interaction among governments and peoples, it must be applied by each nation individually for its survival and defense.

6

More on the energy hypothesis

WILLIAM J. COAD, PE,
Vice President, Charles J. R. McClure
& Associates, St. Louis, Missouri

In the previous chapter entitled "Energy Economics: A Needed Science," the following hypothesis was presented:

"The forces that motivate and strengthen the monetary economics of a society are diametrically opposed to the conservation of energy resources."

This chapter further discusses the significance of this hypothesis as it relates to the energy economics efforts of the engineering community and governmental policies.

The statement was called a hypothesis, not a law, on the basis of the definition of a hypothesis, "an interpretation of a practical situation or condition taken as the ground for action."

Examples that justify the hypothesis are endless. To state a few:

• The monetary economics of the energy industries are such that retardation of ever-increasing energy use could financially destroy them. Consider the commonplace "demand" charges that penalize a consumer for not using more hours (energy) of demand (power); or the rates for either fossil fuels or electric energy, which reduce the unit cost and consequently the *average cost per energy unit as consumption increases.*

• Examples in transportation are numerous. Consider the complex interrelations of the automotive industry, the highway systems, and the building industry. Development of interstate highways into and around major metropolitan areas spurred development of suburban or satellite business communities such as office and industrial parks as well as housing communities and associated commercial developments. The only links between the housing communities, suburban business communities, and "downtown" areas are the super highways. The required use of the family car for transportation to a daily place of work not only increased the per capita gasoline consumption significantly, but also increased demand for the second car, which was now sorely needed for visits to the doctor, trips to the store and shuttling the kids to their extracurricular activities. We became a nation on wheels with no alternative. Yet, no one lost in this conversion: the building industry boomed; the automobile business boomed; and the energy business boomed.

• An example that presents an overview of the hypothesis is shown in Fig. 6-1. Against the time span of 1909 to 1973 is plotted Gross National Product per capita in units of thousands of dollars; superimposed is the annual energy consumption per capita in units of billions of Btus. It is immediately evident that the two curves are very nearly congruent. This phenomenon tends to prove the legitimacy of the hypothesis.

Several conclusions could conceivably be drawn from the above examples, particularly the one illustrated by the graph, and some of these conclusions are most distressing. For example, one might conclude that the two phenomena are so closely interrelated that by some law of economics a downward slope in one would dictate a downward slope in the other, and vice-versa. This conclusion could likely be justified by the current situation in the United States. We have been effectively reducing our energy consumption by various external and internal pressures; and concurrently, we are in the midst of one of the most complex monetary crises of modern times— unemployment and inflation are higher than they should be, shortages prevail, and many of our largest business institutions are in serious financial difficulty.

Another interesting observation is that previously all discontinuities or changes in direction of the curves were caused by other forces affecting GNP, with the energy curve following. But this time, the energy curve is the controlling parameter, and the GNP curve is doing the following. This simple observation, if studied in more depth, could go a long way toward explaining the current economic situation.

But in the study of economics, the laws are not as clearly defined or as hard and fast as the laws of the natural sciences; and in most cases, the recognition of a phenomenon as stated in the hypothesis can provide guidance for intelligent solutions to seemingly overwhelming problems. If, for example, in the development of national energy policies, every effort is made to defy the hypothesis, to separate the two curves— bending the energy curve horizontally if not downward while forcing the GNP curve upward—further crises can be avoided.

There is no question but that this can be achieved, and this is the goal of energy economics. In the design of any energy conversion system or the conception of any alternatives to existing methods, the use of energy economics should include the identification of the "energy product" requirement. This identification should start with the concept and be carried through the entire design development of each and every energy related component.

For example, in transportation, the "energy product" is simply the units of energy required to move an individual or given cargo from Point A to Point B in a given time. From

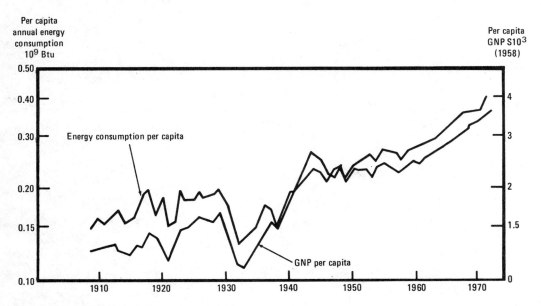

Source: Institute of Gas Technology

Fig. 6-1. Trends of per capita GNP and annual per capita energy consumption in the U.S. 1909–1973.

this point, the burdens of the practical systems such as vehicles are considered. The phase-by-phase degradation continues until we have achieved a method of providing for the "energy product" in the most effective manner. The final measure is to compare the resource utilized to achieve the product requirement. This evaluation parameter is commonly identified in the automotive area as miles per gallon or in public transportation as passenger miles per gallon.

The same approach can validly be applied to building systems, industrial processes, etc.

In the draft of Section 12 of ASHRAE Standard 90-75, this concept is applied to a degree to building systems.

In the development of energy policies, whether they be national, international, or regional, consideration must be given to the energy hypothesis. The question must be answered: Will the policy reduce energy use while not adversely affecting the economic situation? There are numerous ways of achieving this goal either totally or in part. Lack of recognition of this interplay has plagued us so far.

7

An energy outlook

A. A. FIELD
London, England

Events of the past have jolted European thinking out of its groove of complacency toward energy and given the community some fundamental problems in cooperation. Apart from the U.S.S.R., and with few other exceptions, all European countries depend heavily on imported energy. The most committed country is Denmark with no indigenous resources, whereas Holland imports only 26 percent. France imported about 73 percent of its energy in 1974; Italy, 82 percent; Switzerland, 79 percent; and the U.K., 47 percent.

While the search for new supplies through exploration inland and offshore continues, the attitude toward exploitation of new forms of energy is hesitant. Research funds for solar, geothermal, wave and tide investigations are at the moment inadequate to produce the necessary answers within the next decade or so. Nuclear energy, on the other hand, is getting the big boost. Funding is at a rate almost too fast to be used, and this emphasis is setting off reactions from environmentalists that are likely to prove too strong to allow the planned programs to come anywhere near their hoped for contribution.

A brief survey of events in individual countries shows why it is necessary to be so critical. A look at the finances of the whole situation may well expose some unpalatable, but at some stage unavoidable, realities.

Energy finance

The Organization for Economic Cooperation and Development (OECD) warned its 24 industrialized member nations that they will have to live with huge combined annual oil deficits until at least the end of the decade. The current deficit is $39,000 million, and the 1980 deficit is foreseen at $35,000 million, based on an oil price of $10.80 per bbl. The Gulf price is already above this.

Fernand Spaak, director-general of energy of the EEC, has pointed out: "The member states are all committed to purchasing a large part of their energy needs on the world energy market: in 1974, the European community of nine imported over 60 percent of its needs, of which 59 percent was in the form of oil and oil derivatives!" This quote appeared in an article on *Coal and Energy Quarterly*, published by the U.K. Coal Board in 1974. Spaak adds: "For a period that extends at least until around 1980, this situation cannot be greatly altered. The introduction of new sources of hydrocarbons and policies to accelerate the introduction of alternatives to oil, or restrain their decline in the case of coal, can have a significant impact thereafter. Until that time, the only way dependence can be reduced is by the restraint of demand and the conservation of energy, particularly in its imported forms. The crisis that followed the Yom-Kippur War in 1973 was a profound shock, and its implications are far from being worked through yet—not least in terms of economic and social consequences."

The energy commission sees the way out for the community as being a vast development of nuclear energy, and it has further recognized that the member nations may need to buy Russian enriched uranium in the 1980s.

Some 56 nuclear plants currently exist in

EEC countries and have a capacity of 11,500 MW. Estimates for 1981 are set at 119 plants, and some 217 plants are expected for 1985. Capacity is forecast at 54,000 MW in 1980 and 176,000 MW in 1985.

Another of the recent pleas from the commission stressed that the EEC use coal for more of its energy needs. The aim is to mine 300 million tons annually from 1975 to 1985. Unfortunately, the investment needs are prodigious, and there is also an agreement that imports will have to be stepped up, particularly from the U.S.S.R.

The U.K. has not been unaware of its difficulties, but there has been an alarming dependence politically and economically on the prospects of North Sea oil and gas. Engineers have called for a saner and less headlong exploitation while maintaining the coal industry tonnage (Fig. 7-1), but political forces are likely to win in the end.

In the aftermath of the embargoes, there was some reduction in energy use, but by June of 1974, consumption of total energy had recovered to the previous year's level. By September, oil consumption alone was returning to the 1973 levels. Savings in the first 11 months of 1974 attributable to conservation appear to be no greater than 2 percent.

The first U.K. oil production platform in the North Sea was installed toward the end of 1974. Output is expected to reach 250,000 bbl per day in 1976.

On the natural gas front, while the U.K. needs to buy from the Norwegian fields, development of the big supply promise near the Shetland Islands, north of the U.K. has been delayed.

Some restrictions on oil and energy use were introduced at the end of 1974; but apart from the increase in the tax on gasoline, the other measures, such as reduced room temperatures, will be difficult to enforce on a national scale.

On the coal front, the government's Department of Energy has reported that, ". . . an efficient coal industry has an assured future." Unfortunately, the Department of the Environment has caused a delay in the start on the industry's only major new coal field by demanding, ". . . the full planning permission procedures."

The electricity industry is suffering from delays in the decisions for new power stations and from an ever increasing resistance from local groups over the sites to be chosen.

Uranium shortage

In the face of a likely world shortage of uranium in the 1980s, the U.K.'s labor government is being pressured by its left wing to cancel arrangements to buy from South Africa. The only possible substitute is the U.S.S.R.

An indication of the strength of the anti-nuclear lobby, which concentrates on the dangers of every aspect, is the plan to build a chemical plant that would solidify highly radioactive wastes at a cost of at least $60 million. Production is likely to start in 1982.

The French have decided on a crash program to build 40 nuclear reactors (over seven per year).

The French ceiling on oil imports for 1974 was set at 10 percent below 1973 levels and expressed in terms of cash at $11,300 million. The French government is backing conservation measures, and the 1975 expenditure was around $42 million.

France has already made it clear that she has little interest in the oil sharing program among the 11 leading consumer nations. Her own view, according to President Giscard d'Estaing, is for a get-together of producers and consumers to find solutions to the whole energy crisis. This has not prevented France from making independent deals with the Arab countries, seeking more Algerian gas, and concluding a 20 year agreement with the U.S.S.R. for about 10 percent of expected gas consumption. Most French research efforts are reserved for nuclear energy and the fast breeder reactor.

The likely changes in energy pattern in France were outlined recently by P. Bertrand, honorary chairman of the heat service operators association (SNEC). Over the periods 1960–72 and 1972–84, he calculates an increase in energy consumption of 80 percent, instead of 90 percent as originally forecast

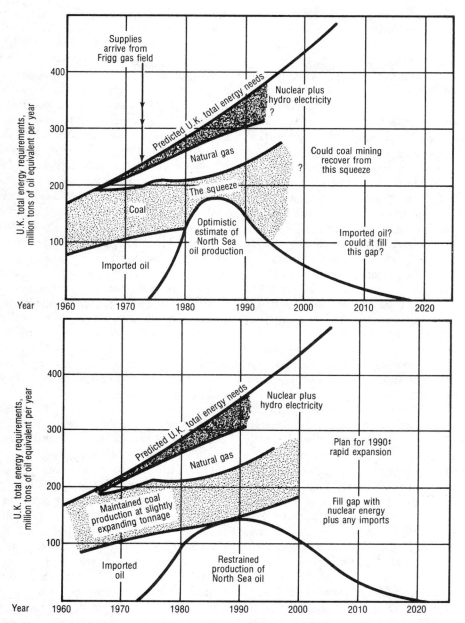

Fig. 7-1. (Top) unrestrained growth of North Sea oil could irreparably damage the coal industry. The bottom drawing shows how to make the best use of both coal and North Sea oil in the U.K.

before the energy crisis. This means an increase in the residential and tertiary sectors of 100 percent rather than the predicted 130 percent. In these sectors, oil and gas will now take 48.5 percent of the energy, compared with 53.5 percent originally predicted; and electricity will rise to 38 percent from 34 per-

cent. All of these figures relate to 1985. In terms of energy in all sectors, the 1985 total of 450 million tons of coal equivalent will be cut to 400 million tons.

French government investments in the coal industry will be shared; roughly two-thirds will maintain the present level of production (about 33 million tons), and one-third will be applied to continued modernization of the industry.

In West Germany, oil supplied more than half the 379 million tons of coal equivalent of energy consumed in 1973. With little chance of having her own oil, the country is determined to both reduce dependence on oil and to spread her dependence on imports as widely as possible. For example, by 1980 gas imports will be about 63,000 million cu meters; half will come from Holland and the rest from Russia, Algeria, Norway, and Iran.

West Germany is also subsidizing coal and seeking to find the maximum use for it. RWE, one of the largest electricity boards in the country, recently switched plans for six new power stations from oil to lignite—a total of 3600 MW capacity. The share of oil for RWE stations will then be 28 percent. Out of the total generating capacity for Germany, only 12 percent is fueled by oil.

Germany is also speeding nuclear development and is creating a major German Oil company to support a buyer's cartel. An unusual deal with the U.S.S.R. is the serious proposal for West Germany to supply nuclear reactors and receive payment in the form of electricity routed via West Berlin. A supply contract for $410 million in uranium in the period to 1990 has also been signed.

In Norway, a member of EFTA (a loose grouping of all the nonaligned European countries), the search to find sites suitable for nuclear stations has been abandoned, and although no decision on whether to go nuclear or not is expected before the spring, it is clear that the protesters have the upper hand. Norway, however, has no immediate energy problems. Some 75 million megawatt hours of electricity are produced by hydroplants, and the North Sea oil and gas discoveries allow plenty of latitude. Nevertheless, Statoil, the national oil corporation, does need some $3500 million capital. Repayment will almost certainly conflict with the current determination to exploit Norwegian reserves at a slow pace. With planned taxation on oil on a scale rising to 90 percent, Statoil could have the problem of being the only explorer-developer in these waters and make nonsense out of the foregoing estimate of capital required—as inflation will do in any case.

Holland has good prospects of using her gas supplies. Already, they account for over 50 percent of consumption and are exported to Belgium, Germany, France, and Italy. But Holland is also running into severe resistance to nuclear plants. The country is introducing measures to cut back on energy and enforce insulation standards. Research and advisory bodies are also being set up. A comparison of dependence on imported energy between 1960 and 1970 for Western European nations is shown in Fig. 7-2.

Switzerland, an EFTA member, is facing the most difficult set of problems. Practically all electricity is produced from hydro stations, and this output is already inadequate—even with three nuclear stations in operation. Opposition to nuclear plants is so strong that the program is already four years behind schedule. Meanwhile, imports of oil are running at about 80 percent of energy use and are likely to stay at that level, even if the nuclear program is kept up to date.

Austria, also a member of EFTA, is seeking answers to energy problems through nuclear power, but plans are meeting stiff opposition by environmentalists.

One of the most fortunate among the EFTA nations is Iceland. With hydroelectric and geothermal power, the country could be more than self-supporting in the future. Since the price of oil imports from the U.S.S.R. has risen more than 150 percent, the motivation to further develop these resources is considerable. Estimates suggest that some 28 million megawatt hours per year are available from hydro power, of which only 8 percent is being used now. Geothermal sources are estimated as in the region of 80 million to 8000 million megawatt hours per year. Already nearly half the houses are heated from direct networks supplied from geothermal wells, and it is planned to raise this to two-thirds soon.

The COMECON nations, the grouping of the U.S.S.R. and the eastern European states, are dominated by the U.S.S.R. and

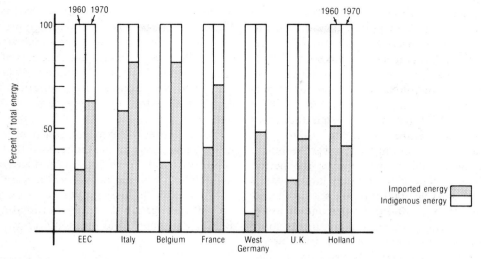

Fig. 7-2. A comparison of dependence on imported energy covering 1960 and 1970. The neglect of indigenous sources of energy has led Western Europe into rapidly increasing dependence on imported fuels.

dependent on her for oil and other energy forms. Under the contracts for the 1971–75 five year plan, oil was bought at $3 per bbl, and the increase for the next planning period will be large, though unlikely to reach the level of $14 per bbl currently quoted for U.S.S.R. sales to the West.

U.S.S.R. oil

Already, it is clear that many of the states will have to buy oil from outside countries, and Russia will aim to keep such purchases to a minimum. But her own urgent need to buy Western technology makes energy an important means of barter. Indeed, the EEC is now recognized by COMECON and is considered to be an indication of the political needs of the U.S.S.R. for consumer goods.

A cooperative agreement already exists between the U.K. and the U.S.S.R. for nuclear research groups, and the U.S.S.R. is further proposing collaboration on pressure type reactors.

Oil output from the U.S.S.R. is very nearly as large as that of the U.S. The idea of long distance power transmission is being actively pursued. Russia already has one line of 75,000 MW capacity at 800 KV.

Yugoslavia, never a willing junior partner

to Russia, is aiming at a high degree of self-sufficiency. She has a joint project with Hungary and Czechoslovakia to develop and bring oil from the Adriatic coast. It is calculated that supplies from the U.S.S.R. will decline during this decade.

Romania is considering buying 20 Candu reactors from Canada over the next 20 years. She is also seeking a direct interest in North Sea oil projects.

Where is the money coming from?

Unfortunately, for all the hopes and promises, a conference held in London by the *Financial Times* to consider the problems of finance for energy indicated that: apart from the fact that few persons had the courage to add up all the estimates of cash needed for the development of the various energy forms, it appeared that inflation was likely to call a near halt to it in the future anyway. T. A. Green, senior international executive with the British National Westminster Bank, told the conference that oil, needing a threefold increase in production to meet world demands in the period 1972 to 1985, remained the most important source of energy and would enjoy a high measure of priority in banking finance. He stated, "It is difficult at the moment to see how much of this money is to be found . . . one of the reasons it is particularly difficult at present to take a view is that the financial markets are undergoing such a shakeout." Funding of oil development had switched from 90 percent in-company in 1960 to 70 percent in 1973, and was still falling. Professor C. F. Karsten, of Design Industries Association, embellished the point: "Estimates of the amount to be invested totaled $1,180,000 million spread over the period to 1985. Assets of the world's largest banks are about $1,901,000 million, so the financing needs were about 17 percent of assets per year. Since banks do not invest more than about 30 percent of a project's needs, they would have certainly been up to their task." But the weaknesses revealed in the banking system may mean they have reached the limits of the funds they can attract. Only with a return of confidence would the banks be able to fulfill their traditional role.

Dr. Denis M. Slavich, senior economist in the Commercial Development Dept. of the Bechtel Corporation, pointed out that one of the constraints on nuclear energy could be the availability of capital. Total energy investment in the period to 1985 would be $2,000,000 million, with $1,125,000 million needed for the nuclear industry. However, all these figures came under a serious doubt when Morton M. Winston of the Oil Shale Corporation told the conference that in March, when the company had been preparing for a big investment in the rapid expansion of the U.S. oil shale fields, the cost of the plans was estimated at $435 million. By October 1974, six months later, inflation had raised the figure by 40 percent to $630 million.

These figures were more than backed by other suggestions at the conference that the capital costs of North Sea oil production set at $2500 per bbl per day in 1973 had trebled to $7500 in 1974. These figures were by Quentin Morris, group finance coordinator of British Petroleum, and he should know.

Energy Outlook

There is a serious situation in Europe. With little in the way of immediately available supplies, an unknown investment factor to recover reserves, and the traditional leaning to a growth economy, the society must either become more and more at the mercy of the Middle East oil producers, or slowly grind to a halt.

Inadequate attention is being given to other forms of energy and to energy recovery, and the prospects of an early contribution from these factors is being lost.

Figure 7-1 is adapted from Fuel Economy Handbook, published by Graham Trotman Dudley, London. Figure 7-2 is based on OECD data and other sources.

8

Energy: crisis and conservation

A. A. FIELD,
London, England

Energy projections made in 1973 for the member countries of the European Economic Community left no room for complacency; supply would just match demand, even with all available resources plus anticipated reserves of North Sea gas and oil. Recent constraints placed on oil supplies by the Middle East countries have brought sharply into focus the need for better management of energy. The most severely hit countries are now suffering a 20 percent cutback. Energy used in the building services sector represents 40 to 50 percent of the total energy used in a country.

Acceptability index

For too long, engineers have judged the acceptability of a system solely on its inherent efficiency—basically, the performance of hardware—whereas a more rational approach would be to minimize energy consumption in absolute rather than relative terms. This closed-loop mode of thinking was forced on engineers because until recently they did not play a part in the overall building concept. The rise of multi-services design, turnkey projects, and the creation of building design teams, which place architecture and engineering under the same roof, have led to integrated design.

A basic "equation" to establish acceptable standards for energy consumption was recently proposed by two Italian engineers,

Mario Costantino and Pierluigi Cattaneo.[1] The equation reads simply:

$$(E - X)/E = I$$

where

E = primary energy entering the system
X = transformed or degraded energy leaving the system
I = numerical value on the acceptability index.

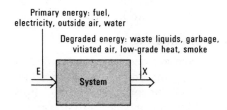

Fig. 8-1. Energy flow in a system. Basic concept of the acceptability index developed by Costantino and Cattaneo.

In Fig. 8-1, primary energy, E, enters a system in the form of fuel, electricity, outside air, or water, and it leaves the system degraded as garbage, vitiated air, low-grade heat, smoke, or waste liquids. The numerical value of the index, I, is a measure of a system's acceptability. When a system is operating at peak efficiency, a value of 1.0 is approached on the index.

[1]Superscript numerals indicate references at end of chapter.

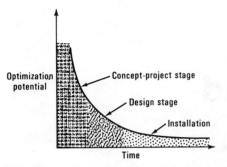

Fig. 8-2. Graph illustrating the decreasing opportunity for basic design changes from project stage to installation. (Based on a diagram by Rusenberg, cited by Costantino and Cattaneo.)

Effluent energy, X, must be kept as low as possible since all such forms of energy damage the environment in terms of pollution—thermal, chemical, and material. As the global value of X increases, so does pollution potential with its insidious effects on life, climate, and buildings.

The amount of primary energy must be limited by optimizing construction and reducing primary energy requirements. This can only be done effectively at the project stage. As the design advances, opportunities for major changes decrease. Figure 8-2 is an idealized representation of this.

Oil is main heating source

All nine member countries of the European Economic Community depend heavily on imported oil for primary energy. Italy is the most dependent since 75 percent of her total energy comes from oil. U.K. and West Germany are the least dependent, relying on oil for 50 percent of their total energy. In all countries, the coal industry is being run down, and nuclear energy—except in the U.K.—is a slow starter. In 1975, France and West Germany produced 2 to 4 percent of their total primary energy with nuclear reactors, with a planned increase to 5 to 9 percent by 1980. North Sea reserves of gas and oil account for 10 percent of Europe's energy supply as of 1975.

In some countries, energy for heating buildings is more heavily weighted toward oil than in the overall balance. France, for example, in 1970 used 62 percent oil for heating builings, whereas in terms of total energy consumption, the figure was 55 percent. The complete reverse occurs in the U.K., where the domestic sector uses only 9 percent compared with a national average of 47 percent.

The prospect for the immediate future in most European countries therefore, is not encouraging. Diminishing supplies of oil and a predictably slow growth in developing other forms of energy leave only one solution: improvement in the acceptability index. This can be done in three basic ways: improving the efficiency of the energy cycle, reducing the energy demand, and reclaiming low-grade heat.

Combustion cycles

Conventional boilers and direct fired heat exchangers have now reached a practical efficiency limit. Their efficiency ceiling is in the 85 percent region and controlled by the condensation factor. However, seasonal efficiency can fall appreciably below optimum efficiency because of part load operation. Correct choice of burner/heat-exchanger will ensure that near peak efficiency is maintained during reduced loads, and a search for the "optimum control policy" (opc)—a term introduced by P. Robertson of Glasgow University—will get the best out of a multiple boiler plant.[2] Overestimating heating demands results in a low load factor that leads to lower seasonal efficiency. Maintenance is another important factor. Dirty heat exchange surfaces can reduce efficiency considerably. Oil firing is associated with a high fouling factor. This has increased interest in the development of a water-oil emulsion burner that produces much cleaner combustion.

A method of breaking the efficiency barrier without any of the accompanying disadvantages utilizes the technique of submerged combustion (burning gas in a cascade heat exchanger). Efficiencies of 95 percent are realized on the gross calorific value; this is equivalent to 105 percent on the net calorific value. A limitation is the relatively low mean water temperatures required to realize high efficiency levels. Return water needs to be maintained at about 90°F and flow water at about 180°F or less.

Generating electricity is the most wasteful energy cycle. The average generating efficiency in the U.K., for example, is 28 percent. One recently commissioned power station has an output of 4000 MW achieved at the expense of an input of 10,000 MW of heat. Power station waste is so extensive that recoverable energy lost in the U.K. in 1970 was almost equal to the whole of the domestic sector heating load in the same year.[3]

Most stations use oil or coal, but some are turning to natural gas. In Belgium, one-third of all electricity comes from gas fired stations. From the environmental viewpoint, this is squandering a form of energy that could be better used for heating where the almost totally innocuous products of combustion could be released at a low level, or for refrigeration and total energy.

Next to Russia, West Germany makes the most use of combined heat power. Over 70 percent of the present district heating networks are supplied from pass-out steam. Germany also has one of the highest proportions of coal fired stations—a total of 67 percent in 1972.

Rational use of primary energy. A 106 MW back-pressure turbine supplies an additional 200 MW (680 million Btuh) of heat to a town in Sweden. (Photo courtesy of Stal-Laval Ltd.)

Nuclear power combined with a heat supply to a district network would seem to be the ideal energy cycle. Sweden is the only country thus far to use this method.

Total energy systems, in the generally understood sense of on-site generation of power and heat, are making slow progress in Europe. One of the largest total energy projects in France—excluding generation from the steam cycle—is the new international center in Paris. Total heating load is 114 million Btuh; one-fifth of this comes from 5 MW reciprocating engine-generator sets. The U.K.'s largest total energy system is at a cigarette factory. It has an output of 9 MW with 70 million Btuh from gas turbine alternators.

Reducing energy demand

Energy consumption predictions for buildings are normally based on rate of new construction, modernization, and improvements in environmental standards; rarely are they based on improved building design. To conserve energy in the future and to aim at a high acceptability index, basic heating requirements and cooling loads must be reduced. This means better insulation, less glazing, and more deep plan structures; more precisely, a reduction in the specific heat load, Btuh per cu ft.

Building insulation standards vary considerably in Europe. The U.K. has the distinction of being at the bottom of the league even though London has the same number of degree-days as several European capitals that require lower outside design temperatures. The Swedish standard, for example, is three times higher than that of the U.K. A recent study of the effects of improved insulation in new domestic buildings in the U.K. used 2 in. rather than 1 in. of mineral wool in roofs of single family houses and resulted in a predicted effective energy saving by 1980 of 1.5 million tons of oil per year. This figure relates to only 30 percent of the total housing.

In France, the higher than normal standards now being used for all-electric systems have had repercusions in the insulation industry; one of the largest manufacturers of glass wool reported a 30 percent increase in sales during 1973.

The importance of insulation can be judged from the fact that a reduction of only 10 percent in energy used in buildings in the domestic sector alone in France and West Germany would amount to over twice the present installed district heating load in both countries.

Considering that the domestic sector is by far the largest consumer of heating energy (between three and four times the nonresidential sector, depending on the country), insulation improvements in dwellings will have the greatest global effect. In the U.K., there is a substantial industry in insulation for existing structures. The most notable is in double glazing; some 200 firms share a market of around $60 million. Another growth area is foamed in-situ insulation for cavity walls, and mineral wool insulation for roofs.

Another factor contributing to the rise in energy consumption is the progressively higher temperatures demanded for living spaces. A decade ago, 68°F was considered sufficient for dwellings; now 70 to 72°F is not unusual. Fanger's recent work suggests that 73 to 75°F is not too much for sedentary subjects.[4] On the other hand, from the psycho-physiological standpoint, Missenard proposes not more than 65°F.[5] In the long term, he says, temperatures of 73 to 75°F will lead to atrophy of the human thermoregulatory system. The question of optimum temperatures, therefore, is by no means settled, and the importance of energy economy cannot be exaggerated.

Outside air supply rate is another important factor in balancing energy consumption. Based on physiological criteria alone, outside air supply could be reduced to one-quarter of the present recommended rate. Odor levels are the governing criteria and not lack of oxygen or buildup of CO_2. Utilizing activated carbon filters plus HEPA filters could reduce mass flow and recycle still usable air.

Reclaiming heat

Apart from the other instances of reclaiming heat (power stations and total energy systems), high grade heat can be directly recovered from refrigerating plant condensers and lighting systems. These are sufficiently established in the U.S. so that no further mention is required.

Exhaust air heat recovery is a technique fast gaining ground in Europe. The principal heat recovery device is a thermal wheel (rotary regenerative heat exchanger). Sweden in particular is using this and a form of heat pipe unit for heat recovery in apartment block ventilation. The heat pipe unit was used in systems for 3600 apartments up to the end of 1971, and its efficiency in terms of waste heat recovered is reported as approximately 50 to 60 percent.

Heat can also be recovered from waste matter. Rising quantities of waste and rising calorific values make this an economic as well as socially desirable proposition. For example, incinerators connected to the Paris district heating network deliver 1.2 million tons of steam a year from an annual 1.6 million tons of waste (specific energy, 1200 Btu's per lb). Another process being researched at the U.K.'s Warren Spring laboratory is pyrolysis, a process that distills combustible gases from waste in the absence of air. A heat release rate equivalent to 1 lb of coal from 10 lbs of waste is claimed.

Future energy sources

Nuclear energy is the principal future energy source, thinking in terms of the end of the century and beyond. However, some interesting ideas are being put forward both for alternative sources and media.

F. Lessing, for example, proposes the use of hydrogen.[6] This would be generated from seawater by using nuclear energy, and the plants would be stationed on large floating platforms near the coast. It is possible to transmit hydrogen through pipelines at about one-eighth of the present cost of transmitting electricity. The advantage of using

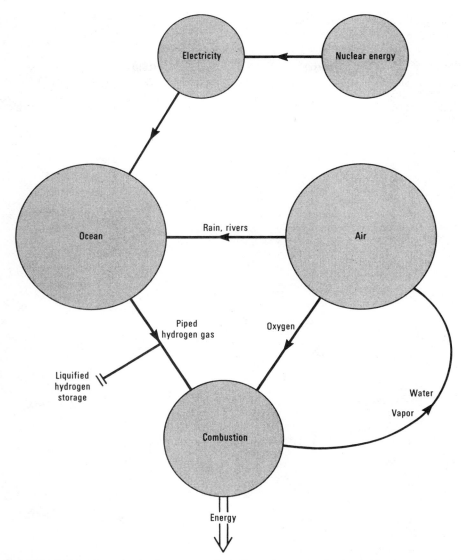

Fig. 8-3. Lessing's idea for using hydrogen as a fuel. One advantage of using hydrogen as a fuel would be the purity of the cycle.

hydrogen is the purity of the cycle; only water is formed at the point of combustion. Catalytic burners would replace jet burners. (See Fig. 8-3.)

A. Bruckner, an American, claims that windmills could be used. Water pumped to elevated reservoirs by 1000 windmills could supply power equivalent to that produced by one nuclear station. Thring suggests onsite generation of heat and power for single family dwellings by using windmills (aerogeneration) and solar collectors.[7] A 4 kw generator plus batteries could provide sufficient power for a three person dwelling. The cost would be around $2000 for the generator. Solar collectors would cover the roof area to supply 30 to 50 percent of space heating, and the remainder would come from a heat pump driven by a windmill.

Whatever the long term solutions to energy, in the next decade or so, the increasing energy famine is only going to be beaten through better management to reduce or at least prevent a rise in per capita demand.

References

1. Costantino M. and Cattaneo P., "L'Impianista e l'Ambienté," *Condizionamento dell Aria*, July 1972.
2. Robertson P., "Economic Sizing and Operation of Boiler Plant," *Conrad*. Vol. 3, No. 1.
3. *Climadata Research International.*
4. Fanger P. O., "Practical Application of the Comfort Equation." *Building Services Engineer*, Sept. 1973.
5. Missenard A., 'L'homme: Son Confort et la Pollution." *Revue Géneralé de Thermique*, Oct. 1973.
6. Lessing F., "L'hydrogene: Energie du futur," *L'Usine Nouvelle*, Feb. 1973.
7. Thring J. B., "Autonomous heat and power," *Heat and Vent. Eng.*, Nov. 1973.

9

The energy intensity of building materials

More energy is consumed by the construction of a building than will be used in many years of operation.

ROBERT A. KEGEL, PE, formerly
Vice President, Hevac Engineering Inc.,
A Division of K.C.M. Engineers,
Wheeling, Illinois, presently with Beling
Engineering Consultants, Joliet, Illinois

Americans have lived in a very affluent society for many generations. We have learned to enjoy and take for granted an over-abundance of energy, food, water, and natural resources. We consume a prodigious amount of commodities during the course of our everyday life. It is part of human nature to be free in spending, in consuming, and, I'm afraid, in misusing such vital commodities as energy. Nations which do not possess these necessities in such abundance have learned (as we have not) to conserve, ration, and use wisely.

Now for the first time in contemporary history, we are beginning to experience shortages in some areas, mainly energy. We recognize some inequitable distribution, allocation problems caused in some cases by the overinvolvement of some government systems, and deliberate tampering with the free market system. These, however, contrary to the statements of ultra-conservative critics of the "establishment" are not the only causes. Other factors include: the prodigious appetite of the American consumer, a tendency to misuse material wealth, and recalcitrance to ration personal expenditure of energy. In short, we are all to blame for the current energy problem!

Those of us who consider ourselves to be realists are convinced that an energy crisis exists—not only exists, but is getting worse. If the problem continues, it will be a real deterent to the policy of continued economic growth. There are those who do have problems admitting that there is a crisis and that the super abudant life enjoyed by the privileged of the world (citizens of the United States and Canada) is threatened. These people wear rose colored glasses.

I am not saying that the "sky is falling in" or that we should revert to a pre-automobile existence—not at all! Depending upon the source of the data, there are still adequate sources of fossil fuel—coal, oil, and natural gas. But, the questions must be asked: How adequate? and at What cost? The answers depend upon the depth of research performed. It is beyond the scope of this chapter to argue the point; however, the fact does remain that there is a limited amount. There was only a given amount created in the beginning, and additional fossil fuels are not being generated, created, or manufac-

we do not conserve today. One does not die from eating one fish contaminated with mercury; however, if it is part of a steady diet, you may become very ill.

In short, we have the responsibility to recognize that the problem exists, thoroughly define it, and collectively solve it. Each of us contributed to the problem; each of us must contribute to the solution.

Energy intensity construction

A great deal has been said, is yet to be said, and should be said about the operation and design of all types of buildings from the standpoint of how much glass, insulation, and lighting should be installed; what type of mechanical system to use; and various ways to save energy, etc. In this chapter, I would like to take a different and new approach to this topic—energy consumed by new construction.

Energy consumed by new construction is over and above that used to heat, light, cool, and ventilate a new building. The materials that a building is constructed of and the equipment used to construct it require considerable quantities of energy during manufacturing assembly and in transportation to the construction site. Construction itself is a consumer of very significant quantities of energy.

Life-cycle cost accounting

A brief review of economics serves as a foundation for this argument. The readers who are economists of some type and/or who are in positions of responsibility in their respective occupations need to be acquainted with the concept of life-cycle costs. Very simply stated, life-cycle accounting is a complete accounting of all expenses incurred with a particular investment over its useful life. For example, in a particular building, an initial investment cost of $40 per sq ft may be realized. The building operating costs, including heating, ventilating, cooling, lighting, and electricity, and maintenance cost for cleaning, repairing, and replacement, plus usury costs are incurred each month and year for the entire life of the

tured. In short, we will run out. Contrary to the rather simplistic and naive attitude of most people toward subjects such as energy or the national debt, it is our responsibility to concern ourselves with the future, as well as the present.

Certainly, the federal government in part recognizes potential problems in the area of energy: note, for example, Project Independence. Whether we can achieve the goal of becoming independent—even by 1985—is questionable.

This is, however, not to say that all forms of fossil fuel will be depleted tomorrow if building. Assume the building has a useful life of 40 years. In that time, it may be necessary to replace a boiler or an air conditioning unit. In that event, the replacement cost must be considered in the total life-cycle cost analysis.

All actual costs and all projected costs should be compared in the same time frame since with the elapse of time, money would earn interest if invested. Future investments can be accounted for by considering the amount of money that would accrue from interest if it had been deposited at an interest bearing rate and withdrawn at a future time to meet interim expenses.

Alternative building systems can be compared by this type of analysis. In true life-cycle considerations, it is within the realm of reason to additionally consider the total use of energy in a given project. This is in addition to both first and owning costs.

Other than the monthly utility charge, regardless of what fuel the building uses, the evaluation of energy expenses has to date been less tangible. Now, however, it is and should be a very real and vital part of the total analysis. A building design should not only be more advantageous from a life-cycle—dollar consideration—standpoint, but it should also be advantageous from a total life-cycle energy viewpoint.

Other energy expenses

Consider that a building is composed of tremendous quantities of steel, aluminum, copper, glass, insulation, dry wall, vinyl tile, carpeting, ceiling tile, asbestos tile, concrete, etc. Consider that it requires approximately 27,500,000 Btu's to fabricate one ton of steel. And this includes only the heat required to convert it from raw ore, limestone, and coke to steel ingots plus rolling to some structural shape. It does not include all the energy necessary for mining, processing, transportation, etc.

Similarly, it requires approximately 82,000,000 Btu's to fabricate one ton of aluminum. Other building components, such as those listed previously, have equally impressive energy requirements. This concept is

Table 9-1. Energy intensiveness of typical building materials.

Material	To fabricate	
	Btu's per lb	Btu's per unit
Aluminum	41,000	
Ceiling materials	1,500	
Concrete	413	
Concrete blocks (8 by 8 by 16 in.)		15,200 per block
Copper	40,000	
Drywall	2,160	
Glass	12,600	
Insulation		
• Duct (1 in., 3 lb density)		51,400 per sq ft
• Pipe (2 in.)		7,700 per ft
• Building (board)		2,040 per sq ft
Paint	4,134	
Roofing		6,945 per sq ft
Steel	13,800	
Vinyl tile	8,000	

Many studies have been initiated by, among others, the Federal Energy Administration to verify this type of data. At the time this research was conducted, much of this information was "classified" by the respective manufacturer. Some of those consulted in obtaining this data are listed here: 1) Aluminum Association of America, 2) American Brick Co., 3) Anaconda American Brass Co., 4) G.D.F. Corp., 5) Gustin Bacon, 6) Gold Bond Corp., 7) Pittsburgh Plate Glass Industries, Inc., 8) Portland Cement Association, 9) Republic Steel Corp., 10) U.S. Gypsum, 11) University of Illinois, Dept. of Forestry.

occasionally referred to as the *energy intensity* of a material. Table 9-1 lists some of these data.

In addition to the energy consumed to properly heat, cool, and light a given space and the energy used to fabricate the various materials, there is a considerable energy expense incurred in erecting the materials—digging foundations, pouring concrete, putting steel in place, hoisting wall panels, pouring floors; installing interior partitions, door frames, windows, glass, carpeting; and assembling a whole host of other building components as the labors of the engineer, architect, and contractor become reality.

There are three major energy components:
• Energy required to fabricate materials and equipment.
• Energy required to erect materials and equipment.
• Energy required to heat, light, cool, ventilate, and other miscellaneous uses for electricity (operating energy). The first two, of course, are expended prior to moving into a building, and the latter encompasses operating energy used after the building is occupied.

These energy components may be looked at as analogous to life-cycle analysis. The exception is that the energy that will be spent in terms of Btu's—not dollars—does not earn money at the present, as invested dollars reserved for future expenditures would earn interest. Therefore, a usury multiplication factor cannot be used as it can with money. In fact, on a monetary basis, the cost of energy will rise with time. We have seen the last days of so-called inexpensive energy in the United States.

To illustrate this concept, we have analyzed a typical project constructed for use as a community college. The building has 432,000 sq ft, three stories, a steel structure, and glass and architectural steel walls. It is located in Chicago. The entire building has been dissected from the standpoints of equipment and materials used, energy consumed for construction, and energy used for operation. Although this building is an institutional structure, a designer may apply the same thought processes in analyzing the energy intensity of any type of building.

During this analysis, it will be noted that familiar buildings are considered, whether they are used for learning, working, playing, health, or living, and more energy was used before they were occupied than will be used for a considerable time after startup. In fact, for this example, construction consumed six years worth of operating energy.

It is very interesting (and something the designer should be aware of) to note the total weight of material and equipment of a building (see Table 9-2). This particular building has approximately 39.5 tons of aluminum, 30,375 tons of concrete, 3518 tons of steel. Table 9-2 summarizes the various components and shows both total quantity as well as equivalent weight in pounds per square foot of the total building. The total weight is 51,779 tons.

Mechanical designers may be interested that the steel alone in mechanical equipment, pipes, and hangers weighs 185 tons. There are 207 tons of steel for plumbing pipes, valves, and accessories; 300 tons of sheet metal and fans; and 24 tons of steel

Typical construction scene at analyzed college.

Table 9-2. Summary of building materials and equipment used to construct the example building.

Material	Total weight, tons	Weight, lbs per sq ft
Aluminum	39.5	0.18
Ceiling materials	200	0.93
Concrete	30,375	141
Concrete blocks	14,040	65
Copper	109	0.50
Drywall	2,157	10
Glass	94.7	0.44
Insulation	27.5	0.1
Paint	17.5	0.1
Plumbing fixtures	100	0.46
Roofing	1,080	5
Steel	3,518	16.3
Vinyl tile	21.5	0.1
Total	51,779.7	240.1

Table 9-3. Summary of energy used to fabricate the materials and equipment used to construct the example building.

Material	Total Btu's $\times$ 10^6	Btu's per sq ft
Aluminum	3,198	7,400
Ceiling materials	600	1,390
Concrete	25,057	58,000
Concrete blocks	3,862	8,940
Copper	8,720	30,190
Drywall	1,492	3,450
Glass	2,484	5,750
Insulation	9,814	22,720
Paint	145	340
Plumbing fixtures	3,971	9,190
Roofing	1,000	2,320
Steel	139,901	323,840
Vinyl tile	1,150	2,660
Total	201,394	476,190

in electrical equipment. This is exclusive of copper, aluminum in the mechanical and electrical equipment and steel in the plumbing fixtures. Electrical switch gear alone has 36 tons of copper. There are 6319 light fixtures; each has copper, plastic, and steel components. Although we think of most insulation as being relatively lightweight, the pipe, duct, and building insulation combined weighs 27.5 tons.

After determining various weights, the values listed in Table 9-1 can be used to translate weight into energy used in fabrication of the various materials and equipment. Table 9-3 shows the totals for these calculations. Remember, these values cover only the manufacture of the material and assembly of equipment, or shipping plus fabrication. Nothing is included for the energy expended in mining and transporting the material to the mills or factories.

Like the data on weight, this information has been itemized on the basis of total energy required to fabricate as well as on a Btu per square foot basis. The numbers expressed in this manner are rather impressive—aren't they? Did you every think that to construct this type of building 476,190 Btu's per sq ft of energy is expended for materials and equipment? Other familiar buildings, regard-

less of construction, would be equally impressive when presented in this way.

Now that the amount of energy consumed by materials and equipment has been determined, let's set this part aside for a moment and define the other component of the energy expended before operation commences—the energy required to erect materials and equipment.

Refer to Table 9-4 and visualize the normal construction activities, such as site clearing, excavation, forms, concrete, backfilling, etc., just to name a few. These have been listed

Table 9-4.   Energy expended on construction.

Activity	Btu's $\times$ 10^6	Btu's per sq ft
1) Clearing of site	7,840	18,148
2) Compaction of materials	200	463
3) Crane operation	1,526	3,532
4) Design energy	2,000	4,630
5) Excavation of footings, etc.	1,045	2,419
6) Fill (backfill)	627	1,452
7) Gasoline powered compressors, etc.	100	231
8) Helicopters to set equipment in place	4,689	10,854
9) Labor traveling to job site	17,500	40,509
10) Material delivery	7,560	17,500
11) Off-site fabrication	10,000	23,100
12) Plumbing and heating, pipe assembly	160	370
13) Roofing	50	115
14) Seeding and planting	100	231
15) Temporary electric power	4,643	10,748
16) Temporary heat	30,600	70,833
Total	88,640	205,135

in a generalized form in Table 9-4. Included are such items as the gasoline used by the designers in commuting to their offices as well as to the jobsite for observation (Item 4). Item 9 includes the expenses of construc-

tion labor traveling to the jobsite. Item 11 includes an estimate of the energy required for off-site fabrication. In this case, it is primarily for architectural steel. Other items listed are self-explanatory.

It is interesting to note that helicopters were used to set certain equipment in place. In this case, two different sizes were used; one has a greater capacity and is used to handle central equipment.

Total energy expended for this effort is 86,640 $\times$ 10^6 Btu's, or the equivalent of 205,135 Btu's per sq ft for the project.

Operating energy

Having determined the energy expended for materials and equipment and to construct the project, the energy required to operate the building can be calculated. The design engineer's calculated heating, ventilating, and cooling loads plus connected lighting and miscellaneous power loads are itemized in Table 9-5. The building is occupied 85 hr per week. This is somewhat greater than for a grade or high school, but it is not unusual for a higher education institution.

Table 9-5.   Summary of energy expended for building operation. Design conditions are: summer: 78° F with 65 percent RH inside and 94° F with 78 percent RH outside; winter: 72° F inside and –10° F outside.

Loads	Energy, Btu's
Heating	
• Building transmission load	3,650,000
• Building ventilation load	17,264,000
Total	20,914,000
Cooling	20,975,000
Lighting and miscellaneous power	2.88 watts per sq ft
Summary	
Heating	
• Day	21,272,934,000
• Night	8,898,709,000
Cooling (Btu's consumed within the building)	6,410,673,000
Lighting and electricity (Btu's consumed within the building)	10,828,728,000
Annual total	47,411,044,000
Total energy per square foot per year	109,748

The ventilation load may seem high, but at the time of construction, state codes dictated very high ventilation quantities.*

Table 9-5 reflects the total annual energy requirement. This was determined via normal ASHRAE calculations. (The writer is aware of the opinion that a computer analysis would be somewhat more complete and more accurate as well.) The result is that the building will require 109,748 Btu's per sq ft per year to provide HVAC, electricity, and a total environment. This works out to be $47,411 \times 10^6$ Btu's per year.

The results of all three of the energy consuming components are totaled in Table 9-6.

The time the building can operate before exceeding the energy consumed prior to occupancy in this example is 6.12 years:

$$(671,325/109,740) = 6.12$$

Other buildings, of course, would have different pay-back periods. The point is, however, that everything requires energy, especially in the construction industry. The designer, mechanical engineer, and architect should all be cognizant of the energy required to produce certain materials, to fabricate equipment, and to erect materials, etc.

*See "Proven Ways to Save Energy in Institutional Buildings," Chapter 54.

Table 9-6. Summary of energy consumed by all three major components of project.

Energy consumed	Base building, Btu's	Btu's per sq ft per yr
Building materials and equipment	201,394,000,000	466,190
Construction energy	88,640,000,000	205,135
Total	290,034,000,000	671,325
Operating energy, annual	47,411,044,000	109,748

Energy intensive materials

It is evident from looking at a list of materials that some require more Btu's per pound or per square foot to fabricate than others. These are the so-called energy intensive materials. Glass, aluminum, and steel, for example, are higher in energy intensity than drywall or paint. Frequently, a designer (particularly where the eventual operation of a building is affected) has a choice of using various materials that have a bearing on the total energy, and life-cycle accountability of building. For example, steel can be used in lieu of aluminum, or wood can be substituted for either of these. The design of a building may be optimized based on the materials selected by the designer and the total energy required to construct it. Operating energy may be decreased in addition.

It is not fair to simply look at materials

on the basis of energy intensiveness but rather what is being gained from the investment of energy required to produce the material.

It is interesting to consider in relation to life-cycle concepts, the pay-back period in terms of energy of a particular material. For example, consider insulation. Insulation has a rather high energy intensity—a high ratio of Btu's per pound to fabricate. However, in spite of the rather large quantity of insulation used in the example building, it really does not weigh very much per square foot. It actually has a very high pay-back rate!

Consider another example. To fabricate a 1 sq ft sheet of $\frac{1}{8}$ in. glass requires 19,500 Btu's (heating only). The U-value for single glazing is 1.13 Btu's per sq ft, °F; and for double glazing, it is 0.64 Btu per sq ft, °F. The difference is 0.49 Btu per sq ft, °F. If we assume an inside design temperature of 72°F, and an average outside temperature of 35°F (typical of northern Illinois), the time required to recover 19,650 Btu's through the investment in a second sheet of glass is (space heating only):

$$19,650/0.49(72 - 35) = 1084 \text{ hr}$$

A typical heating season in the Chicago area is usually put at about 5500 operating hours. At this rate, the second sheet of glass would pay for itself in terms of energy in less than 0.2 years. Certainly, this is a wise investment of energy. Remember, this is on the basis of heating only. If cooling were to be considered also, it would be even less.

Other examples comparing materials to Btu's saved per Btu's invested per year are: glass fiber building insulation, 20:1; duct insulation 15:1; pipe insulation, 47:1; and glass (by above example), 5:1. Similarly, a designer should investigate other examples.

Energy optimization, or energy consciousness, and its uses in our environment are quite evident in these examples. Each of us in our everyday lives should be conscious and aware of the energy required to support our way of life as it exists today and should be in the future. I am not proposing that Americans discontinue their way of life, only that we conserve energy to provide for future generations.

It is our responsibility to use each Btu as wisely, as conscientiously, and as carefully as possible. Then, we can provide for continued growth of the economy.

The building is used by the City Colleges of Chicago and is owned by the Capital Development Board, State of Illinois, with whom Mr. Kegel worked during this project. The architect was Dubin Dubin Black & Moutoussamy; the mechanical engineer was Environmental Systems Design, Inc.

10

Energy conservation and HVAC

E. R. AMBROSE, PE,
Mount Dora, Florida

The ever widening energy gap in the United States between the available supply and the increasing rate of consumption is receiving serious attention from government agencies, as well as from industry and other interested groups. There is cause for concern, and the situation can get progressively worse. Studies indicate the total energy demand in the United States has doubled in the past 20 years, will double again by 1985, and triple by the end of the century if allowed to continue unabated.

Contributing factors

Contributing to this quandary is space heating, which accounts for roughly 18 percent of the total energy consumption, and air conditioning, while representing only about 2.5 percent, seems to be growing at a rate 2.5 times faster than the total usage. When domestic water heating, refrigeration of foods, and lighting (all closely related to heating and air conditioning) are added, the percentage increases to roughly 28 percent of the total energy usage (Fig. 10-1).

What are the possibilities for applying more extensive energy conservation practices in the heating and air conditioning fields? Certainly, no particular effort has been made during the past 30 or 40 years to conserve energy since there seemed to be an unlimited supply at a relatively low cost. There was little or no incentive to conserve. The systems were developed and the equipment selected to produce desired conditions at minimum first cost. The operating cost was usually secondary. Now year 'round air conditioning is an accepted and essential part of every new building of any significance, and the accelerated growth is likely to continue.

Codes and ordinances revisions

Fortunately, attractive possibilities exist for improving the energy consumption (in both new and existing structures) by taking advantage of available energy conservation methods and procedures to eliminate unnecessary waste and inefficiency. To bring this about in a practical and expeditious manner will require revisions in the existing local and state obsolete codes and ordinances, many of which are at least 40 to 50 years old. Also, the standards and practices of the manufacturers, contractors, trade associations, and labor unions, among others, will have to be updated to achieve the desired results.

A start has already been made in this direction. The National Bureau of Standards (NBS) at the request of the National Conference of States on Building Codes and Standards prepared a draft of a standard on energy conservation for new buildings. This draft was referred to the American Society of Heating, Refrigerating and Air-Conditioning Engineers (ASHRAE) in February 1974, with a request that they proceed with the writing of a standard using the NBS draft as a base. The ASHRAE standard,

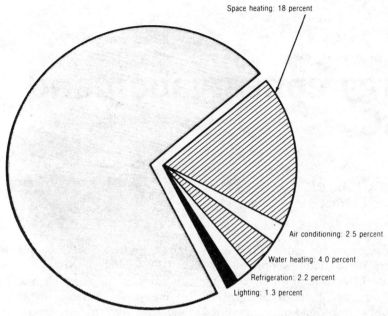

Space heating: 18 percent

Air conditioning: 2.5 percent

Water heating: 4.0 percent

Refrigeration: 2.2 percent

Lighting: 1.3 percent

Total: 28 percent

Fig. 10-1. Estimated percentage of total energy consumption in U.S. allocated to space heating and closely related systems.

90–75, has been adopted and can be constantly updated and serve as a reference for local codes and ordinances and for future federal legislation.

Practices, procedures

It was not an easy task to obtain agreement and acceptance among all interested parties on such standards for energy conservation. A considerable portion of the present practices followed by architects, consulting engineers, and contractors have been acquired over the years. These groups usually have a preconceived preference for a particular type of structure or a given system design, which have proven dependable and satisfactory. They are familiar with the operating and first costs of these designs and are reluctant to change without sufficient justification.

Universal acceptance of the standards may come in several different ways. The attention being given to the critical energy situation is making the government and the public more aware of some of the possible improvements. This situation should cause the heating and air conditioning industries to undergo a self-examination of their position. Proper specifications, codes, and ordinances should make it easier for manufacturers to justify more efficient equipment and for the architects and engineers to convince the owners to select the proper structure and the most economical system for minimum energy consumption.

Undoubtedly, some government legislation will be necessary to outlaw certain wasteful practices and procedures and to establish a national energy conservation policy. Caution must be exercised in this regard. It is desirable to first validate the effectiveness of certain conservation practices before recommendations are enacted into law and become mandatory.

Normally, the building construction as well as the type of system chosen will be governed by economic justification unless there are laws to the contrary. Taxing and banking influences must help to provide the necessary motivation and support. Also, in the final analysis, the actual amount of energy conservation realized will depend to a large degree on the extent and nature of the procedure actually implemented by the building owner and the operator.

11

Energy effectiveness factor

A new evaluation function permitting comparison of alternate integrated conversion systems.

WILLIAM J. COAD, PE,
Vice President, Charles, J. R. McClure
& Associates, St. Louis, Missouri

Throughout the engineering sciences, the concept of evaluation functions is employed as the prevailing technique for comparing alternatives. And in the design of machinery, components, or systems, the iterative process of organized comparisons of alternatives is the fundamental method of ultimately achieving the best result from the alternatives considered. Some examples of evaluation functions include temperature, specific heat, specific weight, enthalpy, modulus of elasticity, thermal efficiency, coefficient of performance, and many others.

This chapter poses the need for, and presents the concept of, a new evaluation function called *energy effectiveness factor.*

The need for the development of a concept such as energy effectiveness factor was created by a combination of the recent awareness of limited energy resources to provide for the needs of an increasingly energy dependent society and the increasing attention being given to the concept of integrated energy conversion plants or energy communities. For the purpose of this discussion, an energy community is defined as a system within defined boundaries having energy needs in one or more of the following forms:

- Shaft power.
- Light.

- Heat loss to surroundings, requiring an equal input to maintain steady state conditions.
- Heat gain from surroundings, shaft power, light, or biological cycles, requiring an equal removal rate to maintain steady state conditions.

The energy effectiveness factor, Ee, is defined as a dimensionless ratio that enables the effectiveness of the conversion of energy from the depletable resource potential form to the final use form to be expressed.

In efforts preceding the origin of the integrated conversion plant concept, evaluation functions directed at comparing or illustrating the effectiveness of the conversion included:

- *Thermal efficiency*—the net work delivered by a heat engine to some external system divided by the heat supplied to the engine from a high temperature source, in consistent units (dimensionless decimal or percent, generally not time integrated; *i.e.*, a power ratio at design output).
- *Heat rate*—usually applied to electric power generating plants and defined as the annual or seasonal fuel input in thermal units (Btu's) divided by the plant electrical product in kilowatt-hours (Btu/kwh).
- *Fuel rate*—usually applied to internal combustion prime movers and defined as the

fuel input in gravimetric or volumetric units divided by the shaft or delivered energy output (lb fuel/hp hr).

• *Coefficient of performance*—a power function relating to refrigeration machines and defined as the ratio of the rate of heat into the refrigeration system from the low temperature sink to the motivating power input, both expressed in consistent units (dimensionless ratio). A dimensional form of *COP* is called the energy efficiency ratio, *EER*, with the numerator expressed in units of Btu's per hour and the denominator in watts (Btuh/w).

• *Conversion efficiency*—a function relating to energy conversion in plants wherein the conversion is from a potential contained in input fuel or resource to a high level thermal output, defined as the useful thermal form output divided by the potential combustion heat input (*HHV*) in consistent thermal units (a dimensionless ratio usually expressed as a percent).

Integrated plants have been developed through efforts to make more effective use of resource energy potential. They include plants with the following combinations of products:

• Electricity and heating.
• Electricity, heating, and·cooling.
• Electricity and cooling.
• Heating and cooling.

In evaluating the effectiveness of these energy service facilities to provide for the needs of any given energy community, a method of comparison of alternative schemes is mandated. If an evaluation function method is used for the comparison, the same method can be employed to compare the performance of the operating entity to the performance anticipated during the early decision making stages of design.

How to determine Ee

To develop the concept of energy effectiveness factor, *Ee*, consider two plants that provide only one form of product energy. The first plant, illustrated in Fig. 11-1 by the first law balance diagram, is an electrical generating plant. Definition of the boundaries is extremely important, and for the purposes of this discussion the boundaries are defined as the plant itself and its distribution network to the point of product delivery. What occurs within the boundaries (the "system") in the form of energy conversion characteristics of the subsystems and components contributes totally to the value of *Ee* but is not relevant in the final analysis of the numerical calculation.

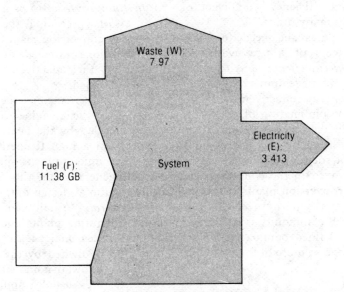

Fig. 11-1. First law balance diagram for electric conversion "system."

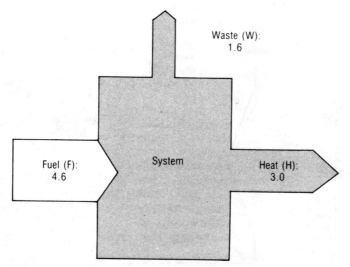

Fig. 11-2. First law balance diagram for thermal conversion "system."

Since the energy flow from the "system," E, and the product delivered, e, are both from the plant to the community,

$$\text{product} = \text{energy flow}$$

$$e = E$$

In the example, the annual product delivered is 1 million kwh, which expressed in equivalent thermal units is 3.43 GB.*

The input fuel energy, E, is expressed as the high heat value of the depletable energy resources (fossil or nuclear fuel) consumed by the plant (or community) annually. Since this energy flow is also in the same direction as the product flow, f,

$$f = F$$

And for the example shown, $F = 11.38$ GB.

The Ee is calculated by dividing the output *product* by the input *product*, or

$$Ee = e/f$$

$$= 3.413/11.38$$

$$= 0.30$$

The analysis to determine the input energy required to generate the useful output con-

*For the purposes of this discussion, $1GB = 10 \times 10^9$ Btu.

sisted of highly sophisticated calculations involving plant machinery characteristics, annual load profiles, part load subsystem performance profiles, plant burdens, combustion losses, distribution system losses, etc.; but once the input required to produce the annual output was determined, the only data required to calculate Ee were the input quantity and the delivered output quantity. Also, the only instrumentation required to determine the Ee of an operating plant is metering of the annual input and product.

The second example is the plant illustrated by the first law balance diagram of Fig. 11-2, which is a thermal plant supplying steam to the energy community. The thermodynamic difference between this and the first example is that we now have a first law conversion plant instead of a second law conversion plant. But as in the first example, both the energy flow, H, and the product flow, h, are from the plant to the community; thus, again:

$$\text{product} = \text{energy flow}$$

$$h = H$$

In the example shown, the product delivered, h, is 3.0 GB, and the input product, f, is 4.6 GB. The energy effectiveness is

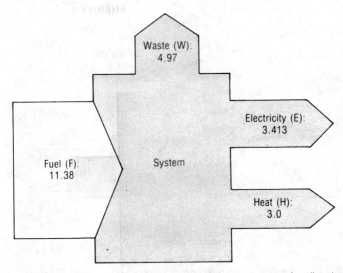

Fig. 11-3. First law balance diagram for electric/heat conversion "system."

then:

$$Ee = h/f$$

$$= 3/4.6$$

$$= 0.65$$

In calculating the energy effectiveness factor, all inputs to the plant must be expressed in thermal value of fuel. Thus, any electric energy input to the thermal plant of Fig. 11-2 must be divided by the Ee of the generating plant that produced the electricity and the quotient added to the thermal value of the input fuel for the thermal plant.

If the products of Figs. 11-1 and 11-2 were to be produced by an integrated plant producing both steam and electricity, the first law balance diagram might appear as in Fig. 11-3, where

$$e = 3.413 \text{ GB}$$

$$h = 3.0 \text{ GB}$$

$$f = 11.38 \text{ GB}$$

and

$$Ee = (e + h)/f$$

$$= (3.413 + 3.0)/11.38$$

$$= 0.56$$

Note that the Ee of this combined plant, compared to the Ee of the two separate plants in serving the *same* community re-

quirements (0.40),* reveals a 40 percent more effective use of the resources.

The Ee thus developed is analogous to the concept of seasonal efficiency as applied to either the first law or the second law plant. The energy effectiveness factor makes it apparent that in an integrated plant, the "seasonal efficiency" for the generation of electric power is identical to the "seasonal efficiency" for the generation of thermal energy, and the "seasonal efficiency" for each is equal to the energy effectiveness factor.

A look at heat removal

Most energy communities have three basic forms of product energy requirement:

• Electric energy—for lighting and power drives.

• Heat addition—to offset heat losses to surroundings maintaining comfort conditions, to heat water, and to provide for process needs.

• Heat removal—to remove heat and moisture gained from surroundings, to provide for low temperature storage, and to meet low temperature process needs.

The heat removal could be accomplished

*The sum of the two outputs, 3.413 + 3.0, divided by the sum of the two inputs, 11.38 + 4.6, = 6.413/15.98 = 0.40.

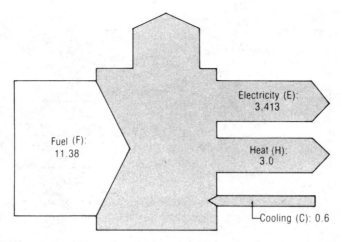

Fig. 11-4a. First law balance diagram for electric/heating/cooling system.

within the community by one or a combination of electric energy to power compression refrigeration machines, fuel to power a prime mover driving compression refrigeration or powering absorption refrigeration, or thermal energy to power either compression or absorption refrigeration. Even the most cursory consideration of the phenomenon of balancing thermal and shaft requirements of a combined plant reveals that the load balance is a function of the product delivered or, stated another way, a function of the *energy community systems*. If these are balanced between the thermal and electric forms such that they optimize the use of salvage heat from the second law generating process within the plant, the sum of *e + h* compared to *f* will increase. As a result, the energy needs of the same community, although remaining fixed, will be provided for with less consumption of resource energy.

In many cases, such as central heating-cooling plants, total energy plants, and recent conceptual developments known as MIUS and TIES* systems, the cooling requirements of the community are integrated into the plant. In this case, in addition to the electric and heat energy supplied to the community, heat energy is removed from the community to the plant through a low tem-

perature fluid system (either single or two-phase). Fig. 11-4a is a first law balance diagram for a plant that includes products of electricity, heating, and cooling. Note in the first law balance that the cooling energy enters the plant boundaries. Despite this, it represents a plant *product* delivered. This phenomenon, recognized by any student of thermodynamics (that the energy flow is negative while the useful product is positive), is the phenomenon that led to such evaluation functions as coefficient of performance. *COP*, and more recently, energy efficiency ratio, *EER*, both conceived to express the relationship between low temperature energy removed and energy consumed to accomplish the removal.

Classically, coefficient of performance is defined as refrigeration effect divided by input energy required to accomplish the effect, both expressed in equivalent units.

As stated above, fuel, electricity, or thermal energy could be converted outside the boundaries of the plant to provide the cooling needs of the community. If this were done, the energy required would be:

$$e, h, \text{ or } f = \frac{\text{cooling required}}{(COP)_s}$$

where

$(COP)_s = COP$ of community refrigeration system(s)

*MIUS is Modular Integrated Utility System and TIES is Total Integrated Energy System.

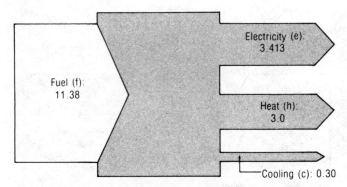

Fig. 11-4b. Product diagram for electric/heating/cooling system.

It is germane to the concept of energy effectiveness factor to recognize that *the consideration here is to establish the value of the product produced by the plant in units consistent with those of the other products, and that how the refrigeration is accomplished within the plant is irrelevant.* Thus, the numerical value of $(COP)_s$ must be fixed at a level commercially available in community systems. Theoretically, the $(COP)_s$ could be the annually integrated Carnot *COP*; in reality, however, the Carnot *COP* cannot be reasonably approached for any system being compared or evaluated.

As an example of a reasonable value relating to articles of commerce, it is suggested that $(COP)_s$ be set at the values established in ASHRAE Standard 90-75, a consensus standard for energy conservation in new buildings. The standard sets minimum values of *COP* for various size categories of machinery, consistent with achievable limits in articles of commerce. Thus, these values, calculated by a weighted average of the connected loads, would serve as useful constants in defining a universal evaluation function.

The energy effectiveness factor for a combined plant that includes cooling product can then be developed by addressing the difference between the first law balance diagram of Fig. 11-4a and the product diagram of Fig. 11-4b. Again, the boundaries of the system being evaluated must be carefully defined, and in the simplest terms must include the conversion plant, distribution system, and any satellite conversion

apparatus connected to the distribution system. In converting from the first law balance concept to the product concept, *the boundaries must be held fixed.*

Referring to Fig. 11-4a, the energy values are:

E = annual electric energy output (kwh) $\times$ 3.413
H = annual heat energy output (Btu)
C = annual cooling energy input (Btu)
F = annual fuel energy input (Btu)
W = annual energy wasted to environment (Btu)

And the product values for Fig. 11-4b are:

$e = E$
$h = H$
$c = C/(COP)_s$
$f = F$

The energy effectiveness factor for the integrated plant of Fig. 11-4b is then:

$$Ee = (e + h + c)/f$$

The numerical value of $(COP)_s$ used to convert the cooling energy of Fig. 11-4a to the cooling product of Fig. 11-4b was 1.8. This value was taken from ASHRAE Standard 90-75, a 1977 value for apparatus with a capacity less than 65,000 Btuh, which would typically represent a residential type energy community. The *Ee* for the combined plant serving the community is:

$$Ee = \frac{3.413 + 3.0 + 0.33}{11.38}$$

$$= 0.59$$

Use Ee for solar, too

The concept of energy effectiveness factor is a useful tool in quantifying the value of using nondepleting energy sources such as solar energy. As an example, if the "system" of Fig. 11-4 were to be designed with a solar collection and conversion system that, with the same product, required only 10 GB of fuel energy (reduced from 11.38 GB by the solar energy utilized), the *Ee* would be:

$$Ee = \frac{3.413 + 3.0 + 0.33}{10}$$

$$= 0.67$$

Thus, there would be a 14 percent increase in the effectiveness of the use of resource energy.

In summary

Energy effectiveness factor is based on the concept of product value ratio rather than energy ratio. This provides for the combination of cooling energy, heating energy, and electric energy in a single evaluation parameter (a combination required for a total evaluation of effectiveness in the use of energy in contemporary energy communities).

If properly applied, it provides a consistently valid evaluation function for both comparing alternative methods of conversion for any given energy community and comparing actual with anticipated performance using a minimum of instrumentation.

Energy effectiveness factor can be applied to any integrated community, from a single residence to an entire living-working-transportation community served by a single conversion plant or by a multitude of plants and direct fuel supply points.

The fundamental requirements for proper application are to convert rigidly from energy to product terms and to define carefully the boundaries of the analysis.

In a recent study of one energy community served in 18 different manners, the energy effectiveness factor was found to vary from 0.28 to 0.416 with the consumption of resource varying in inverse correlation from 360 to 263 GB per year.

SECTION II

Planning for energy conservation

The importance of developing and implementing an energy conservation program is discussed in these 7 chapters along with the following detailed subjects relating to the planning phase:

- A detailed problem solving approach for organizing an energy conservation program.
- The use of the computer for energy analysis.
- A flexible approach to computerized HVAC system analysis.
- The architectural aspects relating to HVAC systems.
- The need for thermal insulation and double glazing.
- The importance of reducing excessive infiltration and ventilation air.

12

How to develop and implement an energy conservation program

A detailed problem solving approach to help organize an attack against the multi-faceted energy conservation dilemma.

JOHN S. BLOSSOM, PE,
President, Ziel-Blossom & Associates,
Inc. Cincinnati, Ohio

The energy crisis and its effects may be with us as engineers the rest of our working lives. Engineers can make a significant short-term contribution to solving the problem by setting up energy conserving programs in new and existing buildings. This is no minor task.

The broad spectrum of ideas, factors, and problems involved in developing and implementing an energy conservation program requires some type of methodical framework to relate and integrate the various complex parts into a unified whole. To achieve this goal, a problem solving approach is needed. This term is often used to refer to management seminar work or management techniques that have been developed on an administrative level in a context that is not ordinarily associated with engineering. The problem solving approach outlined in this chapter is somewhat of a combination of both engineering and man-

This chapter is based on a paper presented by Mr. Blossom at a special series of two-day seminars sponsored and developed by Heating/Piping/Air Conditioning on How to Save Energy in Existing Buildings.

agement methodolgies, and it deals specifically with how to attack the problem of developing and implementing an energy saving program.

In this chapter, two flow charts are shown that outline a problem solving technique. This approach consists of eight basic steps. Each step can be amplified or reduced in scope to fit particular problems, but it is important that all eight steps are followed to some degree.

This type of approach has been developed over several years, and it has been used in a multitude of various problems in facility engineering. We developed it as a method of relating to clients and ourselves three basic factors: what we are doing, where we are going, and how we got there if we succeeded. We have refined this technique to the point that it has become a specific program.

The basic approach

The first step in this approach as shown in Fig. 12-1 is to define the objective.

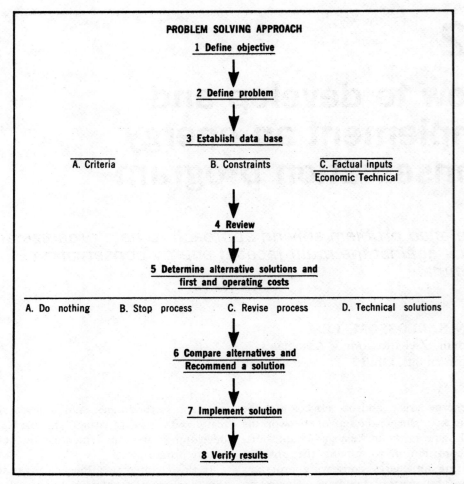

Fig. 12-1. Flow chart illustrating the general basic steps in the problem solving approach.

Do not assume that you know it. Even if it is simple, it is worth writing down. The objective in this chapter, of course, is to explain a method of analyzing energy conservation.

The second step is to define the problem. The problem is not the same as the objective. The problem is why you have an objective in the first place. The trouble spot must be defined before initiating the third step, which is to establish a data base.

A data base consists of the factual inputs that reflect on both the objective and the problem. We have broken this step down into three categories. Category A is termed criteria. In our professional context, criteria usually refers to factors imposed by an owner,

a regulatory agency, or any external considerations that limit our actions in a given situation. For example, an owner might set a budget limit.

Category B, constraints, might be interpreted as being very similiar to criteria in some cases. However, constraints as used here refer to physical limits involved in a particular problem that relate to the building itself, its envelope, or various other factors of a similar nature.

The factual inputs dealt with are both technical and economic in nature and relate to these specific elements. This division makes up category C and examples will be presented throughout the discussion.

The fourth problem solving step is review.

It is one of the most important steps because some feedback is developed at this point. Feedback is real, important, and useful. In some actual problem solving cases, we had proceeded some length before we discovered that the owner's idea of the problem and ours were two different things, and a lot of time and effort had been expended pulling in two different directions. This does not lead to good relationships with clients or owners, and it is really only a common problem in communication. A problem solving approach is a method of communication. Review is basic and should not be skipped.

The fifth step is to determine alternative solutions and associated first and operating costs. The first two choices are extreme. There are really only two limits in any problem. The first one is to do nothing—leave it alone. It is one solution that is always available to you. It may not be any good, but it exists. The second extreme is to quit doing whatever is involved. A building could be shut down or abandoned. Between these two poles, a whole series of solutions exist that revolve around Alternative C, revise the process, which covers a general format for problem solving. This finally leads to the point where various technical solutions and how they relate to your particular problem are reached.

After the alternative solutions are determined, the next step (six) is to compare the various alternatives and to recommend a specific program to solve the problem.

Step seven, then, is to implement the solution by going through the various steps involved. Finally, step eight is to verify the results.

Those are the eight basic steps, and in our way of thinking, none of them should be omitted in any problem solving approach.

Air and water pollution

Table 12-1 was developed as a general approach to air and water pollution problems. In this instance, provisions have been made for noise and solid wastes, but the headings can be changed to reflect any particular environmental problem. Air and water pollution are included in this dis-

cussion because these are elements of energy conservation. Such regulatory problems must be related to how they become energy users, and hence, become part of the total problem. Viewed in this context, the steps in Table 12-1 become a little clearer.

The first of the eight basic steps, to define the objective, in this case is usually to meet federal and state regulations. The second step relates to the system, the equipment, the personnel, or whatever elements enter into a particular air or water process. After the owner's criteria and any pertinent constraints have been determined, technical data must be collected. This can be accomplished through either a water audit in the case of water pollution or an emission inventory, stack sampling, and testing, etc., to gather air pollution data. Now, it is time to review; go back and ask "Do I have all the data relating to the defined problem; and if so, do I feel that I am in a good position to meet the objectives?"

When you feel confident that the answer to both is yes, the alternative solutions in Step five is the next effort—should nothing be done, stop the process, or revise the process? The final alternative in this case becomes the various related technical steps, such as for air pollution, apply scrubbers, filter, collect, incinerate, etc. After you have considered all the alternatives, step six is to recommend a solution; step seven is to construct it and put it into operation; step eight is to verify the results and see if it works.

Verify means to return to steps one and two and see if the objective actually was met and the problem solved. Sometimes you find that neither was satisfactorily achieved.

An expanded approach

In Fig. 12-2, the eight basic steps have been expanded to include more specific considerations involved in energy conservation problem solving. However, it is still somewhat generalized in its selection of the common elements of the problem. It is really impossible to include every item involved in energy conservation; there is almost no place to stop. Figure 12-2 is an attempt to list

Table 12-1. Problem solving approach applied to environmental problems.

Steps in problem solving approach	Air pollution	Water pollution	Noise	Solid waste
1 Define objective: Legal requirements Lower costs Environment Other	Meet federal and state regulations	Meet federal and state regulations		
2 Define the problem: System Equipment Personnel	Survey the plant	Survey the plant		
3 Establish a data base	Emission inventory Stack testing	Water audit, flow and temperature Test sample Flow chart Piping distribution system		
4 Review	Do	Do		
5 Determine alternative solutions and first and operating costs of each	Scrub Filter Collect Incinerate, etc.	Reduce flow Control flow Recirculate Filter Separate Treat, etc.		
6 Compare alternatives and recommend a solution	Check solution against objective and problem	Check solution against objective and problems		
7 Implement solution Objective met? New problems?	Plans and specifications Purchase Install Test Test and monitor	Plans and specifications Purchase Install Test Test and monitor		

the most probable ones, but, of course, each specific problem generates its own list.

The objective is to save energy. A degree of amplification is always necessary because it is important to know what types of energy are pertinent in an existing building. Fig. 12-2 specifies the common ones: kilowatt hours of electricity, tons of coal, gallons of oil, cubic feet of gas, and man-hours. Man-hours are energy. The number of people and how they are used determine whether more or less energy will be consumed, but regardless of this, each person in a process consumes energy and represents a possible opportunity to save energy.

After determining exactly what types of energy are to be conserved, the second step is to define the problem. The current problem really relates to three major elements. The first is the limited availability of energy sources. This varies, of course, with each source. The second element is the astounding rate of cost increases, and the third is the growth of government regulations.

These are the problem elements, and each affects each particular energy source. In other words, availability affects all types of energy, electricity, oil, gas, man-hours, etc. We have done studies where the energy resource in shortest supply was manpower, and some of the required measures were dictated by the shortage of skills in this particular area. Yet in other studies, increasing costs have been pinpointed as the problem, and this, of course, affects every energy source.

Recently, we did a plant study where the

owner was paying $58.10 per ton of coal, which is the highest price that I have encountered. Studies conducted only three or four years ago priced coal anywhere from $7 to $12 per ton and predicted a fairly flat price curve for the next 10 years.

Once the problem has been defined, a data base must be developed. The data base has several major elements that have been divided into two major categories called analytical elements and physical components. The analytical elements are subdivided into two parts: economic and technical. The economic costs are the usual ones: first costs and operating and maintenance costs. Life cycle economic costs, benefit-cost ratio, cost projections, and taxes are the basic influential elements on the economic side.

On the technical side, there are more than engineering considerations involved. Engineering is the first item listed, and this category covers the whole gamut of mechanical, electrical, etc., engineering features of concern here.

This category also covers the need to deal with humanistic factors, both psychological and physiological, in an energy conservation program. Both are very important when changing modes of operation or operating conditions within an environment. These must be done in a climate that considers the human element so that the process is not defeated by those it affects. For example, if an operating solution calls for changing the on-off schedule for lights, people must cooperate or it won't work. The schedule must consider their needs.

The third element is regulatory factors, such as building codes: the pollution regulations, and the energy regulations of some form.

The regulatory category should be carefully tracked as it develops. For example, Section 8 of Senate bill No. 2176 passed by the Senate would require engineers and contractors who design and install central air conditioning systems in previously occupied buildings to furnish the buyer (beginning in about two years) with the estimated annual operating costs. This would be stated in advertisements of specific systems given to prospective customers at the time quotations are offered and included in sales contracts. These estimates of annual operating costs would be based on so-called model calculation procedures evolved by the federal trade commission through rule making proceedings to be conducted within 18 months.

Professional engineering societies and organizations representing the climate conditioning engineering industry would participate in these proceedings. Senator Tunney of California has said that provisions of Section 8 are limited to previously occupied buildings at the insistence of the building industry. This is in the political world and subject to constraints that are neither technical nor logical in nature.

The final technical element covers the availability of resources, such as fuels, manpower, building space, funds, and so on.

The physical components relate to the building itself, the site, the mechanical and electrical systems. The rather obvious elements that occur with each of these are listed.

Step four, having defined the objectives, the problem, and gathered a data base, is to review and develop feedback. If all is satisfactory to this point, proceed to step five, developing alternatives.

As stated previously, alternatives are divided into two categories: operational solutions and engineered solutions. Operational solutions include items that are usually performed in-house, such as setting controls, modifying outside air intakes, operating blowers and lights on schedule, etc. Engineered solutions may require modifications of systems that are beyond in-house capabilities. This might lead to the development of plans and specifications for construction by another organization.

Once the possible solutions are ready, then step six is to decide what to do next, make a recommendation or take a position on the problem. Now is the time to compare each alternative against the defined objective and problem and then recommend a specific energy conservation program. This process

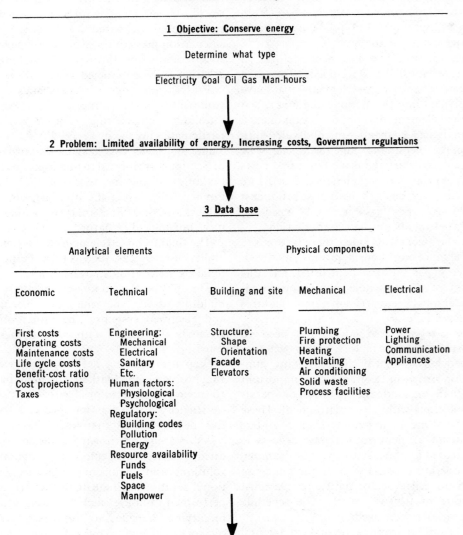

Fig. 12-2. An expanded flow chart of the eight basic steps in the problem solving approach.

can result in two phases: a study report phase and a plans and specifications phase.

A study report is usually a way of writing up operational solutions and spelling them out. Plans and specifications are a method of defining an engineered solution that leads to a construction program. Step seven then is to implement this program. This presumes that the program is presented to a decision making authority and that an agreement is reached on what form of action to take. The next step is to take your ideas and put them to work by following the recommendations in the study report and/or constructing the facilities generated by the plans and specifications. Finally, put the plan into operation, test it, and verify it against the problem definition.

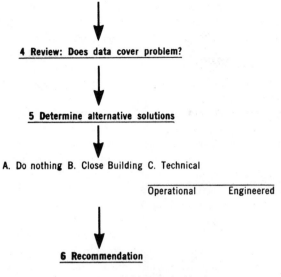

4 Review: Does data cover problem?

5 Determine alternative solutions

A. Do nothing B. Close Building C. Technical

Operational   Engineered

6 Recommendation

Compare alternative against problem and objective

Recommend a specific energy conservation program

Study report   Plans and specifications

7 Implementation

Make operational techniques developed in study

Construct facilities per plans and specifications

Test all work implemented against program definition

8 Verification

Monitor program as operating

Evaluate monitored data and optimize program details

Verify if problem was solved and objective attained

Fig. 12-2. *(Continued).*

Now the energy conservation program is almost complete except for one important facet, gathering feedback. It is not only feedback to the original problem that is important, but also on the complete program in existence now to see if it works as intended. This is an important step!

We do considerable remedial design work involving systems that have never been placed in operation in the manner intended. This is a very common problem that points to the second important facet of feedback, verifying the entire system operation. This requires monitoring the program by collecting data to check actual operating conditions. Then the data must be evaluated to optimize the program details. The program itself limits how far this can proceed.

At this point, the professional can feel that he has done everything possible to make the program work. The last step is to return to the original objective and problem to see if the objective was attained and the problem solved. This is the last feedback step, and it is very critical.

There are some conclusions regarding this approach that we have garnered in the time that we have been using it. These should be emphasized to avoid mistakes.

Avoiding mistakes

It is vital to perform each step in order, even though the problem seems extremely obvious and the data is apparent. Each step should be written down in a clear order. It is very easy to have an incomplete problem definition that results in a communications breakdown among the owner, manager, client, and the engineer.

The most common mistake probably made in problem solving is to immediately jump to step five or six as soon as we hear the problem and to start talking about solutions. We have made mistakes this way, and we have also consistently found that if we go back through all the steps, we will save time and conserve our own energy.

A third cautionary comment is the need for a *factual* data base. It is not always easy to get factual data. It may be necessary to collect your own hard data by monitoring and taking test readings on the electrial side and checking air flow, etc., on the mechanical side—in short, performing yourself whatever must be done to obtain real data. Assumed inputs can cause a great deal of trouble.

The second critical point regarding the data base is to obtain the owner's criteria or constraints, particularly in the economic category. Some owners are very insistent on short payback periods, which can definitely limit the scope of engineering solutions. When this occurs, the owner must be told the meaning of the constraint, and why it will have a certain effect.

A secondary precaution is the need for the development of the energy conservation program as a set of interacting elements. It may be a good idea to split up the whole program into a series of problem solving approaches and run each element (one relating to electrical work, one relating to air or water, etc.) through as individual problems, but each element must be integrated back into the program to form a whole entity.

13

The computer as a tool for energy analysis

The engineer is the technician; the computer is the tool. The engineer provides the judgment and skill, and the computer yields the optimum in quantitative data. Together they can lick many of today's complex energy problems.

WILLIAM J. COAD, PE,
Vice President, Charles J. R. McClure & Associates, and Affiliate Professor of Mechanical & Aerospace Engineering, Washington University, St. Louis, Missouri

Computers have proven very useful for many applications. In the HVAC field, they have greatly increased the capability of the consulting mechanical engineer in designing systems. They have also provided him with the means to study systems in existing buildings to effect substantial savings in energy and operating costs.

The intent of this chapter is to discuss the functions of the computer as a tool. The engineer is the technician; the computer is a tool. How does the computer fit into the problem solving process?

A computer performs the arithmetic necessary to yield reliable data required for problem solving. An engineer recognizes that many assumptions and approximations are required in solving numerous problems. As a field of engineering matures, it becomes

This chapter is based upon a paper originally presented by Mr. Coad at a special series of two-day seminars on How to Save Energy in Existing Buildings sponsored by Heating/Piping/Air Conditioning magazine.

increasingly exacting, requiring a diminishing use of the "approximation" approach.

The computer is the tool that has permitted engineers to reach the nth degree of exactness for many types of problems because it can perform many complex calculations in seconds. Without it, comparable computations would require unreasonable lengths of time. Simulating the performance of a HVAC system to determine annual energy usage could take several months, even on the simplest of buildings. Yet, the result would not be as accurate as a properly designed computer program could provide in a few hours. Even changing one or two variables to examine their effect on consumption could necessitate redoing many or all of the calculations. In the past, this situation prompted many assumptions and approximations to be used in determining a building's energy consumption. To conserve valuable time, this was a necessary alternative. A computer provides the quantitative data that otherwise would have to be approximated; that is all that it does!

When a computer is applied to the area of energy economics, two considerations are necessary. First, the applied program must contain the proper algorithms relating to the problem and all mathematical relationships that affect energy consumption in a building. Second, the output information should not simply be taken as the answer. This was a common error in some initial efforts, but as computerized engineering problem solving developed, it was realized that *the* answer, or final solution, could only be obtained by applying skilled engineering judgment. A computer simply gives good quantitative data that can be used to arrive at better solutions.

The end-to-end computer program, or one in which input data is supplied and a final solution comes out, is generally not a good engineering program. Valuable data generated between these two points, which might indicate that a change in input is desirable, could easily be overlooked. Input should be applied with judgment and experience. Thus, interface programs are more effective in engineering work.

Determining energy flow rate

Physics is a science of exact definitions, and engineering is a subscience of physics. It may be beneficial to review some of the basic concepts common to both. Energy is defined as work or the capacity to do work. The basic unit of energy is foot-pounds. The unit used in the thermal sciences is the Btu, which is by definition, the equivalent of 778.26 ft-lbs (approximately the amount of heat required to raise the temperature of 1 lb of water 1°F at standard atmospheric pressure).

Power is the time rate of energy. Some commonly used power units are Btu per hour, tons of refrigeration, horsepower, and kilowatts.

In energy economic studies, it is imperative that the difference between energy and power is clearly understood and kept in mind. This is significant because power units are generally input functions to an energy program. Clearly, control (minimization) of the power aspects of any system reduces energy use if other factors, such as operating hours, are kept constant. (To attain this condition, a computer is not needed.) If 1 Btuh is consumed for 8760 hours a year, the energy consumed is 8760 Btu. The significance of this simple statement is that if power is reduced at the front end of the program, the output quantity is also reduced. Now, let us explore this relationship by looking at a simple diagram.

Fig. 13-1 is a basic diagram of energy flows in a building system. The largest block on the left represents the space

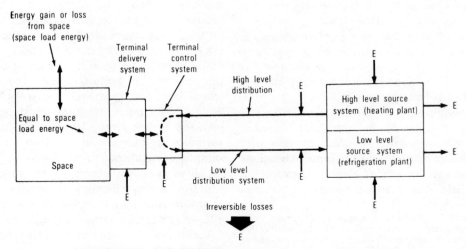

Fig. 13-1. A simplified diagram of energy flows in a building system.

(building, zone, room, etc.) being studied. The rate at which energy enters or leaves this space at any given time is the load—a power function. It is a cooling load if it enters and a heating load if it leaves.

Once a building's envelope is fixed, this energy flow rate becomes the absolute integrated energy requirement. In this analysis, the absolute integrated energy requirement is considered to be the block load, not the sum of the peaks. This function is the starting point in a study of energy economics. Energy requirements at the building ultimately lead to consumption of resource energy—coal, gas, oil, etc.—in a magnified form due to losses and inefficiencies during distribution and conversion processes.

The first point of application for a computer when trying to understand and thus minimize usage is the load study. Load energy is a function of:

• Building design, location, and orientation.
• Ventilation rates based on contaminants.
• Occupancy schedules.
• Weather conditions.

Of these four, occupancy schedules and weather are essentially uncontrollable inputs. The controllable variables of the load are:

1. Fenestration systems, including:
   • Area of the glass
   • Type of the glass
   • Framing systems
   • Interior shading
   • Exterior shading.
2. Lighting systems and appliances.
3. Roofing systems and materials.
4. Wall systems and materials.
5. Excess ventilation rates due to:
   • Infiltration
   • Overdesign of mechanical system ventilation rates
   • Antiquated building code requirements.

A properly designed computerized load program enables a designer to modify any of the controlled variables to determine the effect on the design load. When considering existing buildings, the location,

orientation, and occupancy are, obviously, already established. Hence, the engineer must confine himself to the five controllable load factors listed.

The effects of changes in these five variables, either individually or in any combination, on both the block load and the sum of the peaks can be readily determined by a computer during a load study. Without a computer, these detailed load studies would be approximations at best, in addition to being extremely cumbersome and time consuming.

At this point in an energy economic study it is important to note two items. First, the time integrated block loads—heating and cooling—on a building are the useful product energy requirements. Second, this product energy aspect—input to the mechanical system—is, except for excess ventilation, strictly an architecturally controlled parameter.

System component definitions

Before proceeding further, it is necessary to define the terms used in the following discussion.

• The *terminal delivery system* conveys air to and from the space. It includes the fans, supply and return ducts, supply outlets, grilles, etc.

• The *terminal control system* solves the space psychrometric problems. It contains everything needed to meet the conditioned space needs: cooling and heating coils with associated controls; control and mixing dampers; air terminal units, such as terminal reheat, mixing boxes, VAV terminals, etc.

• The *high level distribution system* conveys a thermal fluid from a source to heat transfer surface (hot water or steam from a boiler to a coil). This process is self-contained in an electric resistance heating coil. Heat is generated by the current flow, and combustion occurs remotely at the generating station.

• The *low level distribution system* conveys a fluid from a refrigeration machine (the source) to a chilled water coil or a direct expansion coil (the heat transfer surface).

• Motivating energy for the terminal control system includes the air energy required to operate the air terminal units. These require a minimum static pressure to assure proper operation. For example, self-contained variable air volume terminals require a minimum of 2 in. WG static pressure. This must be added to the static pressure required to overcome the duct system resistance before an adequate amount of energy is provided to operate the terminal unit.

Building vs. system energy

Referring again to Fig. 13-1, the second block from the left represents the terminal delivery system. All load energy to and from the space flows through it. A first law analysis* of the space clearly shows that the energy flow between the space and the terminal delivery system is equal to the *block* load energy.

Further analysis reveals that load energy flows into or out of the delivery system, while system motivating energy flows into the system. When a net block heating load exists, the delivery system motivating energy flows to the space. Energy analysis reveals that the motivating energy for the terminal delivery system often represents a major contribution to a building's annual energy consumption.

The terminal control system is represented by the smallest box from the left. Here, the space psychrometric problem is solved. Portions of this subsystem could be integrated with the delivery system in actual practice. In this analysis, however, it is important to identify the terminal control system as a separate subsystem.

First law analysis of the terminal control system reveals one of the more interesting aspects of energy consumption in HVAC systems. The energy flow arrows in Fig. 13-1 represent input from the high level distribution system, motivating energy, output to the low level distribution system and either input or output between the terminal control and terminal delivery systems. The terms high and low level refer to temperatures above and below space ambient conditions, respectively, *i.e.*, heating and cooling distribution systems. The high level distribution system generates the "run-around" energy in the overall system. When the net building load is in the cooling mode, this system can receive:

• Space energy.

• Terminal delivery system motivating energy.

• High level distribution system energy.

• Terminal control system motivating energy.

All of these will flow to the low level distribution system. This run-around energy from the high to the low level distribution system is, as the space sees it, wasted energy. This condition often exists at full load as well as part load. As far as the terminal control system is concerned, this run-around energy may be essential to maintain the space conditions.

At full load, many building systems utilize run-around energy to make up the difference between the block load and the sum of the peaks. Figure 13-2 demonstrates this. It shows a single floor of a commercial building

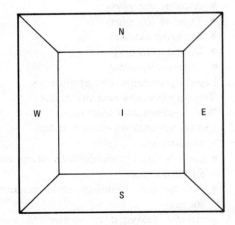

Block cooling load = 36.75 tons

Sum of zone peaks = 48.75 tons

Fig. 13-2. A single floor of a commercial building at net design cooling load conditions.

---

*The first law of thermodynamics states that (classical definition) work and heat are mutually convertible. This definition can be broadened to include all forms of energy; i.e., one form of energy may be converted to another.

at net design cooling load conditions. The block load is 36.75 tons. This is the rate of heat gain into the space at design conditions. The sum of the peak zone loads, however, is 48.75 tons. This is approximately 34 percent higher than the block load.

Many systems are psychrometrically designed to handle the load represented by the sum of the peaks. Many others are designed to provide a cooling plant capacity somewhere between the sum of the peaks and the block loads. The energy difference between the two loads comes from the high level distribution system. Thus, the high level system false loads the low level one. From the standpoint of energy economics, this false loading becomes even more significant. It requires added heat energy and also added refrigeration energy to remove the extra heat energy, hence the term run-around energy evolved.

Typical examples of false loading are: hot decks of dual systems; reheat coils; perimeter radiation in conjunction with cold deck VAV systems; and single fan, multiple zone economizer systems. As previously mentioned, the external energy for the terminal control system often is a significant contributor to energy consumption when high pressure delivery air is required to "control" the terminal units.

Energy is also transferred between the terminal control system and the high and low level source systems through their respective distribution systems. (In an actual system configuration, these systems are often integrated with the terminal delivery systems.) External energy is required to motivate this energy transfer, and significant savings often can be achieved by careful system analysis. For example, with water, or other single phase heat transfer fluids, doubling the temperature range halves the flow rate, which in turn reduces the power (pump horsepower and energy) required.

In existing buildings, this approach would reduce the distribution system energy to as low as one-eighth of its original value.

The first law energy balance also applies to the high level source system where either fossil fuels or electricity are converted (minus losses) to the heat energy supplied to the high level distribution system.

The low level source system follows the second law of thermodynamics: heat will not flow of its own accord from a cold to a warm sink. Thus, it requires external energy to transfer the energy from that system to an available heat sink. This external energy may be in the form of a prime mover or, with absorption refrigeration, heat energy. It is important to remember that the external energy applied to power the refrigeration unit is not the only input. All parasitic loads on the system, such as condenser water pumps, condenser fans, cooling tower fans, control power, ventilation fans, etc., must also be considered.

Building environmental control systems operate at less than design load for a major portion of the year. Therefore, any valid energy analysis should recognize part load performance characteristics of the high and low level source systems. Virtually all energy conversion systems become less efficient (useful power output/power input) as the load decreases from the design capacity. The exponential curve in Fig. 13-3 depicts this common system characteristic. The fixed

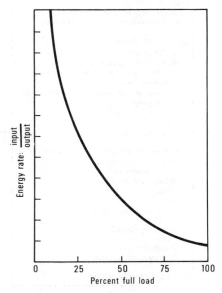

Fig. 13-3. An exponential curve depicting the decline in efficiency as the load decreases.

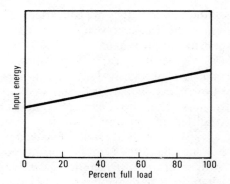

Fig. 13-4. A linearized curve as in Fig. 13-3.

parasitic loads establish the rate of increase of input to output as the load decreases, and the curve approaches infinite input per unit output at zero load. An example of this is a boiler maintained at operation temperature when there is no heating demand. The curve in Fig. 13-3 can be linearized (Fig. 13-4) for computer programming purposes by multiplying the ordinate by the abscissa.

From load to energy analysis

It is with recognition of the proper algorithms and concepts previously discussed that a computer program to analyze energy con-

sumption in a building is employed. Also, human interface with the output of the load program and the input to the energy program is essential in order to arrive at the best energy reduction solution.

Figure 13-5 illustrates, the minimum requirements and output data of an energy analysis report. (This discussion is concerned with the application of these programs rather than an in-depth discussion of the programs themselves.) It was stated at the beginning of this chapter that the computer is a tool. As such, the user applies it to determine what system modifications and/or changes in operating techniques are practical to achieve the desired goal.

Existing buildings provide the opportunity for immediate reductions in energy use compared to projects in the design stage. The procedure for applying computer analysis to existing buildings is as follows:

1. Study the building plans (if available).

2. Make a building survey to check the actual building conditions against the plans and/or to gather what pertinent data are required to perform the study.

3. Run a load program calculation.

4. Input the energy program with the out-

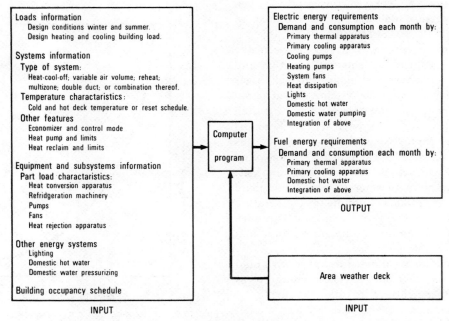

Fig. 13-5. The minimum requirements and output data for an energy analysis program.

put data from the load program and all the necessary data relating to the mechanical systems.

5. Compare the output from the energy program with historical energy consumption records of the building such as monthly and annual fuel and electric consumption from the utility bills.

6. Identify the reasons for any differences between calculated quantities and the historical data. Examples of differences are:

- Improper calibration of controls.
- Damper leakage.
- Operating techniques.
- Installation errors.
- Design errors.

7. With the knowledge of the relevant energy consuming contributors, starting with the building design, explore practical building and system modifications and control and/or operating techniques by processing them through the load and energy programs. From the energy savings developed, determine the dollar savings yielded by each modification.

8. Determine the costs of implementing each of the various conservation steps analyzed.

9. Conduct a quantitative comparison between the energy cost savings and the implementation costs.

10. Determine the combined owning and operating costs and the rate of return for each conservation measure.

11. From all of this information, prepare a feasible course of action in implementing the steps that prove economically viable.

It is the quantitative aspect that is provided by the computer. Reductions in energy consumption can be accomplished in virtually all existing building systems. However, when modifications are undertaken, capital investment is often required, and only a thorough quantitative analysis will determine the wisdom of the investment.

With a valid energy program, the calculated results can be assumed correct, adjusting, of course, for statistical versus actual weather conditions. Obviously, this also depends on all input factors being correct. Many times, factors causing high energy use problems can be readily identified and corrected with little or no capital investment, and engineering costs represent the expense involved.

Proper understanding of the factors affecting energy usage in buildings and of the involved algorithms in computing this energy use will permit the user to intelligently apply the computer to solving most, if not all, of the problems he is likely to encounter.

Figure 13-1, is adapted from the ASHRAE Journal, Vol. 16, No. 5, May 1974.

14

Flexible approach to computerized HVAC system analysis

Basic hardware combines with versatile software to provide a total evaluation of system energy performance.

MILTON MECKLER, PE, formerly President, Meckler Associates, Inc., Consulting Engineers, an Envirodyne Company, Long Beach, California, presently President, The Energy Group, Inc., Los Angeles, California

Much has been written about the energy crisis, and there is hopefully a keen awareness in design offices and among their clients of a new energy ethic. ASHRAE Standard 90-75, developed in response to an earlier National Bureau of Standards draft, attempts in an enlightened cookbook fashion to establish a so-called energy budget. State and federal agencies and legislatures throughout the country are pressing for the adoption of similar prescriptive remedies[1] to avoid, if possible, the dire consequences and real or psychological threats posed by further oil boycotts, brownouts, blackouts, etc.

There are those among us who would have us listen to a different drummer and return to our old wasteful ways with promises of adequate energy supplies. Many of these are people with vested interests or who fear the environmentalists and bureaucratic meddling.

For better or worse, we must be drawn to a better understanding of the energy consequences of our design choices,[2] and the new responsibilities imposed upon us as consultants or designers by these new priorities. Fortunately, many of us read the signs a few years ago and have been active in the building or acquiring of software capabilities to deal with the mass of data and information that must be handled in comparing alternate HVAC building and systems designs on the basis of minimum total energy input.

Unfortunately, those who have continued to rely on past experience, manual computations, etc., may be in for difficult times when faced with rapidly demonstrating meaningful HVAC system comparisons in a period of galloping inflation and design/construct pressures. Far too much emphasis, we believe, has been placed in the hope that computers will save design time by the programming of load procedures, and numerous programs now exist for this purpose alone.

---

[1]Superscript numerals indicate references at the end of the chapter.

We have found that design procedures that reflect the energy interaction of various HVAC system types with a common building load profile can be programmed to deal with these problems by projecting actual monthly energy use billings and other useful parametric system comparisons, as we shall see later.

With all forms of energy in short supply, with money tight and interest rates high, our project designers are under increasing pressure to optimize building energy use. With unlimited lead time, we could explore all the options and consider all the variables before key building decisions are firmed up or dollars committed. Now, our MACP-101 computerized energy program permits us to do this for clients, and do it quickly with an accuracy that has been verified in existing installations.

MACP-101 is a two-part program. Part one calculates energy requirements. It considers hour-by-hour, programmed building usage and occupancy profiles (including nights, holidays, weekends), and also heat losses and gains, weather variations, solar exposures, building orientation, control sequences, structural conditions, and variable electrical and thermal load demands throughout the year.

Part two of this program deals with equipment selection and energy consumption. It compares up to nine commonly used HVAC systems, including dual duct, reheat, induction, fan-coil, variable volume, and various all-electric and/or total energy systems. It also evaluates the performance of each system by simulating its operation to meet the hourly energy requirements established in Part one.

MACP-101 answers many questions. Among them: What is each system's relative thermal performance? How much waste heat is recoverable? Is total or partial power generation feasible? What size units are optimum?

The program printout is a summary of the monthly and annual gas, auxiliary fuel, electricity demand and consumption, and comparable utility billings in dollars.

The validity of this program has been established in hundreds of installations, including office buildings, shopping centers, hotels, and apartments, schools, hospitals, and industrial facilities.

The MACP-101 program analysis also provides information for a broad range of environmental studies. It also permits economic evaluation of total energy plants,[3] comparing effects of interest, insurance and property tax rates, direct labor and maintenance costs, and annual cost of space or buildings to house total energy equipment, and credit for power sold to a system in excess of building requirements, etc.

Flexible approach necessary

We have found that no one program satisfies all of our needs, and as a result, we utilize one or more of four separate programs in energy analysis work. None of these programs duplicates the others. It may prove helpful first to describe some of the key features of these programs, the degree of sophistication, and the general range of their application.

The purpose of the building energy analysis computer program MACP-201 is to provide a means of estimating reasonably, accurately, and quickly the energy consumption and building demands for heating, cooling, lighting, and base loads. Minimum input data are required. This program has undergone extensive field testing and is in everyday use by our staff. It was written primarily to evaluate the energy requirements of smaller buildings with less complex heating and cooling systems. However, it can be used on larger buildings under certain conditions.

MACP-201 represents a computerization of a manual calculator procedure for determining the monthly and annual energy consumption and demands of alternative energy sources. It utilizes the time-temperature duration method with the design heat loss and gain values during occupied and unoccupied periods.

The program is written to consider a specified occupied and unoccupied time period with constant levels of load. However, it can consider each hour of the day for different load levels. In this case, the calculated monthly values must be summed manually. The following types of cooling systems can be evaluated: no-cooling system refrigeration cycle with or without an economizer cycle, and fossil fuel system with or without economizer.

The following types of heating systems can be evaluated: no heating system, resistance heat, fossil fuel heat, reverse refrigeration cycle (heat pump with supplemental resistance heat), refrigeration cycle recovery with supplemental electric heating, or unitary water-to-air heat recovery system with fossil fuel heating.

The following types of energy sources for cooling and/or heating systems and heating domestic water can be evaluated: electricity, natural gas, fuel oil, coal, propane, and steam. Domestic hot water requirements can also be considered, including an option for available usable condenser heat of rejection for preheating the water.

Two miscellaneous uses of fossil fuel other than for heating, cooling, or domestic hot water can be considered: inside or outside the conditioned space, and outside the conditioned space.

MACP-201 is predicated on the availability of the design heat gain and loss calculations for both occupied and unoccupied time periods. The appropriate data for a heating and cooling system is input to the computer. Base load electrical usages and demands are calculated monthly and summed for annual requirements.

The number of hours in each month of each 5° F temperature grouping is calculated. Based on the specified period that the building is occupied, the total hours in the designated occupied and unoccupied time period is calculated monthly.

Based on the design values, the building heat loss and gain are calculated for the occupied and unoccupied time periods at each temperature grouping. According to the type of heating and cooling system and system operation input, a heat balance is then made for each temperature grouping for the occupied and unoccupied time periods.

The amount of heating and/or cooling required at each temperature grouping is then determined.

Energy usage and maximum demands are calculated for the heating and cooling systems for the occupied and unoccupied time periods both monthly and annually. The total available heat of rejection is calculated by temperature grouping on a monthly basis for the occupied time period only. No domestic hot water requirements can be considered during the unoccupied period.

If domestic hot water is needed, the water and heat energy requirements are calculated on a monthly basis. If the condenser heat of rejection is to be used as preheat, the heat energy requirements are compared with the usable portion of the available heat of rejection. The supplemental heat usages for base electrical loads, heating and cooling, domestic hot water, and two miscellaneous fossil fueled items are summed to obtain annual values. The maximum kilowatt demands for all electrical usages, individually and totaled, are shown for each month.

The MACP-202 program is a more advanced version of the MACP-201 program. It permits a more detailed evaluation of constant volume reheat and variable volume systems with and without reheat. In addition to other modifications, the capacity to simulate water-to-air heat recovery systems has been added.

The MACP-301 program is designed to interface with the MACP-100 and 200 series (or independently) to establish more efficient combinations of heating and cooling equipment. The MACP-301 can simultaneously evaluate a number of investments, revenue sources, costs, effects of tax and discount rates, depreciation, and various cash flow options.

A comparison of any two alternate designs can be made with a straight line, sum-of-year digits or double declining balance method of calculating depreciation. When using cash flow option, it is possible to select a standard access method of computing cash flow or an extended method incorporating loan costs.

The above and other useful features,

including heat recovery, are incorporated in the MACP-301 program. This permits concise comparisons of different system combinations to establish lowest first, highest operating, or annual cost, based on selected cost data and financial criteria.

Using the latest cash flow analysis technique, it is possible to readily compute actual payouts for systems being compared that reflect the advantages of more efficient equipment to be gained from lower operating costs. In addition to assuring a thorough review of various heat conservation alternates, MACP-301 prints out a rate of return format that reflects cost differences in system alternates before and after taxes. The program can be used for any kind of financial analysis, including lease or purchase decisions, various engineering or management investment decisions, or various real estate options.

The office programs described utilize IBM equipment, which is stored with our service bureau. Although we presently use a batch mode for our work, the systems can be switched to a time share mode at any time without reprogramming.

These programs are available through contractual arrangements with a service bureau. A service bureau has both the resources and the experience to handle the complete range of data processing needs presented by a customer. Aside from supplying the equipment and time to execute our software programs and providing the people and expertise to develop total systems, a client can jointly provide a cost effective solution by applying the most advanced hardware.

Comprehensive energy planning

Of the four programs described, we feel that MACP-101 program is the most comprehensive. To illustrate its versatility, we have selected a recent office building energy analysis. The project involved the California Coast office plaza, a 17 story office structure constructed adjacent to the Los Angeles International Airport. The building contains approximately 12,000 sq ft gross area per floor. It can best be described by

Table 14-1. General Building and system features and study criteria used to analyze the systems charted in Figs. 14-1 through 14-7.

Building features:
- 17 stories
- 205,700 sq ft
- 50 percent glass uniformly distributed (0.41 shading factor)
- Southeast and Northwest primary exposure

System features:
- Gas heating
- Electric cooling
- Centrifugal chillers, forced draft cooling towers, hot water boilers

Study criteria:
- Hours of building operation: average 11 hr per day (08:00–19:00)
- 0.20 roof U-factor
- 0.12 wall U-factor
- Occupancy: 120 sq ft per person
- 21 cfm outside air per person
- 0.5 watts per sq ft, power
- 4.5 watts per sq ft lighting intensity
- Weather station, Los Angeles International Airport

Assumed weather/system design criteria factors:
1) 1 percent design cooling profile
2) 99 percent design heating profile
3) 65°F maximum outdoor temperature requiring heating
4) 31°F minimum outdoor temperature requiring cooling
5) 50°F minimum outdoor air temperature at which outside air volume entering system is reduced
- Electrical rates: Southern California Edison Co., rate schedule A7 E1173
- Gas rates: Southern California Gas Co., rate schedule G1 A0474

reference to Table 14-1, which also indicates some key computer input design criteria.

It may be of interest to note that computer cost, exclusive of keypunch or employee time for data take-off/input of computer forms, for the entire 17 story analysis for all system options described was only $74.29, applying normal service bureau billing schedules.

Tentative selection of centrifugal type chillers indicated a probable performance ratio ranging from approximately 15.4 Btu per watt at 80°F ambient (dry bulb) temperature to a 9.5 Btu per watt level at 105°F ambient temperature. A total of nine separate HVAC system types were evaluated for our energy study project. They are as follows:

- Single duct terminal reheat
- Four-pipe induction
- Reheat induction
- Fan-coil
- Dual duct with fixed outside air option
- Dual duct with atmospheric cooling cycle
- Single duct variable air volume
- Single duct variable air volume with cold plenum reset option
- All-air cooling induction.[4]

Refer to Figs. 14-1 through 14-7 for a monthly (seasonal) trace of the following key HVAC performance factors for each of the system types described above that were characterized by means of the MACP-101 programmed system logic format illustrated in Table 14-2.

The key performance factors for each of the system types are:

- System cooling in ton-hr versus month of year, Fig. 14-1.
- System electrical energy consumption for fans in kilowatt hours versus month of year, Fig. 14-2.
- System electrical energy consumption for cooling in kilowatt hours versus month of year, see Fig. 14-3.
- System total electrical energy con-

sumption in kilowatt hours versus month of year, Fig. 14-4.

- System electrical cost in dollars per month versus month of year, Fig. 14-5.
- System heating consumption in mega-Btu versus month of year, Fig. 14-6.
- System gas cost in dollars per month versus month of year, see Fig. 14-7.

The result of the studies revealed the ranking shown in Table 14-3 among the six system types covered by Figs. 14-1 through 14-7.

Figs. 14-1 through 14-7 give a comparison of monthly variations that characterized each of the systems presented. For example, in Fig. 14-1, notice the wide variation of system interaction with the same incident cooling load as a typical annual cycle progresses.

In Fig. 14-2, the contribution of cooling tower fans is apparent for the two groupings, *i.e.*, constant volume Systems 1, 2, 3, and 6, and variable volume Systems 4 and 5 during the warmer months. The advantage of an atmospheric cooling cycle is obvious when comparing, for example, Systems 2 and 3. Interesting, however, is the heating penalty that results from the atmospheric cooling option. This is clearly apparent in Fig. 14-6. Furthermore, the analysis points up the often overlooked penalty of resetting system operating control points in response to

Table 14-2.   Programmed system logic format.

| System type | Computer simulated logic | | | |
	Duct distribution: 1 (single duct); 0 (dual duct, multizone)	Outside air: 1 (variable) 0 (constant)	Zone air: 1 (variable) 0 (constant)	Cold plenum temperature 1 (variable) 0 (constant)
Single duct terminal reheat	1	0	0	1
Dual duct	0	1 or 0[1]	0	0[2]
Single duct VAV	1	1	1	1 or 0[3]
Induction (four-pipe)	1	0	0	1
All-air induction	1	0	0	0
Reheat induction	1	0	0	1
Ceiling induction	1	0	1	0
Fan-coil	1	0	0	1

[1] Fixed outside air: 0; atmospheric cooling cycle: 1.
[2] Use 1 for humidity controlled systems or systems employing outdoor air or return air temperature reset option.
[3] Use 1 for systems employing outdoor or return air temperature reset.

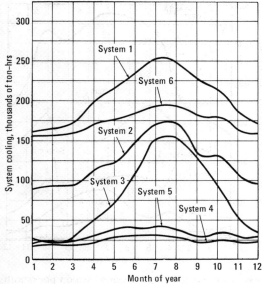

Fig. 14-1. Energy consumed monthly by system for cooling.

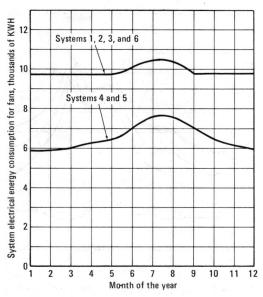

Fig. 14-2. Electricity consumed monthly by fans for each system.

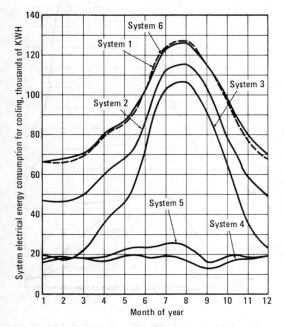

Fig. 14-3. System electrical consumption for cooling.

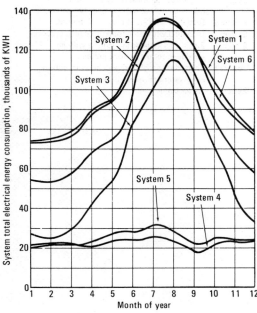

Fig. 14-4. System total electrical energy consumption.

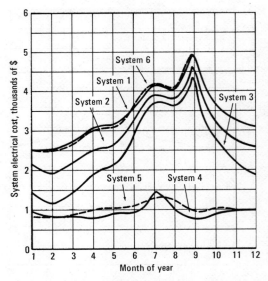

Fig. 14-5. System electrical cost per month of year.

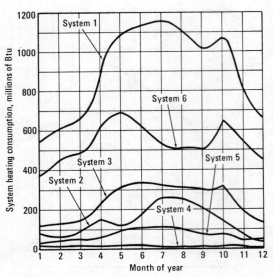

Fig. 14-6. System heating consumption per month of year.

Legend

System 1: Single duct terminal reheat, tour-pipe induction, reheat induction, or fan-coil.
System 2: Dual duct with fixed outside air option.
System 3: Dual duct with atmospheric cooling cycle.
System 4: Single duct variable air, volume with no outside or return air reset option.
System 5: Single duct variable air volume with outside air or return air reset option.
System 6: All air cooling induction.

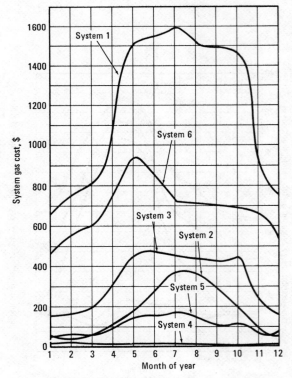

Fig. 14-7. System gas cost per month of year.

Table 14-3.   The results of the programmed studies revealed the following ranking among the six system types in Figs. 14-1 to 14-7. The systems are ranked from lower to higher values on an annualized basis in each category. The systems are numbered as follows: 1) single duct terminal reheat, four-pipe induction, reheat induction, or fan-coil; 2) dual duct with fixed outside air option; 3) dual duct with atmospheric cooling cycle; 4) single duct variable air volume; 5) single duct variable volume with cold plenum reset option; 6) all-air cooling induction.

System cooling	Fan energy consumption		Cooling energy consumption	Total system energy consumption	System operating cost		
					Electricity	Gas	Total
4	4 ⎫		4	4	4	4	4
5	5 ⎬ equal		5	5	5	5	5
3	1 ⎭		3	3	3	2	3
2	2 ⎫		2	2	2	3	2
6	3 ⎬ equal		6	6	6	6	6
1	6 ⎭		1	1	1	1	1

Refer to Figs. 14-1 through 14-7 for the monthly performance of each system in the various categories.

changes in the outside air or return air temperature. This is contrasted in a comparison of Systems 4 and 5.

It should be pointed out that the emphasis here was not in developing optimum conditions from an energy conservation point of view, but to discriminate among various system types available and their order of magnitude energy consequences, which, together with other factors, would assist in choosing the so-called best overall system.

The rising priority of system energy utilization, however, does weigh heavily upon us today when evaluating against the pros and cons of other systems.

Refer to Fig. 14-7 for a comparison and analysis of bottom line energy costs for each of the systems studied. In an attempt to compare our values with the latest available energy cost figures for air conditioned office buildings in Los Angeles, we find that published data[5] suggest that fuel and electricity

Table 14-4   Comparative analysis of computer output annual system energy costs.

System type	Energy costs, $				System load factor, sq ft per ton*	System cost, $ per sq ft
	Electricity	Percent of total energy cost	Gas	Total		
Single duct terminal reheat, four-pipe induction, reheat induction, and fan-coil	42,364.09	75	14,080.37	56,444.46	343	0.27
Dual duct with fixed outside air option	37,111.38	95	2,095.73	39,207.11	343	0.19
Dual duct with atmospheric cooling cycle	31,060.50	89	3,918.61	34,979.11	343	0.17
Single duct variable air volume	12,512.16	99	111.70	12,623.86	343	0.06
Single duct variable volume with cold plenum reset option	13,730.44	92	1,208.24	14,938.68	343	0.07
All-air cooling induction	42,011.39	83	8,403.79	50,415.18	343	0.24

*Nominal 600 ton design requirement established by Part 1 of MACP-101 program.

components, based on 1972 operating average, for example, yields 16¢ per sq ft. This figure, we believe, compares favorably with those listed in Fig. 14-7 when the escalation of 1974 (compared with 1972) energy costs is considered, and the preponderance of System 3 types installed in Los Angeles office buildings as of 1972.

Energy as a parameter

We have found that these programs have instilled in our designers a new energy consciousness since energy use can be readily, and reasonably accurately, quantified. In addition, with an ever increasing inventory of commercially available energy conservation devices, such as heat/enthalpy[6] reclaim wheels, double bundle heat pumps, heat exchange troffers, etc., energy is becoming an essential design parameter[7] in developing guidelines for new and modernized buildings.

Some final comments with respect to variable air volume systems may be in order. In addition to the obvious energy benefits illustrated from the computer modeling study, the application and/or limitations of the selected system must always be considered. On the plus side, variable air volume systems are generally self-balancing, and when equipped with suitable volume control and aspirating diffusers, they permit stable operations over a wide range of conditions[8] and can shut off air to any part of a building not requiring it without unbalancing the air supply to the remainder of the building. Yet, on the other hand, it is essential that terminal devices be unaffected by upstream variations in velocity or pressure. Also, diffusers must be carefully selected to avoid stratification or drafts. On reduced demand for air, care must also be taken to provide minimum ventilation at all times. Furthermore, fans must be equipped with controls, such as vortex dampers, to achieve potential horsepower savings, etc.

In short, energy conservation may require greater sophistication[9] in the design of HVAC systems as we attempt to use the computer to help us model *how* and not just *how much* energy moves or is moved around a building. What is important is that with the new energy priorities, we are now in a better position to demonstrate to clients improvements in operation through the use of time programming system[10] features and (where appropriate) building automation for new or retrofitted existing buildings.[11] Furthermore, the mechanical engineer is now in a better position to significantly influence the architectural and electrical design of buildings within his purview.[12]

References

1. Hess, R. A., "HUD Apartment Study—Thrust for Energy Laws," *Air Conditioning and Refrigeration Business,* Aug. 1974.
2. Stubblefield, R. R. "Energy Conservation in Commercial Buildings," Chapter 53.
3. Emerick, P. O., "A Fresh Look at Total Energy—Benefits from Every By-Product," *Actual Specifying Engineer,* Mar. 1974.
4. Meckler, M., "Evaluating All-Air Cooling Induction Systems," *Heating/Piping/Air Conditioning,* Jan. 1969.
5. Wiggs, R. C., "Operating Costs for Air Conditioned Office Buildings," *Heating/Piping/Air Conditioning,* Jul. 1974.
6. Meckler, M., "Integrated Mechanical Evaporative Cooling System," *Air Conditioning, Heating & Ventilating,* Feb. 1966.
7. Coad, W. J., "Energy Economics: A Design Parameter," Chapter 26.
8. Daryanani, S., "Design Engineers Guide to Variable Air Volume Systems," *Actual Specifying Engineer,* Jul. 1974.
9. Pannkoke, T., "Energy Conservation: Why Not Go All the Way?," Chapter 19.
10. Karman, D. J., "How to Select and Specify Time Controls to Save Energy," *Actual Specifying Engineer,* Apr. 1974.
11. Bodak, L., "Saving and Improvement through Automation Systems," *Actual Specifying Engineer,* Sept. 1974.
12. Shavit, G., "Enthalpy Control Systems: Increased Energy Conservation," Chapter 64.

15

Architectural aspects of energy conservation in HVAC

A proper balance between a building's functional and esthetic features is a necessity.

E. R. AMBROSE, PE,
Mount Dora, Florida

An architect's ability and judgment in keeping a proper economic balance between the functional and esthetic features of a building are very important in controlling heating and cooling requirements as well as corresponding first and operating costs. Table 15-1 lists some significant considerations and recommendations for achieving these objectives.

The geographic location and its prevailing outdoor temperatures and wind and solar intensities are also influential factors in building design. Usually, the greater the average seasonal inside/outside temperature differential, the more attractive energy conservation practices for the heating season become. On the other hand, the greater the solar intensity, the more necessary it becomes to incorporate practical design features to reduce heat gain during the cooling cycle.

Selection of the most favorable orientation and configuration can pay impressive dividends in reducing both solar and conduction loads. Even with these reductions solar gain on windows facing West, Southwest, and South can still represent a large portion of the cooling load, and it is essential to consider minimum glass areas supplemented by available shading devices.

These and other design considerations must be thoroughly compared and evaluated in order to determine the most economic selection.

Evaluation of components

Comparison and evaluation of various materials and components contemplated for the exterior surface of a structure can best be initiated by making heat gain and heat loss computations followed by an annual owning and operating evaluation.

Heat gain and heat loss computations include conduction heat transmission and solar effect, as well as ventilation air and internal heat gain components.

Some idea of the expected size of different components contributing to total heating and cooling loads for a conventional commercial office building of the late 1940s is illustrated in Table 15-2. Solar gain during the cooling cycle can be a very important contributing factor and can represent a considerable portion of the cooling load, particularly if large areas of unshaded glass are involved. The quantities of ventilation air employed, together with the lighting load, also materially influence equipment selection and system design.

Table 15-1. Building design considerations and recommendations for energy conservation of heating and cooling systems.

Design consideration	Recommendation
Orientation	Consider sun and wind effects. Least sun load obtained from Northeast and Northwest. Provide minimum glass exposure to West, Southwest, and South.
Shading devices: use overhangs, exterior louvered sun screens, awnings, interior shades.	Maximum shading required for all glass exposed to sun.
Window: amount and kind of glass.	Keep to absolute minimum. Insulating glass or double glazed windows and doors can generally be justified. Reflective or tinted glazing is a possibility.
Walls and roof: materials with a low thermal transmission value are demanded.	Employ maximum thermal insulation determined by economic evaluation.
Configuration: length to width ratio.	Minimum surface exposed to elements important. A square or circular building has minimum exterior surface for given floor area.
Massiveness of heat storage capacity of structure.	The greater the heat storage effect, the lower the total coincident cooling load.
Space provisions for heating and cooling systems.	Provide sufficient space for practical, acceptable, and efficient design: ducts piping, equipment.
Geographic location.	Heating provisions need special attention in cold winter areas, and cooling warrants priority in warmer areas.

Owning and operating costs

The calculated heat gain and loss of a structure, such as indicated by Table 15-2, serves as a basis for sizing heating and cooling equipment and for designing the associated air conditioning system. It also must serve as a basis for determining the annual owning and operating costs, including such influential factors as:

- Capital cost of the heating and cooling system and of the heat conservation devices considered for the structure.

- Annual fixed charges due to amortization and depreciation of equipment, interest, taxes, insurance, and other fixed charges.

- Annual operating cost, including energy, general maintenance, and service costs.

Unfortunately, a number of variables and assumptions are involved in an owning and operating cost analysis, including:

- The selection of the climatological data for an average year; an average year is difficult to define, however.

- The operating conditions to be maintained: average daytime and nighttime temperatures, average amount of ventilation air, and seasonal efficiency of the heating and cooling system.

- Proper credit for internal heat gain during the heating season.

Table 15-2. Basic components contributing to the heat gain and heat loss of a structure.

Component	Percent of total heat gain	Percent of total heat loss
Conduction and solar effect:		
• Walls	0.2	48
• Windows	19	
• Floor and roof	3.1	
Lights	25	
Ventilation air:		
• Sensible heat	13.7	52
• Latent heat	24	
People:		
• Sensible heat	9	
• Latent heat	6	
Miscellaneous internal heat gain	—	—

Design considerations: Heat loss 70° F indoor outdoor temperature difference. Heat gain: 80° F, 50 percent relative humidity indoor; and 95° F, 48 percent relative humidity outdoor. Lights and Miscellaneous: 214.3 KW, or 2.22 watts per sq ft Floor area: 87,552 sq ft; six stories. Volume 1,150,000 ft. Dimensions: 152 by 96 ft. Construction: 12 in. brick wall; no thermal wall insulation; no exterior shading devices; windows are double glazed, and cover 40 percent of exterior; built in Roanoke, Va., in 1948.

• The period of time to be used: whether for 10 years, 20 years, the normal life of equipment or estimated life span of the building.

It would be most helpful if actual published operating data were available to guide these analyses. As it happens, actual heating and cooling field data for large commercial and industrial buildings are practically non-existent, and it is essential that steps be taken to acquire such data. Architects, consulting engineers, contractors, manufacturers, and trade associations can serve an important role by acquiring and effectively using all data made available. A number of organizations have developed computer programs that can be readily adapted to economic justification studies of various energy conservation devices and heating and cooling equipment selections if proper data are available.

New and existing buildings

About 70 percent of all the nation's buildings to be used during the next 10 years are already standing. Consequently, if significant savings are to be made in the total energy consumption of today's buildings during this period, building owners and designers must be encouraged to make all justifiable improve-ments in these building designs and in the heating and cooling systems.

The suggestions and recommendations made in connection with Table 15-1 apply primarily to new commercial and institutional buildings for human occupancy, but many of the comments on external construction apply also to existing structures, including dwellings. For instance, window areas may be reduced, insulating glass considered, shading devices incorporated, and thermal insulation can perhaps be added. In many instances, however, the first cost of making these changes in existing building designs may not be economically justifiable. Consequently, each installation must be considered individually.

This chapter has been mainly concerned with the exterior or envelope of a building. In cooler climates, heating requirements are usually the controlling factor, and thus thermal insulation is an important consideration. Insulation also has attractive possibilities for economic savings in warmer climates when considered on a year 'round basis. Because of its importance, the economics and acceptable practices of applying thermal insulation and double glazing are discussed in Chapter 16.

16

Thermal insulation and double glazing

Important factors in reducing energy consumption in commercial, residential, and institutional buildings.

E. R. AMBROSE, PE,
Mount Dora, Florida

It has been proven quite conclusively that thermal insulation and double glazing can make important contributions to energy conservation in the consumption of energy for heating and air conditioning systems.

In all probability, federal, state, and local codes and ordinances will take full advantage of this possibility by specifying the maximum allowable winter heat loss and summer heat gain for various types and sizes of buildings in different geographic areas. Actually, such specifications have been used for some time by the Federal Housing administration in its "minimum property standards for one and two living units" and by the American National Standard Institute for mobile homes. From the earliest conception of residential resistance heating, the electric industry recommended the use of the National Mineral Wool Insulation Standards or the National Electrical Manufacturers Association All Weather Perfomance Recommendations for maximum insulation and double glazing. Also, the American Society of Heating, Refrigerating and Air-Conditioning Engineers Standard 90-75, Design and Evaluation Criteria for Energy Conservation in New Buildings hopes that this standard will serve as a guide for all codes and for future federal legislation.

Such standards are not equally applicable to all types of buildings. They can usually be successfully applied to dwelling units because the exterior surface represents the major portion of heat loss and gain. In commercial and institutional buildings, however, heating and cooling requirements cannot be limited to only the exterior surfaces since the ventilation air, internal heat gain, and other such contributing factors play an important part in the total load determination. Consequently, it is desirable to cover the recommended standards and recommendations for energy conservation practices and procedures of residences separately from those applying to commercial and institutional buildings.

Residential conservation efforts

Energy conservation offers attractive possibilities in the residential sector. Based on what appears to be the best authoritative estimates, there were approximately 69 million dwellings in the United States at the end of 1972; 88 million are expected by 1982. The required energy for these installations would represent about 10 percent of the total annual consumption in the United States if current practices are allowed to continue unabated.

With the expected change in federal, state, and local codes and ordinances to limit heat loss and gain, and with cooperation from contractors, trade associations, and financial institutions, installing insulation and double glazing in the expected 19 million new houses by 1982 should not create too much of a problem. The big difficulty will be developing practical methods for reducing energy consumption in the 69 million existing dwellings. Installing storm doors and windows and adding ceiling or roof insulation are practical in many cases, but side wall insulation is not feasible in most instances.

Undoubtedly, a considerable number of insulation and double glazing installations have been made due to the influence of the FHA, ANSI, and NEMA codes. This is particularly true for electric heating systems where a thermal resistance of R-19 in ceilings, R-11 in exterior walls, and R-13 in floors over unheated spaces were specified. The gas and oil industries also recommend thermal insulation and double glazing for houses, but not to the degree followed by the electrical industry. Their recommendation generally consisted of 2 to $2\frac{3}{4}$ in. of ceiling insulation and double glazing. On some occasions, side wall insulation was employed, but this has not been a universally accepted practice and for the most part was limited to cooler climates. Consequently, there are still a number of existing dwellings well below what is considered today's accepted insulation standards. These installations could be considered good prospects for additional insulation if proper incentives and encouragement are forthcoming.

Some idea of the effects of various thickness of insulation and double glazing on heating requirements of a typical ranch type house is illustrated by Fig. 16-1. Adding 2 in. of ceiling insulation (R-7) results in a 11.7 percent reduction in heat loss. An additional 4 in. (R-19) causes another 4.3 percent reduction—a total of 16 percent. Double glazing accounts for 14.9 percent reduction and side wall insulation (R-11) for another

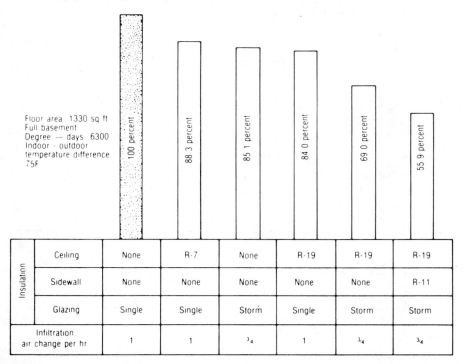

Floor area 1330 sq ft Full basement Degree — days 6300 Indoor - outdoor temperature difference 75F	100 percent	88 3 percent	85 1 percent	84 0 percent	69 0 percent	55 9 percent	

Insulation	Ceiling	None	R-7	None	R-19	R-19	R-19
	Sidewall	None	None	None	None	None	R-11
	Glazing	Single	Single	Storm	Single	Storm	Storm
	Infiltration. air change per hr	1	1	³₄	1	³₄	³₄

Fig. 16-1. Btuh heat loss reduction in percent for a typical residence due to different amounts of insulation. Uninsulated structure with single glazing assumed as base of 100 percent.

13.2 percent—to an attractive overall total energy reduction of 44.1 percent.

A number of factors can influence choice of insulation. The type of construction, for instance, generally limits the thickness of sidewall insulation. With frame wood siding, standard stud space is 3 to $3\frac{1}{2}$ in. Thicker insulation should often be considered in the more northern areas, but special, more costly construction would be required or an insulation more efficient than glass fiber or mineral wool must be developed. Urethane, employing the foaming, pouring, or blowing in place application technique, offers attractive possibilities and justifies further investigation. With the emphasis on energy conservation and with the increase in energy cost taking place, all of these insulation recommendations must be reevaluated and in all probability the recommended thermal resistances should be increased.

Commercial and institutional

Thermal insulation and double glazing are also applicable to commercial and institutional buildings to lower heat loss and gain, with a corresponding reduction in the equipment size and annual energy consumption.

ASHRAE's Guide and Data Book gives the coefficient of heat transmission for a number of different building materials combined with various amounts of insulation, similar to that illustrated by Fig. 16-2. Such data are very useful in comparing the conduction heat flow and in selecting the most practical combinations. It can be noted that the addition of a relatively small thickness of insulation on the exterior surfaces materially reduces the Btuh per sq ft of conduction heat transmission.

In the case of commercial and institutional

Construction	Conduction coefficient of heat transmission, U Btuh per sq ft F	
	Base	2 in insulation
Wall: 4 in brick and 8 in concrete block 	0 16	0 09
Wall metal sandwich curtain 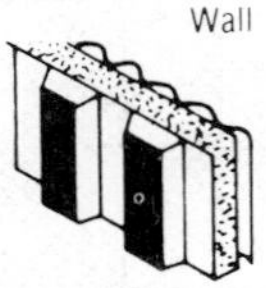	0 55	0 12
Wall wood frame and wood siding 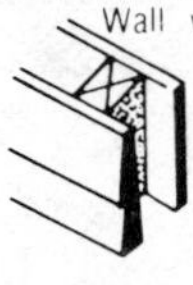	0 19	0 08
Roof metal deck suspended ceiling 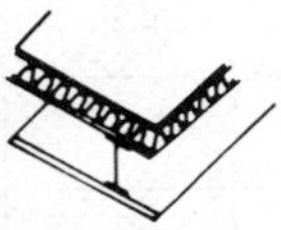	0 29	0 14
Windows Single glazing Double glazing Insulating glass - $\frac{1}{2}$ in air space	1 13 0 56 0 58	

Fig. 16-2. Coefficient of conduction heat transmission for several building materials. Comparison shows advantage of thermal insulation.

buildings, however, many different components contribute to the overall heat gain and loss. Thus, exterior walls, ceilings, and floors may represent a relatively small percentage of the total heating and cooling load. Consequently, in many instances, wall insulation may be difficult to justify economically from an owning and operating viewpoint in both existing and new buildings, particularly if the external wall has a reasonably low transmission coefficient of say 0.16 to 0.20. Double glazing, on the other hand, combined with proper shading devices for reduction in solar gain during the cooling cycle can usually always be justified.

There are many such options in system and design and material selection for a building. It seems essential, therefore, to develop a nationally recognized standard to assure maximum energy conservation. Such a standard would serve as a guide and reference for federal, state, and local codes and ordinances as well as for architects, consulting engineers, mechanical contractors, and others interested in the subject.

To bring this about, ASHRAE Standard 90-75 covers all phases of heating and air conditioning, including the maximum allowable heat flow (U) through the exterior envelope of a structure. Since the economic justification of the energy conservation practices is related to the severity of the climate, the allowable heat flow is made to vary inversely with the heating degree-days. For instance, the maximum allowable heat flow for gross exterior walls at design condition for an area such as Pittsburgh (with 5000 average yearly heating degree-days) is 0.26 Btuh per sq ft, °F for one and two family living units and 0.32 Btuh per sq ft, °F for all other buildings. Similar values have been developed for roofs, ceilings and floors. Some flexibility is left to the judgment of the architect or engineer making the analysis. For example, it would be acceptable to have the total building heat transfer (heating and cooling) at design conditions equal to or less than the allowable U values of the subsections.

Such standards will make all parties aware of the importance of the various factors affecting the energy conservation practices and procedures. They will also cause selections and evaluations to be made on a uniform basis to help eliminate disagreements and misunderstandings.

Insulation and double glazing make important contributions to energy conservation and to the sizing of heating and air conditioning equipment, particularly for new buildings. Outdoor ventilation air reduction, on the other hand, has considerable promise for energy conservation in both new and existing structures.

17

Will less glass save more energy?

The author questions the emphasis that ASHRAE Standard 90-75 places on the building envelope.

PARAMBIR S. GUJRAL, PE,
Associate Partner, Chief Engineer,
Electrical-Mechanical Div., Skidmore,
Owings & Merrill, Chicago, Illinois

ASHRAE Standard 90-75, Energy Conservation in New Building Design, provides prescriptive guidelines for the thermal design of a building envelope; the design of mechanical, electrical, and illumination equipment; and systems that promote the efficient use of energy. The standard was prepared and approved by the American Society of Heating, Refrigerating and Air-Conditioning Engineers in 1975.

This standard is an important step toward reducing energy consumption. It is the first comprehensive energy conservation standard to be developed, and it has been praised. Arthur D. Little, Inc., for example, in a report for the Federal Energy Administration, states that the standard "is very effective in reducing annual energy consumption in all building types and locations."* However, in my opinion, the following analysis, which uses simulations of the thermal criteria, uncovers areas in need of further study and modification before 90-75 can be embraced as the only means for energy conservation in buildings. The results of further study could make the standard even more effective.

Thermal criteria

Section 4 of Standard 90-75 establishes prescriptive criteria for the thermal design of the exterior envelope of new buildings. These criteria are based on specified wall conductance values. It is accepted that the envelope generally accounts for the bulk of energy consumed in a building and that gross wall conductance is the prime variable to control consumption.

The guidelines recommend that a building be designed with a specified combined thermal transmittance value (U value) for the gross area of exterior walls and roof not to exceed certain values given in Section 4 (Figs. 3 and 4 in Section 4 of Standard 90-75). To see these values more clearly, Table 17-1 was developed using the standard's winter heating criteria for various cities. Fig. 17-1 plots the conductance values (Q) against the heating degree-days (DD).

This analysis and Tables 17-1, 17-2, 17-3, and 17-4 are predicated on what the stan-

*Arthur D. Little, Inc., An Impact Assessment of ASHRAE Standard 90-75, Energy Conservation in New Building Design, Report to the Federal Energy Administration, Dec. 1975.

Table 17-1.   Conductance values based on Standard 90-75 winter heating criteria for five cities.

City[1]	DD[2]	Toa	Ur	Uw	dT	Qr	Qw
Chicago CO	5882	1	0.076	0.33	71	5.40	23.43
New York City Central Park	4871	15	0.084	0.36	57	4.79	20.52
Portland CO	4109	29	0.091	0.38	43	3.91	16.34
San Francisco CO	3001	44	0.010	0.41	28	2.80	11.48
Los Angeles CO	1349	44	0.010	0.45	28	2.80	12.60

Key:
  DD = degree-days
  Toa = winter design temperature, °F
  Ur = thermal transmittance roof, Btuh per sq ft per deg F
  Uw = thermal transmittance gross wall, Btuh per sq ft per deg F
  dT = temperature difference between interior and exterior design conditions, °F
  Qr = conductance, roof, Btuh per sq ft (Qr = Ur × dT)
  Qw = conductance, gross wall, Btuh per sq ft (Qw = Uw × dT)

[1] Weather station designations follow the nomenclature used in Chapter 33 of the 1972 ASHRAE Handbook of Fundamentals. Data for stations followed by "CO" came from office locations within an urban area and generally reflect an influence of the surrounding area. American Society of Heating, Refrigerating, and Air-Conditioning Engineers, New York, N.Y.
[2] Degree-day values were taken from Chapter 43 of the 1976 ASHRAE Handbook & Product Directory.

dard terms are for a Type B building that is over three stories high. These are non-residential buildings, such as the educational, business, mercantile, institutional, and warehouse structures covered by the standard.

The allowable window fraction (ratio of glass area to gross wall area) is shown in Table 17-2 for single glazed (P_s) and double glazed (P_d) windows with an opaque wall conductance of 0.06 Btuh per sq ft

per °F. The maximum allowable window fraction (P_{sm} for single and P_{dm} for double) is also computed. These maximum glass areas were computed by assuming a theoretical value of zero for the thermal transmittance of the opaque wall area. Fig. 17-2 shows these as a function of heating degree-days.

The values were calculated in the following manner. The percent window area allowed

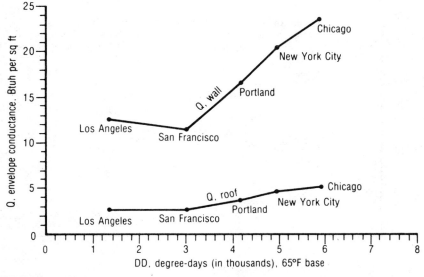

Fig. 17-1. Envelope conductance per ASHRAE Standard 90-75 winter conditions for the five cities shown in Table 17-1.

Table 17-2.   Allowable window area—decimal—based on Standard 90-75 when opaque wall area has a U value of 0.06 Btuh per sq ft per deg F.

City	Ps	Psm	Pd	Pdm*
Chicago	0.28	0.33	0.50	0.55
New York	0.32	0.36	0.55	0.60
Portland	0.34	0.38	0.59	0.63
San Francisco	0.37	0.41	0.65	0.68
Los Angeles	0.41	0.45	0.72	0.75

Key:
Ps   = fraction of window area, single glazed
Psm = fraction of window area, single glazed, maximum
Pd   = fraction of window area, double glazed
Pdm = fraction of window area, double glazed, maximum

*Maximum allowable glass area based on heating criteria if an opaque wall U value of zero could be achieved.

is denoted as P and the window area, A_{wd} equals PA_w. So, the opaque wall area is:

$$A_{wl} = (1 - P) A_w$$

Hence, the specified gross wall transmittance value can be evaluated as:

$$U_w = U_{wl} (1 - P) + U_{wd} (P)$$

The subscripts, w, wl, and wd, refer to gross wall, opaque wall, and window, respectively.

The allowable window fraction is:

$$P = (U_w + U_{wl}) \ / \ (U_{wd} + U_{wl})$$

The window fractions allowed by the guidelines as computed from this equation are a strong function of opaque wall conductance.

The recommended cooling criteria must also be examined, since the more stringent of the heating or cooling requirements of the exterior envelope will govern the design selections.

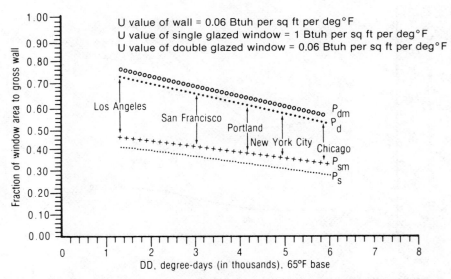

Fig. 17-2. Allowable window fraction shown as a decimal per ASHRAE Standard 90-75 winter conditions for the five cities shown in Table 17-2.

Table 17-3. Summary of summer design criteria for determining allowable window areas when mechanical cooling is used.

City	North latitude	OTTV	Toa	SF	dT	Ps	Pd
Chicago	41° 5w′	33.9	91/76	128.5	13	0.240	0.357
New York	40° 5′	33.7	91/76	127.5	13	0.240	0.359
Portland	44° 3′	34.8	88/68	132.5	10	0.246	0.364
San Francisco	37° 5′	32.8	77/62	125.5	−1	0.266	0.391
Los Angeles	34° 0′	31.8	90/70	123.0	12	0.235	0.350

Key:

OTTV = overall thermal transfer value, Btuh per sq ft
Toa = summer design dry bulb/wet bulb temperatures
SF = solar factor, Btuh per sq ft
dT = temperature difference between exterior and interior design conditions, °F
Ps = fraction of window area, single glazed, expressed as a decimal
Pd = fraction of window area, double glazed, expressed as a decimal

Note: The above values of Ps and Pd are based on an opaque wall U value of 0.06 Btuh per sq ft per deg F, a 0–25 lb per sq ft (TD$_{EQ}$ = 44° F) wall construction, and a fenestration shading coefficient (SC) of 0.93 with single glazing, and 0.64 with double glazing.

Table 17-4. Allowable window areas expressed as a decimal for heating and cooling with an opaque wall U value of 0.06 Btuh per sq ft per deg F.*

City	Psw	Pss	Pdw	Pds
Chicago	0.28	0.246	0.50	0.365
New York	0.32	0.248	0.55	0.368
Portland	0.34	0.246	0.59	0.364
San Francisco	0.37	0.266	0.65	0.391
Los Angeles	0.41	0.235	0.72	0.350

Key:

Psw = fraction of window area, single glazed, winter
Pss = fraction of window area, single glazed, summer
Pdw = fraction of window area, double glazed, winter
Pds = fraction of window area, double glazed, summer

*When mechanical cooling is used, the smaller value must be used.

Any building that is mechanically cooled has an overall thermal transfer value (OTTV) for the gross area of exterior not exceeding the required value given in Section 4 (Fig. 6 in Section 4). OTTV is a function of opaque wall, shading coefficient of the glass, the glass solar factor value based on latitude (given), and temperature difference based on mass per unit area of opaque wall.

Table 17-3 is a summary of summer criteria for the same five cities listed in Table 17-1 and shows the allowable window fraction for single and double glazed windows. Table 17-4 compares the allowable fraction for single glazed summer criteria (P_{ss}); single glazed, winter (P_{sw}); double glazed, summer (P_{ds}); and double glazed, winter (P_{dw}).

Criteria implications

It is obvious that in each case the summer design criteria dictates the fraction of window area when mechanical cooling is involved. If the building is designed for winter heating only, the allowable glass area varies from 28 to 41 percent for single glazing and from 50 to 72 percent for double glazing. If the building is to be air conditioned as well, the allowable glass area is about 25 percent maximum for single glazing and 39 percent for double. It is apparent that the recommended guidelines do not allow "all" glass building construction. (While doing design work on a building planned for Anchorage, Alaska—North latitude, 61° 1', with a 10,864 degree-day heating season—the calculations showed that more glass area would be permitted than if the structure were located in Los Angeles—34° 0' N. latitude, with 1349 degree-days.)

A second major implication of the standard relates to the total energy consumed. All of the previous considerations have been based on reducing thermal requirements. For large commercial buildings with appreciable internal gains from people, lights, fenestration, and equipment, energy requirements for air conditioning may be comparable to that for space heating.

Appreciably lower wall conductance values will markedly increase annual air conditioning loads in many types of commercial buildings; higher values will increase space heating requirements. It would appear that this consideration merits further study.

Beyond the envelope

Standard 90-75 zeros in on the envelope thermal design to the exclusion of other energy related environmental factors. The building's site orientation, aspect ratio, geometric shape, the number of stories for a given floor area requirement, thermal mass, exterior surface color, shading or reflection from adjacent structures, surrounding surfaces or vegetation, opportunities for natural ventilation, and wind direction and speed, all affect the rate of energy consumption. In fact, the optimal use of these variables would have a greater impact than the restriction of the use of glass.

The standard does recommend that a designer consider these factors, but it does not go beyond an endorsement. There are no guidelines for these variables. Standard 90-75 offers comprehensive restrictions on the basic envelope, yet sacrifices these other environmental factors to the whim of the design team. It is conceivable that the savings effected by careful attention to the percentage of glass could be canceled by lack of consideration for these other variables.

Although a performance standard is provided in Section 10 of ASHRAE Standard 90-75, many developers and designers may fall back on the prescriptive values due to a lack of incentive, motivation, or tools to evaluate design alternatives. In the developer's attempt to lower first cost, the true objective, energy conservation, can be minimized in spite of the fact that the Section 10 approach could yield lower life-cycle costs and even lower first cost. Another conclusion of the Arthur D. Little study cited earlier was that smaller engineering offices that do not have experience with and easy access to computer technology would be more likely to follow the prescriptive/performance method outlined in Sections 4 through 9 than the performance method of Section 10.

Compliance with Standard 90-75 would reduce the total energy consumption of space heating on the order of a possible 10 to 20 percent along the coastal areas of the Pacific Northwest (Oregon and Washington). Here, the average winter design temperature is approximately 45° F, and heating degree-days are typically between 4100 and 5200. Also, the internal source heat reclaim (heat pump) system is commonly used when applicable to new commercial construction. (The standard can be expected to save more energy in more severe climates.) However, there will be strong constraints on the allowable building designs. New buildings may be less desirable, both from an energy and land use viewpoint, than if a basis other than envelope areas had been chosen for the standard.

It appears that the design of buildings at present is not controlled by space thermal requirements, but by such factors as space available, function, and cost. The building shape would be governed by these considerations, and a space heating and cooling standard would have little influence.

Since buildings are designed on the basis of function and a desired floor area, it would seem that a standard specifying an allowable conductance based on floor area may be more desirable. Allowable floor conductance values ought to vary with floor area in order to not unduly restrict the design of small buildings and yet take advantage of the energy saving potential of large buildings.

18

Excessive infiltration and ventilation air wastes energy

E. R. AMBROSE, PE,
Mount Dora, Florida

Discontinuing the practice of specifying and using excessive quantities of ventilation air could result in significant reductions in energy usage as well as in the first cost of heating and cooling systems. For example, each 1000 cfm reduction of outdoor air (in a 6000 degree-day area) represents a savings of about 3 tons in cooling equipment and 80 M Btuh in heating equipment. With a heat pump system, this could cause an annual operating cost reduction of approximately 30,000 kwh and an incremental equipment cost saving of about $2500. Comparable savings can be realized with fuel burning heating systems and motor driven cooling systems, depending on the type of system and seasonal operating efficiency.

Unfortunately, to take full advantage of such savings, several practices and procedures discussed in this chapter must have universal acceptance. It will be seen from the discussion to follow on each of the components listed that the suggestions are not new or revolutionary. In fact, they have been used quite successfully in a number of installations to obtain practical and economical systems.[1]

• *Standards for minimum ventilation requirements*—The unanswered question has been, and still is, what constitutes excessive ventilation air; or stating it another way, what are minimum ventilation requirements?

ASHRAE in the past has related ventilation air requirements for human occupancy in general office areas to the occupied space per person and to the activities to be assumed.[2] These outdoor air requirements, based on the minimum odor-free air required to remove objectionable body odors, as shown in Fig. 18-1, Curve C, were established in 1936 under laboratory conditions for the heating season. These recommended quantities vary from 8 cfm per person when occupying a volume of 400 cu ft to 25 cfm per person with 100 cu ft of space. In connection with the cooling cycle, the *ASHRAE Guide and Data Book* also recommends outdoor ventilation air quantities of 15 cfm per person with a minimum of 10 cfm for general office area with some smoking. For private offices with no smoking, the values increase to 25 cfm per person, and with considerable smoking, to 30 cfm or 0.25 cfm per sq ft of floor area, whichever is greater.[3] [4]

It is interesting to note that the above outdoor ventilation air requirements specified by ASHRAE are not governed by oxygen requirements or the carbon dioxide liberated, which for normal activities are about 1.3 and 3.7 cfm per person, respectively (Curves A and B, Fig. 18-1). Instead, they are governed by the odors rising from occupancy living habits, and related factors. The idea is that reduction or control of these odors can generally be accomplished by the introduction of outdoor or purified air in sufficient

---

[1]Superscript numerals refer to references at the end of the chapter.

Table 18-1.   Items needed to realize energy reductions by using minimum ventilation air quantities.

- Standards for different operating conditions that are based on authentic data and give recommended minimum ventilation air quantities to satisfy all safety and health requirements.
- Uniformity in federal statutes and local and state ordinances.
- Fundamental system designs based on sound engineering principles to take full advantage of the recommended requirements.
- Acceptable procedure for using purification equipment.
- Residential installations need recommended infiltration rates.
- Research and development to obtain factual data that can serve as a guide for establishing minimum ventilation rates for health and safety of the occupants.

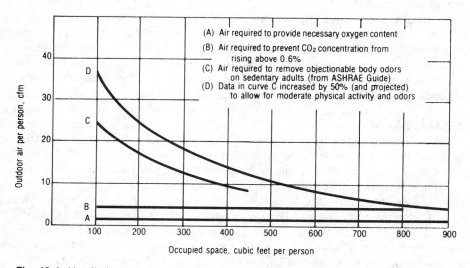

Fig. 18-1. Ventilation requirements related to cubic feet of occupied space per person.

quantities to dilute the odor concentration to a nonobjectionable level.

The acceptable ventilation air quantities are listed in the latest ASHRAE *Standards for Natural and Mechanical Ventilation.*[5] The quantities specified for general office space and similar areas in commercial and institutional buildings are substantially the same as established in 1936 and used during the intervening years. The important difference is that the meaning of the term, ventilation air, has changed. It no longer refers exclusively to outdoor air. The latest recommended definition is "an outside (outdoor) air supply plus any recirculated air that has been treated to maintain the desired quality of air within a designated space."[6]

Ventilation means air quality

This new definition calls for a certain quality of air. This quality is regulated by means of a *Table of Maximum Allowable Contaminant Concentrations for Ventilation Air,* which has been established by the National Air Pollution Control Administration and made a part of the ASHRAE standards. This regulation permits reducing outdoor air quantity to 15 percent of the specified ventilation requirements. In no case, however, "shall the outdoor air quantity be less than 5 cfm per person."

- *Uniformity in federal statutes, local and state ordinances, and among consulting engineers*—The outdoor ventilation air requirements for air conditioning systems found in local codes and ordinances through-out the United States vary widely. They range from 10 cfm per person to 2 cfm per sq ft of floor area, depending on the city and geographic area under consideration and on the type of space involved. The code in one large Midwestern city, for example, specified that 33⅓ percent of the total air circulation must be outdoor air.

In fact, a number of consulting engineers base their ventilation air requirements (when not particularly specified by codes or ordinances) on a certain percentage of the total air circulation. Such a quantity is justified in their opinion to assure proper dilution of the odors and at the same time furnish sufficient makeup air for the ex-haust systems throughout the building. Needless to say, such high rates result in a considerable waste of energy.

The main difficulty facing the air con-ditioning industry is to get an agreement in all these local codes to permit the use of the minimum amount of ventilation air. Undoubtedly, the best way this can be accomplished expeditiously is by federal statute.

• *Basic system design and operation with minimum outdoor ventilation require-ments*—Special consideration must be given to the air conditioning systems in order to effectively use the minimum outdoor ventila-tion quantities. Systems for the larger multi-story commercial or institutional buildings, for example, frequently employ medium or high pressure induction units, using 100 percent outdoor primary air. Such designs may result in ventilation quantities in excess of the minimum required for health and safety with a corresponding increase in energy consumption.

There are several other designs that are more suitable for lower ventilation rates when energy conservation is a prime con-sideration. One possibility is to use re-circulation air instead of outdoor primary air with the induction units. Other possi-bilities are to use fan-coil units or self-contained conditioners as terminal units.

The selection of the outdoor ventilation

air quantities and the design of the air conditioning unit will in turn affect the design of the exhaust system. For the 100 percent outdoor primary air induction units, an equal portion of the recirculated air (to compensate for the outdoor ventilation air) must be exhausted from the building, usually through hallways, kitchen, lavatories, etc. For other air conditioning designs, the venti-lation air and exhaust systems, for other than the office and other occupied areas, can be grouped together and operated inde-pendently of the main system.

It is frequently feasible in such designs to deliver the air directly from outdoors to the space involved. For example, in the case of kitchens or chemical laboratories and the like, vented hoods mounted directly over heat and odor producing equipment can be a practical solution. The hoods can often contain both the unheated supply air and exhaust air circuits. An acceptable hood design must have an effective method for entraining and exhausting the pollutants from localized areas with a minimum con-sumption of room air. Careless and over liberal designs of hoods will result in exces-sive ventilation quantities with corresponding high initial and operating costs.

These air conditioning designs are appli-cable and practical for new buildings still in the planning stage. Similar adjustments, revisions, and alterations, however, can be made to numerous systems in exisitng buildings to materially reduce the ventila-tion air rate. In these cases, it is not usually practical to replace the air conditioning units, but major changes can often be made in the distribution system.

In connection with the ventilation air systems, a factor always present is the difficulty of checking to see if the design quantities are being used in the day-to-day operation. It is possible for the manager or operating engineer to adjust the ventilation air controls and dampers to give quantities below that specified, or on the other hand, it could be considerably in excess of the specifications and not discovered because of the safety factor generally used in sizing equipment. This applies particularly to existing installations and to systems that have been altered and/or modified. Until positive procedures are developed and followed to determine the ventilation air actually being used, the estimated energy saving on a national basis can be very misleading.

• *Purification equipment has attractive possibilities*—Actually, the direct use of even minimum quantities of outdoor air may be unacceptable because of the objectionable irritable, and toxic odors and other pollutants caused by a high percentage of hydrogen sulfide, smog, and other chemical effluents from industrial processes and other such chemical byproducts often found in the atmosphere.

It is because of this possibility that use of physical or chemical odor removing equipment has such an appeal. It has been found for instance, that an absorbing material such as activated charcoal together with an efficient mechanical air filter or electronic filter is quite effective in removing tobacco smoke and other such odor causing substances as well as particulates. Using such units along with a minimum amount of outdoor air in a year 'round system (to satisfy the oxygen and carbon dioxide requirements) can be very practical and will frequently show an attractive saving in both the first and operating costs over a system striving for dilution of the odors by using larger quantities of outdoor air.

• *Residential applications require different considerations*—The discussion so far has been concerned primarily with commercial and institutional buildings. Heating and cooling systems for residences and other such dwelling units do not as a rule provide for the introduction of a predetermined amount of ventilation air into an occupied space. Outdoor air infiltration through cracks and openings is usually considered more than sufficient to satisfy the health and safety living requirements of a normal household. In fact, the problem has usually been to reduce the leakage to an acceptable level.

The instantaneous infiltration can vary with the type of construction, the design of the windows, the wind velocity, occupancy, type of heating system, number of appliances being used, and with the operation of the kitchen and bathroom exhaust fans as well as the clothes dryer. It was found that the infiltration measurements in two research houses varied from 0.12 to 1.34 air changes per hour, depending on the operating conditions.[7]

In electrical heated residences with full insulation, double windows and doors, which are considered criteria of tight construction, an average value of $\frac{1}{2}$ to $\frac{3}{4}$ air changes per hour is often assumed for design conditions. In fossil fuel systems, the amount can be much greater because air for combustion and flue drafts and exhaust systems are involved.

Numerous field tests have been made to measure the infiltration to verify the above conclusions. As a consequence, the same practices used in the early thirties are still being followed in arriving at a quantity of outdoor air for heat loss and heat gain calculations.

The latest ASHRAE standards[8] for single unit dwellings recommend 7 to 10 cfm per human occupant with 200 sq ft of floor area. This is equivalent to about $\frac{1}{2}$ air change per hour. Until now, infiltration has not been considered a major contributing factor. With national attention being directed to energy conservation, however, additional efforts must be made to obtain more realistic values.

• *Research and development necessary*— Additional research is needed to establish factual minimum quantities of outdoor ventilation air for the health and safety of the occupants under various operating conditions for both residential and commercial applications. Such action is mandatory to obtain full agreement by the existing local codes and ordinances throughout the country. Many of these standards have been in existence for 35 years or more, and revisions will be slow in coming without substantiation of the recommendations, even though the energy conservation may be considerable.

A recent advancement in this direction is the establishment of Quality Standards of Ventilation Air.[9] This permits filter equipment and efficient adsorption or other odor and gas removal equipment to be employed to purify the air. Thus the ventilation air can consist mostly of recirculated air with only a relatively small portion of outdoor air required to supply oxygen requirements and take care of the liberated carbon dioxide. In fact, purification of recirculation air as a substitute for outdoor air offers a most promising means of energy conservation. Exploration and development of various equipment designs and procedures are needed to obtain the required air quality and at the same time realize a practical means of obtaining maximum energy conservation during both the heating and cooling cycles.

Air purification systems are applicable in both existing and new buildings. Actually, an attractive potential for immediate energy reduction is realizable in existing commercial and industrial buildings.

References

1. "Heat Pump Performance—A Symposium," Bulletin 65, Annual Meeting of ASHRAE, Philadelphia, Pa., Jan. 28, 1959. "Operation of Large Air Source Central Heat Pump Systems," by E. R. Ambrose, p. 18.
2. *ASHRAE Handbook of Fundamentals—1972*, American Society of Heating, Refrigerating and Air-Conditioning Engineers, New York, N.Y., p. 190.
3. *Ibid.*, "Outdoor Air Requirement," p. 421.
4. *Standards for Natural and Mechanical Ventilation*, 62–73 (1973), American Society of Heating, Refrigerating and Air-Conditioning Engineers, New York, N.Y.
5. *Ibid.*
6. *Design and Evaluation Criteria for Energy Conservation in New Buildings*, Standard 90-75 (1975), American Society of Heating, Refrigerating and Air-Conditioning Engineers, New York, N.Y.
7. *Infiltration Measurements in Two Research Houses*, paper presented by R. C. Jordan and G. A. Erickson, ASHRAE Annual Meeting, Milwaukee, Wis., June 24-26, 1963.
8. *Op. cit. Standards for Natural and Mechanical Ventilation*.

Other sources

1. *Air Conditioning, Heating, and Ventilating*, Philip Sporn and E. R. Ambrose, July 1960.
2. *Heat Pumps and Electric Heating*, E. R. Ambrose, John Wiley & Sons, Inc., New York, N.Y., 1966.

SECTION III

Designing the systems for energy conservation

The need for good, sound engineering design of heating, ventilating, and air conditioning systems including their ancillary services for conservation of energy is discussed in detail in these 19 chapters. Included are the following:

- Selecting design parameters.
- Integrating the system design with the building.
- Selecting an energy source.
- Applying the computer in HVAC design.
- Selecting heat recovery systems.
- Determining thermal insulation for ducts, equipment, and piping.
- Applying heat energy from waste incineration.
- Using life-cycle costing in system selection.

19

Energy conservation: why not go all the way?

Some thoughs on energy conservation, the environment, and the need for HVAC designers to consider factors beyond the building boundary.

T. E. PANNKOKE, PE,
Engineering Editor, Heating/Piping/
Air Conditioning, Chicago, Illinois

A consulting mechanical engineer makes many decisions when designing a system for a client. The problems involved are familiar to most engineers and will not be repeated here. However, another factor has now appeared that promises to affect this decision making process. This is the ASHRAE energy conservation Standard 90-75. Its purpose is to provide energy conservation in new buildings through design criteria and guidelines that significantly influence the architectural, mechanical, and electrical design of buildings encompassed by it.

The standard's form considers efficient utilization of energy within a building only, and it does not deal with energy consumption at its source and accompanying transportation and transmission losses to a building. This decision was based on the rationale that ASHRAE members have the expertise to handle the varied and conflicting problems inherent in energy use within a building. Many felt that it is generally outside their expertise and scope to consider the effects of energy utilization back to its source.

It is not the purpose of this chapter

to argue with the decision of the ASHRAE committee but rather to suggest that consideration of the impact upon energy use and the environment—*i.e.*, considering the energy source—not be abandoned.

Needless to say, much has changed since the world was created. Man's ability to discover scientific and physical laws of the universe and apply them to specific problems has drastically altered the relationship of man to his environment. Man's ability to manipulate the environment has been a mixed blessing, with greater convenience and comfort seeming to go hand-in-hand with increasing environmental degradation.

Environmental pressures have been compounded by the exponential growth of the world's population, which is presently close to 4 billion and could exceed 6 billion by the year 2000.

More people means more land is consumed by housing, roads, commercial centers, etc., and less is available for agriculture.

If the present situation continues, the United States will have to rely more on coal to supply its energy over the next several decades. If the strip mining of coal is substantially increased, as is proposed, by opening up the western lands to wide-spread stripping, the impact on the environment will be profound. Unless the stripped land is carefully reclaimed, it will be useless for agricultural purposes to help feed the expected population increase.

What is the best use for some of these western lands?

• Continue to raise cattle, which consume approximately 8 lb of grain for each edible pound of meat.

• Use the water resources (which are not too plentiful in some areas) to irrigate more land to grow more grain for direct human consumption rather than suffer the losses in fertilizer, fuel, transportation, etc., that are associated with cattle raising.

• Mine the coal and use the water reserves for power plants and substitute natural gas (SNG) plants. Conflicting water needs for coal gasification and farming have slowed the development of coal gasification plans in North Dakota and Montana.

Obviously, no easy answers to this question or other questions involving choices among various alternatives exists. If they did, there probably would be no global concern over energy, mineral, and food supplies. Biologists, sociologists, anthropologists, and others have written extensively on aspects of the problems involved—and probably will continue to do so unless we experience a paper crisis also.

This brief chapter does not presume to solve the problem. It is merely intended to reinforce the fact that in these complex times, decisions often have effects reaching beyond the immediate problem—in this case, energy conservation in new buildings.

Data do exist that supply some of the answers to the question of overall fuel utilization from source to final use. Table 19-1 presents such data for electrical generation. Deep-mined and surface-mined coal, onshore and offshore oil, nuclear systems, and natural gas are compared.

Efficiencies of the different steps from extraction of the raw energy source to delivery to the customer are given. These steps are: extraction, processing, transportation, conversion, and transmission. Extraction efficiency takes into account the percentage of the raw fuel that is recovered from the ground as well as the energy used in the recovery process.

All efficiencies range from 10 to 25 percent. The two most efficient systems utilize strip mined coal—25 percent—and natural gas—24 percent. The first has the greatest negative impact on the environment from the standpoint of strip mining's effects on the earth and air pollution from the combustion process. Furthermore, energy used in reclaiming the land (such as restoring its approximate original contour) could make electricity generated by strip

Table 19-1.   Efficiencies of electric energy systems.

System	Component efficiency, percent					
	Extraction	Processing	Transport	Conversion	Transmission	Total
Coal						
Deep-mined	56	92	98	38	91	18
Surface-mined	79	92	98	38	91	25
Oil						
Onshore	30	88	98	38	91	10
Offshore	40	88	98	38	91	13
Natural gas	73	97	95	38	91	24
Nuclear	95	57	100	31	91	16

Reprinted from "Energy and the Environment—Electric Power," Council on Environmental Quality, U.S. Government Printing Office: Stock No. 411100019, August 1973.

This publication presents considerable data on the component efficiencies shown above (see Table 19-2) and on the environmental impact of each of the operations shown above on the various methods of power generation. Some of these data could be applied to HVAC systems utilizing direct and combustion on premises to obtain the overall utilization efficiency of fuel used and environmental impact.

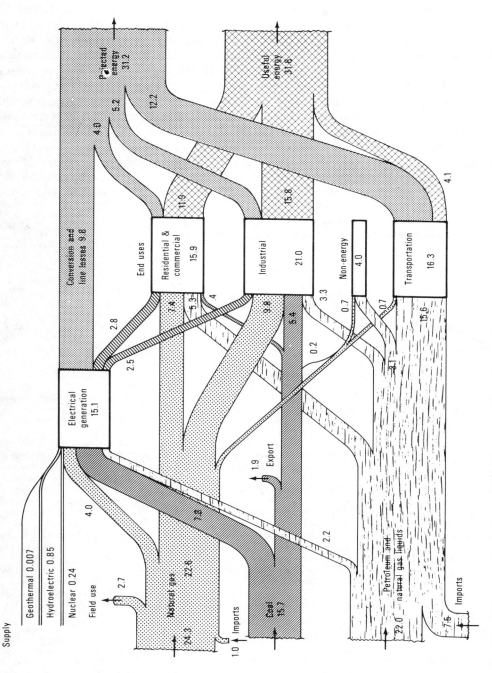

Fig. 19-1. U.S. energy flow patterns, 1970. (Reproduced from Energy: Uses, Resources, Issues. Lawrence Livermore Laboratory, University of California/Livermore, 1972. Prepared for the U.S. Atomic Energy Commission.)

mined coal a less desirable alternative. The second, natural gas, is in short supply.

Table 19-2 presents a more detailed breakdown of the figures shown in Table 19-1. Similar data for the other fuels along with considerable information on the environmental impact of the various steps in the conversion process are found in the source listed under Table 19-1.

Figure 19-1 shows energy flow patterns as they existed in the United States in 1970. Geothermal, hydro-electric, nuclear, natural gas, coal, and petroleum sources are included.

Data for the efficiencies of the extraction, processing, and transportation processes could be applied to the direct conversion of fuel on the premises. Thus, the comparable overall efficiency from the source to the final use for gas, oil, and coal on the building premises could be determined. This could also be done for heat pumps and heat reclamation systems that commonly are driven by electricity. If all major sources and forms of energy for HVAC systems were compared on the same bases, it would be possible to make more intelligent decisions about energy use based on efficiency of utilization, environmental impact, land use, and the other complex and sometimes conflicting factors that must be considered.

The decision of the ASHRAE committee

Table 19-2.   Component efficiencies of coal fired electricity systems.

| | Component efficiency, percent | | | | | | | |
| | Extraction | | | | | | System[1] | |
Factors	Deep	Surface	Processing	Transport	Conversion	Trans-mission	Deep	Surface
Resources recovered[2]	57[3]	80[4]	92.1[5]	99[6]	38[7]	91.2[8]	18.0	25.3
Energy input[9]	0.8[10]	0.8[10]	0.1[10]	0.9[11]				
Net efficiency[12]	56.2	79.2	92.0	98.1	38	91.2	17.8	24.9

[1]System efficiency is the output energy of the last component divided by the total energy input to the system, including external energy.
[2]Percent of energy value of coal energy entering each component which remains after that component.
[3]Lowrie, R. L. "Recovery Percentage of Bituminous Coal Deposits in the United States, Part 1, Underground Mines," Washington, D. C., U. S. Department of the Interior, Bureau of Mines, p. 1, 1968.
[4]Averitt, P., "Stripping Coal Resources of the United States," U. S. Department of the Interior, Geological Survey Bulletin 1322, Washington, D.C., Government Printing Office, pp. 7-8, January 1970.
[5]Total 1969 raw coal production and percentage of raw coal wet cleaned, dry (pneumatically) cleaned, and unprocessed are shown in Figure A-1, U.S. Department of the Interior, Bureau of Mines, 1971. "Minerals Yearbook—1969," Washington, D.C., Government Printing Office, pp. 309, 347, 348. Total raw coal production includes processing plant wastes. Processing removes approximately 12 percent of the energy value, Btu, of the coal, based on Lenard, J. W. and Mitchell, D. R., eds., "Coal Preparation," New York, American Institute of Mining, Metallurgical, and Petroleum Engineers, p. 5-32, 1968.
[6]Approximately 1 percent of the coal is blown away during transport unless it is wetted. Delson, J. K. and Frankel, R. J., "Residuals Management in the Coal-Energy Industry," April 1972. Manuscript to be published by Resources for the Future, Inc., Washington, D.C., R.F.F., p. 1V-45.
[7] A large new powerplant is assumed to have a heat rate of 9000 Btu per KWH, equivalent to a 38 percent conversion efficiency. The best plants have achieved around 8530 to 8900, whereas the national average is around 10,500. Federal Power Commission, "The 1970 National Power Survey," Part I, Washington, D.C., Government Printing Office, pp. 1-5-6, 1-5-7, 1972.
[8]Power sold divided by power produced, 1969. Federal Power Commission, "Annual Report—1971," Washington, D.C., Government Printing Office, pp. 11, 13, 1972.
[9]Expressed as percent of energy content of the coal (or electricity) entering each stage. Based on 11,639 Btu per lb of coal, the 1969 average for coal burned at powerplants. National Coal Association, "Steam-Electric Plant Factors," 1971 ed., Washington, D.C., N.C.A., p. 107, 1972.
[10]Data supplied by the InterTechnology Corp., Warrenton, Va.
[11]Assumes all coal is shipped by rail for an average of 300 miles. Actual 1965 average distance is 297 miles. National Coal Association, "Bituminous Coal Facts—1970," N.C.A., Washington, D.C., p. 91. Two-hundred ton miles of cargo movement requires approximately 1 gal of fuel with an energy content of 136,000 Btu per gal. Rice, R. A., "System Energy as a Factor in Considering Future Transportation," a paper presented at the American Society of Mechanical Engineers Winter Annual Meeting, November 29 to December 3, 1970.
[12]Component net efficiency equals (% resource recovered) — (% energy input).
Component efficiencies for each of the systems shown in Table 19-1 and the environmental impact of each operation can be found in the following publication from which this table is reproduced. "Energy and the Environment—Electric Power," Council on Environmental Quality, U.S. Government Printing Office: Stock No. 4111-00019, August 1973.

not to go beyond the building boundary may well be correct considering the background, knowledge, and expertise of its members and the society as a whole. However, the serious problems facing us to-day—of which the need to use energy more efficiently is only one—defy solution by one group.

Perhaps the question of energy utilization outside a building's boundaries could best be tackled by other organizations that can interface with ASHRAE. The National Bureau of Standards, the Federal Power Commission, the Atomic Energy Commission, and the Bureau of Mines are among several government agencies that have data on fuel reserves and/or energy utilization.

In the private sector, groups such as the American Gas Association, the Electric Energy Association, and the National Coal Association can assist in supplying further data that may be required to ascertain the overall effects on energy resources and environmental impact of building energy use.

It is not possible to incorporate factors involving the overall use of energy in the 90-75 Standard without additional study. However, from the overall standpoint of a desire to make optimum use of our energy, land, and water resources, the ultimate good for society demands that provisions be considered for energy efficiency from extraction of the raw source to the point of end use for inclusion in Standard 90-75.

20
Design parameters

WILLIAM J. COAD, PE,
Vice President Charles J. R. McClune &
Associates, St. Louis, Missouri

Whether it be in operating a business, creating an artistic work, working at a skilled trade, or practicing engineering, there is an inherent tendency to filter out many of the seemingly insignificant bits of basic knowledge upon which our expertise was structured and base our daily functioning and decisions on selected conglomerations of these basics. This developmental process comes with experience and is a prerequisite to effective production.

From time to time, however, with changing technology, social structures, monetary economic conditions, etc., every practitioner should re-evaluate the parameters upon which he is basing his daily decisions. The re-evaluation of parameters, if applied to engineering designs, would produce vastly different results from those we have seen in the past two decades.

To develop this concept, consider that as any new field of engineering technology is born, there is essentially only one design parameter—performance. Examples: that the machine convert thermal energy to shaft work; that the generator convert shaft work to usable electric energy; or that the bridge span the river and support the weight required. Then, as each field of engineering technology matures, other design parameters inevitably are required. Take, for example, the heat engine, for which the original parameter was simply to convert energy from a thermal to a mechanical form. Subsequent parameters that evolved included improving the heat rate (or energy efficiency), decreasing the weight-to-horse-power ratio, reducing the maintenance requirements, refining or improving the automation and safety systems, and decreasing the production or manufacturing cost per unit of power required.

As the practicing engineer embarks upon any phase of design, he is well advised to stop for a moment and compile a list of the design parameters that he will attempt to satisfy. In the field of building systems design, as in any discipline of systems engineering, such an examination of relevant parameters must be undertaken at numerous phases throughout the design development. A typical listing of such phases relating to building environmental systems would be:

- Establishment of indoor conditions.
- Calculation of the loads and load profiles.
- Selection and design of terminal control systems.
- Selection and design of terminal distribution systems and methods.
- Selection of type(s) of thermal distribution systems and subsequent design.
- Selection and design of ranges, rates, and thermal levels of thermal distribution systems.
- Selection and design of high level (heating) primary conversion system.
- Selection and design of low level (cooling) primary conversion system.
- Selection of high level (heating) energy source or sources.
- Selection of low level (cooling) energy source or sources.

Consider, for example, the first phase

as is shown—establishment of indoor conditions. If this question were addressed hastily regarding air conditioning for human comfort, one might state a specific dry bulb temperature and relative humidity (such as 75 °F DB and 50 percent RH). But upon reflection, we all realize that the basic parameter is human comfort. Thermal comfort, in turn, is a physiological phenomenon achieved by a balance among metabolic heat rate (input), work produced (output), and heat dissipated (rejected). In oversimplification, the human machine can be described thermally by the classical block diagram used in elementary thermodynamics to illustrate the second law (see Fig. 20-1). The designer of the thermal environment is concerned chiefly with the dissipation. Considering the heat transfer phenomena effecting this exchange—radiation, convection, evaporation, and conduction—establishes the following comfort parameters:

- Dry bulb temperature.
- Relative humidity.
- Mean radiant temperatures (enclosing and "exposed" surfaces).
- External radiant effects (solar, direct and reflected).
- Contact surface temperatures (chairs, desks, etc.).
- Air velocity.

Obviously, this is quite an expansion from simply dry bulb temperature and relative humidity. However, thousands of buildings enclosing millions of sq ft have been constructed without the ability to provide for thermal comfort because of

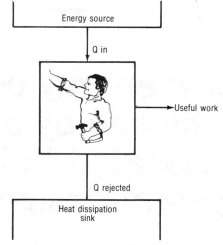

Fig. 20-1.

a failure to recognize one or more of these as relevant parameters. Space does not permit a discussion of some of these typical failures, but every practicing engineer can provide his own.

Another interesting aspect: It becomes immediately evident that the thermal environmental system is not *added* to the building; rather, it is an integral part thereof. The design engineer, therefore, cannot avoid involvement in the basic building design, or at least in the aspects related to heat exchange.

Each designer should compile his own listings; in addition to improving the wisdom of design decisions, this will make it increasingly evident why the engineer must participate in the architectural aspects of building design decisions.

21

How energy conservation is achieved by applying integrated building and system design

PAUL K. SCHOENBERGER, PE,
Energy Management Consultant
Columbus, Ohio

When an integrated approach between the construction of a building and the selection of the systems used to control the environment is applied, definite energy consumption of the mechanical and electrical systems can be minimized. By designing the building envelope to minimize heat transmission and solar effects and by reducing the lighting levels, the size of the mechanical equipment can also be reduced. The smaller sizes of mechanical equipment in turn will reduce the size of the electrical power distribution system and will result in a net reduction in first cost for both the mechanical and electrical systems. The additional costs that may be incurred for constructing the building envelope to minimize the transmission and solar effects may not be as much as the savings obtained in the first cost of the mechanical and electrical systems, but the net result in energy conservation will be experienced in the reduction of the lifecycle cost of operating the building.

Three design examples prove results

The building used to show the effect of an integrated energy conservation design by three examples is a two story office building in Columbus, Ohio. Design parameters for the facility are as follows:

1. Maximum summertime conditions: 92° F DB and 77°F WB outside and 76°F DB and 64°F WB inside. Building is located 40° north latitude, and design wind velocity is 7.5 mph.
2. Maximum wintertime conditions: 0°F outside and 73°F inside. Design wind velocity is 15 mph.
3. Occupancy: one person per 120 sq ft of floor area.
4. Flourescent lighting fixtures are used throughout the building.
5. Building is two stories high with 20,000 sq ft on each floor, 250 ft long by 80 ft wide. Orientation is with long dimension running north and south. Roof area is 20,000 sq ft, and perimeter area of wall and glass is 14,000 sq ft.
6. The maximum heating and cooling loads have been calculated using data from the "1972 ASHRAE Handbook of Fundamentals". Judgment indicates that the roof and west glass will combine to produce the maximum cooling load at 4:00 PM, sun time. The radiant portion of the instantaneous heat

gains for building designs "A" and "B" are averaged over a 3 hr period up to and including the time of maximum load conditions. The building design "C" is averaged over an 8 hr period.

Building design "A" construction details

Building "A" reflects an envelope construction and an electrical lighting system similar to the design of many recently completed facilities where conservation of energy has not been of primary concern.

1. Wall construction: 8 in. lightweight concrete block, 4 in. brick veneer and $\frac{5}{8}$ in. sand aggregate plaster. The "U" value is 0.29 for winter and summer.
2. Glass construction: $\frac{1}{4}$ in. heat absorbing plate/float glass (gray, bronze, or green tinted) with light colored interior drapes or blinds. Shading coefficient is 0.53 when the fenestration is sunlit as well as when it is on the shaded side of the building. The "U" value is 1.13 for winter and 0.81 for summer.
3. Roof construction: metal deck with 1.5 in. of preformed roof insulation, built-up roofing and $\frac{1}{2}$ in. suspended acoustical tile ceiling. The "U" value is 0.13 for winter and summer.
4. Minimum outside air ventilation rate: 30 cfm per person.
5. Interior lighting levels are maintained using an average of 4.5 watts per sq ft including the wattage of the fixture ballasts.
6. Exterior walls: 70 percent glass on all exposures.
7. High velocity air distribution systems supply air to all spaces. System fans increase supply air temperatures by 4° F.
8. The desired ventilation rate is used for determining the ventilation component of the heat gain. Infiltration is assumed to be one air change per hour for determining the maximum heating load.

Building "B" construction details

1. All construction details, ventilation rates, and lighting levels used in Building "A" will be used in Building "B" design.
2. Glass construction: $\frac{1}{4}$ in. heat absorbing plate/float glass outside and $\frac{1}{4}$ in. regular plate/float glass inside with light colored interior drapes or blinds. Shading coefficient is 0.51 when the fenestration is sunlit as well as when it is on the shaded side of the building. The "U" value is 0.65 for winter and 0.52 for summer.
3. Glass areas are reduced from 70 percent used in Building "A" to 20 percent.

Building "C" construction details

1. Wall construction: 8 in. lightweight concrete block, 4 in. brick veneer, studs with 3 to 3.5 in. of mineral fiber blanket insulation having a resistance of 11 and $\frac{1}{2}$ in. gypsum plasterboard. The "U" value is 0.07 for winter and summer.
2. Glass construction: $\frac{1}{4}$ in. heat absorbing plate/float glass outside and $\frac{1}{4}$ in. regular plate/float glass inside with light colored interior drapes or blinds. The "U" value is 0.65 for winter and 0.52 for summer. Minimum amounts of glass will be used in all walls; 120 windows, 1 ft–9 in. wide by 3 ft high will be used to permit at least one window in all perimeter rooms, requiring 630 sq ft of glass or 4.5 percent glass area on all exterior exposures.
3. Roof construction: metal deck with 4 in. of preformed roof insulation, built-up roofing and suspended $\frac{3}{4}$ in. acoustical tile ceiling. The "U" value is 0.06 for winter and summer.
4. Minimum outside air ventilation rate is 5 cfm per person. Interior lighting levels are maintained using an average of 2.5 watts per sq ft including the wattage of the fixture ballasts.
5. Low velocity air distribution systems

supply air to all spaces. System fans increase supply air temperature by 2° F.

6. The desired ventilation rate is used for determining the ventilation component of the heat gain. Infiltration is minimized due to tight construction and minimum amounts of fixed glass.

Tables summarize effects of insulation

Tables 21-1, 21-2, and 21-5 show that the maximum heat gain and heat loss of the roof and wall areas is significantly reduced when the areas are well insulated. In addition to reducing the maximum loads, insulation has supplementary benefits. During wintertime operation, the inside surface temperature of the well insulated wall is higher, causing the occupants to lose less heat by radiation to the warmer wall surfaces. During summertime operation, the addition of insulation to the wall and roof areas causes a lag between the time the sun hits the exposure and the time that this heat travels into the building. Additional insulation can make those areas peak during evening hours when the facility may be unoccupied. Insulation of the walls will greatly reduce the transmission and solar load of these areas during summertime peak load conditions. The maximum load imposed upon the air conditioning systems will be reduced appreciably by the addition of insulation to the roof because the time is changed that the roof load peaks.

Table 21-1.   Instantaneous convective and radiant heat gains.

Section	Net area sq ft	"U"-Value	Heat gain at 4:00 PM		Heat gain 3 hr ave.	
			°F Temp. Diff.	MBH Heat Gain	°F Temp. Diff.	MBH Heat Gain
Building "A"						
Roof	20,000	0.13	73	189.8	69	179.4
South wall	509	0.29	13	1.9	11	1.6
East wall	1,591	0.29	23	10.6	22	10.2
North wall	509	0.29	7	1.0	7	1.0
West wall	1,591	0.29	12	5.5	12	5.5
TOTAL				208.8		197.7
Building "B"						
Roof	20,000	0.13	73	189.8	69	179.4
South wall	1,358	0.29	13	5.1	11	4.3
East wall	4,242	0.29	23	28.3	22	27.1
North wall	1,358	0.29	7	2.8	7	2.8
West wall	4,242	0.29	12	14.8	12	14.8
TOTAL				240.8		228.4
Building "C" (Heat gain is 8 hr ave.)						
Roof	20,000	0.06	40	48.0	21.8	26.2
South wall	1,620	0.07	11	1.2	11.0	1.2
East wall	5,064	0.07	20	7.1	16.6	5.9
North wall	1,620	0.07	7	0.8	7.6	0.9
West wall	5,064	0.07	14	5.0	16.4	5.8
TOTAL				62.1		40.0

Note: Since ave. temp is 80° F and room temp. is one degree greater than 75° F, 6° F has been subtracted from the tabulated values.

Table 21-2.   Transmission cooling load.

Section	60% of 3 hr ave. heat gain	40% of 4:00 PM heat gain	Cooling load MBH
Building "A"			
Roof	107.6	75.9	183.5
South wall	1.0	0.8	1.8
East wall	6.1	4.2	10.3
North wall	0.6	0.4	1.0
West wall	3.3	2.2	5.5
TOTAL	118.6	83.5	202.1
Building "B"			
Roof	107.6	75.9	183.5
South wall	2.6	2.0	4.6
East wall	16.3	11.3	27.6
North wall	1.7	1.1	2.8
West wall	8.9	5.9	14.8
TOTAL	137.1	96.2	233.3
Building "C" (60% of 8 hr ave. heat gain)			
Roof	15.7	19.2	34.9
South wall	0.7	0.5	1.2
East wall	3.5	2.8	6.3
North wall	0.5	0.3	0.8
West wall	3.5	2.0	5.5
TOTAL	23.9	24.8	48.7

The heat absorbed in the insulated walls and roof areas will dissipate into the building during the unoccupied summertime evening hours. Fan systems can be operated using 100 percent cool outside air during the early morning hours to purge the facility of the heat buildup without the use of mechanical refrigeration, thereby conserving energy.

Calculations reflect effect of glass areas

Tables 21-3 and 21-5 show that double insulated solar reflective glass has less effect on the maximum load than does the use of minimum glass areas. Glass is not all bad, just the quantity of it. Glass on any exposure has more undesirable effects than good effects on the environmental control system.

When direct sunlight hits glass areas, interior conditions adjacent to the glass become extremely hot without the use of reflective shades or draperies. When the draperies are closed, it is usually necessary to operate the interior lighting. The orientation of glass areas with respect to the sun can be used to minimize these undesirable effects. In northern latitudes south glass exposed to sunlight during the wintertime will provide heating to the building. One way considered to be ideal to minimize the undesirable effects of glass would be to install exterior insulated shutters that would provide shading and could be closed to convert the areas to insulated walls whenever desirable.

Lighting levels show effect on loads

Table 21-4 shows that reducing lighting levels will significantly reduce the maximum cooling load on the building. Lighting generates sensible heat, and occupants add both sensible heat and moisture to the building

Table 21-3.   Cooling load through windows.

Location	Area sq ft	42% of 4:00 PM Heat gain MBH	58% of 3 hr ave. Heat gain MBH	Air to air heat gain MBH	Cooling load MBH
Building "A"					
S. windows	1,188	10.6	28.6	15.4	54.6
E. windows	3,712	18.2	31.2	48.1	97.5
N. windows	1,188	6.1	10.1	15.4	31.6
W. windows	3,712	178.5	213.7	48.1	440.3
TOTAL					624.0
Building "B"					
S. windows	340	2.9	7.9	2.8	13.6
E. windows	1,060	5.0	8.6	8.8	22.4
N. windows	340	1.7	2.8	2.8	7.3
W. windows	1,060	49.0	58.7	8.8	116.5
TOTAL					159.8
Building "C" (58% of 8 hr ave. heat gain)					
S. windows	76	0.7	2.1	0.6	3.4
E. windows	239	1.1	6.5	2.0	9.6
N. windows	76	0.4	0.7	0.6	1.7
W. windows	239	11.1	6.5	2.0	19.6
TOTAL					34.3

year around. Occupied spaces on the interior of the building which have no roof heat loss require cooling year around. Occupied spaces with only roof heat losses may also require cooling year around when well insulated roofs are used. The heat generated from occupants and lighting must be offset with cooling from the environmental control system during summertime operation; however, this heat should be used to offset building heat losses during wintertime operation. To conserve energy, systems must be used that move the heat generated in the occupied interior spaces to the perimeter areas that require heating. Excessive lighting of areas will increase the maximum cooling load, increase equipment sizes, increase first cost and cause waste in energy consumption.

Use "free" cooling with outside air

"Free" cooling using outside air in lieu of mechanical refrigeration can be provided for spaces with excessive heat gains during wintertime operations. When outside temperatures permit, outside air may be mixed with return air to provide mixed air at a temperature adequate to cool the spaces. Extreme care must be taken to prevent this mixed air economizer from falsely loading the system and causing an increase in the heating requirements of the building due to excessive ventilation rates. It has been common practice in the past to install economizer controls that maintain 55°F mixed air to the systems for "free" cooling of the spaces with excessive heat gains. If the return air temperature is 75°F and the outside air temperature is 35°F, the system should mix one part of outside air with one part of return air to maintain the 55°F mixed air temperature. This means that the economizer is causing the system to operate using 50 percent outside air while the minimum ventilation rate may be 10

Table 21-4. Summary of maximum calculated cooling loads.

Item	Bldg "A" cooling load MBH	Bldg "B" cooling load MBH	Bldg "C" cooling load MBH
Transmission			
Roof	183.5	183.5	34.9
South wall	1.8	4.6	1.2
East wall	10.3	27.6	6.3
North wall	1.0	2.8	0.8
West wall	5.5	14.8	5.5
Windows			
South	54.6	13.6	3.4
East	97.5	22.4	9.6
North	31.6	7.3	1.7
West	440.3	116.5	19.6
Internal loads			
Infiltration sensible	2.9	2.9	2.9
Lights	613.8	613.8	341.0
People sensible	83.3	83.3	83.3
Ventilation sensible	176.0	176.0	29.3
Fan loads	305.2	216.0	51.0
TOTAL SENSIBLE LOAD	2,007.3	1,485.1	590.5
Latent loads			
Infiltration	5.4	5.4	5.4
People	66.7	66.7	66.7
Ventilation	324.3	324.3	54.1
TOTAL LATENT LOAD	396.4	396.4	126.2
GRAND TOTAL COOLING LOAD	2,403.7	1,881.5	716.7
(Tons of refrigeration)	(200.3)	(156.8)	(59.7)

Table 21-5. Summary of maximum calculated heating loads.

Item	Bldg "A" heating load MBH	Bldg "B" heating load MBH	Bldg "C" heating load MBH
Roof	189.8	189.8	87.6
Wall	88.9	237.1	68.3
Glass	808.4	132.9	29.9
Slab on grade	31.7	31.7	31.7
Ventilation	788.4	788.4	131.4
Infiltration	473.0	473.0	
TOTAL	2,380.2	1,852.9	348.9

percent of this value. When the outside air temperature is -5°F, the system should mix one part of outside air with three parts of return air to maintain the 55°F mixed air temperature. This means that the system will be using 24 percent outside air while the minimum ventilation rate may be 20 percent of this value. If this 55°F mixed air is supplied to any spaces that do not have cooling requirements, the air must be reheated to space conditions through false loading. If the economizer remains in operation when the building is totally unoccupied the entire air delivery of the fan system must be neutralized through false loading before the heat losses can be offset. "Free" cooling economizers must supply cooling air to the spaces with cooling loads only, and if a constant temperature of mixed air is used, the volume of air supplied to the spaces must be varied in direct proportion to the space cooling requirements. Minimum ventilation rates should not be exceeded in areas that require heating to offset heat losses.

Calculations prove reductions in loads

It is apparent from Tables 21-4 and 21-5 that construction of the two-story building can significantly affect the maximum heating and cooling loads. When comparing building calculations "A" and "C", the maximum cooling load is reduced by 70.2 percent and the maximum heating load is reduced by 85.3 percent. It is also apparent that half-measures will accomplish very little in the reduction of these loads. Reducing the perimeter glass areas from the 70 percent single thickness glass used for building "A" to 20 percent of double thickness glass used for building "B" reduced the maximum heating and cooling loads by 22 percent each. Constructing buildings to minimize loads can greatly simplify design of the environmental control systems; however, to assure minimum energy usage, it is necessary to install systems that will respond to actual usage and swing in solar loads without false loading. If heating and cooling loads are minimized and responsive environmental control systems are used, minimum energy usage will be achieved.

22

Thermal effectiveness of a vapor compression cycle

WILLIAM J. COAD, PE,
Vice President, Charles J. R. McClure &
Associates, St. Louis, Missouri

The second law of thermodynamics states essentially that heat cannot be made to flow from a region of low temperature to one of higher temperature without the input of energy from an external source. This sets the ground rules: a building enclosed in higher temperature surroundings cannot be cooled without consuming energy. The corollary to the Carnot principle takes us one step closer to quantifying the dictates of the second law by stating that no refrigeration machine operating between a constant temperature source (To) and a constant temperature sink (T) can have a coefficient of performance (COP) higher than a reversible cycle. A simple thermal analysis of the Carnot cycle (a reversible cycle) further reveals that the COP of the Carnot cycle is:

$$COP = \frac{To}{T - To}$$

This is the Carnot or Ideal COP. All too often, the tendency is to blame Carnot for our second law energy conversion losses or nonproductive burdens. Consider a simple analysis of how well we are doing compared to the ideal energy requirement. Starting with 1 ton (12,000 Btuh) of cooling and proceeding through an analysis of the useful (product) and nonuseful (process) energy flows through a simple refrigeration system from the cooling coil (input) to the ambient surroundings. For the analysis, conditions are set at 75°F DB and 50 percent RH indoors and a 95°F ambient sink. The Carnot energy required to accomplish heat pumping from 75 to 90°F is 449 Btu; the Carnot energy required to dehumidify, reducing the low temperature sink to 55°F dewpoint, is 483 Btu.

The sum of these two ideal energy inputs is the theoretical or Carnot energy required to produce 1 ton of cooling: *i.e.*, 0.273 KW. Another interesting observation is that theoretically, dehumidification requires 7.5 percent more energy than sensible cooling.

Continuing through the analysis, the energy losses or burdens are:

• *Cooling coil heat transfer burden*—The increase in Carnot energy required to provide for a 10°F temperature differential between conditioned air dewpoint and 45°F entering water temperature is 256 Btu.

• *Chiller heat transfer burden*—The increase in Carnot energy required to provide for a 5°F temperature differential between the leaving (45°F) chilled water and the evaporating refrigerant at 40°F is 132 Btu.

• *Condenser heat transfer burden*—Current product catalog literature was scanned to find a "typical" condensing temperature for a 95°F ambient, air cooled chiller-condenser combination. This was found to be 121°F condensing temperature. Thus, the heat transfer burden imposed is represented by the temperature differential 26°F (121°—

111

95°F). Incorporating this in the Carnot COP produces a condenser heat transfer burden of 624 Btu.

• *Fluid and thermodynamic cycle burden*—The ideal vapor compression cycle closely approximates the Carnot cycle; the deviations being the superheat resulting from isentropic compression diverging from the vapor dome and the constant enthalpy expansion in lieu of an isentropic expansion. When these deviations are considered with a given refrigerant, the burden of that particular fluid can be determined. For this analysis, the most common air conditioning fluids, R12 and R22 were considered. The R12 burden was found to be 596 Btu and R22 was 637 Btu.

• *Mechanical and electrical burdens*—At this point in the analysis, Carnot necessarily yielded to available hardware. Again, a product catalog scan and selection of one of the low KW per ton chiller condenser combinations revealed that the burden imposed by mechanical and electrical cycle motivation was 1486 Btu.

• *Condenser motivation energy*—In addition to the Carnot heat transfer burden, energy is required to move the cooling fluid through the condenser heat exchanger. Product catalogs were again consulted for a "typical" but conservative value, and the condenser fan resulted in a burden of 471 Btu.

• *Fluid distribution*—Since a water chiller was selected for the analysis, the fluid motivation energy should be considered. Based on a 50 ft pumping head, it is calculated to be 122 Btu.

Table 22-1 shows Btu per ton-hr values for each energy use and burden, and the percentage of input represented by each. Summing all the energy componenets reveals an input of 4660 Btu per ton-hr or 1.36 KW per ton. If the Carnot COP energy requirement is considered as the ideal, and the thermal effectiveness is defined as the ideal energy requirement divided by the actual, the system is found to be 20 percent effective. Fig. 22-1 illustrates the same data in a classical first law thermal flow diagram.

Table 22-1.   Thermal requirements per ton-hr.

	Btu	%
Carnot sensible cooling	449	9.64
Carnot latent cooling	483	10.34
Cooling coil heat transfer	256	5.50
Chiller heat transfer	132	2.83
Condenser heat transfer	624	13.38
Fluid and thermodynamics	637	13.67
Mechanical and electrical	1486	31.88
Condenser heat rejection	471	10.11
Cooling fluid distribution	122	2.65
Total energy input	4660	100.00

The purpose of this brief excursion into the thermal effectiveness of a refrigeration cycle is to illustrate the theoretical potential for energy consumption reduction in a major subsystem of the building environmental system. If, for instance, the energy effective-

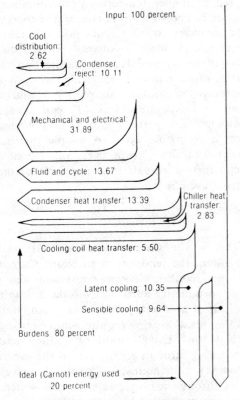

Fig. 22-1. Thermal balance flow diagram.

ness in heat transfer technology and application could be doubled, this would decrease the consumed energy in refrigeration systems well in excess of 10 percent. Other areas of concentration indicated are in mechanical and electrical technology and thermal fluids.

This analysis has considered only the design capacity energy use and burdens. As the building system operates at reduced loads, the seasonal energy effectiveness is seen to generally reduce radically.

Thus, the question might be asked. Would it be more cost-energy effective if we devoted some of the vast sums of money now being spent on such projects as solar energy conversion to more mundane efforts, such as heat transfer, fluids, mechanical concepts, and electrical convertors?

23

Picking an energy source and conversion system

WILLIAM J. COAD, PE,
Vice President, Charles J. R. McClure &
Associates, St. Louis, Missouri

Many decisions are required as the design of a building thermal environmental system develops. This chapter suggests some parameters to apply in the selection of energy sources and primary conversion systems.

The state of the art in available hardware has resulted in two forms of energy being needed to motivate and supply the HVAC needs of a space; these are thermal energy and electrical energy. The forms available for consideration for building systems as supplied are: electricity, fossil fuels, district heating and cooling, and solar energy.

Although there is some interrelationship between the use form and the supplied form, it is quite minimal. For example, available technology dictates that electricity is universally used to provide the energy for such auxiliary system drives, such as fans and pumps for any number of uses, leaving the interrelationships between use and source to the primary motivating energy of the primary high level (heat) and low level (cooling) source systems. Thus, except for these two selections, the remainder of the subsystems are best designed irrespective of the source.

The selection of the building primary conversion system, however, cannot be separated from the source. The suggested parameters are:

- *Availability of source matter or energy-* It would be ridiculous to consider natural gas in an area where it is not available (a rather obvious parameter).

- *Unit cost of matter or energy*—Probably the most obvious parameter, and one which many times has been considered without regard for any other parameters.
- *Integrated efficiency of conversion to the required use form*—The term integrated is of paramount importance since the building energy system is a time integrated entity; and over a given time span (such as a calendar year), it operates at less than design load the vast majority of the time. This realization coupled with the inherent reduction in efficiency at less than full load mandates consideration of the time integrated efficiency.
- *Investment cost of storage, handling, and conversion apparatus*—Like the energy unit cost, this parameter appears obvious. However, in this context, it is emphasized that investment cost considerations of other subsystems, such as *distribution and terminal conversion or control, are to be considered separately.*
- *Environmental considerations of the space*—This is an indirect parameter in that it relates to a consideration of the basic need for energy. For example, in a warm climate, if thermal source energy is being considered for the purpose of dehumidification only, perhaps alternative methods of satisfying that need could be developed. Another example might be the reconsideration of utilizing high level lighting heat when a fuel source might prove more energy resource effective.

• *Environmental considerations of the community*—Although it may seem a bit idealistic to suggest considerations beyond statutory requirements, such idealism has relevance both morally and potentially economically. From the moral side, all engineers are sworn to consider the impact of their designs on the public welfare. Economically, future legislation could have significant financial impact on a building's primary conversion system (whether on- or off-site) if the system is considered detrimental to the environment.

• *Energy consumption and demands* (*power*)—This parameter, coupled with the unit cost schedule is required to determine the ultimate annual comparative energy cost.

• *Cost and availability of maintenance and service for the conversion apparatus*—Although this may appear to be an obvious ingredient to those who support the concept of life-cycle cost analysis, it is the least easily quantifiable ingredient thereof. The hourly cost may be easily identifiable, but the hours required are the most difficult to assemble or predict. The second consideration, availability, is the most overlooked parameter in energy conversion system selection. The availability of skilled maintenance and service personnel is highly regional in nature, particularly for sophisticated machinery and controls.

• *Cost and availability of replacement components*—Keeping in mind that most buildings are long term investments, the conversion apparatus and its controls must be selected with the consideration that replacement components will be readily available for years to come.

• *Reliability of the source and conversion apparatus*—Reliability of the conversion apparatus, whether it be a boiler or a transformer/switchgear combination, is interrelated with cost; *i.e.*, redundancy could achieve the needed reliability. Reliability of the source, however, is an often overlooked parameter. If there is any question of future availability or future stabilized cost, the system should provide for conversion to an alternate source.

Not included are sources dictated by code or legislation. This is not an oversight, but rather an expression of faith in a free economy. Hopefully, well founded application of design parameters to energy source selection as well as other building conversion systems will result in the most effective use of energy resources; such that if and when federal or regional policies affecting energy use are adopted, the engineering profession will have arrived there first.

24

The computer and HVAC system design

WILLIAM J. COAD, PE,
Vice President, Charles J. R. McClure &
Associates, St. Louis, Missouri

Energy and computers are probably the two most written about topics in the HVAC field today. Although building systems have been utilizing energy for centuries, it was not until the increased use in recent years resulting from product and system advances that did not consider energy economics as a design parameter, and the recent awareness of energy source limitations by the public that the energy parameter has come forward.

The digital computer, on the other hand, is a relatively recent development, which can handle, process, and store vast amounts of information in microseconds. It found an almost instant market in processing readily quantified variables, such as cost accounting, bookkeeping, etc. Properly programmed, it can "solve" complex problems in a small fraction of the time it would take to do manually. In many cases, the time required for a manual calculation would be so prohibitive that an exact solution would be impossible.

Design engineers in the HVAC discipline were not quick to utilize these advantages for several reasons, thus the 15 to 20 year delay. Some reasons for this time gap are:

• Designers lacked training and skill to author or use programs.
• The need was not evident. "Successful" systems had been designed for years utilizing manual calculations.
• Fee structures discouraged improved designs utilizing computer techniques at added cost.

• Catalog data on components and subsystems in tabular or graphic form were accepted as accurate.
• Most design firms were relatively small and could not financially afford the necessary hardware.

Numerous changes, however, have occurred in the past five to ten years that have created a vital interest in computerized techniques. They are:

• Recent engineering graduates were educated in computer programming and use techniques.
• Shared-time networks availed even the smallest firm of large capacity computers for a relatively small terminal lease charge.
• First cost pressures accompanied by significant advances in systems concepts initiated an element of need.
• Energy awareness by the non-engineering public has opened a market for energy calculations (calculations so complex and extensive that manual solutions would be prohibitively expensive).

Enormous sums of capital have been invested throughout the past decade in seeking methods to use the computer advantageously in this field; many efforts have failed. However, at this time a plateau has been reached from which computer technology will doubtless move into a lasting and functional pattern.

One lesson learned was that there are significant differences between engineering computer programs and so-called data processing. Data processing programs are based on a well defined set of equations, numerical relationships, or algorithms, and can be effectively written by professional computer programmers. Most programs related to HVAS, however, are so complex that *most valid programs are written by skilled systems engineers*, for whom computer programming is a secondary skill. Programs thus developed are a bit more cumbersome but are vastly more valid.

Another significant difference is twofold. First, hardware capacity limitations are most detrimental, and second, logistics are such that the scheduling of the computer for performing a calculation cannot be predetermined. When used as a design tool, computerized techniques are most cumbersome if access to the hardware is not readily available. The reason for this is that a good design program does not provide the *answer; it provides valid information based on the input.* Thus, when the output is obtained, it is often necessary to study it, and then change the input as dictated thereby until the output reflects the most desirable design. (This concept has been challenged by some, but the majority concurs.)

Another lesson learned is that HVAC system design and system analysis have been less than exact sciences. For example:

• Load calculations formerly used provided an "adequate" answer. However, when the computer was employed to quantify many judgments and approximations of the manual techniques, it was found that considerable basic research in building load phenomena was required.

• Elementary energy calculation techniques previously used, such as the degree-day and simplified bin calculations, were quickly seen to relate poorly with actual energy usage in commercial and institutional buildings because of the consumption of auxiliary systems, control techniques, and ventilation and occupancy schedules. Again, considerable research efforts to identify the algorithms were dictated.

• Fluid systems designs, classically based on standardized constants applied to the Darcy Weisbach equation and Moody's friction factors, were found to be useful only for sizing piping systems on the basis of full design flow in all branches. Again, extensive research into the dynamics of the system under varying operating conditions was dictated.

For detailed discussions of computer applications in energy and HVAC system analysis, see Chapters 13 and 14.

25

Integrated loop system for campus cooling

Solution maximizes utilization of existing chiller capacity and upgrades overall campus air conditioning while reducing machine operating hours and operating and maintenance costs.

WILLIAM J. COAD, PE,
Vice President, Charles J. R. McClure
and Associates, Consulting Engineers,
and Affiliate Professor in Mechanical
and Aerospace Engineering, Washington
University, St. Louis, Missouri

The campus of the University of Missouri at Rolla, like many others, reflects the changes in building technology that have occurred over the past half century—a period in which the mechanical and environmental disciplines of that technology were changing at an exponential rate. The problem posed by such a situation is one of periodically re-evaluating older buildings from a functional point of view and arriving at decisions as to whether capital should be invested to upgrade these spaces environmentally and functionally or whether they should be abandoned to obsolescence.

The Rolla campus currently consists of 30 buildings including the power plant, as shown in Fig. 25-1. Ten are served by modern central chilled water cooling plants; nine of these systems have absorption chillers while the remaining one has an electric motor driven centrifugal unit. Total installed chiller capacity is approximately 1817 tons. Steam for the absorbers is supplied from the power

plant, which also provides steam for space heating and domestic hot water heating. The low pressure steam is available as a byproduct of power generation. Table 25-1 lists the buildings having central chiller systems.

Cooling in the remaining buildings is provided by unitary equipment, window units, or both. Generally, only partial cooling is provided in these structures. In service are 53 pieces of unitary equipment totaling 424 tons of capacity and 246 window units with a combined capacity of 325 tons. Tables 25-2 and 25-3 show the distribution of this equipment throughout the campus.

A study was undertaken to evaluate the present and future cooling needs of the Rolla campus and determine the feasibility of developing a single chilled water plant. The objectives were to reduce energy consumption, operating costs, and maintenance costs while increasing reliability and maximizing the useful life of the equipment.

Obviously, the existence of approximately

118

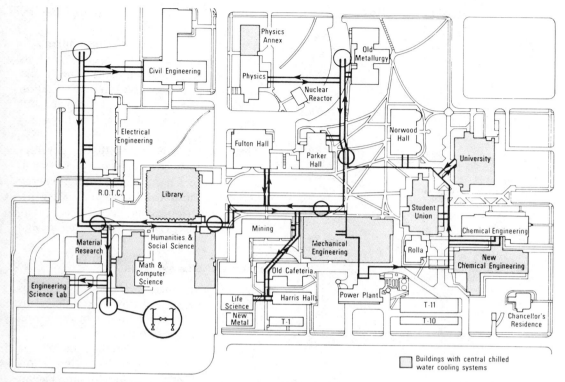

Fig. 25-1. Campus of the University of Missouri at Rolla. Buildings in shaded areas have central chilled water refrigeration plants. Shaded piping shows Loops 1 and 2, Loop 1 at left and Loop 2 at right, in phased construction for providing cooling to most of the campus from the existing chiller systems. Together with the black piping, they make up the Sigma Loop which was planned for the third phase of development. Installed capacity will handle virtually the entire campus because of diversity in building loads. Future piping connections for the initial loops and final loop are shown by the circles on the piping layout. Buildings marked with asterisks are those without central cooling for inclusion in the integrated Sigma Loop system.

1817 tons of chiller capacity could not be ignored because of the capital investment involved and the long service life which this type of equipment provides. The challenge was therefore to develop an integrated system making maximum use of the existing central plant type equipment. This led to a *decentralized* central plant concept in which the

Table 25-1.   Buildings served with central chilled water systems.

Building	Area, sq ft	Equipment	Installed tons
Physics Annex	14,800	Absorption	71
New Chemical Engineering	78,600	Absorption	360
Library	85,600	Absorption	274
Mechanical Engineering	38,922	Absorption	200
Student Union	17,900	Centrifugal	115
University Center	38,400	Absorption	204
Materials Research	28,600	Absorption	120
Math & Computer Science	35,900	Absorption	123
Humanities & Social Science	30,600	Absorption	155
Engineering Science Lab	42,400	Absorption	195
Totals	411,722		1817

Table 25-2.   Unitary cooling inventory.

Building	Installed tons
Rolla Building	10.0
Physics Auditorium	25.0
Norwood Hall	15.5
Parker Hall	60.0
Fulton Hall	15.0
Harris Hall	37.5
Old Cafeteria	12.5
Old Metallurgy	2.0
Civil Engineering	35.5
Chemical Engineering	18.0
Chemical Engineering Addition	15.0
Electrical Engineering	23.0
Reactor	7.5
Life Science	17.5
T-10	5.0
T-11	20.0
T-C	17.5
Math and Computer Science	87.0
Total tonnage	423.5
Total number of units	53

T denotes temporary buildings

Table 25-3.   Window unit inventory.

Building	Units operating	Total tons
Rolla Building	15	21.4
Physics	5	7.8
Norwood Hall	32	54.2
Parker Hall	1	0.8
Fulton Hall	30	37.2
Harris Hall	12	12.2
Old Cafeteria and T-6	2	3.2
Old Metallurgy	6	5.2
Civil Engineering	9	15.5
Chemical Engineering	41	66.0
Mechanical Engineering	2	3.2
Electrical Engineering	21	24.2
Reactor	5	4.1
Life Science	3	3.0
T-10	4	7.8
T-11	4	5.9
T-1	4	2.7
T-2	9	10.7
T-4 (Military Classroom)	4	5.0
T-7 (Military Supply)	3	2.8
Chancellor's Residence	8	7.5
Mining	26	24.4
Totals	246	324.8

existing chillers would be tied together by chilled water loops.

An inventory of the existing chiller systems was followed by a thorough analysis of their operating characteristics. The system evaluations provided revealing information that emphasizes the need to take into account energy utilization in environmental systems design.

This chapter will deal with the development of the integrated chilled water system. The following Chapter 26 explores energy economics concepts and shows how the work at Rolla underscores the need for considering these concepts in all building design programs.

Existing chiller systems

Descriptions of the central cooling systems on campus are given below. Flow diagrams showing major system components were prepared for all of these buildings.

• *Physics Annex*—Its system includes a 71 ton absorption chiller and a crossflow cooling tower with an indoor pump. The air side system includes a 12 ton multizone unit, heating-cooling unit ventilators on a two-pipe distribution system. Perimeter finned tube radiation is provided in some spaces. Space and piping arrangements have been provided for the future addition of another absorption unit with associated pumps and cooling tower.

• *New Chemical Engineering Building*—This building is served by a single 380 ton absorption unit. The air side consists of two high velocity central station type air handling units supplying dual duct variable volume terminal devices. The two central station coils are wild-flow devices and represent the only two "loads" on the chilled water system. Computer rooms, which require close control of relative humidity, are provided with self-contained air cooled units, each of $7\frac{1}{2}$ ton capacity.

• *Library*—The library is served by a single 275 ton absorption unit located in a penthouse equipment room. The air side consists of two high velocity dual duct air handling units located in a room immediately adjacent to the chiller room. The cooling coil banks in the built-up units, running wild with an uncontrolled cold deck, are the only chilled water users. Heat dissipation is through multiple forced draft cooling towers located on the roof adjacent to the equipment room.

• *Mechanical Engineering Building*—Only the 38,920 sq ft addition to this building, built in 1967, has been provided with central air conditioning. The cooling source is a 200 ton absorption unit. On the air side are 36 ceiling mounted four-pipe air handling units with chilled water coils supplying terminal reheat coil zones. Supplemental perimeter heating is done with wall hung convectors.

• *Student Union*—This is the building that employs the electric motor driven centrifugal chiller. It is a 125 hp hermetic packaged unit. The distribution system is a two-pipe hot and chilled water network serving three central station air handling units and 44 terminal fan-coil units.

• *University Center* (new Student Union) —Chilled water to cool the building is provided by a 200 ton absorption unit located in a ground floor mechanical equipment room. Condenser water is cooled by a roof mounted cooling tower. The chilled water distribution is separate from the hot water distribution (four-pipe) and supplies three

multizone air handling units and one small draw-through conditioner. The kitchen make-up air unit is not provided with a cooling coil.

• *Materials Research*—A 120 ton absorption chiller located in a ground floor equipment room supplies chilled water to a two-pipe hot-chilled water distribution system serving room terminal units, which include a combination of unit ventilators and fan-coil units. Outside air is supplied to those units toward the building interior by a non-pressurized central outside air intake duct system. The cooling tower is located on the roof.

• *Math and Computer Science*—Cooling is provided by a 170 ton absorption unit located in a penthouse equipment room. The cooling tower is roof mounted. The terminal system is supplied by three high pressure central station air handling units (zoned south-interior-north), which utilize terminal reheat units for space control. A computer area is served separately by five water cooled self-contained units for year 'round temperature and humidity control. Computer room units are supplied with condensed water from the same tower serving the absorption unit, but separate pumps are provided. Flow diagram and equipment data are presented in Fig. 25-2.

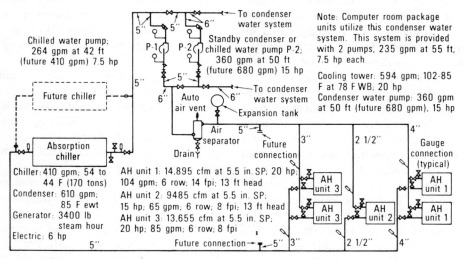

Fig. 25-2. Flow diagram is for central cooling plant in Math & Computer Science Building. It is representative of those prepared for all buildings with central plants.

• *Humanities and Social Science*—Building cooling is provided by a 135 ton absorption unit located in a central grade level mechanical equipment room. A terminal reheat medium pressure central station air handling unit is located in the same equipment space. This terminal reheat system provides conditioning for the entire first floor and interior spaces and corridors on the second and third floors. A two-pipe unit ventilator system provides conditioning for the second and third floor classrooms, and the same piping loop serves two-pipe fan-coil units in the offices on these floors. Supplementary heat by convectors and finned tube radiation is provided where required in lobbies, entries, etc.

• *Engineering Science Laboratories*—The completely air conditioned building is provided with chilled water by a 200 ton absorption unit located in a basement mechanical equipment room. The chilled water is circulated through three central station air handling units that provide space conditioning through high velocity terminal reheat distribution systems. The condenser cooling water is provided by a roof mounted cooling tower. Air handling units and all other major equipment items are located on the lower level.

Discussions with the university staff and inspections of the buildings not currently provided with permanent central year 'round air conditioning systems yielded helpful information with regard to long range planning.

Buildings considered to be of sufficient continuing value to campus operation to warrant upgrading through the addition of cooling apparatus (though remotely in many cases because of building age) were included in the initial integrated campus cooling source analysis. These buildings are so indicated by asterisks in Fig. 25-1.

The Old Metallurgy Building was dropped from the final analysis because of its age and type of structure, which definitely rendered a major investment in central cooling impractical from an economic standpoint. Connections to the loop were shown for this

and some other buildings not included in the analysis because of their proximity to the proposed campus distribution routing, however. Added chiller capacity would have to be installed to handle buildings not included in the final analysis.

Load analysis

After the installed air conditioning equipment was surveyed, building cooling loads were calculated to determine how they compared with installed capacity. A detailed computerized load calculating program, available from a time-sharing computer service, was used to generate the load figures. The results are shown in Table 25-4. The buildings that are served by machinery capable of being tied into a loop system have a combined peak load of 1403 tons against an installed capacity of 1817 tons. The arithmetic sum of the peak building loads is nearly 25 percent less than the available tonnage.

It is apparent from these figures that the chillers can handle much more space than they do now since one of the major advantages of supplying a group of buildings from either a district plant or an integrated loop system is the noncoincident nature of the various peak building loads. Load diversity would permit the chiller equipment to serve additional buildings whose individual peak loads added to the present arithmetic load sum would total more than 1817 tons.

The major advantages of the primary loop system are:

• Existing machinery is used as components of the basic system.

• The distribution system (loop) allows for unlimited addition of building loads and chiller sources with campus expansion. This *open end* design flexibility cannot be achieved with any other concept.

• Investment in piping is minimized because mains sized to carry the summary load are not required.

• Pumping horsepowers are minimized because water flow rates sufficient to carry the summary load are not required.

Table 25-4.   Calculated cooling loads versus installed capacities.

Building	Area, sq ft	Cooling load, tons	Installed capacity, tons	Load, sq ft per ton
Buildings with central cooling systems—				
Physics Annex*	14,800	37	71	400
New Chemical Engineering	78,600	268	360	293
Library	85,600	144	274	594
Mechanical Engineering (1967 addition)	38,922	200	200	194
Student Union	17,900	122	115	147
University Center	38,400	166	204	231
Materials Research	28,600	101	120	283
Math & Computer Science	35,900	87	123	413
Humanities & Social Science	30,600	123	155	249
Engineering Science Lab	42,400	155	195	274
Totals/average	411,722	1403	1817	293
Buildings with packaged and window units—				
Parker Hall	25,500	54	60.0	473
Mechanical Engineering (1949)	17,200	84	0	205
New Metals Building*	4,050	20	20.0	202
Electrical Engineering	46,200	135	23.0	342
Civil Engineering	61,900	150	10.5	412
Harris Hall	18,250	72	37.5	254
Fulton Hall	29,300	98	15.0	300
Old Chemical Engineering	46,302	141	33.0	328
Mining	25,000	97	0	258
Physics	25,400	98	25.0	260
Norwood Hall	50,900	140	15.5	364
Rolla Building*	12,620	30	0	420
Life Science*	4,704	15	17.5	314
Totals/average	367,326	1134	257	330
Total load, all buildings, used in analysis		2435		
Total installed chiller capacity used in analysis			1746	

*Not included in final system analysis. Contribution of Physics Annex chiller considered too small to warrant expense of typing it into proposed loop. Old Metallurgy Building not listed in this table because of prior decision that it would be impractical to install central cooling in it.

Loop analysis

In view of the above, the sizing and routing of the piping loop are important factors. The basic concept of the campus loop system is to connect the loads and the refrigeration sources in series in such a way as to allow the loop temperature to rise as it satisfies the loads of several buildings. Then the temperature is again reduced by another chiller to provide for the cooling needs between that connection and the next cooling source.

To best take advantage of this concept, a thorough analysis of the loads on each build-

ing and their characteristics of coincidence with other building loads at each hour of the year is needed. The *design* condition, the time of year that the loop *sees* the maximum load, is then determined.

After the maximum diversified load was determined, a trial and error method was employed to select the loop routing that would result in the minimum decrement of the cooling capability of the loop. The analysis was then performed at the various reduced load conditions to determine which combination of load/source locations would result in the minimum number of chillers to satisfy the maximum number of loads.

Our own computer programs for developing energy consumption and part load performance data were used, via a time-sharing computer, to determine the various load data and annual hours of refrigeration required for the different loops. It was necessary to modify these programs to obtain the desired information. The programs were run for each building, and the results were then run against one another for the alternate combinations of loop configurations.

The development of the loop was split into three phases for construction budgeting purposes. The first two called for the interconnection of adjacent buildings into two separate loops and placing the chillers into loop operation. The third phase called for the connection of the two loops and extending the loop to

pick up buildings proposed to be served by the existing chillers. The loops in the first two phases are denoted as 1 and 2, and Sigma Loop is the ultimate design (see Fig. 25-1).

The results of the loop performance studies are shown in Figs 25-3, 25-4, and 25-5, which are actually warped bar charts. The vertical axes represent tons of refrigeration capacity; on the right hand side are temperature ranges encountered in 8 and 10 in. diameter pipe loops. The horizontal axes represent building cooling loads in tons. The charts represent full load building quantities because system diversity applies to the loop as a whole and not to individual buildings.

Temperature of water entering a building from the primary loop is taken at the tail of the appropriate diagonal arrow. Temperature in the loop after the water has returned from a building that does not have a chiller is taken

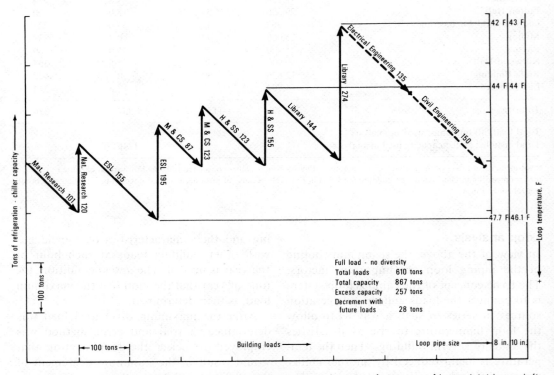

Fig. 25-3. Temperature-capacity gradient diagram for Loop 1 shows performance of Loop 1 (at lower left of Fig. 25-1). Starting at left, chilled water from the loop enters Materials Research Building, absorbs building load of 101 tons (angled line), then gives up 120 tons of heat to the building chiller (vertical line), and re-enters the loop at a lower temperature than when it entered the building. Cycle is repeated throughout the remaining buildings in the loop. Ranges of loop temperatures for 8 and 10 in. piping are shown at right. Dotted lines show effect on loop temperatures if the Electrical and Civil Engineering Buildings were added to the loop and cooled by the excess chiller capacity in the other buildings.

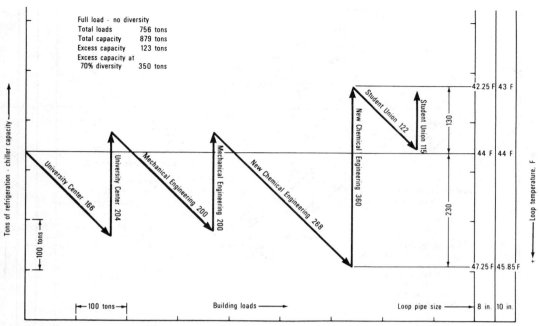

Fig. 25-4. Temperature-capacity gradient diagram for Loop 2 shows performance of Loop 2 (lower right in Fig. 25-1). Again, angled lines represent building cooling loads, and vertical lines represent installed building capacity.

at the head of the diagonal arrow. Temperature in the loop after the water has returned from a building that does have a chiller is found opposite the head of the vertical arrow. Cooling of such a building is accomplished with water from the primary loop before it enters the chiller so that the water entering the chiller is at a sufficiently high temperature to gain maximum capacity from the chiller.

Examination of Fig. 25-3 reveals that even without allowing for diversity, there is an excess of capacity in Loop 1 since the temperature of water leaving the chiller in the library is lower than the temperature of the water entering the Materials Research Building. Adding the Electrical and Civil Engineering Buildings to the loop utilizes the excess capacity, assuming all individual peak loads are additive. With the 70 percent diversity factor calculated for this loop, however, there is an excess capacity of 240 tons.

Figure 25-4 reveals that at full load, without allowing for diversity, there is an excess capacity of 123 tons in Loop 2. Allowing for diversity, the excess is 350 tons.

Sigma Loop is shown in Fig. 25-5. The 70 percent system diversity changes the capacity decrement of 689 tons when all peak loads are added arithmetically to a surplus chiller capacity of 42 tons. It should be noted that on a diversified basis the temperature leaving the Old Mechanical Engineering Building will be approximately equal to that entering the New Chemical Engineering Building.

Examination of Sigma Loop routing and loading in Figs. 25-1 and 25-5 respectively, shows that the selected routing provides for the higher loop temperatures at buildings not having central chilled water systems. In these buildings, designers need only select cooling coils and other apparatus on the bases of the available water temperature at the point of connection and the allowable temperature range in the building loop. While these are not difficult restrictions, it is important that they be met if the integrated primary loop concept is to be successful.

It should be stressed that the limiting factor is the maximum temperature at which chilled water can be supplied to a building load from the loop and still provide adequate dehumid-

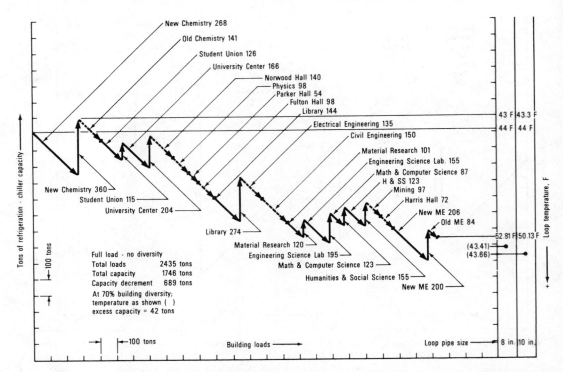

Fig. 25-5. Temperature-capacity gradient diagram for Sigma Loop—combined Loops 1 and 2 plus other connections (Fig. 25-1). Angled lines are building cooling loads, and vertical lines represent installed chiller capacities in buildings that presently have central plants. Temperature rises in 8 and 10 in. loops are shown at right. Closed loop temperatures with system diversity accounted for are shown in parentheses. Dotted lines show buildings that would be cooled from chillers in other buildings. Not all buildings that have connections to the loop in Fig. 25-1 are included in the above analysis since the age of some renders economic feasibility of adding central cooling marginal or impractical.

ification and sensible cooling in the building. Of the two, the latent or humidity control parameter is the more readily affected. The survey of the facilities revealed that those areas on campus requiring close humidity control are provided with independent environmental control systems whose inclusion in the integrated campus system is not contemplated. Thus, the acceptable temperature rise of the loop is directly dependent on the tolerable ranges of relative humidity for normal occupant comfort. Based on the ASHRAE comfort standards, loop interconnection design should be based on a 44°F inlet to the coil with a 16°F range to achieve 75°F DB and 50 percent RH with maximum entering water temperature and coil dew point swings up to 49° and 60°F respectively, to maintain 75°F DB with a maximum rise in RH to 60 percent. Any future critical humidi-

fication applications should *not* be connected to the integrated system. All loads that represent comfort cooling should be connected.

Temperatures of water entering buildings having chillers are kept as low as possible to minimize changes in existing apparatus. In some instances, additional rows of coils are required. In other cases, there is currently excess coil capacity that can be utilized.

Loop pipe sizing

As stated above, one of the advantages of a loop type campus system is that the series connected load and source concept results in minimum pipe sizes for an infinite growth of summary load. Two major considerations in pipe sizing are:

• What is the largest single load or source that will be connected without an accompanying source or load, as the case may be?

• What is the anticipated full load and part load temperature decrement that can be tolerated?

In the study, four basic sizes were considered, 6, 8, 10, and 12 in. The assumed maximum flows of these four sizes are 800, 1700, 3000, and 5000 gpm, respectively. Preliminary analysis led to rejection of the 6 and 12 in. loops, leaving the 8 and 10 in. for further consideration.

The tonnage represented by a load or source connection to the primary loop is calculated by dividing the product of flow rate (in gpm), 500, and temperature range by 12,000. A maximum temperature range of 16°F in the secondary circuits leads to the following limitations for a *single* load or chilled water source:

• 8 in. primary, 1700 gpm: 1133 tons.
• 10 in. primary, 3000 gpm: 2000 tons.

The flow rates represent the maximum amounts of water that can be economically circulated.

In applying these limitations to the system, there is yet an additional restriction: the chiller source (if the limit is reached) must be applied at full capacity (16°F). The same

limitation applies to a load, calculated as above, except that it must be imposed at a point of minimum loop temperature (43 or 44°F). Thus, in such cases the maximum tonnages above would have to be reduced by a factor depending on the physical location of the connection to the loop.

Figure 25-6 is a diagramatic illustration of the method of interconnecting a building system with the primary loop. Although differences in the existing chilled water systems will necessarily dictate some custom modifications of this scheme in the various buildings, the proposed connection method can be closely approached in all of the existing systems. Draining and shutdown of the systems during installation will be required in most cases, however.

Any future buildings that are planned should include the general piping arrangement proposed, complete with valving and blind piping openings for the eventual loop connection.

As the final engineering design is developed for each or any incremental phase of system development, some changes to the proposed ideal connection scheme could be dictated by the loop configuration. The basic concept is

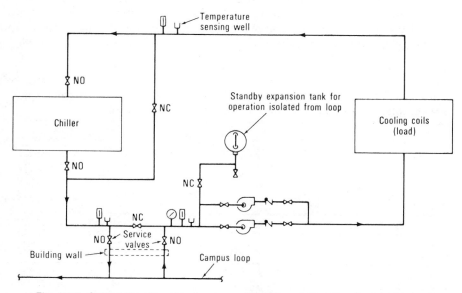

Fig. 25-6. Schematic diagram of proposed building connection to primary loop.

the relative arrangement of the chiller, load, and loop connection to achieve minimum water temperature to the load and maximum water temperature to the chiller.

Summarizing, the previously stated objectives of integrating the presently isolated building cooling systems into a single system are as follows:

- To optimize utilization of machinery.
- To minimize machine operating hours.
- To optimize energy utilization.
- To enhance reliability.

Inherent in the above is accomplishing campus cooling at minimum cost. In this context, cost or economic values are to be applied to virtually all measurable quantities to achieve a common denominator. The quantities involved are:

- Useful machinery life.
- Machinery replacement cost.
- Maintenance costs (planned and unanticipated).
- Energy consumption and cost.
- Water.
- Expendable materials (water chemicals, lubricants, etc.).

To apply a quantitative evaluation to the Rolla campus, the three basic loop developments established were evaluated, each in two different ways: first, with each building sys-

tem functioning independently of the others—assuming that a central cooling plant would be added to buildings currently not fully air conditioned; and second, with the existing cooling systems integrated into the loop in the three phases of development. This was accomplished by hourly load comparisons incorporated with the reduced or part load characteristics of each chiller system. The following two sections show the energy and operating cost savings achieved.

Energy requirements and savings

Energy consumption was considered in terms of thermal energy expressed in pounds of steam per hour and electrical energy expressed in kilowatt hours. The part or reduced load steam consumptions for the absorption units were taken from the manufacturers' cataloged performance curves. Although experience in operating the units indicates that these consumption rates are seldom realized in the installed machines, there was no documentation of this and manufacturers' data were therefore employed.

The comparison was then made for the isolated unit operation and the integrated loop system in the three phases of development, and the results are given in Table 25-5. As can be seen, the annual energy cost saving that can be attributed to total campus interconnection on the Sigma Loop is approximately $10,000.

Table 25-5.   Amounts and costs of electricity and steam for both isolated unit and integrated loop operation in the three phases of development planned.

Condition	Electricity		Steam	
	Quantity, KWHR	Cost, $	Quantity, Mlb per hr	Cost, $
Loop 1 isolated	530,668	9,021	18,821	30,114
Loop 1 interconnected	404,507	6,877	16,291	26,065
Savings	126,161	2,144	2,530	4,049
Loop 2 isolated	524,435	8,915	11,817	18,907
Loop 2 interconnected	339,504	5,772	10,650	17,040
Savings	184,931	3,143	1,167	1,867
Sigma isolated	1,499,343	25,489	45,236	72,378
Sigma Loop	1,055,831	17,949	43,800	70,080
Savings	443,512	7,540	1,436	2,298

Machine operation hours and costs

As stated earlier, one of the major advantages of the integrated loop vis-a-vis individual systems is the capability of operating only some of the units under conditions of reduced loading. The value of machine operating hours is less readily identified than that of energy consumption and varies as a function of plant operating and maintenance techniques. The monetary value of machine operating hours is composed of the following values:

• *Capital cost of machinery*—All but one small unit in the proposed system are lithium bromide type absorption units. The basic motivating energy in these units is high level thermal energy, which achieves the refrigeration effect through the evaporation of water at extremely low absolute pressures, motivated by a chemical process of brine absorption. Thus, the "wear and tear" on such a unit is chemical rather than mechanical. It is reasonable to estimate the life expectancy of this type of unit at 20 years, based on 3000 hr of operation per year. Thus, the machine can be estimated to have a useful operating life of 60,000 hr. For simplification, the value of operating hours was developed neglecting value of monies interest on sinking funds, etc., and the comparisons were made strictly on a cash value basis. Assuming an investment and replacement cost of $250 per ton, average machine size of 181.7 tons, and the 60,000 hr life expectancy, the depreciated value of 1 operating hr of a unit was set at $0.76.

• *Maintenance and operation cost*—Current experience on the campus indicates that the labor and materials costs for the existing units are approximately $28,000 and $12,000 per year, respectively. Calculations indicated that the present operation results in approximately 26,417 machine hr per year; thus, the maintenance and operation value was set at $1.51 per machine hr.

• *Water and water chemical cost*—Although the use of water for condensing purposes is theoretically related to ton-hours rather than operating hours, research indicates that very few data have been compiled

relating to water consumption to reduced load demands. It is reasonable to assume, however, that there is a drift loss and controlled bleed loss that exists in a cooling tower system regardless of the load on the system, even assuming a constant temperature-load relationship. This combined loss thus represents a parasitic burden to system operation. In the context of the study, however, the quantitative value of this parasitic loss was not determined, and no value was assigned to this consideration.

Adding the capital cost value component to the maintenance and operating cost value component results in a machine value rate of $2.27 per hr. This rate was applied to the calculated operating hour schedule shown in Table 25-6 for the three suggested stages of development. The overwhelming significance of the integrated loop approach becomes evident at a glance when the value of the machine hour saving for the ultimate or Sigma Loop is seen to be $111,430 per year. Another interesting correlation is that the machine hours and associated value for the Sigma Loop are *appreciably less* than for the existing non-connected machinery (by approximately one-half), which indicates that the existing machinery essentially comprising the Sigma Loop apparatus would operate for fewer hours to handle the entire campus than they do now, handling only 10 buildings.

Table 25-6. Hours and value of machine operation for both isolated machine and integrated loop operation in the three phases of development planned.

Condition	Hours per yr	Value, $
Loop 1 isolated	12,278	27,871
Loop 1 interconnected	5,500	12,485
Savings	6,778	15,386
Loop 2 isolated	11,809	26,806
Loop 2 interconnected	3,700	8,399
Savings	8,109	18,407
Sigma isolated	62,264	141,339
Sigma Loop	13,176	29,909
Savings	49,088	111,430

It must be pointed out, however, that although the Sigma Loop maintenance and operating costs are predicted to be less than the costs currently incurred, the *total* air conditioning maintenance costs can be expected to increase significantly. The reason for this is the maintenance costs associated with the distribution and terminal systems. These costs will increase in a linear step fashion with the increased number of building systems and will not be affected by the loop integration.

The value savings developed above are not without some degree of inaccuracy. The unit value developed is felt to be as accurate as could be generated with the data known, and it can readily be adjusted if desired with further refinement of cost history. The machine operating hours data were developed by calculating the loads for each indicated building, evaluating the energy and part load profile for each building, and then calculating machine hours. To calculate loop machine hours, all buildings on the loop were summarized as though they responded as a single variable load; then the part load and machine hours were determined. This method of computation resulted in two sources of inaccuracy:

• Because of the loop capacity decrement, under conditions of extremely low loading, during which the program anticipated only one machine, the temperature rise limitation could be exceeded, necessitating the operation of an additional machine.

• The computerized calculations did not take into account operating tolerances beyond the mathematically ideal condition.

The coupling of these two potential inaccuracies could result in a significant difference between the foregoing predictions and operating experience. If system performance is adequately monitored, however, the deviation should not exceed a 100 percent increase in actual loop machine hours over the number predicted. This correction, applied to the Sigma Loop, would decrease the value of the savings to $81,520—still a very significant amount.

It should also be considered that just as construction costs can be expected to increase in coming years, energy and operating costs will likewise escalate. An analysis of current market trends indicates that both of these costs will increase at a more rapid pace than construction costs during the coming decade because maintenance personnel costs are presently lagging behind construction personnel costs and the current energy shortage will undoubtedly result in significant cost increases in the near future.

Monitoring system

To assist in gaining optimum value from the integrated loop and to retain manageable proportions over the extensive machinery that will ultimately be gathered on campus, installation of a monitoring system was recommended, beginning with the first phase of loop development.

Minimally, direct transmission to a central panel should provide temperature indication at the entrance to each load and each chilled water source; loop pump operation; loop flow indication; and loop pressure indication. Each building could be provided with a failure or "out of tolerance" indicating panel, with a signal for each building being transmitted to the central console.

Experience has shown that future construction planning for campuses of this nature is at best a fluid situation. Any provisions in a central or integrated energy system that impose a restriction on planning or system revisions can prove economically disastrous. The future growth potential was therefore studied thoroughly, and the results yielded another plus for the integrated loop concept; *ie.*, since both chilled water sources and loads can be added to the system and their effects controlled completely by the individual design engineer, the loop system developed will be completely compatible with any future growth. The only primary loop restriction would be if any such future load or chilled water source, when translated from Btuh to gpm, exceeded the total flow capacity of the loop at the connection point.

26

Energy economics: a design parameter

A comprehensive study of campus cooling points up the necessity of incorporating the energy economics parameter into the design of building systems, and specific recommendations tell how to go about it.

WILLIAM J. COAD, PE,
Vice President, Charles J. R. McClure and Associates, Consulting Engineers, and Affiliate Professor in Mechanical and Aerospace Engineering, Washington University, St. Louis, Missouri

In Chapter 25, entitled, "Integrated Loop System for Campus Cooling," the reasoning behind the computerized design was explained that resulted in the recommendation to install an integrated loop chilled water system on the campus of the University of Missouri—Rolla.

In the process of gathering the information needed to make this study (heating and cooling loads, types and sizes of refrigeration machinery and auxiliaries, types of building environmental systems, etc.), significant data were developed. These data serve as a report of a comparison of building energy economics. The information is presented to accent the differences in energy demands and consumptions among various buildings with similar uses and occupancy schedules. The excesses of energy required by some indicate that although the buildings were all relatively successful designs architecturally and environmentally, energy economics was not considered as a design parameter.

This is not a unique situation. For the past quarter century in the United States, the building environmental sciences have been undergoing probably a more dynamic progression than any other field of engineering in history, save space technology. As a result, design practitioners have had all they could cope with in keeping pace with progress in building materials, building systems, fenestration, mechanical system concepts, and new machinery. In addition, they have had to meet the demands of the marketplace from the standpoint of ever-increasing requirements for multiple control zones. The additional pressure of satisfying these demands within the budget restrictions established for earlier, less sophisticated systems posed a challenge of almost astronomic proportions to the systems design profession. With these burdens, and with readily available and competitively priced energy sources, design practitioners understandably did not consider energy economics as a design parameter.

It might be emphasized at this point that building environmental systems include lighting as well as HVAC systems. Thus, a major portion of the total output of electric power generating plants serves building environmental needs. Other major areas of energy consumption are the categories of transportation and industrial processes, and neglect of energy economics is as manifest in these areas as it has been in the building industry.

Building loads compared

As was described in Chapter 25, the Rolla campus includes 10 buildings with central chilled water systems. For the sake of continuity, some of the data appearing in Chapter 25 are repeated here in a slightly different format. Table 26-1 shows the comparative loadings of the 10 buildings. They are seen to range from 147 sq ft per ton (82 Btuh per sq ft) for the Student Union to 594 sq ft per ton (20.2 Btuh per sq ft) for the Library. This comparison, however, is not especially revealing since occupant loading densities and functions are considerably different for these two buildings. If these two buildings and the University Center are discarded, the remaining buildings are similar in function, occupancy densities, and occupancy schedules. These range from 194 sq ft per ton (62 Btuh per sq ft) for the Mechanical Engineering Building to 413 sq ft per ton (29 Btuh per sq ft) for the Math and Computer Science Building.

An understanding of the differences in specific capacity requirements is essential if one is to make use of this information in the development of methods for minimizing energy usage. Although specific capacity is only one area of concern, it is the area that often is completely out of the control of the mechanical systems designer (except for the ventilation rate contribution). This specific loading is calculated coincident full load that the building imposes on the system (plus ventilation). Not included in this form are any inherent system inefficiencies, such as reheat, refrigeration and heat energy in reduced load zones at the time of maximum building coincident load, distribution system losses, etc.

Lighting levels range from 1.63 watts per sq ft for the Library to 3.64 watts per sq ft for the Materials Research Building. (The values are based on total installed lighting per gross area.) Again, if the nonsimilar use buildings are discarded, the range is 2.44 watts per sq ft for the New Chemical Building to 3.64 watts per sq ft for the Materials Research Building. The correlation between lighting power density and specific cooling load, however, while not overly impressive, is significant; a calculation for the seven similar use buildings reveals a positive correlation of 0.4. This is not surprising since any building design feature that adds to the cooling load should obviously be scrutinized if energy economics are being considered.

Another feature that has an appreciable effect on cooling load but was rejected from a quantitative analysis is the fenestration of the different buildings. A correlation between fenestration and specific cooling load for these buildings was similar to the lighting

Table 26-1. Calculated cooling loads versus installed capacities for buildings with central chilled water plants.

Building	Area sq ft	Cooling load, tons	Installed capacity, tons	Load, sq ft per ton
Physics Annex · .	14,800	37	71	400
New Chemical Engineering	78,600	268	360	293
Library	85,600	144	274	594
Mechanical Engineering	38,922	200	200	194
Student Union	17,900	122	115	147
University Center	38,400	166	204	231
Materials Research	28,600	101	120	283
Math & Computer Science	35,900	87	123	413
Humanities & Social Science	30,600	123	155	249
Engineering Science Lab	42,400	155	195	274
Totals/average	411,722	1403	1817	293

level correlation ($r = 0.36$). But since all relevant variables were not considered (such as reflectivity of glass, interior shading methods, and exterior shading versus orientation), this feature was not included.

Effect of auxiliaries

Thorough understanding of the building features that contribute to a specific cooling load is the very starting point in designing with energy economics as a parameter. If the total cooling load increases, the base power requirement to drive compressors (kw) or absorption machines (Btuh) increases, the power requirement to drive all necessary auxiliaries increases, and the resulting energy consumption increases. The Mechanical Engineering Building, at 64 Btuh per sq ft, has more than twice the specific cooling load of the Math and Computer Science Building, at 20 Btuh per sq ft.

The mechanical systems designer is seldom in a position to control the specific cooling load. He is in sole control of the energy usage per unit of building requirements. Specific system power requirements, *SSP*, is defined as the power per unit of cooling capacity as required by system design, expressed in such units as Btuh per ton or kw per ton. Specific system energy, *SSE*, is the seasonal or annual energy consumption per unit of system capacity requirement, expressed in such units as Btu per ton or kw per ton.

All of the buildings studied have absorption chillers, except for the Student Union with its electric centrifugal. Specific system power requirements and specific energy requirements were analyzed for the 10 building

systems, and Table 26-2 illustrates the comparative mechanical system power requirements. The refrigeration auxiliaries include chilled water pumps, condenser water pumps, cooling towers, absorption unit pumps, and control power. The fan auxiliaries include only the supply and return air fans (fans necessary to affect space conditioning). These power loads are then brought to a specific requirement by referring to the common denominator of installed tons.

Figure 26-1 illustrates these data in bar graph form.

It is important to note that if these power units were brought to the base of kw per building ton, they would be higher in many cases since installed capacities exceed building loads, partly because of system parasitic loads (required for control purposes) and partly because of available machine sizes. This is illustrated in the last column, which expresses total system auxiliary power requirements per square foot of building area. (Note that this combines the specific load with specific power.) In this evaluation, building function does not have the relevance it did in comparing specific cooling loads, as long as the comparison is made in kw per system or installed ton.

Brief descriptions of building systems rudimentary information is given in Table 26-3. Before any value can be assigned to the data, however, one must understand the systems so as to determine the sources of deviations. Contributing factors included:

• Of the four buildings with the highest specific power requirements, three employed high pressure air distribution systems.

• The cooling tower fan motors ranged

Table 26-2. Comparative mechanical system power requirements.

Building	Area sq ft	Installed capacity, tons	Fan auxiliary power KW	Refrigeration auxiliary power		Total auxiliary power		
				KW	KW per ton	KW	KW per ton	w per sq ft
Physics Annex	14,800	71	8.8	22.1	0.3112	30.9	0.4352	2.0878
New Chemical Engineering	78,600	360	67.5	80.0	0.2222	147.5	0.4097	1.8765
Library	85,600	274	84.0	115.5	0.4215	199.5	0.7281	2.3306
Mechanical Engineering	38,922	200	42.1	66.5	0.3325	108.6	0.5430	2.7901
Student Union	17,900	115	12.0	27.7	0.2408	39.7	0.3452	2.2178
University Center	38,400	204	27.0	52.4	0.2568	79.4	0.3892	2.0677
Materials Research	28,600	120	7.5	24.0	0.2000	31.5	0.2625	1.1013
Math & Computer Science	35,900	123	54.3	44.4	0.3609	98.7	0.8024	2.7493
Humanities & Social Science	30,600	155	23.1	28.1	0.1821	51.2	0.3303	1.6732
Engineering Science Lab	42,400	195	76.0	45.6	0.2338	121.6	0.6235	2.8679
Totals/averages	411,722	1817	402.3	506.3	0.2786	908.6	0.7787	2.2141

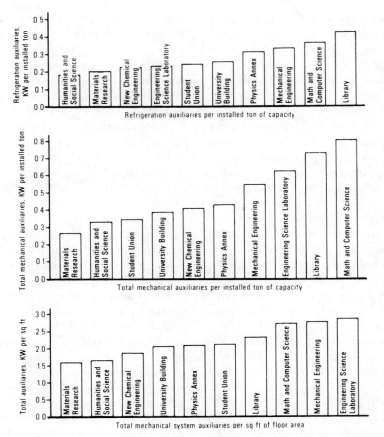

Fig. 26-1. Power inputs for mechanical system auxiliaries in the different buildings compared. Total mechanical auxiliaries include refrigeration auxiliaries and supply and return air fans.

Table 26-3. Total annual electrical and thermal energy requirements of buildings studied. Absorption refrigeration is used in all buildings except the Student Union, which has an electric motor driven centrifugal.

| Building | Area sq ft | System[1] | Annual energy consumptions | | | | ΣSBE[2] Btu per sq ft |
			KWH $\times 10^{-3}$	KWH per sq ft	Btu $\times 10^{-6}$	Btu per sq ft	
Physics Annex	14,800	MZ	220.3	14.9	535	36,200	200,200
New Chemical Engineering	78,600	DD/VV	1201.1	15.3	3876	49,200	217,200
Library	85,600	DD	1590.2	18.6	3525	41,200	246,200
Mechanical Engineering	38,922	RH	622.8	16.0	3427	88,000	264,000
Student Union	17,900	HCO	348.5	19.5	106	59,200	273,400
University Center	38,400	MZ	681.1	17.7	4769	124,000	319,000
Materials Research	28,600	HCO	390.4	13.6	1250	43,700	193,000
Math & Computer Science	35,900	RH	749.7	20.9	5289	147,000	377,000
Humanities & Social Science	30,600	RH,HCO	439.3	14.3	4506	147,200	304,000
Engineering Science Lab	42,400	RH	971.9	23.0	7649	180,000	433,000

[1]MZ = multizone; DD = dual duct; DD/VV = dual duct with variable volume mixing boxes; RH = reheat; HCO = heat-cool-off.
[2]SBE = specific building energy consumption.

from 0.065 to 0.275 hp per ton. The primary difference appeared to be higher horsepower requirements for less real estate. That is, when space limitations prevailed, power requirements increased.

• Chilled water pumps ranged from 0.02 to 0.21 kw per ton.

• Condenser water pumps ranged from 0.032 to 0.15 kw per ton.

Although the absolute values for the above

specific requirements are small, they are significant in their deviations: 420 percent for cooling towers, 1050 percent for chilled water pumps, and 470 percent for condenser water pumps. When extended to system capacity, these deviations are *most* significant.

Annual energy consumption

Power, the time rate of energy production (or consumption), is a significant consideration in the cost of energy since power dictates the investment in generating plant and distribution system. When considering basic resources, however, the energy unit becomes the primary target.

Annual energy consumptions for the included buildings were determined using a computerized calculation technique that has been verified in many buildings. The results are shown in Table 26-3, together with specific building energy requirements, *SBE*, for both electrical and thermal forms and an identification of the type of terminal system control. (A combined result of specific energy and specific loading, each for the time period of one year, in units of kwh per sq ft and Btu per sq ft, is defined as the specific building energy consumption.)

Inspection of the results leads to the following observations:

• There is a positive correlation between systems with higher specific electrical energy consumptions and those with higher specific power consumptions. (This correlation, though only on the order of 0.11, is positive and can be considered as having some significance.)

• Of the four buildings with the highest specific building electrical energy consumptions, three have high pressure fan systems.

• The building with the electrically driven chiller does not have the highest specific building electrical energy consumption (although it does rank third).

• The three similar use buildings with the highest specific building thermal energy consumptions all have reheat systems.

These results are not significantly different from what one would expect. The contribution of a high pressure fan system to specific electrical energy consumption is obvious when one realizes that fan horsepower is directly proportional to pressure and that the fan operates for all the occupied hours of the building (*i.e.*, it does not unload at reduced capacity, at least not in the systems installed in these buildings).

Analysis of a reheat system reveals that absolute energy consumption increases with decreasing building load. In dual stream and heat-cool-off systems, on the other hand, energy consumption decreases with decreasing load. (With any energy conversion device, however, the consumption per unit of demand increases in an exponential manner with a decrease from full load.) Unfortunately, the humidity control capabilities of these three alternatives are inverse functions of their inherent system specific energy consumptions.

The last column in Table 26-3 titled summary specific energy (ΣSBE), is intended to illustrate specific building energy consumption as a single quantity; it was developed using the arbitrary value of 11,000 Btu per delivered kwh for electrical energy. (The calculated thermal energy and electrical energy consumptions include all conversion losses and thus are the quantities delivered to the buildings.) From the standpoint of pure energy economics, this is the most relevant comparison. Since the conversions are arbitrary and subject to challenge, however, they are presented simply for interest.

Energy economics

The Rolla study, which shows a differential of 123 percent between the similar use building with the lowest summary specific building energy consumption and that with the highest, exemplifies the need to include energy economics as a parameter in building environmental system design. The following considerations are offered as a positive approach to incorporating energy economics into the design of all building systems.

• *Energy consciousness in building design*—This consideration is not normally under the direct control of the mechanical engineer on the building design team. Since it has such a direct effect on investment cost as well as on energy consumption of environmental systems, however, he has an obligation to make architects and electrical engineers aware of the implications of their designs with regard to the energy aspects of the building. It is at this stage of design development that the very significant building load is established.

• *Intelligent definition of design parameters*—Available analytical data have been published for some time to illustrate the specific energy requirements of the three basic system control methods. At full cooling load conditions, all three might have the same specific energy requirements. But at reduced load conditions (which an analysis of any building system will show to exist during a majority of the operating hours), the requirements of these systems differ significantly. In order of increasing energy consumption at reduced loading, they are, heat-cool-off and variable volume, dual stream (double duct or multizone), and reheat.

Unfortunately, if one lists the same three basic systems in order of their ability to maintain space humidity control, the order is reversed. Thus, it is the responsibility of the systems designer considering energy economics to establish intelligently the *real* building requirements vis-a-vis humidity ranges or limits and then design a system to provide that level of control and no more. This approach is quite revealing, and it leads to the concept of incorporating energy units into the ASHRAE comfort chart.

Additionally, the single fan "economizer" type system must be scrutinized carefully in the process of systems selection. In many applications (those involving multizone or dual duct installations), specific energy consumption with such a system is higher than when some reduced capacity refrigeration machinery is operated in cold weather periods. The logic is quite obvious: with a single fan "economizer" system, the heating deck becomes a reheat device, and thermal energy consumption increases.

• *Detailed load calculation and part load analysis*—As stated previously, any energy conversion system is less efficient at reduced load operation than at design loading. Thus,

if a system is sized to a "safe" load calculation, it will operate at a lower percentage of full load, and a lower efficiency, at all times. The latest computerized techniques of calculating heating and cooling loads should therefore be employed. As an example, if a system or component is oversized by 25 percent, at peak load it will never operate above 80 percent of design capacity, a point of higher specific energy consumption than at full capacity.

The part or reduced load analysis is a very tedious one to perform, but computer programs are available to assist designers in this effort. (See Chapters 13 and 14 for detailed discussions of using computers for design and analysis.) A part load analysis will reveal the hours per year or the percent of total operating hours that the system will "see" various increments of the full load. By using this analysis in selecting machine size increments, the designer can assure the greatest number of operating hours at the greatest machine load ratios, thereby greatly decreasing specific energy consumption.

• *Flow rate analysis*—Although two-phase steam systems are still widely used for primary heat transfer, and for good reason, the majority of systems employ the single phase hydronic concept for terminal fluids. These systems use the concept of temperature differentials to achieve thermal capacities. In the design of any such system, the systems engineer should employ the maximum possible temperature differential. As an example, if a system temperature differential of 60°F can be achieved in a heating system, the water flow rates and consequently the pumping horsepower can be reduced to one-third of those for a system design based on a 20°F temperature differential. There are several techniques available for achieving this, among them coils designed for large drops and series connection of loads. Primary-secondary circuits are sometimes required, but the total pumping energy is still reduced.

It is more difficult to achieve larger temperature differentials with chilled water, but consider that a 15°F rise in lieu of 10°F will result in a 33 percent reduction in flow rates, and consequently, in energy expended for pumping. An additional benefit is usually lower investment cost in pumping and distribution systems.

• *Air system pressures and operating schedules*—Although some control schemes require high air pressures, the use of high pressure, high velocity air distribution systems should be avoided unless techniques are employed to reduce fan energy consumption at reduced building loads. If a variable volume refinement is not incorporated, fan energy will often exceed the energy requirement at the basic refrigeration machinery because of the greater number of hours of fan operation.

An additional consideration in the selection of fan systems is the building occupancy schedule. The possibility of operating a central fan system a number of hours per year with greatly reduced occupancy rates should be prevented.

• *Primary conversion machinery selection*—Primary energy conversion apparatus such as boilers, chillers, and condensing units should be matched in size modules to the system's part load profile. Each module of machinery should be sized so that it will rarely operate at less than 50 percent of design or rated load. A study of any commercially available energy conversion system will reveal that the energy input rate (per unit output) increases exponentially as the output decreases. The reason for this is the basic parasitic loading that contributes to mechanical inefficiencies at full load. Most of these do not decrease as the output decreases, and with the loss remaining constant, apparatus efficiency decreases.

Capacity reduction modes constitute an additional consideration in the application of refrigeration machinery in smaller (reciprocating) sizes. Analogous to terminal system control, the capacity reduction mode that provides the best or most consistent result is the one that consumes the most energy. Compressor control modes, in order of increasing part load energy consumption, are: on-off, cylinder unloading, hot gas bypass.

• *Refrigeration and boiler system auxiliaries*—Energy consuming auxiliary devices such as condenser water pump motors, cool-

ing tower fan drives, forced draft fan motors, feedwater pump drives, etc., should be scrutinized thoroughly to determine the combination of devices that imposes the lowest power burden on the system and thus results in minimum energy consumption. Cooling tower fan drives can be very low in specific power requirements, but selection without concern for this parameter commonly results in requirements as high as 0.25 kw per ton. In applying condenser water pumps, one must exercise care to arrange the tower, cold water pump, and pump to minimize static lift. In larger systems, multiple cells and variable pumping provide means of optimizing energy efficiencies.

• *Energy conservation devices*—There are numerous devices and components available that have been specifically developed and marketed to conserve energy. Some examples are heat reclaim wheels, double bundle heat pumps, "heat of light" systems, so-called total energy systems, water source incremental heat pumps, etc. These devices cannot be overlooked in any system design; they must be considered. Unlike the measures recommended in the previous considerations, use of these devices may often penalize investment cost, returning the increment out of operating energy cost savings.

• *Energy source selection*—The selection of an energy source or sources should be made independently of the above considerations. If specific building energy consumption has been minimized, the source selection is relegated to owning and operating cost economics, not energy economics.

Possibly the major pitfall of designers has been to initiate an energy economics evaluation with a study for energy source selection, a technique that clearly is analogous to the tail wagging the dog. When the above considerations are applied to the energy economics parameter, power and energy consumption are minimized; and thus the cost will be minimal, regardless of the source. A suggested checklist for energy source selection is:

1. Cost
2. Availability

3. Efficiency of conversion, at full and part load
4. Investment cost for storage, handling, and conversion apparatus
5. Environmental requirements of space
6. Environmental requirements of community
7. Demands and consumption of system
8. Availability of apparatus involved and its maintenance and service availability
9. Reliability of source
10. Reliability of conversion apparatus

Owner's design guidelines

In applying the parameter of energy economics to building systems, the designer must take care not to introduce complexities that cannot be understood by the maintenance and operating staffs. Complete understanding is necessary if the design intent is to be carried out. Many efforts at achieving energy economics without consideration of this point have led to failure of a system to operate efficiently and, in many cases, failure even to satisfy performance requirements. Thus, the designer must always be guided by the rule that simplicity in design will result in successful performance. This consideration certainly affects the specific building energy requirement, and it offers more assurance that design performance will be achieved.

The University of Missouri—Rolla, as a result of the study, has developed a set of system design guidelines to be applied to future buildings on campus. Those guidelines involved in energy considerations are:

- Cooling coils for general space cooling shall be designed for a temperature range of 44 to 60° F.
- Whenever possible, cooling coil capacity control shall be by throttling valve, resulting in load response by variable flow.
- No loads that are humidity-critical, because of process requirements, shall be connected to the building chilled water system. All loads representing normal occupancy or human comfort shall be connected to the chilled water system.
- Special attention shall be given to minimizing all system auxiliary motor loads.
- If the following limitations are exceeded, special permission must be obtained from the University: refrigeration auxiliaries, 0.25 kw per installed ton; system auxiliaries, 0.25 kw per installed ton.
- Every effort shall be made in building and system design to minimize the energy requirements of the environmental systems. A thorough analysis of the energy requirements shall be performed and reviewed with University authorities prior to final design development. Areas of special attention shall include: lighting levels, light switching techniques, ventilation requirements, system selection and control logic, and fenestration.

These design guidelines could be considered a method of energy rationing, but it is hoped that they will simply serve to make design teams conscious of the energy economics parameter.

The author wishes to acknowledge Raymond A. Halbert of the University of Missouri and Victor E. Robeson of the University of Missouri—Rolla for their initiation of, and their role in arranging funding for, the study that served as the basis for this detailed analysis.

27

Heat reclaiming systems

A review of systems and factors governing their use.

E. R. AMBROSE, PE,
Mount Dora, Florida

The successful use of energy conservation practices in commerical and industrial structures is influenced by a number of design and application factors. These include architectural aspects, geographic location, outdoor temperatures, thermal insulation, ventilation air quantities, and other components that contribute to heating and cooling loads of a structure. In turn, practical heat reclaiming systems are governed by orientation and the relative amounts of perimeter and internal areas of a structure, together with accompanying internal heat gain from lights, people, and heat producing equipment.

To provide complete indoor comfort throughout a structure, an air conditioning system may be called upon to furnish heating to exterior zones and mechanical cooling to the interior, even on colder winter days. One of the significant objectives of reclaiming cycles therefore is to limit the common, wasteful practice of removing excess heat from the internal area while simultaneously furnishing heat from an outside source to satisfy the perimeter portion.

A number of heat reclaiming systems can effectively transfer heat from one area to another. Before making a selection, however, it is desirable to make a complete evaluation of the structure's heating and cooling requirements at various outdoor temperatures.

Heating and cooling evaluation

A method of analyzing heating and cooling requirements at various outdoor temperatures is illustrated by Fig. 27-1. Everything below the zero abscissa line represents addition heat required by the structure, and all above the zero line represents available heat gain from the interior space plus the heat equivalent of the work of compression of the refrigeration equipment. The net heat loss of the perimeter space, Curve C, consists of the heat loss, including ventilation air, minus the useful internal heat gain in the area.

During occupied winter heating periods, the useful heat gain from the interior space, Curve B, balances the net heat loss of the exterior space, Curve C, at $19\frac{1}{2}°$F outdoor temperature. From $19\frac{1}{2}$ to $53°$F outdoor temperature, the interior useful heat gain exceeds the exterior heat loss by the amount shown by Area A. Above $53°$F outdoor temperature (called the changeover temperature), the system changes to the cooling cycle. Below the $19\frac{1}{2}°$F outdoor balance point, exterior heat loss exceeds interior useful heat gain by the amount shown by Area D. During this period, heat can be taken from a storage system and/or supplemental heat can be used.

A similar study must be made for unoccupied winter periods when a large portion of the heat from lighting and other heat generating sources is not available. As a partial counter balance, ventilation air requirements can usually by lowered to reduce heat loss somewhat during this period.

With an analysis as illustrated by Fig. 27-1.

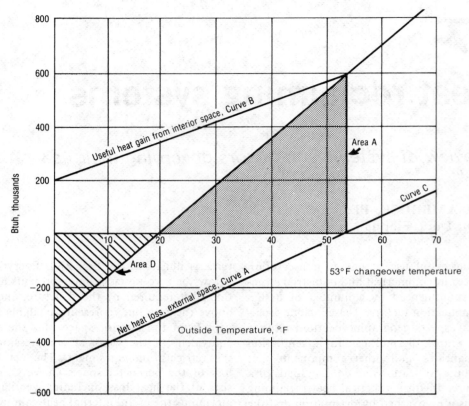

Fig. 27-1. Heating and cooling requirements of a structure during occupied periods. Heat available for storage (Area A) and useful internal heat gain (Curve B) consists of the excess internal heat gain plus 20 percent for the heat equivalent of the work of compression from the refrigeration compressor.

it is more feasible to select the most practical heat reclaiming system. Systems to be considered include:

- Air-to-air heat exchangers, consisting of rotary devices, glycol/water runaround coils, heat pipes, and parallel plate heat exchangers.
- Double circuit condenser systems.
- Domestic hot water/vapor compression refrigeration combinations.

Air-to-air heat exchangers

Air-to-air heat recovery exchangers appear to be practical for many commercial and industrial applications where considerable quantites of supply and exhaust ventilation air are required. Several designs are discussed in Chapter 28.

- *Rotary cylindrical drum or wheel*— The revolving heat exchanger, consisting either of a metal mesh or of a non-hygroscopic

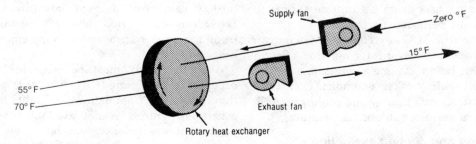

Fig. 27-2. Diagram of system using recirculated air exhaust and outdoor air supply during heating cycle. Approximate temperature transfer is shown.

material impregnated with desiccant, is divided into two separate counter flow air circuits as illustrated by Fig. 27-2. The metal mesh type is suitable for sensible heat recovery only; the second type is employed where both sensible and latent heat recovery is desired. An exhaust fan draws air through one half sector of the heat exchanger wheel, and a supply fan delivers outdoor air through the other sector to the conditioned space.

A temperature exchange as indicated by Fig. 27-2 can be expected between the two air streams depending on the operation cycle. During the cooling cycle, the revolving wheel absorbs heat from warmer ingoing ventilation air and rejects heat to the cooler exhaust air. Conversely, during the heating cycle, the wheel transfers heat from warmer exhaust air to the cooler ingoing ventilation air. The efficiency of such a device can vary from 60 to 90 percent, depending on the type of surface employed, the face velocity, and other design features.

• *Heat pipes*—Heat pipes have no moving parts. They are reversible isothermal devices strictly for sensible heat recovery.

A heat pipe consists of a sealed metal tube lined with a "wick" that is saturated with a relatively volatile working fluid. Heat applied at one end of the pipe vaporizes the fluid. The vapor flows to the opposite or cold end where it condenses and is absorbed by the wick. The liquid returns to the warm end through capillary action.

For air conditioning systems, multiple tubes are arranged in parallel and have fins, having an appearance similar to a conventional finned coil. A portion of the unit is located in the exhaust air stream, and the remainder is in the supply air stream.

Typical efficiencies range from approximately 60 percent with a face velocity of 600 fpm to 67 percent at a face velocity of 300 fpm. Figure 27-3 illustrates a typical heat pipe.

• *Runaround coils*—Another method of heat recovery is provided by so-called run-

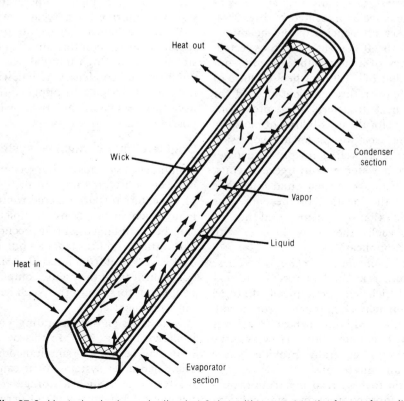

Fig. 27-3. Heat pipe is shown inclined at 0 deg with respect to the force of gravity.

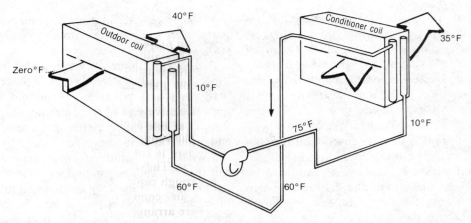

Fig. 27-4. Air-to-air heat exchanger employing runaround coils. Water or non-freezable solution is used as circulating medium. Possible temperature gradients are indicated for the heating cycle.

around coils, as illustrated by Fig. 27-4. Heat transfer coils are located in the exhaust and supply air streams. Heat is then transferred from one air stream to the other by means of water or a non-freezable liquid that is circulated between the surfaces.

Practical designs usually result in a temperature exchange as indicated by Fig. 27-4. Heat recovery efficiencies in the neighborhood of 40 to 60 percent can be expected. The efficiency of a runaround system is not as high as that of a rotary wheel because of the larger temperature gradient needed for acceptable heat transfer. Also, provisions must be made for the possibilities of moisture condensation, frost formation, and corrosion of the surfaces, as well as for pressure losses in the air and water or anti-freeze circuits. The main advantage of runaround coils is the flexibility in the location of the coils. The supply and exhaust streams need not be adjacent to each other as is the case with rotary wheel applications.

• *Parallel plate heat exchangers*—These provide another method of recovering sensible heat. Efficiencies are typically 40 to 65 percent. Units may be fabricated from metal, plastic, or fiber. Maximum heat transfer is obtained with counter flow. However, the cross flow type generally provides more convenient air convections.

In addition to heat recovery from the exhausts of the more conventional air condition-

ing installations, the rotary wheel, heat pipe, runaround coil systems, and parallel heat exchanger offer possibilities for many special applications where mechanical refrigeration is not employed. This includes the reclaiming of outgoing heat from ovens, driers, kitchens, and other such high temperature exhaust systems requiring 100 percent ventilation air.

When mechanical refrigeration is a basic requirement for comfort air conditioning installations with high internal heat gains, the double circuit condenser system, with its multiplicity of design and application features, usually offers more practical heat recovery possibilities.

Double circuit condenser systems

A number of systems permit reclaiming internal excess heat to supplement the demands of the external space on cold winter days. In many such installations, complete comfort can be maintained during occupied hours with little or no heat from an outside source. This excess heat from internal areas may consist of heat from the lights, computer rooms, people, motors, and other such heat producing equipment.

Such a design (incorporating a heat pump) is shown in Fig. 27-5. In this design, excess heat may be used to simultaneously supplement perimeter systems, or it can be stored for future use. Such a storage arrangement has the further advantage of permitting off-

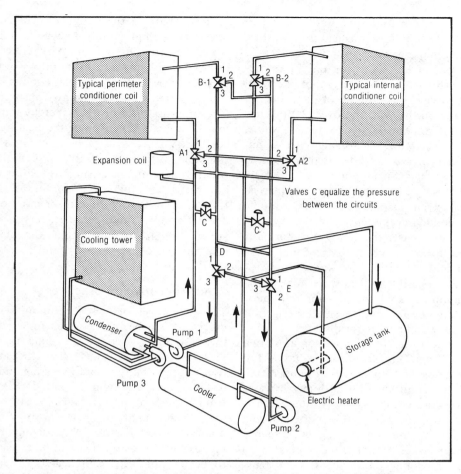

Fig. 27-5. Heating cycle: position Valves A-1 and B-1 to path 1-3. Pump 1 circulates warm water in a closed loop through condenser, Valve A-1, perimeter coils, Valve B-1, and Valve D. Cooling cycle: (simultaneously) position Valves A-2 and B-2 to Path 1-2. Pump 2 circulates cold water through cooler. Valve A-2, internal conditioner coils, Value B-2, and Valve E. Intermediate cycle: when heating is no longer required, condenser Pump 3 can dissipate the excess heat to a cooling tower or other heat sink. Modulating Valve D maintains a predetermined water temperature, generally between 100 to 120°F in condenser circuit. Similarly, Valve E maintains a predetermined temperature of about 45 to 55°F in the cooling circuit. Storage tank cycle: supplemental heat can be supplied from the storage tank during the heating cycle when the water temperature is below that suitable for direct heating; it can be used as a heat pump system down to a temperature of about 40 to 45°F to further improve the overall efficiency. Similarly, supplemental cooling can be supplied from the storage tank during the cooling cycle.

peak direct electric heating and off-peak cooling during summer periods in order to take full advantage of any special energy rates that may be available.

Heat reclaiming cycles for larger commercial and industrial buildings, where a considerable amount of internal heat gain is involved, are particularly adaptable to a heat pump system. It is usually feasible, however, but not always practical, to use fossil fuel equipment to supply additional heat require-

ments over and above the useful heat gain obtained from interior areas.

Domestic hot water heating

Frequently overlooked in energy conservation evaluations are the possibilities of using reclaimed heat for domestic water heating. Energy consumed for water heating in small commercial and residential structures represents a considerable portion of total annual usage. For instance, it is not unusual for a typical family of four (two adults and two children) with clothes washer, dish washer, and other major appliances to use an average of 100 gal. of hot water per day. This represents from 16 to 30 percent of the total annual energy consumption of the residence depending on the type and overall efficiency of the system. In the United States as a whole, domestic water heating for all types of applications represents an estimated 4 percent of the total energy consumption.

An attractive possibility for commercial and industrial installations is the double circuit condenser system depicted by Fig. 27-5. In this case, the hot water normally diverted to the cooling tower could serve instead as the domestic hot water supply.

It is also possible to heat water for domestic use in most vapor compression refrigerating systems by installing a small refrigerant water heat exchanger in the discharged refrigeration line from the compressor as shown in Fig. 27-6. This arrangement takes full advantage of the high superheat temperature of the refrigerant gas to heat domestic water whenever the compressor operates. In the case of the heat pump, this will be during either the heating or cooling cycle. Since hot water usage is not directly related to heating or cooling requirements of a structure, some type of storage must be provided or else a supplemental heating system must

be made part of the design in order to be sure that an ample supply is always available.

Such a combination unit by effectively using the condenser heat normally wasted to the atmosphere during the cooling cycle can materially reduce energy consumption for domestic water heating, thereby improving the overall efficiency of the heating and cooling systems. With a heat pump, a considerable portion of available condenser heat will also be available during the heating cycle at other than full load conditions. Field tests indicate that 65 percent or more of the domestic hot water requirements can be supplied on an average summer day, 45 percent for the entire summer, and 50 percent for the entire year.

Other conservation practices

The four types of heat reclaiming systems described can be incorporated into most of the commonly used year around air conditioning systems. In fact, their chief attraction is adaptability to both existing and new installations. It should be noted that heat from industrial processes can also be recovered and used for space heating and/or domestic water heating. At least one manufacturer offers a unit that can be considered to be an "un-fired" unit heater. Instead of combustion products, high temperature exhaust gases circulate through the fire side

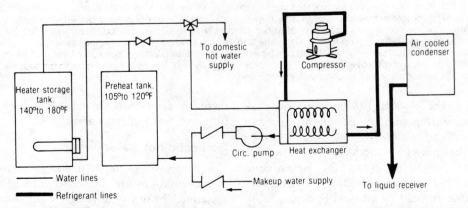

Fig. 27-6. Heating domestic water with conventional cooling cycle. The compressor discharges the high temperature, high pressure refrigerant gas through the heat exchanger to the air cooled condenser (or other heat sink). From the condenser the liquid refrigerant goes through the cooling coil back to the compressor in the usual cycle. In the heat exchanger, the refrigerant gas heats the domestic water that is circulated from the storage tank in a closed loop.

of the heat exchanger. Air-to-air recovery devices can transfer heat from process to space heating systems; heat pumps can recover process heat from space heating and/or domestic water heating.

Other energy conservation practices related to the possible efficiency improvements in the design and application of the basic heating and air conditioning systems are high pressure duct systems, the reheat cycle, the use of terminal units, and the variable air volume system.

References

1. Ambrose, E. R., *Heat Pump and Electric Heating*, John Wiley & Sons, Inc., New York, N.Y., 1966.
2. Ambrose, E. R., "The Heat Pump: Performance Factor, Possible Improvements," *Heating/Piping/Air Conditioning*, May, 1974.
3. Bonilla, C. F., PE, "Rotary Heat Exchangers," *Actual Specifying Engineer*, May, 1964.
4. Bowlen, K. L., "Energy Recovery from Exhaust Air," *ASHRAE Journal*, April, 1974.

Figures 27-1 and 27-5 are reproduced by permission of the publisher from Reference 1, pages 82 and 49, respectively.

28

Heat recovery devices for air conditioning

Guidelines to options in heat recovery devices: how they work and how their costs compare.

RICHARD K. FERGIN, PE,
Formerly of San Diego, California, and Staff Scientist and Principal Investigator, Geoscience Ltd., Solana Beach, California, currently Hugh Carter Engineering Corp., Consulting Engineers, La Jolla, California

Heat recovery devices are being used in a wide variety of HVAC, snow melting, and industrial applications. Simple devices used for various methods of heat recovery (or rejection) are shown in Figs. 28-1 through 28-6 and discussed below. In each, only one row of condenser or evaporator tubes is shown, but more can be used where necessary. Air is referred to in the figures and discussion, but with appropriate modifications for controlling conductances, the recovery methods would apply to any fluid.

Figures 28-1 through 28-3 show evaporator-condenser heat recovery units using gravity liquid return. Greater heat transfer rates can be attained by using internal fins. The simplest unit is that in Fig. 28-1 where the cold air condenses vapor in the upper section of the unit. The liquid drops by gravity to the bottom section, where it is evaporated by hot air.

The gravity return in Figs. 28-1, 28-2, and 28-3 makes it compulsory that the cooler air stream be at the top. If an exchanger is treat-ing outside air for environmental control, this restricts it to a seasonal operation unless there is a duct and damper arrangement to alternate air streams from top to bottom depending on whether heating or cooling is desired.

The operation in Fig. 28-2 is similar to that in Fig. 28-1 except that the evaporator-condenser tubes are inclined instead of vertical and the condenser and evaporator sections can be placed side by side instead of over and under.

Figure 28-3 shows a closed loop evaporator-condenser heat recovery unit, again with gravity liquid return but without the liquid and vapor flows in opposite directions as in single tube units. In this system, higher transfer rates can be attained, as can greater physical separation of the condenser and evaporator sections.

Design methods for gravity units are the same as those for individual, conventional evaporators and condensers using proper flow directions. If desired, an overall *U*-factor

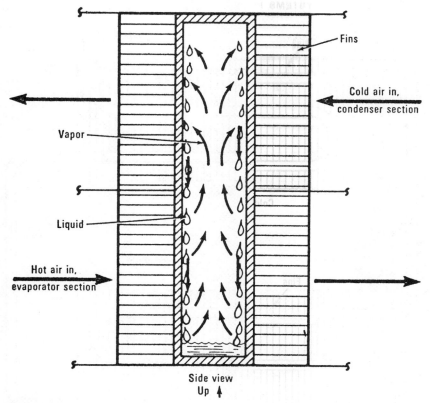

Fig. 28-1. One-pipe evaporator-condenser heat recovery unit with gravity liquid return and over-under air streams.

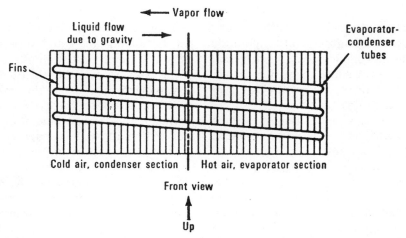

Fig. 28-2. One-pipe evaporator-condenser heat recovery unit with gravity liquid return and side-by-side air streams.

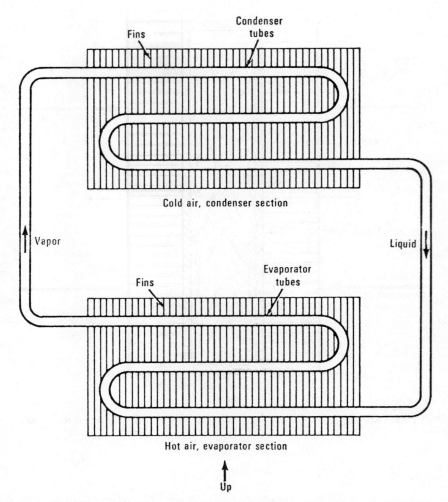

Fig. 28-3. Closed loop evaporator-condenser heat recovery unit with gravity liquid return.

can be developed for both sections, including both airside conductances and evaporation and condensing conductances. A refrigerant liquid pump can be added to assist or replace gravity, and relative locations of evaporator and condenser can be varied.

Figures 28-4, 28-5, and 28-6 show the heat pipe, heat wheel, and plate heat exchanger, respectively. All are reversible, regenerative devices. When a section is exposed to a temperature gradient with respect to another section, the cycle commences, attempting to eliminate the gradient. This allows year around operation in treating outside air.

In Fig. 28-4, capillary attraction in the wicking of the heat pipe is used in lieu of gravity to return the liquid to the evaporator section. Careful evaluation of the properties of the capillary wick and the fluid are necessary to determine whether this unit will overcome the effect of gravity if it is installed in other than a level position, with the evaporator section above the condenser section.

The rotary air-to-air regenerative heat exchanger or "heat wheel" in Fig. 28-5 alternately heats and cools the air streams by its matrix of heat absorbing material. (The packing material absorbs heat from the hot stream, rotates, and rejects the heat to the cold stream once each revolution.) Design

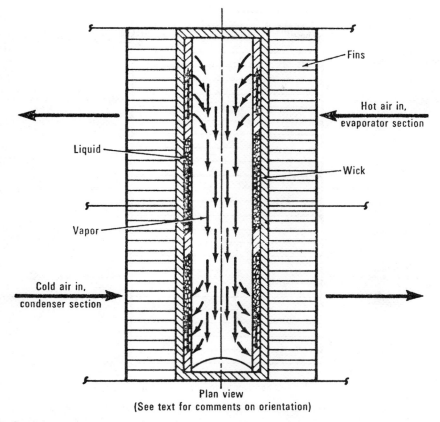

Fig. 28-4. One-pipe evaporator-condenser heat recovery unit (heat pipe) with capillary wick liquid return.

criteria for several heat wheel matrix configurations are available.

The air-to-air plate heat exchanger in Fig. 28-6 may be of concurrent, counter, or cross flow design. No intermediate heat transfer fluid or material is utilized, except to separate the two streams.

Costs of the various methods vary and should be evaluated on the basis of energy recovered. If recovery of sensible heat alone is required, the air-to-air plate heat exchanger is the least expensive; and the gravity liquid return units are next, with a slight increase in the order of Figs. 28-1, 28-2, and 28-3. If recovery of latent as well as sensible heat is desired, however, a rotary air-to-air exchanger may be the least expensive.

In a given application, none of the systems may appear to be usable, but rearranging system components usually can make one or more so.

Maintenance and operating costs depend on the degree of filtration required, the frequency of cleaning, and the fan energy needed to overcome the static pressure loss through the equipment. The heat wheel has an additional cost in that it requires a motor, usually of fractional horsepower, for rotation.

As indicated above, the heat wheel is generally not economically feasible for sensible heat recovery in air conditioning applications, but it may be at an advantage if latent heat is recovered also.

All of the units shown, except the heat wheel, can have zero mixing of the air streams. This can be a prime factor in selection if contamination is important. The finned units shown normally have aluminum or copper fins, but the air-to-air plate heat

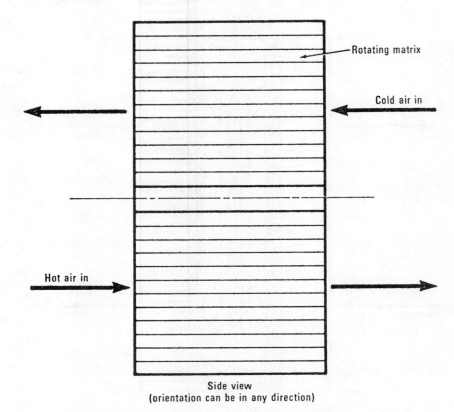

Fig. 28-5. Rotary air-to-air regenerative heat exchanger (heat wheel).

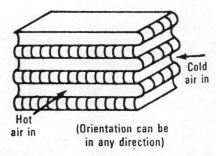

Fig. 28-6. Air-to-air plate heat exchanger (cross flow type shown).

exchanger may be of metal, plastic, or fiber construction.

Following are some references that contain design information on heat recovery devices.

References

Gravity return units

ASHRAE Handbook of Fundamentals, American Society of Heating, Refrigerating and Air-Conditioning Engineers, Inc., New York, 1972.

Threlkeld, J. L., *Thermal Environmental Engineering*, 2nd ed., Prentice-Hall, Inc., Englewood Cliffs, N.J., 1970.

Heat pipes

Schwartz, J., *Performance Map of an Ammonia (NH₃) Heat Pipe*, ASME Paper 70-HT/SpT-5, presented at Space Technology and Heat Transfer Conference, Los Angeles, June, 1970.

Eastman, G. Y., "The Heat Pipe: Space Spinoff for Heat Transfer," *Heating/Piping/Air Conditioning*, December, 1969.

Harbaugh, W. E., and Eastman, G. Y., "Applying Heat Pipes to Thermal Problems," *Heating/Piping/Air Conditioning*, October, 1970.

Mason, H. E., "Heat Pipe: Economical Snow Melting Device?," *Heating/Pipine/Air Conditioning*, November, 1971.

Plate, finned, and regenerative

Kays, W., and London, A. L., *Compact Heat Exchangers*, 2nd ed., McGraw-Hill Book Co., New York, N.Y., 1964.

ASHRAE Handbook and Product Directory, Equipment. American Society of Heating, Refrigerating and Air-Conditioning Engineers, Inc., New York, N.Y., 1975.

29

Air-to-air energy recovery

A large variety of heat transfer components and packages provides many options.

T. E. PANNKOKE, PE,
Engineering Editor, Heating/Piping/Air Conditioning, Chicago, Illinois

Industrial rates of electricity for the United States as a whole increased 75 percent from January 1972 to April 1975, according to the United States Department of Commerce. The average price of natural gas sold by interstate pipelines *directly* to industrial users increased 30 percent from December 1973 to December 1974. Prices charged by gas distribution companies have also increased over the past few years, and some classes of customers have experienced curtailments in their supply in various parts of the country. The Commerce Department reported a rise in average utility charges to customers of 11.8 percent during 1976, compared with a 17.5 percent rise experienced in 1975. (These figures are based on data for all types of public utilities, not just energy companies.)

The American Petroleum Institute reported that for the week ending March 12, 1976, imports of crude and refined products exceeded U.S. domestic production for the first time in history. Average daily figures for this period are approximately 8.2 million barrels of imported oil versus 8.05 million barrels of domestic oil.

The incentives to conserve energy are increasing, at least for the industrial sector of the economy. The cost of energy is becoming a larger part of the cost of finished products. Also, eliminating energy waste is sometimes the only way a firm can obtain additional energy to boost production.

Applications of various types

The first patent for a heat recovery system by preheating combustion air was issued in Britain in 1816. Table 29-1 shows the extent of fuel savings obtainable by preheating combustion air required by industrial furnaces—savings of up to 50 percent. Table 29-2 shows some outdoor air load reductions that can be achieved in various locations by means of air-to-air heat exchangers used to precool or preheat ventilation air.

Air-to-air recovery systems are typically used in conjunction with the following general applications.

- Space heating and/or cooling (HVAC).
- Recovery of heat from a process to return the heat to the process.
- Recovery of heat from a process to return the heat to another process.
- Recovery of heat from a process to use it for makeup air heating or space heating.
- Recovery of heat used in fume incineration for use in a process.
- Lowering of process exhaust temperatures to a level compatible with emission control equipment.

Table 29-1.   Percent of fuel savings for various combustion air preheat temperatures. The fuel is natural gas at 10 percent excess air.

Furnace outlet temp., °F	Preheat temperatures, °F										
	400	500	600	700	800	900	1000	1100	1200	1300	1400
2600	22%	26%	30%	34%	37%	40%	43%	46%	48%	50%	52%
2500	20	24	28	32	35	38	41	43	45	48	50
2400	18	22	26	30	33	36	38	41	43	45	47
2300	17	21	24	28	31	34	36	39	41	43	45
2200	16	20	23	26	29	32	34	37	39	41	43
2100	15	18	22	25	28	30	33	35	37	39	41
2000	14	17	20	23	26	29	31	33	36	38	40
1900	13	16	19	22	25	27	30	32	34	36	38
1800	13	16	19	21	24	26	29	31	33	35	37
1700	12	15	18	20	23	25	27	30	32	33	35
1600	11	14	17	19	22	24	26	28	30	32	34
1500	11	14	16	19	21	23	25	27	29	31	33
1400	10	13	16	18	20	22	25	27	28	30	—

Courtesy of the American Schack Co.

36 in. diameter industrial shell and tube air-to-air recovery package. Exhaust air flows in tubes, makeup air in the shell. This type is well suited to high exhaust particulate concentrations. (Courtesy Allied Air Products Co., Oregon.)

Basic equipment types

Air-to-air heat recovery equipment falls into two general categories.

Regenerative units have alternating air flow—hot and cold—over the heat transfer surfaces. This category can be subclassified into stationary and rotary types. Flow of hot and cold air streams are periodically reversed in the stationary regenerator. Rotary regenerative units maintain a constant flow direction for both air streams. The heat transfer media rotates through a supply and an exhaust stream which flow in opposite directions. Rotary units are more widely used than stationary ones since they are suitable for HVAC applications as well as for many industrial uses over the past 17 years.

Recuperative units in the strict sense are devices in which continuous unidirectional flow across the heat transfer surfaces of the separate hot and cold streams is maintained across the heat transfer surfaces of the heat exchanger.

Table 29-2.   HVAC system load reduction per 1000 cfm of outdoor air with air-to-air heat recovery.[1]

City	Summer operation (tons)				Winter operation (Btuh $\times 10^3$) sensible recovery only[3]		
	Total	Recovery	Sensible	Recovery			
	85%	75%	75%	65%	85%	75%	65%
Albuquerque, N.M.	1.39[2]	1.22	1.08	0.94	50.5	44.5	38.6
Atlanta, Ga.	2.93	2.59	0.95	0.82	25.7	22.7	19.6
Boston, Mass.	1.40	1.24	0.68	0.59	56.9	50.2	43.5
Chicago, Ill.	2.30	2.03	0.95	0.82	69.8	61.6	53.4
Des Moines, Iowa	2.62	2.31	0.95	0.82	68.8	60.8	52.6
Fairbanks, Alaska	—	—	—	—	112.0	98.8	85.6
Los Angeles, Calif.	1.23[2]	1.08[2]	0.81	0.70	25.7	22.7	19.6
New York, N.Y.	2.30	2.03	0.81	0.70	51.4	45.4	39.3
St. Louis, Mo.	2.93	2.59	1.15	0.99	58.8	51.8	44.9
Rochester, Minn.	2.02	1.78	0.68	0.59	68.8	60.7	52.6

[1]Based on ASHRAE Standard 90-75 criteria; Indoor design, summer, 78°F DB, 60% RH except as noted in footnote 2, winter, 72°F DB, outdoor design, summer, 2.5 percent values, winter 97.5 percent values.
[2]Lower indoor RH used due to relationship to outdoor design conditions: 30% RH for Albuquerque, 40% for Fairbanks, 50% for Los Angeles.
[3]Units having both sensible and latent capability will recover additional energy during the winter, particularly when humidification is provided.

Heat transfer within heat exchanger units is accomplished primarily by conduction—plate types—or by change of state of a "working fluid"—heat pipes.

Under a broad interpretation, air-liquid-air systems may be considered to be recuperative *systems* since the hot and cold air streams always flow in the same direction across or through the exchange media. With these systems, the hot and cold *sides* of the heat transfer surfaces are remotely located from one another. The coil runaround cycle is an example of such a system. The runaround cycle is a sensible heat transfer system. An air-liquid-air enthalpy exchange cycle employing a hygroscopic liquid is also available. In some respects, it is similar to the more familiar coil runaround cycle.

Efficiency and effectiveness

Air-liquid-air cycles do not require that supply and exhaust ducts have to be connected to the same apparatus. Therefore, design flexibility is enhanced.

Efficiency is a measure of the maximum amount of heat that a heat transfer can recover at a given set of conditions and depends to a great extent on the mass flows and the velocity through the unit. Catalog efficiencies generally are based on equal mass flows (usually stated in terms of standard cubic feet of air per minute, scfm) through the inlet and exhaust air sides.

At a specified flow velocity, the device will have a certain efficiency value. A particular unit's efficiency will vary as the velocity varies, being higher at lower velocities because of the longer time exposure to the heat transfer surface.

It is not unusual for heat exchangers to operate with unbalanced mass flows, however. With HVAC systems, the exhaust stream mass flow often is less than the supply. In other cases, the exhaust mass flow may be a larger quantity. Unbalance between the supply and exhaust streams has a definite effect on the performance of heat exchange units with regard to the total amount of heat that can be transferred.

Efficiency is the ratio of the energy recovered to the total energy available:

$$\eta = \frac{W_s(t_2 - t_1)}{W_e(t_3 - t_1)}$$

where

η = efficiency
W_s = mass flow of the supply stream
W_e = mass flow of the exhaust stream

t_1 = entering supply stream temperature
t_2 = leaving supply stream temperature
t_3 = entering exhaust stream temperature

Enthalpy values instead of temperatures are used for total heat calculations.

Effectiveness is a term currently being applied to air-to-air heat recovery devices. Effectiveness is defined as:

$$\epsilon = \frac{W_s(t_2 - t_1)}{W_{min}(t_3 - t_1)}$$

where

ϵ = effectiveness
W_{min} = the lower of W_s or W_e

The amount of heat recovered by an air-to-air unit is measured by $W_s C_p(t_2 - t_1)$ where C_p is the specific heat. The amount of heat available in an exhaust stream is measured by $W_e C_p(t_3 - t_1)$. The limits on what the supply stream can recover is set by $W_{min} C_p(t_3 - t_1)$ where W_{min} equals W_s. Therefore, when the mass flow of the supply stream is equal to or greater than the mass flow of the exhaust stream, efficiency is equal to effectiveness. When the mass flow of the supply stream is less than that of the exhaust stream, efficiency is less than effectiveness. At the condition of $W_s < W_e$ for a given unit, the effectiveness of the device is the ratio of the recovered heat to the theoretical capability defined by $W_{min} C_p(t_3 - t_1)$ rather than the maximum heat available, which is defined by $W_e C_p(t_3 - t_1)$. Stated another way, at $W_s \gtreqqless W_e$ the unit theoretically could pick up all of the heat in W_e; at $W_s < W_e$, the unit no longer can.*

Most manufacturers' literature currently provides rating data in terms of efficiency. It is expected that as catalogs are revised, performance data will be presented in terms of *effectiveness*.

As a matter of interest, ASHRAE is working on the development of a standard (proposed Standard 84P) that provides the criteria

*For more information on this subject, the reader is referred to Gawley, H.N., and Fisher, D., "The Effectiveness and Rating of Air-to-Air Heat Exchangers," presented at the symposium on heat recovery at the ASHRAE Annual Meeting in Boston, June 22-26, 1975.

by which air-to-air recovery devices/systems are to be rated for heat transfer capacities, thermal effectiveness, and any contaminations of the supply air stream by the exhaust.

Nonstationary component systems

Some air-to-air recovery devices or systems require a driving force—drive motors or pumps—to permit the cycle to function, while others contain no moving components and rely on external sources, *i.e.*, supply and exhaust fans, to supply all the required mechanical operating energy.

Rotary regenerative heat recovery units or heat wheels perhaps are the most familiar air-to-air heat recovery devices, a type of which is shown in Fig. 29-1.

They are available with various nonmetallic and metallic heat transfer media to provide total heat transfer or sensible heat transfer only.

Either desiccant impregnated (lithium chloride), nonmetallic media is used when sensible and latent recovery is desired or an aluminum oxide coated woven aluminum wire medium is used.

Moisture from the air stream having the higher humidity ratio is absorbed directly by the desiccant in the nonmetallic wheel and given up to the stream having the lower humidity ratio. Frosting on the wheel gen-

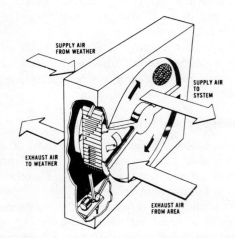

Fig. 29-1. Cutaway of rotary regenerative heat exchanger. (Courtesy of Cargocaire Engineering Corp.)

erally will not occur until temperatures fall well below freezing.

With the metallic media, latent transfer takes place via condensation. Condensation will occur when the exhaust air dry bulb temperature is considerably below the dew point temperature of the outside air.

The desiccant impregnated media has greater latent capability.

Some media present a corrugated or honeycomb appearance when viewed from the wheel face. Others appear similar to metal wool. The corrugated material provides for laminar or straight through flow of the air streams, while the woven metal variety provides random path or "turbulent" flow through the media in which air flow simultaneously occurs axially and radially.

Seals and purge sections are used to reduce and control leakage between the supply and exhaust air streams. Up to the 300 to 400° F temperature range, elastic materials such as neoprene may be used. Above 400° F, some manufacturers use a molded refractory insulation material for seals for temperatures up to 1500° F. Others rely on minimum clearances to control leakage at high temperatures.

Purge sections for heat wheels used in HVAC applications can reduce carryover from the exhaust stream to acceptable limits for most installations. Carryover can be reduced to as little as 0.04 percent in some units.

The purge section will operate properly if the static pressure on the exhaust side is lower than on the supply side of the wheel. The supply fan may be located on either side of the exchanger as long as a sufficient Δ P across the supply and exhaust sides of the wheel is maintained.

Heat wheel assemblies may be obtained in various ratings from approximately 750 to 72,000 scfm. Exhaust stream temperatures up to 1500° F may be accommodated. Approximate capacity ranges and maximum exhaust temperatures for various energy exchange media are summarized below:

• Desiccant impregnated: 1000 to 75,000 scfm; maximum temperatures 120 to 400° F depending on the manufacturer.

• Woven aluminum mesh: 750 to 72,000 scfm; 150° F for aluminum, 300° F for aluminum oxide surface.

• Stainless steel mesh: 750 to 72,000 scfm; 400 to 600° F.

• Corrugated aluminum: 750 to 72,000 scfm; 300° F.

• Corrugated aluminized steel: 750 to 72,000 scfm; 600° F.

• Corrugated alloy steel: 400 to 43,400 scfm; 750° F.

• Corrugated stainless steel: 400 to 18,400 scfm; 1500° F.

• Honeycomb ceramic: 1000 to 15,000 scfm; 1500° F.

The heat transfer efficiency of heat wheels generally range from 60 to 85 percent depending on the installation, type of media, and air velocity through the units.

• *Desiccant spray units*—Figure 29-2 depicts a layout of an air-liquid-air enthalpy transfer cycle. It provides year round energy recovery on HVAC applications by preheating and precooling ventilation air.

The transfer solution is lithium chloride, which is hygroscopic. Energy transfer takes place in towers in which both the supply and exhaust air streams are in contact with lithium chloride solution sprays. The supply and exhaust air towers are connected by piping and circulating pumps as shown. The washing action of the sprays also cleans the air, and NBS dust spot test efficiencies as high as 35 percent are obtainable. Bacteria removal efficiency of 95 percent as measured by the six plate Anderson sampler may also be realized.

Tower capacities range from 7700 to 92,000 scfm. Multiple towers may be used for larger installations. Annual enthalpy recovery efficiency is in the range of 60 to 65 percent.

Three basic system variations are obtainable depending on winter design conditions and how much indoor design relative humidity is desired or required.

1. *Basic system*—This operates without the winter solution heater. The enthalpy exchange is in balance between the two towers. Most applications are in areas where the outdoor temperature does not go below freezing;

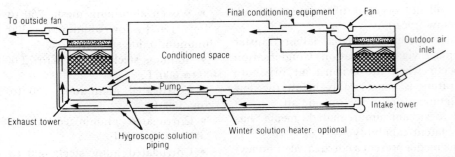

Fig. 29-2. Sensible-latent recovery system. Energy recovery is accomplished through the contact of a lithium chloride solution with air. (Courtesy of Kathabar Systems.)

it may be used in colder climates if space humidities below 35 percent RH are acceptable.

2. *Extra humidification*—This is applied in areas having cool, dry winter climates. During cool weather, the lithium chloride solution is maintained at a fixed dilute concentration by automatic addition of water to the system.

3. *Extra humidification and extra heating*—Water is automatically applied to the cycle to maintain a fixed dilute concentration.

Air leaving the supply air tower is maintained at a fixed temperature by a solution heater. It can take the place of preheat coils. The lithium chloride solution is a satisfactory antifreeze at the solution concentrations and outdoor ambients involved.

• *Coil runaround cycle*—This is a recovery concept that has been in existence for at least 40 years. It may be used to preheat and precool ventilation air, and it may be applied to industrial heat recovery. Figure 29-3 shows the basic system. In its simplest form, two finned coils are connected together with a piping loop. A pump circulates a heat transfer fluid—usually water or a glycol solution—through the coils. Heat is picked up from the warm air stream and released to the cooler one.

Up to now, this cycle has not been widely used. Coil selection depends on a number of variables. Among these are:

• The volume of flow through the coil.
• Face velocity through the coil.
• Entering air and liquid temperatures.
• Number of coil rows.

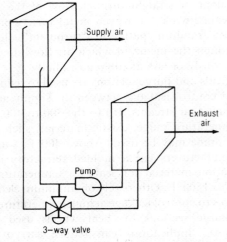

Fig. 29-3. Coil runaround cycle for sensible heat recovery in HVAC and process applications.

• Number of fins per inch.
• Air side and water side pressure drops through the coils.

Prior to the advent of computerized coil selection procedures, a tedious trial and error solution was required to match two coils and to determine whether during the winter the entering liquid temperature to the cooling coil would become low enough to cause condensate freezing and plugging of the coil.

Presently, several manufacturers have computer programs capable of optimizing coil selections for many types of coil runaround applications.

Minimum gpm flow through the system provides greater temperature differences and higher efficiency. Lower liquid flow also increases the probability of condensate freezing. A three-way valve is recommended to avoid this. If the entering water temperature to the cooling coil (EWTc) falls to a level where the coil surface freezing can begin to occur—30 to 32° F—the valve bypasses sufficient solution so that the EWTc remains at a level to preclude condensate freezing. The three-way valve generally opens when outdoor ambients are in the range of 0 to 15° F, as also explained later in the chapter under "Temperature control."

With the three-way valve in the system, recovery steadily increases as the temperature difference between the inlet exhaust and inlet supply air streams increases *until* the valve starts to open. Then, as the Δt between the air streams increases further, the quantity of recovered heat remains essentially constant. On an annual basis, the decrease in recovery in efficiency is negligible due to the few hours involved.

If maximum recovery at the winter design condition is desired instead of maximum efficiency, coils can be selected for higher gpm flows in order to raise the EWTc. Depending on climatic conditions, higher flow rates alone may be sufficient. If not, the three-way valve will open at a lower outdoor ambient temperature.

Sensible recovery efficiency is approximately 60 to 75 percent on a seasonal basis.

Stationary heat recovery units

Numerous variations of recuperative stationary, sensible heat recovery devices are available.

The *heat pipe* is unique in that it employs a "working fluid" to provide heat transfer from one end of a sealed, evacuated tube to another as explained in Chapter 28.

Heat pipes are reversible devices and will operate as long as there is a temperature difference between each end. A heat source at one end vaporizes the fluid. The vapor flows to the colder end where it condenses, releasing the heat of vaporization. A wick, which is in contact with the inside of the tube or pipe, provides for the return of the condensed fluid to the evaporator end by capillary action. Depending on the orientation of the heat pipe, gravity can help or hinder fluid flow.

Heat pipes can be made in many shapes and sizes and out of many materials. Those used in air-to-air applications closely resemble in appearance a finned heat transfer coil. Representative data listing sizes, and dimensions for exhaust temperature ranges up to 600° F are shown in Table 29-3. Unlike a conventional coil, there are no pipe connections, and the assembly has a divider to separate the hot and cold air streams.

Tube and fins may be aluminum, copper, carbon steel, or stainless steel, depending on the operating temperatures encountered. Generally, aluminum is used for applications up to 400° F, copper to 800° F, and steel to 1300° F. Protective coatings may be applied to aluminum and copper surfaces for use in corrosive atmospheres. Coil housings may be aluminum; galvanized, aluminized steel; or other suitable materials.

Capacity may be varied by tilting the heat pipe or by means of face and bypass control. With tilt control, flexible connections must be applied to all duct connections at the unit, as also explained later in the chapter under "Temperature control."

Packaged HVAC heat pipe recovery units are available, and nominal capacities of these assemblies range from 750 to 12,500 scfm (see Fig. 29-4).

Recovery efficiency of heat pipes applied to air-to-air applications generally fall into the 55 to 70 percent range.

Plate heat exchangers are made in a variety of sizes and configurations. Units used in HVAC and industrial recovery applications usually are fabricated from metal. (Materials such as fiber and plastics have also been used to make plate exchangers.) Materials used for the heat transfer surfaces commonly are aluminum for temperatures up to approximately 400° F, aluminized steel up to 1000° F, and stainless steel for service up to 1600° F. Housings typically are aluminum, galvanized steel or aluminized steel.

Most available units are designed to operate in counterflow conditions in order to

Table 29-3.   Representative size range of heat pipe units used in mechanical systems.

Exhaust temp., °F	No. of standard sizes available	Minimum and maximum dimensions, in.			No. of rows	Fin spacing per in.
		Height	Length	Depth		
Up to 600°F	76	13.5 to 54	24 to 192	8 to 15.5	3–7	8, 11, 14

Units are made from aluminum tubes and fins or copper tubes and fins. Seven standard heights and 15 lengths are available. Intermediate sizes are also available. (Courtesy of Q·Dot Corp.)

Air-to-air recovery unit assembled from five standard modules. Air flow: exhaust enters bottom and exits at the top on the facing side. Outdoor air enters top and exits bottom on the reverse side. (Courtesy Des Champs Laboratories, Inc.)

Pollution control installation. The five vertical counterflow modules condense exhaust stream vapors. (Courtesy of Allied Air Products Co., Oregon.)

maximize heat transfer efficiency. One type of unit employs cross flow operation. While efficiency is somewhat lower, duct connections for some installations can be simplified.

Rated heat transfer efficiencies are in the 60 to 75 percent range, although one manufacturer lists an efficiency of 85 percent for an extended surface unit when the flow velocity is down at 200 fpm.

Several models may be combined to form

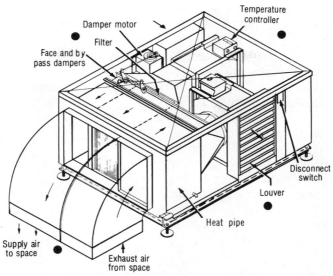

Fig. 29-4. Heat pipe air-to-air recovery package. (Courtesy Isothermics Inc.)

larger capacity assemblies. Individual module capacities range from approximately 1000 to 11,000 scfm, depending on flow velocity through the unit.

Units may be obtained having built-in cleaning apparatus such as water wash assemblies.

Evaporative cooling can be an effective adjunct to the installation. Units may also be obtained with evaporative cooling as an integral part.

Some manufacturers supply packages designed for specific applications such as: supplying kitchen makeup air (Heating coils are an option offered by at least one manufacturer.); cooling sealed electronics enclosures; and controlling air pollution by condensing out pollutants.

Other air-to-air devices that are available include units designed for stack installation to capture heat from the flue gases to preheat combustion air.

Heat exchangers basically designed for combustion heating equipment such as unit heaters are being used to capture process heat and use it for makeup air or building heating. Combined with a blower assembly, they become an "unfired" unit heater.

Many manufacturers of packaged air handling equipment for heating, ventilating, and air conditioning equipment and industrial equipment are incorporating the various heat recovery devices into packages. These packages often include such options as filter racks, supply and/or exhaust fans, damper assemblies, preheat and/or reheat coils, and chilled water or DX coils. Gas or oil-fired heating sections, steam coils, or electric resistance coils may be specified.

Use air filters or collectors

Dirt buildup on heat transfer surfaces reduces heat transfer effectiveness and increases the pressure drop through the unit. Therefore, air filters installed in the supply stream ahead of the heat exchanger are recommended for all HVAC applications and for process applications when conditions warrant them.

When industrial effluents contain considerable amounts of particulate matter, the use of filters or dust collectors in the exhaust stream ahead of the heat exchanger may be advisable.

Air filters are less expensive than dust collectors, but they also have much less dirt removal capacity. Most filter media are not designed for operating temperatures above

200° F, although some can operate at somewhat higher temperatures. Dust collectors can be designed for service at air temperatures in the 500 to 750° F range, and special units can be constructed for higher temperature service.

When high particulate content exhausts are encountered at temperatures above 500 to 750° F, it generally is more economical to cool the effluent than to construct a special dust collector capable of operating at the higher temperature. Effluent can be cooled in various ways: dilution with outdoor air, passing it through a water spray (evaporative cooling) or by passing it through a section of water cooled ductwork.

Plan for moisture buildup

Another factor that affects both HVAC and process installations is condensation of moisture and frost buildup on heat transfer surfaces on the exhaust air side. Whenever the surface temperature of the heat transfer element is at or below the dew point temperature of the exhaust air temperature, moisture will condense on the surface. Therefore, condensate drain pans and drain lines should be provided for all installations, unless it can be definitely determined that condensation cannot ever occur.

When the surface temperature of a metal heat transfer element in the exhaust side is at or below freezing temperature, frost accumulates on the surface and restricts heat transfer capability and air flow. Frost formation on HVAC and lower temperature industrial applications is possible whenever winter design temperatures below freezing are encountered. Whether or not freezing will occur is a function of the dry and wet bulb temperatures of the outdoor supply air and the indoor exhaust air or gas.

During periods of outdoor ambient temperatures at which freezing could occur, the supply air can be preheated prior to entering the heat exchanger so that the temperature of the exhaust side of the exchanger will be high enough to preclude frost formation. It should be noted that frosting occurs at lower temperatures with nonmetallic, desiccant impregnated total heat transfer wheels, since latent transfer occurs in the vapor phase. Nevertheless, preheating inlet air may be required with these devices also.

Temperature control

In many HVAC applications, control of the leaving supply air temperature is mandatory in order to prevent overheating or overcooling the space. Varying the speed of a heat transfer wheel will vary the rate of transfer; consequently, the leaving air temperature will vary.

Heat pipes can be tilted. The effect of gravity can be used to increase or retard the rate of heat transfer. Another method of control is to use face and bypass dampers.

When a coil runaround cycle is employed, the leaving supply air temperature is easily controlled by using a three-way valve ahead of the coil in the supply air stream. The same valve that is used for winter freeze protection may be employed. Two sensing elements are used: one for freeze protection and one for leaving air temperature control. The outdoor air temperature governs which sensing element controls the valve.

With a desiccant runaround spray cycle, the leaving supply air temperature is controlled by modulating the volume of the sprays in the supply and exhaust air towers. Minimum spray volume occurs when the outdoor ambient temperature equals the design leaving supply air temperature.

Face and bypass control can be applied to stationary, plate heat exchangers.

In industrial applications, the problem may be one of controlling the exhaust air temperature so that it does not overshoot and damage the heat exchanger. Or, the problem may be how to control the leaving supply air temperature when process exhaust heat is used to provide space heating.

HVAC applications

Air-to-air recovery systems often are applied in HVAC installations where 100 percent outdoor air is required because of the major reduction in energy use that heat recovery

Heat recovery and water scrubber units being installed at a paper mill. Unit serves both as an air-to-air recovery device and as a water scrubber. Additional heat is recovered by the scrubber operation. (Courtesy of AER Corp.)

can achieve when large volumes of air are heated and cooled.

These systems can also be effective on installations that do not require 100 percent outdoor air, particularly during periods when mechanical cooling is required. In many buildings that operate with a mixture of outdoor and recirculated air to maintain a constant cold deck temperature, there is an excess of heat during occupied periods for much of, or even all of, the heating season.

On new HVAC installations, air-to-air heat recovery can result in appreciable savings in heating/cooling plant costs. When latent as well as sensible heat is recovered, there may be a reduction in total system cost, including the cost of the heat recovery system. Refrigeration and attendant heat rejection capacity, heating capacity, and heating and cooling coil capacities can be reduced, because the outdoor air component of the heating and cooling loads is reduced. When estimating investment cost savings attributable to a heat recovery installation, incremental costs must be applied. Using a budget figure of X dollars per ton for a total system will yield false savings. Air side system costs will remain the same, because space cooling loads depend on the space sensible heat load. Likewise, systems designed to overcome transmission heat gains and losses will require the same capacity whether or not air-to-air recovery is applied.

Industrial applications

Industrial air-to-air heat recovery applications offer a greater opportunity for energy conservation. Industry, however, is production oriented. For years, energy was both cheap and plentiful. The drive in many industries was to get the product out and not worry too much about energy efficiency, because fuel costs usually were a minor factor in the cost of the finished product. Also, keeping production lines operating is a paramount responsibility of production personnel. Installation of systems that require additional service and maintenance without boosting production have not been welcomed.

In certain parts of the country, industry is restricted as to how much natural gas—a highly desirable industrial fuel—it can obtain. Also, the prices of the major industrial fuels have significantly increased. Finally, signs of economic recovery from the business recession appear to be remaining steady, giving promise of continued spending by industry for new equipment and expansion. Therefore, hopefully, the time is ripe to exploit the potential industry offers for conservation.

Air-to-air recovery can be applied to a host of industrial applications. There are many opportunities to save in metal working operations—melting, heat treating, etc., and where materials are baked, cured, dried—electrical manufacturing, plastics, textiles, paper, printing, etc. Large amounts of heat may be liberated, often at high temperatures that facilitate recovery.

Commercial and institutional applications should not be overlooked either. Kitchens and laundries both are good applications for properly designed and installed recovery systems.

Industrial recovery considerations

Dr. Barry Commoner in his book *The Closing Circle* told of the potentate who demanded that his advisors present the laws of economics to him in a concise, simple form. After considerable wrestling with the problem, they succeeded in satsifying their master. The results of their efforts read: *There is no such thing as a free lunch.* Perhaps the laws of engineering can be so summarized as well.

While major energy savings can be achieved with industrial energy recovery, successful, economically viable installations often require considerable study and effort to implement. Unlike HVAC installations, which generally are straight-forward, industrial applications often require the solving of problems posed by high temperatures, corrosive effluents, and dirty effluents. Sometimes, all three problems are present on a given application.

High temperatures can be accommodated by using suitable materials. Aluminized steel, which resists scaling, may be applied at temperatures up to 1000° F. Types 304 and 316 stainless steel are commonly used, and they resist scaling up to approximately 1600° F. Series 400 stainless steels also may be used. Type 446, for example, has been used in high temperature rotary heat wheels. This material withstands temperatures up to 1800° F.

Ceramic materials are successfully applied in rotary heat wheel installations. One type of ceramic is designed for a maximum exhaust gas temperature of 1000° F; another type can be used at temperatures up to 1600° F. Ceramics have a very low expansion coefficient, hence they are dimensionally stable, and they are chemically inert in the presence of most corrosives. Ceramic wheels are, however, more expensive than their high temperature metal counterparts (no free lunch).

Corrosive effluents cause damage primarily if cooled below the dew point temperature, which causes them to condense out on heat exchanger (and other) surfaces. Many effluents contain sulfates, fluorides, chlorides, or metal salts that—in the presence of water—are highly corrosive. Chlorinated hydrocarbons are other potentially corrosive effluents that are particularly difficult to handle.

Choose correct materials

Several options are available to mitigate corrosion problems. The choice of suitable materials is an obvious one.

Copper (used in many industrial heat pipe applications) offers excellent corrosion resistance in the presence of many chemicals. The same can be said for the stainless steels and the ceramics previously mentioned.

It must be remembered that galvanic corrosion occurs if dissimilar metals contact one another in the presence of an electrolyte. Therefore, it may be necessary to specify housings of the same material as the exchanger, or the exchanger should be mounted in a manner that minimizes the possibility of galvanic corrosion occurring. Also, it may be advisable to isolate the energy exchanger unit from the ductwork with nonmetallic connections when practical to do so.

Platings and coatings applied to heat exchange surfaces can also provide protection from corrosion. Copper heat pipes may be nickel-plated to provide added corrosion protection in certain environments.

Phenolic coatings also are applied to copper and steel surfaces to retard corrosion. These can be effective in mildly acidic environments. Since coatings tend to be porous, normally more than one coat should be applied to provide satisfactory coverage. Care must be exercised in handling, shipping, and installing coated units; each scratch in the coating is a spot where corrosion can begin.

Dirty effluents were discussed previously from the standpoint of particulate contamination. However, baking, curing, paint drying, and moisture drying operations liberate organic materials, such as resins, varnishes, waxes, thinners, etc., that can condense on heat exchange materials.

At lower operating temperatures, these materials generally form a soft sludge or grease. Above 300 to 400°F, many organic compounds bake onto heat exchange surfaces, forming a hard coating. These materials are consumed at temperatures above 750 to 800°F; so the problem of organic deposits generally is not significant on higher temperature installations when the heat transfer media itself is at or above this temperature range.

Remove organic deposits

Organic deposits as well as dust and dirt must be periodically cleaned off to assure efficient operation. Failure to do so can result in complete plugging of the heat exchanger.

Dirt and sludge buildups often can be removed by washing or steam cleaning, with or without the addition of a cleaning agent. Washing and steaming assemblies can be built into many installations. Suitable drain provisions must be incorporated into the system to handle the washdown products.

Removal of heat exchanger elements for remote cleaning may be required in some instances. Depending on the size of the heat exchange elements and the type of cleaning agent used, cleaning may be done by immersion in a solvent or a cleaning solution at either room or an elevated temperature or by boiling it in a cleaning solution. The method used depends on the type of material to be removed and the type of cleaning agent. Remote cleaning by pressurized water washing or steam cleaning with or without cleaning agents is another option.

When aluminum is to be cleaned, a cleaner that is not aggressive to this metal must be used.

Effluent from cleaning operations must be disposed of in a satisfactory manner, since it is a pollutant. Some cleaners contain such compounds as phenols, cresols, chromates, and/or methylene chloride; these can be toxic if not properly handled and can present disposal problems.

A phenolic coated heat exchanger surface must be cleaned with a cleaner or solvent that will not attack the coating.

Consider using incineration

Incineration is another method used to clean exchanger surfaces. Remote incineration involves placing the exchange surfaces in an oven at 800 to 1000°F. The contaminants are consumed, and what is left is an easily-disposed-of ash.

In-place incineration can be used to dispose of organic deposits when the design of the heat exchanger lends itself to this method. A burner installed in the exhaust stream ahead of the heat exchange unit is fired to raise the exhaust stream temperature when cleaning is to be initiated.

The higher the incineration temperature, the faster the oxidation. A buildup that requires one or two hours to remove at 800° F may be removed in a manner of minutes at 1000° F. Removal time also depends on the composition of the material, its thickness, and how hard it is baked on.

Incineration can be difficult to control. The material being oxidized can ignite and enter into the combustion reaction, raising the temperature further.

Heat exchanger materials must be selected to withstand incineration temperatures if this method is chosen. Incineration cannot be used with aluminum or mild steel surfaces, for example, since heat exchangers made of these materials have maximum operating temperatures of approximately 400 and 800° F, respectively. To date, incineration cleaning probably has been applied most often to ceramic heat wheel installations.

Industrial recovery control

Overheating recovery systems must be avoided to avert possible damage. Air-to-air exchangers generally are rated for service at temperatures lower than what the material can actually withstand. This margin of safety, however, can be exceeded if a process gets out of control.

Ductwork can be damaged, metal parts warped, and the cements used for sealing heat exchanger joints can be weakened or ruined if overheating occurs.

Heat transfer fluids in heat pipes often break down when subjected to excessive temperatures.*

Finned heat transfer coils used with the coil runaround cycle can also be damaged from overheating. Likewise, damage can occur to heat wheel assemblies and all types of heat wheel exchange media.

The necessity of avoiding overheating is not unique to heat recovery installations.

*Heat pipe assemblies applied to air-to-air recovery usually have several rows of tubes or pipes (just like HVAC coils). On many industrial applications, different heat transfer fluids, each suitable for specific temperature ranges, are used in succeeding rows. Therefore, it is important that the unit be installed with the inlet side facing the oncoming exhaust stream.

Sophisticated and complex monitoring and control systems often are applied to industrial processes to achieve maximum production with minimum waste while protecting expensive equipment. By contrast, the controls applied to air-to-air recovery systems usually are relatively simple and straightforward in application.

Overheating of the exchanger device can occur in a number of ways, such as:

• Run-away heat source—if the process equipment overheats.

• Closing off dilution air supply air on installations where high temperature exhaust is mixed with outdoor air. (This is done to lower exhaust temperature to an acceptable level for a given heat exchanger design or to control the supply air temperature on a process to space heating application.)

• Failure of supply air fan.

• Fires caused by particulate matter or flammable vapors. (These may also result from any of the three preceding occurrences.)

Whether or not heat recovery is involved, high limit temperature controls should be installed on all heat producing equipment, such as furnaces, ovens, boilers, etc. A second limit control installed in the exhaust stream ahead of the heat exchanger is recommended to specifically protect the recovery system.

Exhaust fans should not be operated when supply air fans are inoperative. Both should be interlocked so that the exhaust fan cannot operate by itself.

Air flow switches should be installed in the supply air stream. These will warn of supply air failure when the fan motor is running, but the fan itself is not due to a broken belt or some other cause.

Exhaust bypasses are recommended on process applications. A bypass will permit the process to operate in the event of a malfunction in the heat recovery system. It will also permit shutdown of the recovery system for routine maintenance.

Automatic bypass systems can be interlocked with other controls so that the process can continue uninterrupted in the event that it is necessary to shut down the recovery system. Consideration should be given to

Combined makeup air and heat recovery unit installed at a manufacturing facility. (Courtesy of Weather Rite Div., Acrometal Products, Inc.)

installing a separate exhaust fan for the bypass system to provide even greater reliability for the process when this can be economically justified.

Finally, fire dampers may also be required if the potential for the accumulation of flammable materials within the system exists.

Maintenance is important

Heat recovery apparatus should be installed where it is readily accessible for inspection and maintenance. Sufficient working space should be provided to assure that it is possible to perform maintenance and, where required, cleaning operations. An alternative is to install the equipment in such a manner that the entire unit or heat exchanger components can be readily removed for repair or cleaning.

Maintaining a stock of spare parts can be a good investment. Fan belts are easily stocked. A splice kit to repair the drive belt of a heat wheel or an entire belt are other items that can be kept on hand. Some types of air-to-air recovery units have sectional heat recovery modules. Spares can be kept for replacement of damaged sections. Bearings and seals are other items to consider stocking if downtime of a recovery system is a critical concern.

Heat recovery justifiable

Air-to-air heat recovery is justifiable on many HVAC and process installations. Fast payback and high rates of return on invested capital are the norm at today's energy costs. And energy costs are expected to rise.

Many installations—both HVAC and process—are relatively easy to design, install, and operate. Complex installations are also often good investments for the owners. When they do exist, the potential problems discussed in this chapter usually can be satisfactorily resolved if they are considered in the conceptual stage.

Manufacturers have obtained considerable experience in the application of their products. Engineers who have been involved in the design of air-to-air recovery systems also are knowledgeable in the correct application of this equipment. Further, there is a reservoir of knowledge that has been accumulated by many industries that are good candidates for heat recovery relative to what types of materials to use when conveying corrosive or dirty exhaust products.

Cooperation among all concerned will result in good installations that will benefit all involved parties. What is needed is a total commitment to energy conservation by all who are in a position to implement conservation measures.

30

A winning combination: heat recovery and evaporative cooling

A report on available methods for increasing effectiveness of evaporative cooling with multiple-effect techniques for a fortunate alliance of economy and comfort.

RICHARD K. FERGIN, PE,
Sc.D, formerly, NRG Technology Inc., and, Staff Scientist and Principal Investigator, Geoscience Ltd. Currently, Hugh Carter Engineering Corp., La Jolla, California

Heat recovery devices used for air conditioning in sensible heat recovery[1] operations described in Chapters 27, 28, and 29 were: one-pipe evaporator-condenser with gravity liquid return, closed loop evaporator-condenser with gravity liquid return, one-pipe evaporator-condenser with capillary wick liquid return (heat pipe), rotary air-to-air regenerative heat exchanger (heat wheel), and the air-to-air plate heat exchanger.

This chapter focuses on heat recovery combined with evaporative cooling processes.

The author originally became interested in evaporative cooling processes while doing graduate work in New Mexico, where single-effect evaporative cooling is relatively satisfactory. However, as one moves to regions with higher wet bulb temperatures—not necessarily very much higher—comfort diminishes quite rapidly.

In this chapter, extreme outside air con-

[1]Superscript numbers refer to references at the end of the chapter.

ditions have been used in all of the examples to illustrate the increased capacity of multiple-effect evaporative cooling. Three different systems are described, and each is accompanied by a psychrometric chart that demonstrates the inside condition achieved under these deliberately severe conditions.

For the sake of simplicity, an 82°F DB room temperature and a sensible heat ratio of unity are used in all of the examples, but both can be changed to any desired value. The only reason 82°F DB was selected for a room condition was so that simple evaporative cooling could also be shown for comparison, using extreme outdoor conditions.

Most of the work in this chapter is a report of available methods for increasing effectiveness of evaporative cooling with multiple-effect techniques. These are very important, however, since such systems not only can have lower first costs than refrigeration air conditioning systems, but also operating costs can be an appreciable order of magnitude apart.

Figures 30-1, 30-2, and 30-3 are simple schematics of three evaporative air cooling systems and the respective psychrometric process paths for each. Any of the heat recovery devices listed previously can be substituted into the schemes.

The recovery processes, except where noted, are compared on an 80 percent heat exchanger effectiveness basis. The effectiveness method of analyzing a heat exchanger offers many advantages when comparing various types. (Reference 2 contains a very complete discussion of this area.)

Outside conditions of 95° F DB and 78° F WB are used in all of the examples.

A simple evaporative cooling system is shown in Fig. 30-1. Outside air is cooled by water spray to the desired design supply air condition.

The second method, Fig. 30-2, consists of dual evaporative cooler air saturating devices such as sprays, air washers, cooling pads (or towers) combined with a heat recovery unit. Entering outside air (top left) is cooled by a water spray. The cooled air then absorbs heat from a second outside air stream (lower right), via the heat recovery unit. This second air stream is then further cooled by a second

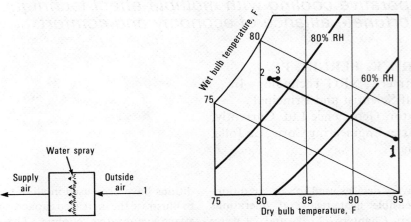

Fig. 30-1. A simple evaporative cooling system and its psychrometric path. The following conditions exist at the numbered points: 1) outside design condition, 95° F DB, 78° F WB; 2) supply air condition, 81.4° F DB, 77.8° F WB; 3) room condition, 82° F DB, 78° F WB.

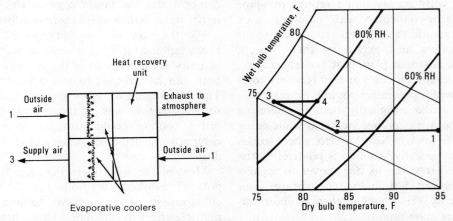

Fig. 30-2. System combining dual evaporative coolers and heat recovery unit and resultant psychrometric path. The following temperatures correspond to the numbered points: 1) outside design condition, 95° F DB, 78° F WB; 2) intermediate condition, 84° F DB, 75.2° F WB; 3) supply air condition, 77° F DB, 75.1° F WB; 4) room condition, 82° F DB, 76.4° F WB.

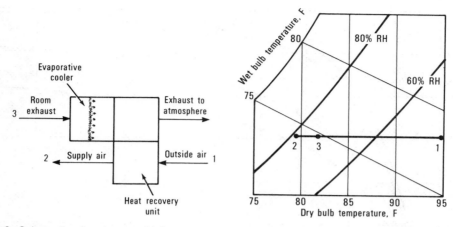

Fig. 30-3. Schematic of system combining evaporative cooler and heat recovery unit and the psychrometric path for the system. The following conditions exist at the numbered points: 1) outside air condition, 95° F DB, 78° F WB; 2) supply air condition, 79.8° F DB, 74° F WB; 3) room condition, 82° F DB, 74.6° F WB.

water spray (lower left) and then introduced into the conditioned space. An alternate method is to use cooled water from the top unit to cool the incoming bottom air.

The system depicted in Fig. 30-3 capitalizes on the room exhaust air and realizes an 80 percent saturation effect and heat exchanger efficiency. It consists of an evaporative cooler combined with a heat recovery unit. Room exhaust air is cooled by a water spray and then absorbs heat from the outside air stream (lower right) via the heat recovery unit. This lowers its temperature prior to introduction into the conditioned space.

By scanning these flow diagrams and the associated psychrometric process plots, the possibility of reducing the temperature of the cooling air without adding moisture until after it has been exhausted from the room becomes evident. This would result in a considerable increase in comfort capacity offered by a simple evaporative system.

The key to this boost in capacity for the system shown in Fig. 30-3 is taking advantage of the lower wet bulb condition of the exhaust room air to help cool the outside air via the heat exchange unit.

There are many additional combinations of dry and wet processes with and without regeneration and cascading that can be examined and optimized. Reference 3 covers several methods that use dry heat exchangers.

An interesting application of these principles may be found in Australia, developed by the Commonwealth Scientific and Industrial Research Organization (CSIRO). (More interest in evaporative processes is evident in Australia than anywhere else in the world, with the possible exception of the Middle East where considerable work has been done on the

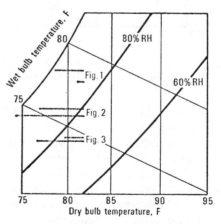

Fig. 30-4. Supply air paths of the three systems depicted in Figs. 30-1, 30-2, and 30-3 compared for 80 and 95 percent effectiveness. At 80 percent effectiveness, the supply air and room conditions are the same for each system as detailed in the respective figure captions. At 95 percent effectiveness (denoted by longer lines), the values change as follows: Fig. 30-1, supply air at 78.8° F DB, 77.8° F WB, room condition, 82° F DB, 78.6° F WB; Fig. 30-2, supply air, 74.3° F DB, 74° F WB, room condition, 82° F DB, 76.1° F WB; Fig. 30-3, supply air, 76.8° F DB, 73.2° F WB, room condition, 82° F DB, 74.4° F WB.

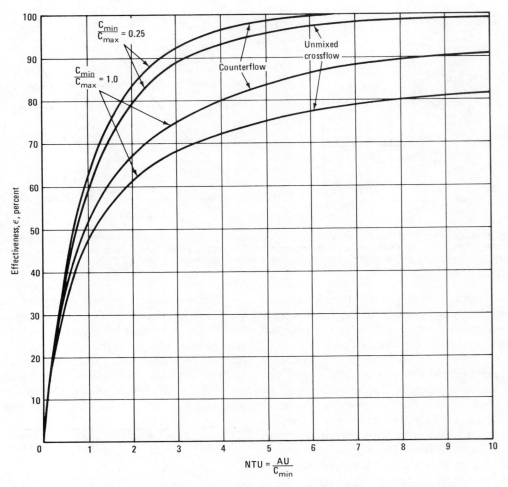

Fig. 30-5. Heat exchanger effectiveness compared with number of transfer units.

dual cooling towers plus heat recovery unit combination.) The Australian device consists of rock (or other suitable heat absorbing materials) beds that are alternately cooled by evaporatively cooled exhaust air, at an air flow rate of 2000 cfm and a water flow rate of 4.6 gpm, followed by sensible heating by outside air. One of the rock beds is cooled and exhausted while the other simultaneously conditions the supply air. Very small temperature changes occur in the beds; the effect is similar to a fast moving heat wheel with an effectiveness of the system illustrated in Fig. 30-3.

The psychrometric chart shown in Fig. 30-4 depicts only the supply air paths for each of the three systems at heat exchanger effectivenesses or saturation efficiencies of 80 and 95 percent. Corresponding increases in comfort and cooling potential are readily apparent.

Effectiveness, ϵ, is defined as the ratio o. the actual heat transfer to the maximum possible. The number of transfer units (NTU) is a measure of the relative size of a heat exchanger. This relationship can be expressed by the following:

$$NTU = AU/C_{min}$$

where

NTU = the number of heat transfer units
A = heat transfer surface area
U = overall coefficient of heat transfer
C = capacity rate, which is equal to the mass flow rate times the specific heat of the fluid on one side of a heat exchanger unit

In several of the processes, the evaporative effects are kept separate from the sensible heat transfer in the heat exchanger core. In heat transfer effectiveness calculations, this corresponds to $C_{min}/C_{max} = 1$ (see Fig. 30-5). C_{min} and C_{max} are determined for the warmer and cooler fluids, respectively, or vice versa, depending on the particular heat exchanger design. If a wetted film is used for evaporation on one side of an exchanger, the ratio is approximately 0.25, which is the ratio of sensible energy to total energy change along the saturation curve in the air conditioning temperature range under consideration.

The curves also show a comparison of cross- versus counterflow design on the number of transfer units required for a given level of effectiveness. The advantage of counterflow sprayed surface design is apparent.

In addition to boosting heat transfer efficiency, sprayed surfaces also reduce the resistance of the sprayed side. These effects are not shown in the illustrations.

The rock bed cooler designed by CSIRO achieves saturation and sensible heat exchange effectiveness exceeding 90 percent.[4] Another of the organization's designs is a cross-flow unit cooler with sprayed surface that reaches an overall effectiveness of 83 percent at normal design flow rates.[5,6] Low cost plastic plate heat exchangers are employed in this design.

Another multiple-effect evaporative cooling system uses all of the processes discussed here plus desiccant wheels to further dry the air. Hygroscopic fluids can also be used in a separate loop to lower the dew point temperature of the processed air. Hybrid units, combining refrigeration and evaporative processes, are the next step in sophisticated design.

The outside conditions used in these illustrations are just about as severe as exist in the United States for wet bulb and dew point temperatures. It is possible, of course, to calculate the air flow per ton requirements for any locality, depending on the individual desired room dry bulb condition.

It should be noted that with increased air velocities, the comfort capacity can be increased considerably. For a look at the effects of air velocity, activity, etc., refer to the Fanger charts in References 7 and 8. Reference 9 deals with the effects of changes in the air velocity only.

References

1. Fergin, R. K., "Heat Recovery Devices for Air Conditioning," Chapter 28.
2. Kays, W., and London, A. L., *Compact Heat Exchangers*, 2nd ed., McGraw-Hill Book Co., New York, N.Y., 1964.
3. Dunkle, R. V., "Regenerative Evaporative Cooling Systems," *Australian Refrigeration, Air Conditioning, and Heating*, January, 1966.
4. "The Australian Rock Bed Cooler," *Australian Refrigeration, Air Conditioning, and Heating*, December, 1971.
5. Pescod, D., "Unit Air Cooler Using Plastic Heat Exchanger with Evaporative Cooled Plates," *Australian Refrigeration, Air Conditioning, and Heating*, September, 1968.
6. Pescod, D., "How Far with Evaporative Cooling," *Australian Refrigeration, Air Conditioning, and Heating*, December, 1971.
7. Fanger, P. O., *Thermal Comfort*, Danish Technical Press, Copenhagen, 1970.
8. *ASHRAE Handbook of Fundamentals*, American Society of Heating, Refrigerating and Air-Conditioning Engineers, Inc., New York, N.Y., 1972.
9. Crow, L. W., "Weather Data Related to Evaporative Cooling," *ASHRAE Journal*, June, 1972.

31

Thermal insulation guide for ducts, equipment, and piping

ROBERT W. ROOSE, PE,
Senior Editor, Heating/Piping/Air Conditioning, Chicago, Illinois

The need for insulation to conserve heat is obvious. However, rising fuel costs and the need to save energy make the selection of insulation for mechanical systems in a building more critical a choice than ever before. Therefore, it is very important to know as much as possible about the available types of insulation, the economical amounts to apply, and how best to apply them. Good insulation of mechanical systems within a building is certainly one of the important methods of conserving energy.

How heat flows

Since it is the purpose of insulating materials to conserve energy by retarding heat flow, it is important to know how heat transfer occurs. When a temperature difference exists between the hot and cold surfaces of any material, heat transfer takes place by means of conduction, convection, and radiation.

Conduction is a molecular transmission of heat; the material in question transmits the heat from particle to particle of its own substance. This conductive heat transfer occurs only between two sections of the material that are at different temperatures; the heat will always flow from the higher to the lower temperature. Time is required for conduction to take place, and the rate of heat transfer varies with the distance between the sections, the temperature difference, and the character of

the material. Poor conductors or insulators permit a very slow rate of heat flow.

Convection is the transmission of heat by the circulation of a fluid or a gas over the surface of a hotter or colder body. The molecules of the moving substance come into close contact with the hotter body and are actually heated by conduction during the period of this contact, but immediately pass on, carrying what heat they have acquired along with them, and cooler molecules succeed them. This circulation may be caused by natural forces or may be produced by mechanical means. Heat transferred by convection depends on the velocity of the moving substance, on the form and dimensions of the body, and on the temperature difference between the moving substance and the body.

Radiation always takes place in straight lines, obeying the same laws as light; so its intensity or amount per unit of surface varies inversely as the square of the distance from the source of radiation to the surface. Radiant heat continues to travel in the same straight line until intercepted or absorbed by some other body. Radiant heat is also similar to light in that it is reflected from various materials, and those substances possessing a high power of radiation have a low reflecting power. The amount of heat emitted by a surface radiating equally in all directions depends only on the nature of the surface, the differ-

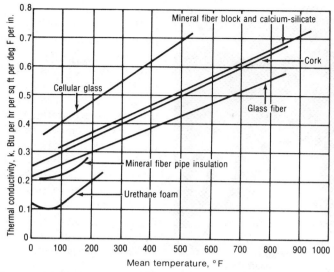

Fig. 31-1. Thermal conductivity is influenced by temperature of application and material.

ence in temperature between the surface and surroundings, and the absolute temperature.

Mass insulations

Thermal insulating materials of the mass type rely on the presence of a large number of small pockets of still air to limit the flow of heat through material. The heat transmission through the air pockets by natural convection is small because of the low conductivity value of still air and the size of the air pockets. Therefore, it is important in the manufacture of insulation to have the air pockets neither too large nor too small. If the air pockets are large, the convective heat flow within them will be too high; if the pockets are too small, the conduction through the solid parts containing the air pockets will offset the insulating value of the pockets themselves. It is for these reasons that the specification of a thermal conductivity for a particular insulating material should include information about its density and the temperature range in which it is most effective. As shown in Fig. 31-1, the temperature of application and the type of insulation material determine the conductivity.

Mass insulations are produced in many forms. Some of these are: rigid boards, blocks, sheets, semi-rigid boards, flexible boards,

Photo courtesy of Johns-Manville

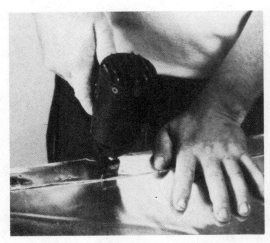

For all types of glass fiber duct, the first step in application is to staple the closure flap (top). Pressure sensitive tape is pressed down firmly to complete the job (courtesy of TIMA).

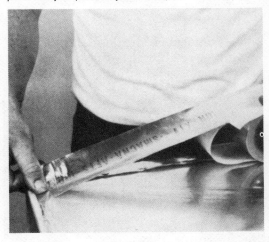

blankets, batts, preformed shapes, tapes, and loose fill.

Reflective insulations

Reflective insulations restrict heat transfer by radiation since the surfaces have a high reflectivity and a low emissivity. These insulations can be a single sheet or multiple layers of metal foil. In a single sheet of metal foil, most of the impinging radiant energy is reflected; only a small amount passes through the reflective layer by conduction to be emitted from the back.

The ability to reflect heat varies with the temperature in accordance with the laws of radiation. The reflective capacity has, as its counterpart, the ability to emit heat slowly from the surface away from the heat source. The two together represent 100 percent; if the reflectivity of a material, such as bright aluminum foil, is 95 percent, then its emissivity is 5 percent. Thus, if foil is mounted on some nonreflective material that first receives the heat, its low emissivity is just as effective as its high reflectivity to heat approaching its exposed face by radiation.

Reflective insulations are not commonly used as the primary insulation for mechanical systems. However, by applying bright coverings over insulation on hot surfaces, a low radiant loss will occur because of the low emissivity of the surface.

Insulation selection factors

When considering the selection of an insulation system, it is the cost of the completed system that is of primary importance, not merely the cost of the insulation alone. Naturally, the first cost of any insulation system is important, but it should not be the major factor in the selection process. The life of the system and the benefits that will be derived from the insulation for the life of the system must be considered. In all cases, the type of insulation to be used should be decided during the design or retrofitting stages, because afterthought insulation is invariably more expensive to apply.

Numerous insulation materials are available, and the choice of one for a particular application is inevitably a compromise dependent on cost and many other factors, which may include:

- Conductivity of mass insulation.
- Effective temperature range.
- Density.
- Ease of application.
- Cost.
- Resistance to combustion.
- Resistance to moisture.
- Resistance to damage and deterioration.
- Resistance to distortion and shrinkage.
- External finish.
- Corrosivity, odor, and health hazards during installation.
- Ability to support a surface finish.

- Ability to prevent vapor condensation on surfaces having a temperature below the dew point of the surrounding atmosphere.

Which insulation to use?

Maximum temperatures that insulation materials can withstand serve as a common basis for determining the type to use for various applications. Since temperature is the first element to consider in designing the mechanical system insulation requirements, it is a practical and logical consideration.

High temperature insulations may be applied where temperatures between 200 and

First step in installing glass fiber pipe insulation is to slip it over the pipe (top). The closure flap is stapled down after the insulation is in place (courtesy of TIMA).

2300° F are to be experienced. High temperature insulation is used on heat exchangers, boilers, steam and process piping, high temperature water systems, and stacks.

Low temperature insulations range from 200° F down to as low as −400° F. Naturally, some insulation materials will be applicable in both ranges, depending on their composition. Since heat flow is inward in low temperature applications, condensation on the surface or in the insulation is a major concern. If water is permitted to condense in the insulation, it will reduce the insulating value of the material. Therefore, it generally pays to install thicker insulations for low temperatures than for high ones. If water vapor cannot be prevented from entering the insulation by its thickness, the insulation should have good moisture resistance, or a good vapor barrier should be applied. Resistance to vapor absorptivity in a low temperature insulation is expressed as vapor permeability, which is measured in terms of the amount of water vapor that passes through one inch thickness of the material. An insulation with excellent moisture resistance may not possess the highest insulating qualities; so selection must result in a compromise.

High temperature insulations

Mineral wool insulation has a recommended temperature limit of 1200° F in the form of blankets and batts. Mineral wool consists of fibers formed from fused limestone or furnace slag. The fibers are bonded with an asphaltic compound and molded to specified shapes. Also, a flexible covering made of loose mineral wool may be used, or slabs may be secured to metal lath or wire netting with an outer casting. The recommended temperature limit for mineral wool preformed in blocks is about 1900° F.

Rock wool insulation is formed from molten rock of a siliceous nature into short fibers, which are bonded into slabs, pipe covering, and flexible blankets. Limiting temperatures are 450° F for nonbonded rigid rock wool material and 1400° F for loose fill.

Glass fiber insulation has a maximum operating temperature of 850° F. It is produced

Lightweight glass fiber duct sections are lifted into place and connected to other sections (courtesy of TIMA).

by blowing steam through streams of molten glass. The composition of the material and the fiber diameters determine the type of service, temperature, and user specifications for which various types of glass fiber insulation can be recommended. Semi-rigid boards can be as thick as 8 in.

Glass wool insulation is available for service to 1000° F. The flexible type in rolls is easy to apply over irregularities. Moisture absorption of these insulating glass wools is low.

Cellular or foam glass is produced by grinding glass to a very fine powder that is passed through high temperature ovens to cause the cells to bond. A rigid, close-cell insulation permits cutting to flat or curved sections. The temperature limit for cellular glass is 1200° F.

Low temperature insulations

Polyurethane is a two-component synthetic resin material. The two component plastics may be mixed at the site or in the manufacturing plant. Carbon dioxide is used to expand the plastic and the resins to produce a tough, cellular material of good insulation and mechanical properties. Limiting maximum temperature is 200° F, while the minimum is −60° F. The use of this insulation on pipes, tanks, and vessels may present a fire hazard unless coated with an approved thermal barrier.

Polystyrene is a transparent, hard, and relatively brittle material possessing good insulation properties and dimensional stability. It is resistant to dilute acids and alkalis, but it is attacked by many solvents. It is manufactured by incorporating a suitable low boiling point substance in the polymer. Upon heating, the volatile substance boils, and the resultant vapor expands the softened polymer, giving a noncommunicating cell structure. The maximum application temperature is 175° F. Both molded and expanded polystyrene will burn if ignited, but they are self-extinguishing when the flame is removed. Nonflammable types are available.

The mineral fiber type of insulation and the glass fiber type are applicable in the low temperature range down to −60° F. Also, cellular glass, which is impervious to moisture, is a good low temperature insulation. Where joints meet, however, a vapor barrier needs to be used.

Expanded silica is mainly used as a fill, or in molded form, with a protective coating or jacket.

Aluminum foil has a low emissivity and rate of reflection of radiant heat. It is usually applied in such a way that an air space is enclosed between the foil and the hot surface. Improved performance is obtained by use of several layers of aluminum foil with an air space between each adjacent layer.

Insulation for duct systems

Insulation of duct systems is not only necessary to conserve heat from heating ducts but to prevent condensation on duct exterior surfaces in cooling systems. Insulations with vapor barriers are available to provide this needed thermal insulation and condensation protection when applied to the exterior surface of sheet metal ducts.

As suggested in ASHRAE Standard 90-75,

all air handling systems delivering conditioned air (both supply and return) and installed in nonconditioned spaces (−20 to 160° F limits) shall be thermally insulated to provide a minimum resistance of R-6 overall from exterior to interior surfaces. Required insulation thickness can be calculated, or thicknesses in Standard 90-75 may be used.

The rigid fire hazard requirements of the National Fire Protection Association Standard 90A for the installation of air conditioning and ventilation systems must be met for all duct insulations. Although adoption of NFPA Standard 90A is not universal, most building codes require compliance with it.

Ducts can be insulated with duct wrap, liner, or any of three types of self-insulating ducts—rectangular, round, or flexible glass fiber ducts can be used.

Flexible glass fiber blankets (duct wrap) are used to insulate the outside surfaces of sheet metal ducts. Duct wrap with vapor barrier facings is supplied in $1\frac{1}{2}$ and 2 in. thicknesses. Unfaced duct wrap is available in thicknesses ranging from 1 to 4 in. Duct wrap generally is designed for use at operating temperatures from 40 to 250° F, although some unfaced blankets can be used at operating temperatures up to 350° F. A good vapor barrier should be applied over the insulation, especially where the ducts carrying cooled air are installed in unconditioned spaces. Joints and laps in the vapor barrier must be tightly sealed to prevent condensation from collecting in the insulation.

Flat glass fiber duct board is used to fabricate rectangular, nonmetallic duct systems. Boards are supplied in standard 1 and $1\frac{1}{2}$ in. thicknesses in 4 by 8 ft and 4 by 10 ft sizes. Duct boards are furnished with precut or molded male and female shiplap edges or with plain edges. They are designed for low velocity applications of up to 2400 fpm at 2 in. WG static pressure, and at temperatures up to 250° F.

Rigid round glass fiber ducts are used for entire air handling systems and as run-outs from main supply ducts, return ducts, and from mixing boxes to diffusers. Standard 6 ft lengths are manufactured in 4 to 36 in. (inside diameter) sheet metal sizes for use at velocities from 2000 to 5400 fpm at 2 to 8 in. WG, depending on the product. Sections are factory finished with or without male and female shiplap ends for joining sections. The ducts are jacketed with glass reinforced aluminum foil vapor barriers and are designed for use at temperatures up to 250° F. Semi-round ducts can also be fabricated from duct board.

Duct liner is used to insulate the inside of sheet metal ducts during fabrication. The surface exposed to air flow is designed to withstand erosion and minimize friction loss. The glass fiber liner is designed for use at temperatures up to 250° F and at velocities of 2000 to 6000 fpm, and should not delaminate. Liner is supplied as flexible rolls or as rigid boards in thicknesses from $\frac{1}{2}$ to 3 in.

Most glass fiber air handling insulation products also provide sound attenuation by helping reduce noises associated with the operation of equipment and rushing air.

Preinsulated flexible ducts should also be considered. These ducts have the insulation as an integral part of the duct and are used as connectors—usually where sharp bends and offsets preclude the use of rigid duct. Flexible ducts are manufactured with or without wire and vinyl air barrier cores and have exterior vapor barriers. Sections come in 7 and 25 ft lengths, the latter compression packed. Flexible ducts are manufactured in 4 to 18 in. (inside diameter) sheet metal sizes.

Methods and materials for applying duct insulations vary from product to product. Therefore, manufacturers' application instructions always should be followed. The most common methods are noted below.

Common duct application methods

Duct wrap is attached to sheet metal ducts with adhesive, mechanical fasteners, outward clinch staples, wire, or tape. Combinations of these materials also are used.

Duct liner is installed with mechanical fasteners and adhesives. It should be installed according to the requirements of the Sheet Metal and Air Conditioning Contractors'

National Association (SMACNA) Duct Liner Application Standard, second edition.

Closures are used in the fabrication of glass fiber ducts and to install rigid round ducts and flexible ducts. Closures are the sealing devices that provide structural integrity in a duct system and also block air leakage at seams and joints. When used as part of a UL 181 Class I air handling system, closures must be tested as part of that system. The various closures are not always interchangeable with the glass fiber air handling systems on the market.

Four types of closure systems have passed UL 181 testing with fiber glass products:

• Pressure sensitive tape—the most commonly used closure system.

• Glass fiber and joint mastic—closures formed by stapling glass fiber fabric along duct joints and applying mastic over the tape.

• Thermally activated closures—closures that, when heated, melt into the pores of the duct facing, chemically bonding to it.

• Mechanical extruded aluminum closure strips with channels that slip over the edges of plain duct board and are used to form rectangular duct sections and to connect sections.

Contact adhesives under the closure flaps also are used, as are combinations of systems.

To minimize application problems associated with some pressure sensitive tapes, SMACNA has developed standards to upgrade tape systems—Performance Standard AFTS-100-73 and Application Standard AFTS-101-73. Both are included in the 1975 edition of SMACNA's Fibrous Glass Duct Construction Standards.

All manufacturers of nonmetallic duct board recommend that when the closure system is pressure sensitive tape, only tapes bearing the AFTS-100-73 designation be used. Other closures that are part of UL 181 listed systems and that meet the manufacturer's recommended fabrication practices are acceptable also.

Reinforcement requirements for glass fiber ducts are also contained in the duct construction standards. Schedules for the size and placement of reinforcements, based on duct size, static pressure, and board type, are pro-

Closed cell insulation on chillers, tanks, and piping also serves as vapor barrier to prevent condensation (photo courtesy of Armstrong Cork Co.).

vided. Sheet metal channels or tee bars are specified as reinforcements.

For the first time the use of tie-rod reinforcing with 12 gauge or heavier wire and washers is permitted by the standard—but only for positive pressure systems.

Trapeze hangers with 1 by 2 by 1 in. channels are recommended as supports for glass fiber ducts. Supporting straps should be 1 in. wide and should be made from 22 gauge or heavier material. Rods $\frac{1}{4}$ in. in diameter also can be used instead of strap hangers. Other supports, such as bar joists, ceiling joists, etc., may be used, provided they meet the hanger specifications listed in the duct construction standards. All support systems must be capable of withstanding a load three times the anticipated load.

Table 31-1 shows representative types and characteristics of insulation for several applications. The ranges of conductivities, densities, and temperatures are broad and do not represent any single product; they are shown here to illustrate the varieties of products for various types of ducts.

Insulation for equipment

Insulation for boilers, tanks, chillers, heat exchangers, and breechings is normally in the form of flat or curved blocks, blankets, and

Table 31-1. Representative types and characteristics of insulation for duct systems. The ranges of conductivities, densities, and temperatures are broad and do not represent any single product.

Insulation types	Temperature range, °F	Conductivity, Btu per hr per deg F per sq ft per in.	Density, lb per cu ft	Application
Boards and blankets with vapor barrier on one side	Up to 250° F	0.23 to 0.36	0.75 to 6.0	Ducts—hot or cold
Thermal and acoustical duct liner blankets and boards	Up to 250° F, 4000 to 6000 fpm	0.20 to 0.48	1.5 to 3.0	Ducts—hot or cold—and acoustical treatment
Glass fiber boards, 1 in. thick	Up to 250° F, 2400 fpm, 2 in. WG	0.21 to 0.29	1.5 to 6.0	Rectangular ducts hot or cold
Rigid and flexible preinsulated ducts	Up to 250° F, 2400 to 5400 fpm, 1.5 to 4 in. WG	0.23 to 0.26	5.0	Round ducts

sprayed-on types. The method of securing the insulation may include banding around the insulation on the equipment, the application of a wire mesh over the insulation, and the application of the insulation to various types of welded studs and angle iron supports on the equipment.

In selecting an insulation for equipment, the degree of flexibility of the blocks and blankets and their shapes are of primary concern so as to minimize the need for cutting and fitting on the job. Availability of insulation in flexible materials is increasing to meet this need. As an example, the scoring of blocks increases the flexibility of rigid insulations for easier application to irregular shapes.

Large flat and curved calcium silicate blocks are commonly used on large tanks and boilers. For above ambient temperature conditions, the insulation should have a finish covering with a wire mesh tightly stretched and secured. A finish coating of hard cement is generally used to fill the joints and to seal the wire mesh covering the blocks. Paint may be applied over the cement.

Mineral wool or glass fiber blankets may be used on surfaces with odd shapes. A metal mesh or wire ties are commonly used to hold the blankets in place. A metal jacketing may also be used to protect the blanket.

Spraying of insulation is one method that is used to cover large surfaces. Urethane and polystyrene foams are applied in this way. It is most important that the temperature of the surface onto which the foam is being sprayed will permit the foaming reaction to be completed. A desirable wall surface temperature of 90° F will achieve this desired foaming reaction. If the tank or equipment is in service, the surface temperature may be controlled by varying the temperature of the contents to reach the 90° F outside surface temperature. Obviously, it is essential that the spraying technique be correct and that wind and moisture be at a minimum. A reinforcing wire mesh should be applied on the surface to serve as a bond for the sprayed-on foam.

For temperatures below ambient, the insulation should have a high resistance to moisture; this could require some sacrifice in thermal conductivity. Cellular glass blocks, plastic foams, and cellular rubber shapes are common materials used because of their moisture resistance. If condensation could develop on the surface of the insulation, a vapor barrier must be applied. The vapor barrier should provide the necessary degree of vapor sealing to avoid entry of moisture from the surrounding air.

Vapor barriers include sheets of aluminum, reinforced plastic, metal foils, treated papers, or metal jacketing. Fastening may be accomplished by using bands, or by sealing the joints with tape. Coatings of asphaltic or resinous materials may also be used. Damage to the vapor barrier during application must be prevented. If a vapor barrier is punctured, its effectiveness is lost, and therefore extreme care must be exercised on the job.

For dual temperature service where equipment is alternately hot and cold, the insulation and the vapor barrier seal must be carefully selected for withstanding expansion and contraction and still maintain the vapor seal.

Representative types and characteristics of insulation for several applications are given in Table 31-2. The ranges of conductivities, densities, and temperatures are broad and do not represent any single product; they are shown here to illustrate the varieties of products for various types of equipment.

Table 31-2. Representative types and characteristics of insulation for equipment. The ranges of conductivities, densities, and temperatures are broad and do not represent any single product.

Insulation type	Temperature range, °F	Conductivity, Btu per hr per deg F per sq ft per in.	Density, lb per cu ft	Applications
Urethane foam	−270 to 225	0.11 to 0.14	2.0	Tanks and vessels
Glass fiber blankets	−270 to 450	0.17 to 0.60	0.60 to 3.0	Chillers, tanks (hot and cold) process equipment
Elastomeric sheets	−40 to 220	0.25 to 0.27	4.5 to 6.0	Tanks and chillers
Glass fiber boards	Ambient to 850	0.23 to 0.36	1.6 to 6.0	Boilers, tanks, and heat exchangers
Calcium-silicate boards, blocks	450 to 1200	0.22 to 0.59	6.0 to 10	Boilers, breechings, and chimney liners
Mineral fiber blocks	to 1900	0.36 to 0.90	13.0	Boilers and tanks

Table 31-3. Representative types and characteristics of pipe insulation. The ranges of conductivities, densities, and temperatures are broad and do not represent any single product.

Insulation types	Temperature range, °F	Conductivity, Btu per hr per deg F per sq ft per in.	Density, lb per cu ft	Applications
Urethane foam	−400 to 300	0.11 to 0.14	1.6 to 3.0	Hot and cold pipes
Cellular glass blocks	−350 to 500	0.20 to 0.75	7.0 to 9.5	Tanks and piping
Glass fiber blanket for wrapping	−120 to 550	0.15 to 0.54	0.60 to 3.0	Piping and pipe fittings
Glass fiber preformed shapes	−60 to 450	0.22 to 0.38	0.60 to 3.0	Hot and cold pipes
Glass fiber mats	−150 to 700	0.21 to 0.38	0.60 to 3.0	Piping and pipe fittings
Elastomeric preformed shapes and tape	−40 to 220	0.25 to 0.27	4.5 to 6.0	Piping and pipe fittings
Glass fiber with vapor barrier jacket	−20 to 150	0.20 to 0.31	0.65 to 2.0	Refrigerant lines, dual temperature lines, chilled water lines, fuel oil piping
Glass fiber without vapor barrier jacket	to 500	0.20 to 0.31	1.5 to 3.0	Hot piping
Cellular glass blocks and boards	70 to 900	0.20 to 0.75	7.0 to 9.5	Hot piping
Urethane foam blocks and boards	200 to 300	0.11 to 0.14	1.5 to 4.0	Hot piping
Mineral fiber preformed shapes	to 1200	0.24 to 0.63	8.0 to 10.0	Hot piping
Mineral fiber blankets	to 1400	0.26 to 0.56	8.0	Hot piping
Glass fiber: field applied jacket for exposed lines	500 to 800	0.21 to 0.55	2.4 to 6.0	Hot piping
Mineral wool blocks	850 to 1800	0.36 to 0.90	11.0 to 18.0	Hot piping
Calcium-silicate blocks	1200 to 1800	0.33 to 0.72	10.0 to 14.0	Hot piping

Insulation for piping systems

Proper selection of insulation for piping systems within the building must take into account not only the thermal properties but the mechanical properties as well. Also, the chemical properties of the insulation must be considered to be sure the piping materials will be compatible with the insulation. Basically, insulation installed on steel piping should be neutral or slightly alkaline, and that installed on aluminum should be neutral or slightly acidic. However, these are very broad guides, since chemical attack may result if salts or other components leak out of the insulation.

The types of insulation that are available for piping system applications are glass fiber, cellular glass, calcium silicate, mineral wool, and rigid urethane foams. Representative types and characteristics of various insulation materials are presented in Table 31-3. The ranges of conductivities, densities, and temperatures are broad and do not represent any single product; they are shown here to illustrate the varieties of products for various applications.

Recommended thicknesses of thermal in-

Chiller equipment and piping. Pipe is covered with glass fiber insulation with canvas and metal jacketing. Urethane foam blanket insulation is used on tanks (courtesy of Johns-Manville).

Pipe size, in.	Mineral fiber Above 35 F	Mineral fiber 35 F to 0 F	Mineral fiber Below 0 F to -30 F	Cellular glass Above 35 F	Cellular glass 35 F to 0 F	Cellular glass Below 0 F to -30 F	Flexible unicellular Above 35 F
Under 1½	1	1½	1½	1½	2	2½	½
1½ to 3	1	1½	2	1½	2	2½	Not to be used
3½ to 5	1½	2	2½	1½	2½	3	Not to be used
6 to 10	1½	2	2½	2	2½	3	Not to be used
Over 10	1½	2½	3	2	3	3½	Not to be used

Fig. 31-2. Recommended insulation thickness for low temperature pipe insulation from the General Services Administration.

sulation for hot piping have been increased about 50 percent in the revision of the General Services Administration Guide Specification PBS 4-1516. The specification applies to insulation for piping within buildings built or maintained by GSA. These thicknesses generally conform to those determined by the ECON method developed by the Thermal Insulation Manufacturers Association (TIMA). The accompanying Figures 31-2 and 31-3 for recommended pipe insulation thickness have been reproduced from PBS 4-1516. Copies of the specification may be obtained from GSA, Washington, DC 20415.

The forms of insulation for piping applications consist of preformed sections, wrapping blankets, and insulating tapes. Adhesives with banding and/or jacketing are common methods of attaching the insulation. Before applying the insulation to the piping system, all connections should be inspected for leaks. Also, it is suggested that draining of the piping during installation should be avoided to eliminate the possibility of moisture entering the insulation. Lengths of pipe insulation are

Insulating cement is troweled into place. In this instance, the insulation is being applied to a breeching in a utility shaft (courtesy of TIMA).

Pipe size, in.	Mineral fiber Up to 450 F LP	Mineral fiber Up to 450 F MP	Mineral fiber 451 to 1000 F LP	Mineral fiber 451 to 1000 F MP	Mineral fiber 451 to 1000 F HP	Calcium-silicate and thermal pipe covering LP	Calcium-silicate and thermal pipe covering MP	Calcium-silicate and thermal pipe covering HP
Under 2½	1½	2	1½	2	3	2½	3	4½
2½ to 3	2½	3	2½	3	4	4	4½	5½
3½ and over	3	3½	3	3½	4½	4½	5½	7

KEY: ■LP. Low pressure steam is up to 15 psig, or temperatures to 250 F.
■MP. Medium pressure steam is from 16 to 75 psig at temperatures from 251 to 320 F.
■HP. High pressure steam is from 76 to 200 psig at temperatures from 321 to 400 F.

Fig. 31-3. Recommended insulation thickness for steam and hot water piping systems from the General Services Administration.

available to provide the most advantageous value for field application. This optimum length is generally 2 to 3 ft, which eliminates the need for frequent cutting as the piping changes direction. Pipe fittings may now be insulated with preformed shapes that provide a snug fit; however, where odd shapes, such as valves, are to be insulated, wrapping tapes may be used to achieve the desired results.

Pipe insulation should have true concentric cylindrical surfaces, and it should have the necessary strength to withstand considerable handling. Insulation materials that have good tensile strength in all directions will meet this requirement and are also more suitable for cutting and fitting into the desired finished shapes. If the ends of the pipe insulation are not square and true, the gaps must be plugged with insulating cement or the end recut to fit. When the insulation is secured by bands or wires, the material must resist the tendency to crack along the wire or bands, because a strapping tool will exert a 600 to 800 lb tensile pull on a strap to draw the joints up tight.

Expansion and contraction of piping can cause serious damage to thermal insulation of preformed shapes and the coverings. Most high temperature insulations shrink as the temperature rises, while the metal pipe expands. Therefore, some provision must be included to allow for these dimensional changes. If cracks in the insulation occur with these temperature changes, the size of the crack is not important. The problem that must be prevented is that water or moisture may enter the crack and cause the insulation to lose its insulating value. Therefore, it is imperative that expansion joints in the insulation be provided under these conditions.

Insulations of the flexible type that are slit for ease of installation on piping should be sealed at the butt joints with an adhesive and/or a sealing tape. Pipe fittings and valves should be insulated with the same material in sheet form of the same thickness. The joints should be sealed with an adhesive in the same manner.

Flexible blanket pipe insulation may be covered with metal jacketing that locks along the longitudinal seam when the jacket is drawn up tight. Other types of insulations

that also rely on the jacketing for long-term protection of the insulation include scored blocks and preformed shapes.

Metal jacketing can be a part of the insulation as it is delivered from the manufacturer, or it can be applied on the job. Corrosion resistance of metal jacketing is important if used in corrosive atmospheres. The most corrosive services may require stainless steel jacketing. Another jacketing material commonly used is asphalt saturated asbestos felt that is reinforced with a glass threaded fabric.

Plastic jacketing may be used in applications where ambient temperatures and other environmental conditions will not affect the life of the material.

Pipe elbows may be jacketed in metal or in plastic. Bands and tapes used with adhesives permit the connection of the jacket to the insulation of the piping for the prevention of heat loss and entry of water.

The insulated piping should be supported by hangers and metal protection shields. For piping 2 in. and larger, an insert consisting

Typical equipment room systems scene shows hot and chilled water piping and hot and cool air ducts. Chilled water lines are covered with glass fiber pipe insulation and all-purpose jacket; hot water pipe is covered with calcium silicate with canvas covering. Air ducts are insulated with a glass fiber duct insulation 500 and 800° F (courtesy of Johns-Manville).

of a piece of mineral fiber, cork, or wood block should be placed between the shield and the piping. The insert should be shaped to fit not less than half the pipe circumference. Protection saddles prefabricated to include the insert material and a vapor barrier may also be used. Flexible cellular types of insulation should be protected from compression at all pipe hanger locations by using compression resistant insulation inserts, of the type mentioned, and protective metal shields.

The chapter by Robert Curt of York Research Corporation, which immediately follows this chapter provides a comprehensive analysis of how to determine the best insulation thickness for piping systems.

The author wishes to thank the Thermal Insulation Manufacturers Association (TIMA) for their help and cooperation in providing information for the duct insulation section of this chapter.

Bibliography

1. *Duct Manual—Fibrous Glass Construction for Ventilating and Air Conditioning Systems*, Sheet Metal and Air Conditioning Contractors' National Association, Inc., Vienna, Va.
2. *Duct Liner Application Standard*, Sheet Metal and Air Conditioning Contractors' National Association, Inc., Vienna, Va.
3. *Standard 90A, Standard for the Installation of Air Conditioning and Ventilating Systems and Other than Residential Types*, National Fire Protection Association, Boston, Mass.
4. *Public Buildings Service Guide Specification, Section 1516, Thermal Insulation (mechanical)*, General Services Administration, Washington, D.C.
5. *Thermal Insulation*, by J. F. Malloy, Van Nostrand Reinhold Co., New York, N.Y.
6. *ASHRAE Standard 90-75, Energy Conservation in New Building Design*, American Society of Heating, Refrigerating and Air-Conditioning Engineers, New York, N.Y.
7. *ECON, A Method for Determining Economic Thickness of Thermal Insulation*, Thermal Insulation Manufacturers Association, Mt. Kisco, N.Y.

32

Economic thickness for thermal piping insulation above ground

Nomographs and calculation procedures are explained.

ROBERT J. CURT,
Assistant Manager, Engineering and Applied Research, York Research Corp., Stamford, Connecticut

The increase in energy costs and the sophistication of today's designers are causing more emphasis to be placed on thermal insulation as an energy conservation technique. Energy conservation techniques are investments and as such are expected to provide the investor with a return on his investment. If a plant owner can realize a 15 percent profit on a product he sells, why should he put his money into an energy conservation technique that returns 10 percent per year? He would be better off expanding his plant capacity to produce more of the product. The problem, then, is determining the return and comparing this to other available investments. Or, since insulation is often a predetermined necessity, how much insulation should be applied to provide an owner with his required rate of return?

In the past, the cookbook approach to specifying thermal insulation thickness was the accepted method for many designers—and justifiably so. It was not practical for an engineer to take great pains in custom designing insulation thicknesses for each pipe size and narrow temperature range when energy was cheap and plentiful. Additionally, since insulation was the last installed item, it often

was specified based on space allowances in previously designed pipe racks or fell victim to budget overruns in the later stages of construction.

Recognizing the value of thermal insulation to industry as a profitable investment, and to both industry and the nation as an energy conservation technique, the Federal Energy Administration sponsored the development of a manual for the calculation of insulation thickness based on the relevant financial criteria. The contractor for this project was York Research Corporation, and the resulting manual is called *ETI—Economic Thickness for Industrial Insulation.*

The ETI manual consists of eight sections. The first two are:

1. Introduction
2. Economics of Insulation

The remaining six sections are:

3. Cost of Energy
4. Cost of Insulation
5. Economic Thickness Determination
6. Condensation Control
7. Retrofitting Insulation
8. Sample Problem

An appendix follows in which the mathe-

(Photo courtesy of Johns-Mansville)

matical derivations, input parameter sensitivity analysis, and piping complexity factor derivations are presented.

Sections 3 through 7 are the working sections of the manual. Each contains a worksheet that provides the step-by-step procedure for the calculation. All calculations are accomplished with nomographs, sparing the user the necessity of performing the math.

Sections 3, 4, and 5 are used to find the economic thickness. Section 6, Condensation Control, allows the user to find the thickness required to prevent condensate formation for cold applications. This thickness can then be compared to the economic thickness, and the greater of the two thicknesses should be specified.

Section 7, Retrofitting Insulation, allows the user to determine whether he is justified in adding more insulation over existing insulation, or if he should insulate bare surfaces in an existing plant.

Perhaps the best way of explaining how the manual works is to present an example problem, along with the actual worksheets and nomographs as used to solve for the economic thickness with the manual. The example progresses as follows.

First, using Section 3, the cost of heat in the steam line is found. As seen on the Cost of Heat Worksheet, not only the fuel cost is used, but also the boiler efficiency, expected annual increase in fuel cost, operation and maintenance expenses, and the capital equipment cost associated with each million Btu in the line. The total average cost of heat over the life of the insulation project is $1.63 per million Btu.

Next, using the insulation installed prices quoted by the contractor, the insulation cost slope, m_c, is arrived at. By determining the cost slope value, in terms of dollars per inch thickness per linear foot, the user is spared the need for providing the actual cost for each thickness. The cost slopes for both single and double layer thicknesses are found because at this point we do not know in which layer range the economic thickness will fall. The cost slope is determined by a simple formula:

$$m_c = PC \, (P_2 - P_1 / L_2 - L_1)$$

where

m_c = Incremental cost for piping, dollars per linear foot per inch thickness; for vessels, dollars per square foot per inch thickness

P_c = piping complexity factor, which is contained in a table in the manual

L_1 = thickness of lower end of layer range, in.

L_2 = thickness of upper end of layer range, in.

P_1 = installed price for L_1

P_2 = installed price for L_2

In the sample problem, it cost more per inch to install the second layer of insulation than the first ($2.72 versus $4.57).

Having found the two necessary costs (the cost of heat and the cost of insulation), Section 5 is then used to determine the economic thickness. Economic thickness is defined as that thickness which produces the lowest total of lost heat and insulation costs over the given life of the insulation.

The procedure is first followed using the single layer insulation cost slope, m_{c_1} = $2.72, (Step 7 on the Economic Thickness Determination Worksheet). However, the anwer obtained by using this cost (8 in.) is beyond the range of a single layer application. Therefore, we return to Step 5 on the worksheet and repeat procedure from that point on using the double layer insulation cost slope, m_{c_2} = $4.57. Using this cost slope, the answer becomes 6 in., which is within the double layer thickness range and is therefore correct.

Financial design criteria

The method of calculating optimal economic thickness, as presented in the ETI manual and

shown in the example problem, arrives at that thickness which produces the minimum cost over the assigned insulation project life (n_1), which is a variable input. The concept to remember for choosing an insulation project life when using the manual is that the last increment (usually $\frac{1}{2}$ in.) of insulation in the economic thickness will just recover its cost in energy savings in the assigned n_1 period, at the assigned cost of money, which is the i_3 term on the Economic Thickness Determination Worksheet.

By varying the n_1 and i_3 terms, therefore, the last $\frac{1}{2}$ in. of insulation in the economic thickness can be designed to pay for itself in a specific time (discounted payback) or return a certain rate (return on investment) over the life of the insulation. This latter case is also the life-cycle costing procedure, as the lowest cost solution for the service life of the insulation is found.

The life-cycle approach was used in the example problem where the n_1 term was 15 years—the entire expected service life of the insulation. The i_3 value used was 15 percent, the required return on investment rate for the utility. However, had we been dealing with a situation that required the investment to meet a payback criteria, possibly an ammonia plant, the n_1 and i_3 values would be different. In this case, the n_1 term would be the payback period (e.g., 3 years), and the i_3 rate would be the simple cost of money (about 10 percent) not including any profit percentage.

The same principles applied to insulation—life-cycle costing, payback, etc.—are being applied more often by engineers and management in evaluating investment alternatives during this period of inflation and high money and energy costs.

The manual is available from the Federal Energy Administration, Office of Industry Programs, 12th St. & Pennsylvania Ave. N.W., Washington, DC 20461.

Sample problem

High temperature steam pipe

A new electric utility installation will have a 1000 ft long, 16 in. nominal pipe size high pressure steam line leading from the steam generator to the turbogenerator building. The steam temperature will be 1050°F. The line will be outside, where the average ambient temperature is 62°F.

The utility will have a net continuous output capacity of 565 megawatts. Total electrical generation per year is expected to be 2772 million KWH (plant factor of 56 percent), and the line is expected to be in service 8500 hours per year. Total expected capital investment for the steam plant (boiler, piping, condenser, etc.) is $55 million. The first year fuel cost will be $18 per ton of coal, which has a heating value of 12,500 Btu per lb. The plant will have a depreciation period of 30 years; however, the insulation is expected to have a useful service life of only 15 years in the outdoor environment. The boiler efficiency is 92 percent, and it requires 10,100 Btu of fuel to produce each kilowatt-hour of electricity.

The return on investment requirement for the company is 15 percent. The bonds issued by the company to finance the plant pay a 9 percent dividend, with flotation and administrative costs adding another 1 percent to this cost over the 30 year life. Fuel cost is expected to increase an average annual rate of 7 percent.

The insulation to be used on the pipe is calcium silicate with an aluminum jacket. The average of insulation contractor unit price estimates for this insulation on a 16 in. pipe at the time of construction are:

Thickness	Dollars per linear foot
2 in. single layer	11.86
4 in. single layer	17.29
5 in. double layer	27.36
7 in. double layer	36.50

The worksheets that follow display the input data and resulting economic thickness

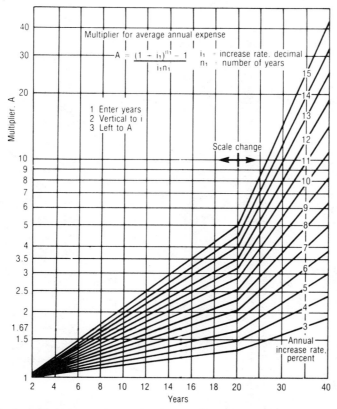

Fig. 32-1. Multiplier to apply to present costs for determining the average annual costs when uniform cost increases occur in future years.

for the above problem, which is 6 in. As shown, the calculation was first made using the single layer incremental cost (m_{c_1} equals $2.72 per in. per linear foot). The thickness using m_{c_1} was in the double layer range; so the procedure was repeated using the double layer slope (m_2 equals $4.57 per in. per linear foot). This incremental cost produced the economic thickness of 6 in., which is in the double layer range and is therefore the correct solution.

Worksheets and nomographs follow

Cost of Heat Worksheet

Step 1, Fig. 32-1.
• Find the multiplier for average annual heat cost, A, using Fig. 32-1.

a. Enter insulation project service life, years. $n_1 = 15$
b. Enter annual fuel price increase, percent. $i_1 = 0.07$
c. Find multiplier A. $A = 1.67$

Step 2, Fig. 32-2.
• Find the first year cost of heat, Ch, using Fig. 32-2. (Fig. 32-2 calculates the cost of heat for coal as a fuel source, but the manual contains nomographs for other fuels as well.)

a. Enter heating value of fuel, Btu per lb. $H = 12,500$
b. Enter efficiency of conversion fuel to heat, percent. $E = 0.92$
c. Enter first year price of fuel, dollars per ton. $P_c = 18$
d. Find first year cost of heat, dollars per million Btu. $C_h = 0.78$

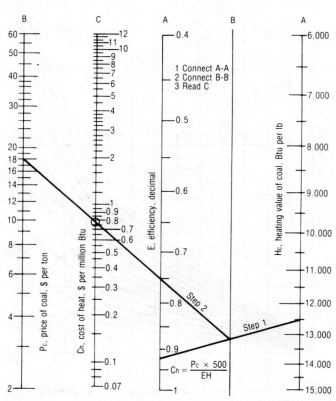

Fig. 32-2. The cost of heat, Ch, for coal as the heat source. Other charts are included in the manual for other fuels.

Step 3, Fig. 32-3.
- Find the average annual heat cost, using Fig. 32-3.
a. Find the average annual cost of heat for purchased steam and electric heat plants (no operating or maintenance costs), dollars per million Btu. $ACh =$
b. Find average annual heat cost for coal, oil, and gas plants (10 percent operation and maintenance costs), dollars per million Btu. $ACh = 1.40$

Step 4, Fig. 32-4.
- Find the compound interest factor, $(1 + i_2)^{n_2}$, using Fig. 32-4.
a. Enter life of facility, years. $n_2 = 30$
b. Enter annual cost of money to finance plant, decimal. $i_2 = 0.10$
c. Find compound interest factor $(1 = i_2)^{n_2}$ $= 17.0$

Step 5, Fig. 32-5.
- Find the annual amortization multiplier for capital investment, B, using Fig. 32-5.
a. Find B using i_2 and $(1 + i_2)^{n_2}$ from Step 4. $B = 0.105$

Step 6, Fig. 32-6.
- Find the annual capital cost of heat, Ck, using Fig. 32-6.
a. Enter expected average annual heat production, Q, millions of 10^6 Btu. (10,000 Btu per KWH $\times$ 10^6 KWH per yr $\times$ 0.92) $\times 10^{-6}$ $Q = 25.76$
b. Enter capital investment in heat plant, millions of dollars. $P_f = 55$

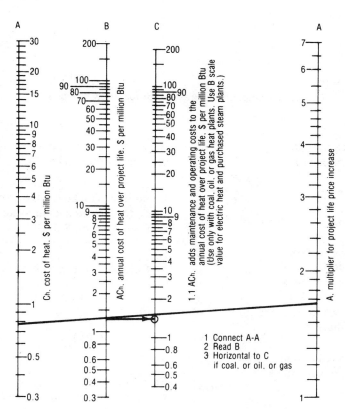

Fig. 32-3. The average annual value of heat cost including operating and maintenance costs at the heat producing facility.

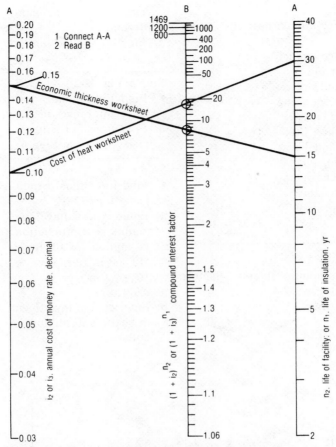

Fig. 32-4. Compound interest factor for use with Fig. 32-5.

c. Find annual capital cost of heat, dollars per million Btu. $C_k = 0.23$

Now proceed to the economic thickness of determination worksheet.

Step 7

- Find the project cost of heat, M. M = $ACh + Ck$

a. From Step 3, we know that ACh including maintenance and operation is, dollars per million Btu. $ACh = 1.40$

b. From Step 6, we know that Ck is, dollars per million Btu. $Ck = 0.23$

c. Find the projected cost of heat, M, dollars per million Btu. $M = 1.63$

Economic Thickness Determination Worksheet for hot and cold systems (if cold system, see Section 8 also).

Step 1

- Calculate mean insulation temperature, °F.

a. $tm = \dfrac{(tp + ta)}{2}$

$\dfrac{(1050 + 62)}{2}$ $tm = 556$

Step 2, Fig. 32-7.

- Enter insulation thermal conductivity

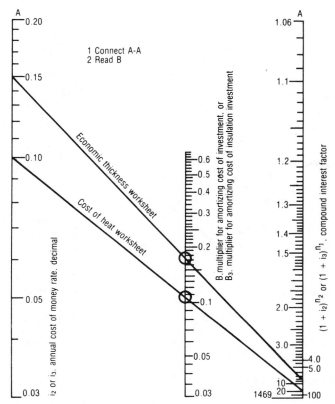

Fig. 32-5. The B multiplier, $i_2 (1 + i_2)^{n2}/(1 + i_3)^{n1}$, for amortizing an initial cost over a period of time with equal increments.

using Fig. 32-7—which is for high temperature insulation materials. Another chart for low temperature materials is included in the manual.

a. Using the mean temperature found in Step 1, find thermal conductivity, k, Btuh per sq ft per deg F per in. thickness. $k = 0.6$

Step 3
• Calculate temperature difference, °F.
a. $\Delta T = (tp - ta)$
    $(1050 - 62)$ $\Delta T = 998$

Step 4
• Enter annual hours of operation, hr.
a. From the sample problem description, we know the plant operates, hr: $Y = 8500$

Step 5, Fig. 32-8.
• Find Ds for a flat insulation installation of Dp for a pipe insulation installation.
a. The example problem deals with pipe insulation, so find the value for Dp on Fig. 32-8. $Dp = 4.3$

Step 6
• Find B_3, the multiplier for amortization cost of insulation investment using Figs. 32-4 and 32-5.
a. i_3, the annual cost of money rate, decimal $i_3 = 0.15$
b. n_1, the life of the insulation, years $n_1 = 15$
c. The compound interest factor, $(1 + i_3)^{n1}$ $= 8.1$
d. The multiplier for amor-

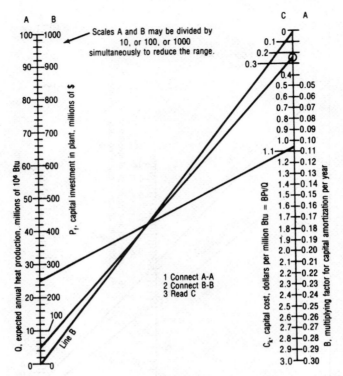

Fig. 32-6. The distribution of the heat production capital cost, Ck over the energy output on an annualized basis.

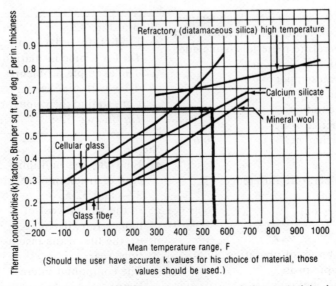

Fig. 32-7. Thermal conductivity factors for high temperature insulation materials. Another chart for low temperature materials is included in the manual.

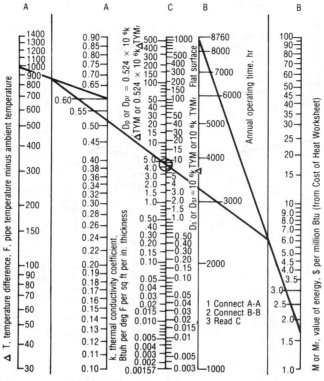

Fig. 32-8. Annual cost of heat lost or gained.

tizating cost of invest-
ment, B_3 $B_3 = 0.171$

Step 7
- Use the incremental cost from Section 6, m_{c_1}
a. m_{c_1} single layer price is determined by $(P_2 - P_1/ L_2 - L_1)$ $(17.29 - 11.86/4 - 2)$ $m_{c_1} = 2.72$
b. m_{c_2} double layer price, dollar per in. thickness per linear ft is determined by $(36.50 - 27.36/7 - 5)$ $m_{c_2} = 4.57$

Step 8, Fig. 32-9.
- Find Zp, the factor for round pipe surface using Fig. 9.
a. For a single layer of insulation, m_{c_1} $Z_{p_1} = 8.5$
b. For a double layer of insulation, m_{c_2} $Z_{p_2} = 5.0$

Step 9
- Calculate kRs
a. $k \times 0.7$ (R_s value of 0.7 is typical). $kRs = 0.42$

Step 10, Fig. 32-10.
- Use Fig. 10 to determine the economic thickness.
a. Use the value found for Z_{p_1} in Step 8, in. $w_1 = 8$

Step 11
- If the economic thickness found in Step 10 is within the single layer range (corresponding to the single layer slope, m_{c_1}, used in Step 7), the thickness is correct. If the thickness is beyond the single layer range, repeat the procedure from Step 7 on, using the double layer slope, m_{c_2}.
a. You would use the value for $Z_{p_2, w_2} = 5$, or 6 in.

Step 12
- If the economic thickness, using the double

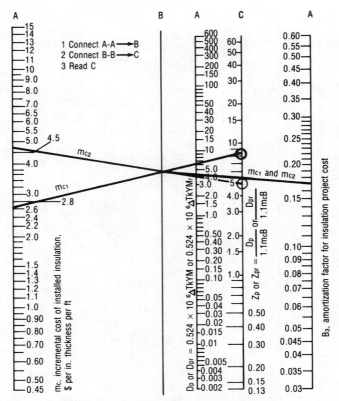

Fig. 32-9. Zp factor for round (pipe) surfaces. Zp includes 10 percent annual maintenance charge for insulation.

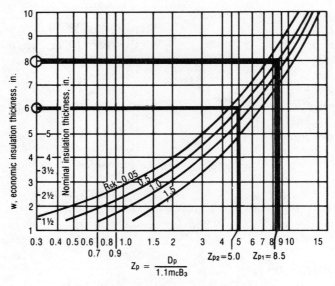

Fig. 32-10. The economic thickness chart.

Photo courtesy of Johns-Manville

layer slope, is in the triple layer range, repeat the procedure from Step 7 on, using the triple layer slope, m_{c_3}.

Note
- It may occur that the economic thickness found with a single layer slope falls in the double layer realm and that the subsequent thickness found with a double layer slope falls in the single layer realm; or that the economic thickness found with a double layer slope falls in the triple layer realm and the subsequent thickness found with a triple layer slope falls in the double layer realm. Should either of these conditions occur, the proper economic choice is the thickest single or thickest double layer, respectively.

33

Thermal insulation for buried piping

ROBERT W. ROOSE, PE,
Senior Editor and

T. E. PANNKOKE, PE,
Engineering Editor, Heating/Piping/Air
Conditioning, Chicago, Illinois

Underground insulated piping systems have been used for many years with varying degrees of success. The earliest known successful venture to supply heat to a group of buildings from a central source via buried pipes was at Lockport, N.Y. in 1877.[1] Some insulated underground systems have been in service for many years. Others, which have had the benefit of modern technology, have deteriorated badly within a few years of installation.

While it is generally more expensive to install piping below grade level than above, buried installations do offer important benefits. Piping and appurtenances are less subject to vandalism. The earth acts as an insulator, so the pipe and its contents are not exposed to as wide a seasonal temperature variation. Therefore, heating and cooling energy may be saved. Also, the possibility of freezing or the solidification of viscous fluids is reduced. At industrial sites, burying some lines makes above ground space available for other services. Finally, burial may be the only feasible design alternative because of esthetic considerations.

Buried and insulated piping systems are used for space heating and cooling and/or process applications via steam, hot water

(either high or low temperature), and/or chilled water or brine from a central plant. These systems also find wide use in industry for transporting viscous liquids, cryogenic liquids, etc.

Federal agency interest

The United States Government, through various federal agencies, is perhaps the largest purchaser of insulated underground piping systems.

Since World War II, many installations have been made to serve various federal facilities. The initial investment was considerable. During the 1950s, system failures became a matter of major concern. Therefore, a Federal Construction Council (FCC) task group was formed in 1957 to determine the reasons for the failures and to develop design and installation criteria that would produce more reliable systems.[2]

The National Academy of Sciences-National Research Council published the findings and recommendations of the FCC in 1958 as Technical Report No. 30, *Underground Heat Distribution Systems.** This

---

[1]Superscript numerals refer to references at end of chapter.

*The Building Research Advisory Board (BRAB) is a unit of the National Academy of Sciences. It undertakes to advance the art and science of building through a broad spectrum of activities. The resolution of specific technical problems is such an activity. Over the years, the BRAB Federal Construction Council has been

report was revised and updated twice—the last time was in 1964.[3] FCC Technical Report No. 39, *Evaluation of Components for Underground Heat Distribution Systems*, was issued in 1960 and revised in 1964.

These two reports provided the basis for the construction specifications of various federal agencies. In 1964, the first interagency specification based on the work of the FCC was published. This is the Tri-Service Specification used by the Army, Navy, and Air Force.

In 1963, FCC Technical Report No. 47, *Field Investigation of Underground Heat Distribution Systems*, was issued.[4] This covered 121 field investigations of 15 different types of buried, insulated heat distribution systems. Both prefabricated and field fabricated systems were covered. The age of these installations ranged from 2 to 46 years.

The specification criteria developed through the efforts of the FCC reversed the

very active in formulating recommendations for solving the varied problems that have been associated with underground heat distribution systems. The purpose of the National Academy of Sciences is to further the use of science for the general welfare of the nation. By the terms of its charter, it is required to act as an official, yet independent, advisor to the federal government. The Academy is not a federal agency, however, and its efforts are not restricted to government activities.

failure trend of the early post WW II period.[2] To take advantage of new developments in materials technology, however, and to reduce costs where lower temperatures and pressures might be safely handled with materials other than steel, another FCC task group was formed to prepare underground heat distribution system design and evaluation criteria based upon current technology and the experience gained through use of the criteria developed previously. The recommendations of the task group may be found in FCC Technical Report No. 66, *Criteria for Underground Heat Distribution Systems*, published in 1975.[5]

An FCC Guide Specification, Section 15705, *Underground Heat Distribution Systems (Prefabricated or Pre-engineered Type)*, has been prepared. This guide specification will be used when a minimum of three systems suppliers have been qualified under the criteria requirements. When issued, it will be used by members of the Federal Interagency Group (which has superseded the Tri-Service Committee) and which currently consists of the three armed services, the General Services Administration, and the Veterans Administration.

Many construction projects are outside the realm of the federal government. However, the technology and the availability of systems that are improved, modified, or developed to qualify for federal contracts can be expected to have a definite effect on designs and construction practices followed on private and other nonfederal projects. Therefore, further reference will be made to FCC reports in the remainder of the chapter.

Since this chapter deals generally with underground insulated systems and not specifically with federal requirements, the reader should be aware that what are made as recommendations in the following text may be requirements for federal installations.

System classifications

Many types of insulated underground piping system concepts are in use. They may be classified in various ways, such as prefabricated, pre-engineered, and field fabricated.

Thermal distribution systems may also be grouped by temperature. Three ranges are generally accepted. These are: above 250° F, steam and high temperature hot water; 180 to 250° F, steam and low temperature hot water; and 35 to 180° F, chilled and dual temperature water.[6]

There are a number of piping materials commonly in use that are restricted by allowable pressure ratings to applications within the lowest or the two lower temperature ranges.

Systems also can be classified by types as shown below:[7]

- Pressure testable conduit systems.
- Nonpressure testable conduit systems.
- Insulating envelope systems.

Because moisture in the form of ground water or pipe leaks has been the largest cause of insulation and pipe failure (by corrosion), emphasis today is on the development and/or use of systems that are either *drainable and dryable or that are capable of confining water to a limited section.*

Pressure testable conduit systems

These systems, sometimes referred to as air gap systems, are drainable and dryable when properly installed. They consist of a carrier pipe, pipe insulation, spacers, and an outer casing pipe or conduit. An annular air space around the insulation provides added resistance to heat flow and also provides the means to leak test the conduit (with pressurized air). It also permits the system to be drained if the conduit or pipe develop leaks (Fig. 33-1).

The outer casing may be of steel, galvanized steel, or cast iron. The steel and galvanized steel conduits are generally covered with either a glass reinforced coal tar enamel having an outer wrap of glass fiber reinforced pipeline felt or with a glass fiber reinforced epoxy resin. Cast-iron casings are not coated.

Pipe insulation is usually calcium silicate or preformed or molded glass fiber. The systems can be designed to handle fluid temperatures ranging from below freezing to 800° F or higher.

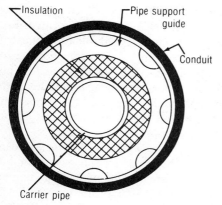

Fig. 33-1. Section of a pressure testable, drainable, and dryable metal conduit.

The carrier pipes are joined by welding, and the joints are covered with split preformed insulation sections. Steel conduits are joined by welding also. Cast-iron ones are connected by sleeves (plain end conduits), bolted together (flanged end conduits), or connected with mechanical joint fittings (mechanical joint conduit).

Steel conduit assemblies are available in 20 and 39 ft lengths. Cast-iron ones come in 13 and 18½ ft lengths.

Pressurized monitoring alarm systems are available that maintain the conduit at a pressure above atmospheric—typically, 5 to 8 psig. The unit signals any loss in pressure if a leak develops or if the casing is accidentally ruptured by other construction. In the case of smaller leaks, the pressure source can restore pressure and prevent water from entering the conduit. Either an air compressor or compressed nitrogen gas may be used.

Manufacturers of pressurized conduit systems also offer prefabricated, pressure testable manholes to facilitate system installation and to provide watertight construction.

Nonpressure testable conduits

These are available in a variety of material combinations to meet the requirements of many types of applications.

Many conduits are factory fabricated and consist of a carrier pipe surrounded by insulation that in turn is enclosed in an outer nonmetallic casing. The area between the carrier pipe and outer casing is completely filled with insulation; and generally the ends of the insulation are sealed with a watertight enclosure, thus limiting the spread of moisture if the pipe or outer covering fails (Fig. 33-2).

Other types of nonpressure testable conduits are field fabricated. These may be a factory engineered system consisting of components specifically manufactured for underground heat distribution components, or they may be concrete trenches or other nonproprietary designs.

Prefabricated conduits of the type shown in Fig. 33-2 can be obtained with a variety of components. Carrier pipes may be made of copper, carbon steel, stainless steel, polyvinylchloride (PVC), fiber glass reinforced plastic (FRP), or epoxy lined asbestos-

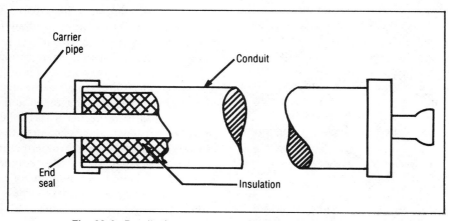

Fig. 33-2. Detail of a water spread limiting conduit is shown.

cement. Outer casings generally are PVC, FRP, or asbestos-cement. Insulations commonly used include foamed polyurethane, calcium silica/asbestos, and preformed foam glass.

PVC carrier pipes are suitable for lower temperature applications, such as chilled water and brine service. The maximum allowable working pressure decreases rapidly above 73° F, and its use above 120° F is not recommended.

Copper, FRP, and epoxy lined cement carrier pipes are commonly used below 250° F. Copper can be used with fluid temperatures up to 250° F. FRP pipe generally is not used above 225° F, and some carries a maximum rating of 150° F. Asbestos-cement pipe has an allowable operating temperature range of 35 to 210° F. Steel pipe is used exclusively on hot water and steam installations operating above approximately 250° F because of the pressures and temperatures encountered.

Stainless steel pipe is selected for any operating temperature when the characteristics of the fluid conveyed require this material.

The insulation selected for prefabricated insulated conduits depends primarily upon the operating temperature of the pipe.

Foamed polyurethane insulation may be used in applications between approximately −320 to 260° F.* Polyurethane resists water penetration. Nevertheless, it must be covered with a vapor barrier of some type. Ground water under a sufficient head may rupture the cells. Damage during construction or pipe flexing due to ground settling after installation may cause cracks to develop in the foam. Also, if the pipe is used for chilled water service without a vapor barrier, the difference in vapor pressure resulting from a pipe temperature lower than the ground water temperature may draw moisture to the pipe.

Composite insulations, consisting of an insulation● suitable for higher temperature

*Discussions with insulated pipe manufacturers on the subject of maximum allowable pipe temperature with polyurethane material and a review of manufacturer's literature indicate that 250° F is an accepted maximum value. At least one manufacturer rates his system at 260° F. Several stated that development of polyurethane derivatives suitable for 300° F or higher was in process. One manufacturer stated that they could supply urethane foam that would withstand 300° F.

service next to the pipe, which in turn is covered with a lower temperature rated insulation, are also used. A combination of calcium silicate or calcium silicate and asbestos covers the pipe and in turn is surrounded by polyurethane. Pipe systems having this insulation construction are rated for temperatures to 450° F and a maximum operating pressure of 500 psig.

Formed cellular glass insulation is also used with prefabricated pipe construction. The insulation itself will withstand temperatures between −450° F up to 800° F. Prefabricated piping with sufficient insulation thickness to protect the outer jacket is available for this temperature range.

Prefabricated insulated pipe lengths vary with the manufacturer, carrier pipe material, and pipe size. Typical lengths are 20 ft for copper, steel, PVC, and FRP carrier pipe sizes up to 12 in. Asbestos-cement carrier pipe sections typically come in 10 and 13 ft lengths with carrier pipe sizes ranging from 4 to 16 in. One manufacturer offers 55 ft lengths with carrier pipe diameters (steel) up to 36 in.

Joining methods

Various methods are used for joining prefabricated pipe sections. For high temperature service, most systems are joined by welding, although one system uses couplings designed to operate at 450° F and 500 psig. Below 250° F, couplings generally are used with steel and copper pipe. The coupling forms a leakproof joint, while it permits sufficient movement at the joint to allow for expansion and contraction within the operating temperature range of the system, often eliminating the need for expansion loops.

Nonmetallic pipes may be joined with couplings; or they may have bell and spigot ends, which are connected with rubber sealing rings; or they may be cemented together, depending upon the carrier pipe material.

Asbestos-cement pipes are coupled together. PVC pipes—bell and spigot types—use sealing rings; plain end PVC pipe is joined with rubber gasketed couplings. FRP pipe with bell and spigot ends is cemented together.

The adhesive needs to cure before pressure can be put on the line. Depending upon the outdoor ambient temperature and the adhesive formula, approximately 20 to 30 hr may be required. To speed up the curing process or to assemble an FRP system at ambients below 45° F, electric heating collars (115 v AC) designed for this application may be used. Heaters cut the cure time down to less than an hour and in some cases to only 10 minutes. Pipe size, FRP material, and adhesive type govern the cure time when heaters are used.

Heating lines and heating/cooling lines are insulated at the joints. (The construction of some prefabricated piping systems joined with couplings precludes the need to insulate joints.) One method utilizes a sleeve that is slipped over the conduit prior to making up the joint, and foam is applied in the field between pipe and sleeve. The ends of the sleeve are sealed with a double wrap of coal tar enamel pipeline wrap. Another method utilizes a sleeve that shrinks when heat is applied to it, forming a watertight covering over the joint insulation.

The joints of insulated chilled water lines generally are not insulated. The heat loss to the ground or the gain from an adjacent hot line is insignificant when compared to the overall ratio of insulated to uninsulated areas.

Tees, elbows, reducers, and other fittings are available in steel, copper, PVC, FRP, and asbestos-cement. PVC fittings generally are not insulated, since they are used in lower temperature service applications for the reasons given in the preceding paragraph.

Field fabricated conduit systems

Field fabricated pipe conduits are nonpressure testable and may or may not have an air space around the pipe. They may be preengineered in that the conduit consists of an assembly specifically designed to house underground insulated pipelines. One type consists of a poured concrete structure formed so that it has a trough for water—either ground water or pipe leakage—to collect and be drained away. Cast-iron pipe supports, rollers, etc. are mounted on the base. Vitrified

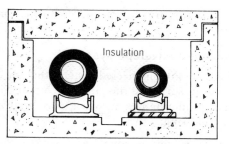

Fig. 33-3. Concrete trench system.

clay side pieces and half-round covers cemented together complete the conduit. The cemented joints are covered with a waterproof mastic to seal the conduit.

Concrete trenches can also be applied as shown in Fig. 33-3. These may be considered mini-tunnels, since they are similar to walk-through tunnel systems. The bottom and sides are formed and poured. The top consists of concrete slabs. After the pipes are installed in the trench, the top is set in place. Tar or other sealing material is used to provide a barrier to ground or surface water entrance at the joint. The top of the trench often is at grade level and serves as a walkway. This facilitates access to the pipe, since the ground does not have to be dug up when checking for leaks, adding a line, replacing a line, etc. It is recommended that the bottom of the trench be set on a vapor barrier and the sides be coated with a mastic material to provide greater resistance to moisture penetration.

Vitrified clay tile and concrete sewer tile also are sometimes used to construct pipe conduits. These and other field fabricated conduits may serve well where drainage is good and the system is installed above the water level.

Years ago, concrete trenches and field fabricated conduits sometimes were completely filled with insulation. However, examination of such systems after they were in operation for a period of time revealed that it was not unusual for the insulation to be wet and the pipes corroded. Therefore, it is recommended that the pipe be insulated and an air space be left between the pipe and conduit. The air space around the pipe permits water to drain from the system, and the heat from the line can dry the insulation if it becomes wetted.

Insulating envelope systems

These are pre-engineered systems that are designed to surround the pipe with a poured in place insulating media, which may be any of the following:

- An insulating concrete.
- A granular hydrocarbon fill—both non-curing and heat curing types are available.
- A treated calcium carbonate fill.
- An insulating granular perlite fill.
- Field installed polyurethane foam.

Insulating concretes are available with two types of aggregates, vermiculite or expanded polystyrene beads. It is the aggregate that provides the insulating value to the concrete.

Figure 33-4 shows the recommended form of an installation. A structural concrete base pad is poured to the desired grade. Precast insulating blocks and drain vents are set on a waterproof membrane that covers the structural slab and lines the trench or forms. The pipes rest on top of the insulating blocks, and the insulating concrete is poured over the slab, pipes, and internal drains. The waterproof membrane is wrapped around the top of the concrete and sealed.

If required, cathodic protection can be achieved by the installation of a continuous ribbon zinc anode parallel to the pipes inside of the concrete.

Internal electrical sensors can be installed inside the internal drain channels to detect and locate leaks.

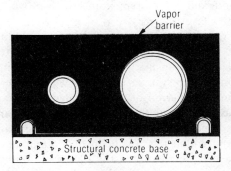

Fig. 33-4. Insulating concrete insulation around pipe.

Added drainage can be accomplished if conditions warrant it by installing an external drain tile.

In addition to providing thermal insulation, insulating concrete provides continuous support and alignment of the piping and resists heavy loads.

Granular hydrocarbon fills are installed in an open trench. A bed of the material is placed in the bottom of the trench. The pipe is installed and more fill is placed around and over the pipe. The material is then compacted to the proper density. The material resists wetting; however, installation of drain tiles along the pipe path may be required, depending on the natural drainage characteristics of the earth and ground water conditions, to avoid build-up of hydrostatic heads that would penetrate the fill.

Natural hydrocarbon mineral material does not have to be heat cured. It is suitable for temperatures between 35 to 500° F.

Heat cured granular fills are derived from petroleum residuals. These are available for various temperature ranges between 150 and 520° F. An asphaltic binder in the fill melts and bonds to the pipe during the curing process. Surrounding the pipe and binder is a sintered zone, and the outer layer of the fill is unaffected by the curing process.

Since these materials are used with metal piping systems, cathodic protection should be installed whenever material selection and soil conditions dictate.

Powdered calcium carbonate, which is treated to resist water penetration, is another insulating material. It is rated for temperatures up to 480° F. Side forms in the trench are all that is required. The material is installed around and below the pipe to the proper dimensions, compacted, and covered on the top with a plastic film. The manufacturer states that due to the hydrophobic properties of the material and its high electrical resistivity—$R = 10^{14}\pi/\text{cm}/\text{cm}^2$—cathodic protection is not required.

Expanded perlite particles are also used as a buried insulation. Powdered perlite is heated to a high temperature, causing it to pop or burst into granular air-celled particles. These particles are mixed in the field with an asphalt binder. A base pad, consisting of the insulating material and binder, is poured in a formed trench. The pipes are laid on the base, tested as required, and then the pad and pipes are primed with a corrosion resisting compound. The installation is then completed by pouring more material over the pipes to the depth of the forms.

Operating temperatures range from 15 to 800° F, and the material is said to have a high electrical resistivity.

Polyurethane foam may be poured or blown around pipes in the field. The systems are ready to be backfilled within an hour after the foam has been placed around the pipes. Forms are required for the froth-pour type of insulation, and a vapor barrier liner and covering generally are used to seal off ground moisture. The direct spray method does not require side forms. Again, vapor barriers are recommended.

Buried tunnels

Buried walk-through tunnels generally are the most expensive way to house underground piping. However, they greatly facilitate piping maintenance, and when properly drained and ventilated, they greatly minimize the development of corrosion on pipe and appurtenances.

Concrete tunnels constructed by forming and pouring are the most expensive. To reduce the cost of tunnels, prefabricated steel tunnels have been developed. These are complete with pipe racks and hangers. The sections are dropped into the excavation and joined together. The steel is factory coated and cathodic protection can be applied to provide additional protection against corrosion.

Large diameter concrete sewer pipe has also been successfully used to construct tunnels. The joints must be sealed and caulked, and as with other tunnel systems, provisions for drainage should be made as job conditions require.

Preliminary design considerations

Many factors must be considered when selecting an underground piping system. Among most important is knowledge of soil conditions along the proposed route, with respect to soil type(s), ground water conditions, corrosivity, soil stability, alkalinity, and drainage patterns. All have a major bearing upon what types of systems should be considered for a specific installation.

FCC Technical Report No. 66 goes into considerable detail on soil types, their relation to ground water conditions, and corrosiveness.[8] Four ground water classifications are set forth in the document; these are: severe, bad, moderate, and mild. Briefly, these classifications are based upon the frequency that the water table will or *will not* be above the bottom of the piping system combined with the length of time that accumulated surface water will remain in the soil around the pipe. Table 33-1 shows the relationship of these two criteria with the four classifications. This report recommends that only drainable and dryable, pressure testable systems be used with the severe classification. Both these systems and water spread limiting ones are considered suitable for the remaining classifications.

A soil survey should be made to determine the conditions along the pipeline route. It should include the determination of the soil

Table 33-1. Underground water conditions classification.*

Classification	Water table above bottom of system	Duration that surface accumulation remains in soil
Severe	Frequent or occasionally	Long
Bad	Occasional or never	Long
Moderate	Never	Short
Mild	Never	Not expected to remain

*This table is based upon criteria set forth in Reference No. 8. Additional information regarding soil types encountered with above classifications, precipitation, and irrigation practices may be found in this reference.

pH, since the acidity or the alkalinity of the soil may adversely affect some materials. For example, asbestos-cement may be harmed by soils having a pH less than 5.5.[9]

The services of a corrosion engineer should be utilized when metal conduits and/or piping are contemplated. Corrosion of buried ferrous structures is very common unless they are properly protected.

Soil corrosivity is most commonly determined by the *soil resistivity test*. This measures the electrical resistance of the soil—the lower the resistance, the higher the corrosivity. Wet soils, organic soils, and soils having a high soluble salt content generally are corrosive.[10]

Corrosion problems can sometimes occur with high resistance soils, so soil resistivity, while a good guideline, does not always give absolute results when higher resistance soils are encountered. Stray electrical currents also may cause corrosion. Stray currents may emanate from direct current transmission lines and electrified rail systems and from industrial operations that require direct current power. It is almost impossible to know whether a stray current problem exists without checking the ground along the job route.

Some system design considerations

Chilled water lines often are installed without insulation. Earth temperatures often are assumed to be relatively uniform year around and low enough not to impose a significant load on the system. Often the latter assumption is correct, particularly at the temperatures normally encountered with comfort air conditioning applications. However, ground temperatures can vary widely on an annual basis as illustrated in Fig. 33-5. In some areas of Florida and Texas, for example, earth temperatures higher than 80° F can be encountered at depths that pipelines are commonly installed.[11]

It is also not unusual to install chilled water lines in the same trench with heating lines.* Earth temperatures in the vicinity of heating

lines will be raised, even though these lines are insulated, when they are in operation. Methods have been developed to determine the heat transfer effects of the earth on buried lines and of parallel buried lines operating at different temperatures to one another.[12] The results can aid in the decision whether or not to insulate chilled water lines.

The anticipated temperature of the earth when heating lines cross buried electrical lines, telephone cables, etc., should be determined to assure that the operation of these facilities will not be impaired. More insulation may be used at these points, or the line elevation may be raised or lowered to provide greater clearance.

Pipe expansion can be handled with expansion loops, expansion joints, and/or ball joints. Prefabricated expansion loop assemblies are available for both pressure testable and non-pressure testable conduit systems. Expansion chambers can be designed into the encasement at corners and loops of insulating concrete installations. Expansion loops may be used with various types of loose-fill insulation, usually with some limitation on the amount of lateral movement that may be accommodated.

Ground water level is not as great a concern with either pressure testable drainable and dryable conduits or water spread limiting conduits as it is with some other types of systems. Some types of soil will retain surface water for extended periods. Also, the trench that the pipe is installed in may disrupt normal drainage patterns and become a catch basin for surface water. Therefore, installation of drain tile along the route may be necessary to assure satisfactory performance of insulating envelopes and some types of field fabricated conduit systems.

Steel and cast-iron conduit systems should be cathodically protected unless a corrosion survey determines otherwise. Cathodic protection should also be applied to piping in loose-fill hydrocarbon type insulations when warranted.

Some other types of insulating fill are said not to require protection due to their high dielectric constant and resistance to water penetration. However, it must be remembered that

---

*Chilled water lines can be and are installed in prefabricated conduit systems with heating lines also. With this form of construction, the chilled water lines should always be insulated.

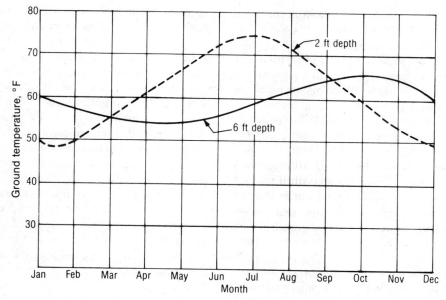

Fig. 33-5. Seasonal variations in ground temperatures. Variation decreases as depth increases. Actual temperatures vary with location, type of soil, etc.

the possibility exists for dirt to contaminate the insulation during installation, thus providing a path for current to flow to the pipe, or a washout could occur later and expose the pipe to ground water. Either condition could create a hot spot where localized corrosion could occur.

Cathodic protection systems should not be designed for built-up areas without consulting with the local utility companies to learn what installations they have. They in turn may have records of other cathodic protection systems. Cooperation among all parties utilizing cathodic protection is essential to assure that all systems will be adequately protected.

Insulation should be specified for all pipes, valves, and other appurtenances in heat distribution manholes. The insulation should be covered with sheet metal jacketing to prevent mechanical damage. Manholes should also be vented. Excessive manhole temperatures have been found to be a major cause of inadequate inspection and maintenance of underground heat distribution systems.[13]

Drainable and dryable systems should be sloped so that water will drain from the entire length of conduits, concrete trenches, etc. Provisions should be made for water removal from manholes should a major inflow occur.

Installation and backfill

All piping should be installed in accordance with applicable codes and industry standards. The conduit or pipe should be firmly supported along its entire length by virgin earth, compacted sand, or insulating fill so as to minimize the possibility of excessive strain due to ground subsidence and the possibility of washouts that could remove protective fill from around the pipe.

Pipe laid in insulating backfill should not rest on bricks, timbers, etc. These provide a moisture path to the pipe, and severe concentration cell corrosion may occur.[9] Remove all temporary pipe supports as backfilling occurs.

All fill type installations must be installed to the proper depth and properly compacted in accordance with the manufacturer's requirements. Care must be taken to avoid contaminating the fill with dirt or other construction debris.

Insulating concrete should be permitted to dry prior to covering with a waterproof membrane and backfilling.

Any damage to protective coatings incurred during shipping or construction must be repaired in accordance with methods compatible to the original coating.

All welding should be done in accordance with the ASA B31.1 Code for Pressure Piping. Manufacturer's requirements for joining coupled pipe should be followed to avoid leaks on low pressure lines and damage to high pressure systems. When plastic pipes are to be joined with solvent welding techniques, the method should be practiced prior to actual pipeline construction. This is no place for on-the-job training!

All lines should be tested for tightness prior to complete backfilling. All joints and fittings should remain exposed until the test is completed. Pressure should not be put on lines before concrete anchors and thrust blocks have cured sufficiently to withstand the stress.

Manufacturers of plastic pipe *do not* recommend testing with air. Hydrostatic testing should be employed with these materials. It is recommended that lines conveying cryogenic fluids also be tested with the fluid to be handled, since at low temperatures, contraction rather than expansion will occur.

Care should be taken during backfilling to prevent damage to protective coatings, conduits, and vapor sealing membranes.

System maintenance

Careful inspection and regular maintenance will reduce the likelihood of premature failure of buried lines. Procedures to be followed should include the following:

• Monitoring of steam pressures or water temperatures at points of send out and use. This will warn of excessive heat loss caused by damaged or wet insulation.

• Periodic patrolling of the line. Burned grass over the route, water or steam coming to the surface, and melted snow indicate problems.

• Periodic inspection of manholes and vaults. Water in the vault or signs of moisture in insulation or in the ends of conduits are further signs of problems.

• Maintaining a regular program of checking the performance of cathodic protection systems. This includes taking anode readings, monitoring rectifier installations, and checking all electrical insulating fittings.

Central heating/cooling plants offer many advantages. They can be more efficient in the use of resources than individual building heating/cooling plants. They provide a means to alleviate the solid waste disposal problem by disposing of combustible trash. And it is easier to control air pollution at a large central installation than at many small ones.

Underground insulated piping systems are an integral part of the central plant concept. In the past, they sometimes have been the weakest link. With the vast amount of past experience to draw from, however, and present technology, buried heating and cooling distribution systems can be installed that will give reliable, economical service for any application that is not beyond the limitations of the system itself.

References

1. *District Heating Handbook*, 3rd ed., International District Heating Association, Pittsburgh, Pa., 1951, Chapter 1, p. 1.
2. Irvin, Jr., L. V., "Federal Agency Specification for Underground Heat Distribution Systems," *Underground Heat and Chilled Water Distribution Systems*, NBS Building Science Series 66, U.S. Department of Commerce, National Bureau of Standards, Washington, D.C., 1975, pp. 73–77.
3. *Underground Heat Distribution Systems*, FCC Technical Report No. 30R-64, National Academy of Sciences—National Research Council (Pub. 1186), Washington, D.C., 1964.
4. *Field Investigation of Underground Heat Distribution Systems*, FCC Technical Report No. 47, National Academy of Sciences—National Research Council (Pub. 1144), Washington, D.C., 1963.
5. *Criteria for Underground Heat Distribution Systems*, FCC Technical Report No. 66, National Academy of Sciences, Washington, D.C., 1975.
6. *ASHRAE Handbook & Product Directory—Systems Volume*, American Society of Heating, Refrigerating, and Air-Conditioning Engineers, Inc., New York, N.Y., 1976. Chapter 14, p. 2.
7. *Criteria for Underground Heat Distribution Systems*, FCC Technical Report No. 66, p. 17.
8. *Criteria for Underground Heat Distribution Systems*, FCC Technical Report No. 66, pp. 4–5, 49-51.
9. Fitzgerald, J. H., "Corrosion Control for Buried Piping," *Heating/Piping/Air Conditioning*, March 1974, pp. 83–88.
10. *Piping-Handbook*, 5th ed, R. C., King, ed., McGraw-Hill, New York, N.Y., 1967, Chapter 9, p. 12.

11. *Criteria for Underground Heat Distribution Systems*, FCC Technical Report No. 66, pp. 64–74.

12. Kusuda, T. "Heat Transfer Studies of Underground Chilled Water and Heat Distribution Systems," *Underground Heat and Chilled Water Distributions Systems*, NBS Building Science Series 66, U.S. Department of Commerce, National Bureau of Standards, Washington, D.C., 1975, pp. 18–41.

13. *Field Investigation of Underground Heat Distribution Systems*, FCC Technical Report No. 47, p. 3.

34

Energy saving warehouse/office design

Energy and life cycle cost benefits can be derived through application of ASHRAE Standard 90-75 to this common building design.

JAMES ROSS,
Manager, Mechanical Engineering
Dept., Inducon Consultants of Canada
Ltd., Don Mills, Ontario, Canada

Various shapes, sizes, and configurations of warehouse or light manufacturing facilities can be viewed in industrial subdivisions and in the suburbs of most municipalities. This type of building is being erected throughout North America at an increasing pace. Generally, an air conditioned office space is attached to or included in the structure.

A large percentage of these warehouse/ offices are constructed or owned by small business ventures with minimum available capital. Each building must be erected at the least possible cost and therefore, provides only the absolute necessities required for occupancy.

The building construction materials usually include bare concrete block walls in the warehouse or manufacturing section with a token quantity of insulation in the roof structure. The walls are usually not insulated.

As the office area or appendix is normally at the front or street elevation, it is often dressed up to include brick or curtain wall with a quantity of fenestration.

A large number of these structures have been and still are being thrown up with a complete disregard for the annual energy costs required for heating and air conditioning.

Prospective owners of single warehousing or light manufacturing facilities cannot be expected to know the economics of energy conservation as related to their planned buildings. The builder, developer, contractor, architect, engineer, or whoever is responsible for the design of the mechanical system is also responsible for making the owner aware of the economics of energy conservation. The owner should be presented with estimates of the construction costs of energy conservation methods and should be made fully aware of the continuously increasing annual energy costs that he can expect unless proper measures are taken during the design stages of his building.

ASHRAE Standard 90-75, Energy Conservation in New Building Design, establishes the minimum requirements for thermal design of the exterior envelopes of new buildings and HVAC system requirements for effective utilization of energy. The standard also deals with component efficiency and efficient electrical distribution and lighting system design.

Standard 90-75 is being adopted by different levels of government throughout the United States. A great deal of interest is also being generated across Canada.

This chapter demonstrates the results of applying Standard 90-75 to a typical warehouse/office structure in the suburbs of Toronto, Canada.

Building construction and systems

The building floor plan is shown in Fig. 34-1. The warehouse section has a floor area of 15,400 sq ft (1430 m²) and is used for storage and handling of metal products, which will be stored on racks to a height of 15 ft (4.6 m). It is 20 ft (6m) from the floor to the top of the steel deck roof. Four 8 by 10 ft (2 by 3 m) wooden high lift shipping doors are at the rear of the warehouse.

The warehouse walls are constructed of 12 in. fluted three core concrete blocks. The roof is a metal deck with built-up roofing and insulated with 1½ in. of rigid glass fiber. The walls are not insulated. This is a typical envelope for this type of building.

A 2400 sq ft (223 m²) office area projects from the front of the building and has considerable single sheet glass on the eastern exposure. The balance of the wall is made of 12 in. hollow core concrete block with ¾ in. thick rigid insulation and additional drywall finish in the exposed areas. The roof deck contains 1½ in. of rigid glass fiber insulation and is 13 ft (4 m) above floor level. The offices have a suspended T-bar ceiling at 9 ft (2.7 m) above floor level.

The heating and air conditioning system is shown in Fig. 34-2. In areas where gas is available, this is the normal, and one of the most economical, methods of heating and air conditioning such structures.

Multiple propeller gas-fired unit heaters are installed around the warehouse exterior perimeter walls. These heaters are normally installed at as low a level as practical or as permitted by the occupancy operation to assure heating at floor level. However, in this case, they are installed at a mounting height 10 ft (3 m) to the bottom of the heaters. Each is cycled from an individual wall mounted room thermostat. High level, high volume, multi-speed propeller fans are installed at the shipping doors to maintain warm air circulation from roof level down to the floor in these areas.

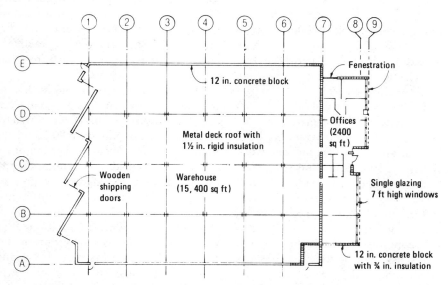

Fig. 34-1. This floor plan shows standard construction features for this warehouse/office.

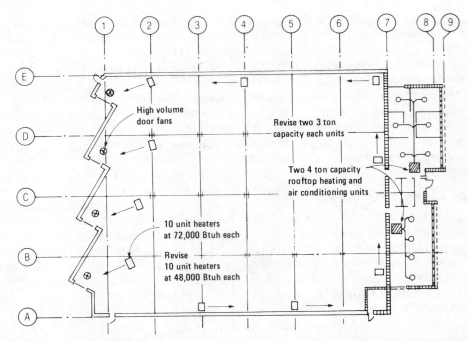

Fig. 34-2. The heating system for the warehouse and the heating and air conditioning system for the office area as designed for standard construction is shown in black. The revisions made in accordance with Standard 90-75 are shown as shaded areas.

The office area is heated and air conditioned by roof mounted, packaged gas-fired heating/electric cooling units with centrifugal fan evaporator and air cooled condenser sections.

Each unit is regulated from a room thermostat with automatic summer/winter changeover. The quantity of roof top units is normally selected to provide comfort control zoning for each major exposure, and two units are provided for this project.

Since the ceiling space is only 4 ft (1.2 m) high, return ductwork is not provided. The ceiling is used as a return air plenum.

The elimination of return air ductwork and the deletion of insulation on supply air ductwork amounted to a considerable cost savings.

The rooftop units handle a predetermined quantity of outside air. Due to the small size of these units, economizer cycles, which permit cooling with outside air between seasons, are not provided. Standard 90-75 specifies that economizer cycles are not required on units with capacities below 5000 cfm (2.36 m^3/s).

These units each handle a maximum of 1600 cfm (0.75 m^3/s).

Design conditions and envelope

Table 34-1 is a compilation of the design conditions and the envelope construction materials for the warehouse section and the office area. In each category, typical practice is compared to that recommended by Standard 90-75.

The first section of Table 34-1 compares the standard system design conditions that are currently used in the Toronto area to those recommended by Standard 90-75 for this latitude and degree-day area. The major changes are a lower outside air ventilation rate and lower summer cooling and winter heating indoor/outdoor temperature differentials.

Standard 90-75 recommendations will result in lower comfort conditions for building occupants occasionally during the peak periods of the heating and air conditioning seasons, and this must be clearly explained to prospective owners and tenants.

The outside air ventilation quantity is re-

Table 34-1. Comparison of typical construction practices to those specified by Standard 90-75 for a typical warehouse/office building at Toronto, Canada.

DESIGN CONDITIONS		
	Typical practice	Standard 90-75
Winter: outdoor	−5 F	1 F (97%)
interior	68 F warehouse*	
	75 F offices	72 F offices
Summer: outdoor	87 F dry bulb	87 F dry bulb
	75 F wet bulb	75 F wet bulb
indoor	75 F dry bulb (offices)	78 F dry bulb
Ventilation rate (outdoor air)	0.25 cfm per sq ft of floor area	0.1 cfm per sq ft of floor area**

* A warehouse minimum design temperature of 68 F is required for this structure in accordance with local codes.
** Using the reduction to 33 percent as specified by ASHRAE Standard 62-73, the outside air ventilation rate would be 0.05 cfm per sq ft. Washroom exhaust ventilation, however, is in excess of 240 cfm, therefore, a rate of 0.1 cfm per sq ft of floor area must be used.

ENVELOPE CONSTRUCTION MATERIALS—WAREHOUSE		
Roof	• Steel deck with 4 ply asphalt and gravel.	• Similar but 4 in. of glass fiber insulation. U value is 0.068 (maximum permitted).
	• 1½ in. of glass fiber insulation with a U value of 0.15.	
Opaque walls	• 12 in. hollow core concrete blocks with fluted face.	• Similar but with granular vermiculate insulation in cores. U value is 0.25.
	• Uninsulated.	
	• U value is 0.47.	
Doors	• U value is 0.51	• U value is 0.51.
Average thermal transmission for gross wall area (Uo)	• 0.48.	• 0.255 (maximum permitted is 0.26).

ENVELOPE CONSTRUCTION MATERIALS—OFFICE		
Roof	• Same as for the warehouse.	• Same as for the warehouse.
Opaque walls	• 10 in. hollow core fluted concrete blocks with ¾ in. rigid glass fiber insulation.	• Similar but with 4 in. thick rigid glass fiber insulation. U value is 0.07.
	• U value is 0.21.	
Fenestrations	• Single glazed.	• Double glazed.
	• U value is 1.13.	• U value is 0.68.
Average thermal transmission for gross wall area (Uo)	• 0.37 double glazed.	• 0.24 (maximum permitted is 0.26).
	• 0.56 single glazed.	
Overall thermal transfer value (OTTV)	• 39.87	• 34.2 (maximum permitted is 34.6).

duced from 0.25 cfm per sq ft of floor area to 0.1 cfm per sq ft. This revised volume is quite sufficient and provides an excess of 15 cfm of outside air per person for this project.

Section two of Table 34-1 describes the warehouse envelope. The U factor for the selected 12 in. concrete block is 0.47, while the wooden shipping doors have a U factor of 0.51.

The average thermal transmission (Uo) for the gross wall area of 9400 sq ft (873 m^2) is 0.48 Btuh per sq ft (1.5 W/m^2). As Toronto has 6827 annual degree-days, Standard 90-75 requires that the average transmission for the gross wall area does not exceed 0.26 Btuh per sq ft (0.8 W/m^2).

Granular vermiculite insulation in the cores of the 12 in. blocks provides an average thermal transmission of 0.25 Btuh per sq ft and is within the recommended rate.

A U factor of 0.15 is provided by 1½ in. of rigid glass fiber roof insulation. To meet the Standard 90-75 requirement of 0.068, a thickness of 4 in. is required.

The increase in construction cost for the additional roof and wall insulation was $12,864.

Section 3 of Table 34-1 describes the office area envelope. The exterior section consists of single glass 6 ft, 6 in. high around the perimeter as shown in Fig. 34-1. The balance of the office wall is of 10 in. fluted block with a layer of $\frac{3}{4}$ in. rigid glass fiber insulation.

Standard 90-75 requires that the average wall thermal transmission value for this building be 0.26 Btuh per sq ft and that the minimum overall thermal transfer value not exceed 34.6 (109 W/m²). The wall must be changed to meet these requirements by the addition of double glazing and the use of 4 in. thick insulation instead of $\frac{3}{4}$ in. on the block surfaces. This provides an average wall thermal transmission value of 0.24 and an overall thermal transfer value of 34.2 Btuh per sq ft.

The rigid glass fiber roof insulation must be changed from 1½ in. to 4 in. to provide a U value of 0.068.

The additional construction cost to provide double glazing and additional insulation was $3900.

Heating and air conditioning

Fig. 34-2 shows the warehouse heating system and the heating and air conditioning system for the office area. Standard construction design is shown in black, and the revisions made in accordance with Standard 90-75 are shown in shaded areas.

Standard design for the warehouse requires the selection of 10 unit heaters, each having an output of 72,000 Btuh (21 kW) and mounted 10 ft from floor level. Using Standard 90-75 permits the reduction in size of units to 48,000 Btuh (14 kW). A smaller quantity of 72,000 Btuh units could have been substituted but would not have provided proper heat distribution or coverage.

The savings in construction cost realized by designing the heating system for the warehouse to Standard 90-75 was $2000. This includes reductions in unit heater sizes, flue sizes, piping sizes, etc.

The standard construction shown for the

office heating and air conditioning system as shown in black in Fig. 34-2 required the selection of two 4 ton (14 kW) refrigeration capacity rooftop units. The reduction in cooling and heating loads realized by adopting Standard 90-75 permitted the substitution of two 3 ton (11 kW) capacity units with an accompanying reduction in unit heating capacity.

The total system air volume is reduced from 3200 cfm (1.5 m³/s) to 2400 cfm (1.1 m³/s). The savings in installed equipment cost $1400. This includes the reduction in size of units, ductwork, diffusers, electric supply, and lower labor costs.

Energy savings

The energy cost savings realized by applying Standard 90-75 to the warehouse section of this building was $1523 per year. This is based on current electric and natural gas fuel costs. The extra cost of additional insulation was $12,864. The net extra cost was $10,864.

Using Standard 90-75, the estimated savings in fuel and electric costs for the office heating and air conditioning system amounts to $606 per year. The largest portion of this savings is from winter heating costs, which show a savings of $534 per year.

Table 34-2. Unit construction cost figures for the office/warehouse.*

Revision	Unit cost difference
Roof insulation boosted from 1½ to 4 in. thick.	Extra 60 cents per sq ft of roof surface
Add granular vermiculite to 12 in. block cores	Extra 40 cents per sq ft of wall surface
Single to double glazing	Extra $3 per sq ft of fenestration area
Reduction in heating/air conditioning system for office area using Std. 90-75	Reduction of 58 cents per sq ft of floor area
Reduction in warehouse heating system using Std. 90-75	Reduction of 12 cents per sq ft of floor area

*These unit prices were obtained for this particular building and cannot be applied to other projects.

Table 34-3. Unit energy cost savings for office/warehouse building using Standard 90-75.*

Area	Energy cost savings per sq ft of floor area per year
Warehouse heating	9 cents
Office heating	22 cents
Office air conditioning	3 cents
Complete building heating and air conditioning	11 cents

*These unit cost savings apply to this particular building and cannot be applied to other projects.

The resulted savings from all the installed equipment amounts to $1400. The extra cost for double glazing and insulation was $3991. This results in a net extra cost of $2591.

Life cycle cost analysis

The total net extra cost to provide a structure and mechanical system for this warehouse/office building in accordance with Standard 90-75 is $13,455 or 75 cents per sq ft of floor area. The total energy savings amounts to $2129 per year.

The life cycle cost analysis is based on the following criteria:
• 6 percent interest rate after taxes.
• An energy and maintenance cost escalation rate of 15 percent.
• A 20 year period.
• Original year's annual energy savings of $1065 after taxes ($2129 before taxes).
• A tax rate of 50 percent.
• Extra construction costs of $13,455.

A benefit cost analysis weighs the benefits in total present value of operating and maintenance savings against the additional capital cost. The cost ratio of this building is estimated at 3.912. (A benefit cost ratio of more than 1 indicates an economic advantage.)

The direct comparison of long term savings on present worth basis indicates that the present worth of the energy savings is $47,920.

The payback period required to recapture the additional construction cost of $12,248 based on the energy savings is estimated at 8.8 years.

The interest rate at which the present equivalent of revenues is just equal to present equivalent costs is 19.8 percent after taxes and 39.6 percent before taxes.

Summary

Previous cost studies have examined fuel savings realized by the application of additional layers of insulation to building envelopes. Due to the long payback periods being realized by this single measure, older and some recently constructed buildings are gluttons for energy.

Standard 90-75, Energy Conservation in New Building Design, recommends that design temperature differentials be reduced and that outdoor air quantities and unnecessary lighting levels be reduced. In addition, an increase in the thermal resistance of the building envelope components is recommended. This, together with increased equipment efficiencies, can result in very short payback periods.

These energy cost savings can be easily demonstrated and should be extremely attractive to prospective building owners.

Most of the various levels of government will eventually enforce or enact Standard 90-75, or similar energy saving standards. This process, however, will probably take considerable time. During the interim period, building systems designers should use every opportunity to adapt Standard 90-75 to their designs and demonstrate to prospective building owners the advantages of energy conservation in new building design.

35

Heat energy from waste incineration: cash for trash

Incineration is a practical and economical method of disposing of our mountains of solid waste, and recovery of heat offers a partial solution to fossil fuel shortages. Here is an approach that can be used to arrive at budget figures for the initial investment and owning and operating costs involved in typical applications.

MAURICE J. WILSON, PE,
I. C. Thomasson & Associates, Inc.,
Consulting Engineers, Nashville,
Tennessee

The crisis in energy follows our life style. It is not the high cost of living but the cost of high living that is of so much concern.

Present sources are yielding fossil fuels with greater and greater reluctance. Each gallon or unit of energy costs more to find, mine, and refine. Nearby sources are becoming exhausted, and transportation drives up the cost of new sources.

Efficiencies of combustion and power generation have improved steadily over the years but are now asymptotic to ideal.

World competition from the developing countries is driving the price of prime energy up to the level of French wine. The Middle East has discovered that it can now charge users much more than the cost of extraction for greater profits.

Conservationists are battling the offshore exploration efforts, the installation of new pipelines, and the construction of refineries and nuclear plants, making the job of satis-

fying our ever-increasing demand for energy more difficult.

The cost of energy as a percent of gross national product dropped steadily for decades, but the trend reversed in 1966. No one expects to see this ratio decline at any point in the future.

Meanwhile, we are generating an ever-increasing quantity of solid waste, approximately half of which, by weight, is a replaceable resource such as paper with a relatively high Btu content. If all of the 250 million tons of residential and commercial waste disposed of yearly were incinerated and its heat recovered, we would obtain less than 3 percent of our total prime energy needs, or less than 10 percent of the heat energy required to heat and cool our residential and commercial buildings.

This, then, is certainly not the solution to the rising cost of energy and the apparent crunch in certain fossil fuels. It is, however, a practical and economical method of disposing of our solid waste—one in which we can get "cash for trash" by utilizing its heat energy for many different applications once served by some of the low priced, nonrenewable fossil fuels.

Solid waste can be a blessing in disguise.

This chapter will evaluate the markets, or applications, for the heat energy available in solid waste, with emphasis on the following:

- Initial investment.
- Fixed charges.
- Operating costs.
- Load factors by market and their impact on production costs.
- The types and temperature levels of fluid used to convey the heat.

There are several markets or applications for the heat recovered from solid waste. These include:

1. Manufacturing processes.

2. Industrial parks in which process and/or comfort heating or cooling are required.

3. Urban areas in which new as well as existing offices, stores, banks, hotels, etc., require year around climate control.

4. College and university campuses with existing central heating and cooling plants serving academic, dormitory, administration, student union, library, and other buildings.

5. Medical centers with diverse needs for laboratory, research, surgical, patient, and administrative areas.

6. Airport complexes with central heating and cooling facilities to provide environmental control for multiple buildings.

7. Shopping centers in which coolant and medium temperature water are supplied for control of temperature and humidity in sales areas, malls, and miscellaneous spaces.

8. Apartment complexes including several high rise structures, often as part of an urban development program.

9. Power generating plants where solid waste heat energy may serve as a supplement to the prime energy for either base load or peak shaving use.

Load factor must be considered

Each of these applications or markets requires a different amount of heating and cooling per unit of demand. In other words, each imposes a different load factor on the incineration plant supplying it with recovered heat energy. Load factor, as used herein, refers to the ratio of an incineration plant's yearly sales to its potential yearly production.

The potential yearly production must allow for scheduled and unscheduled outages of the incineration equipment. Expressed more succinctly, load factor is steam sales, in pounds, and/or equivalent pounds of steam represented by coolant sales, divided by steam production, in pounds. In-plant usage, line losses, and miscellaneous losses are thus excluded as factors.

The total amount of steam produced may be that produced only from solid waste incineration or that produced from incineration and supplementary fuel firing.

The load factor ranges imposed by these applications are shown in Table 35-1. Specifically, a manufacturing process requiring a fixed quantity of steam each operating hour offers the highest load factor of any of the markets analyzed. The demand is af-

Table 35-1. Load factors imposed on solid waste incineration facility by various markets for recovered heat energy.

Application	Load factor range
1) Manufacturing process	
• Process use only	0.70–0.80
2) Industrial park	
• Heating only	0.25–0.35
• Heating and cooling	0.50–0.60
3) Urban area	
• Heating and cooling	0.60–0.70
4) University campus	
• Heating and cooling	0.40–0.50
5) Medical center	
• Heating and cooling	0.47–0.57
6) Airport complex	
• Heating and cooling	0.55–0.65
7) Shopping center	
• Heating and cooling	0.35–0.45
8) Apartment complex	
• Heating and cooling	0.37–0.47

fected solely by the production process; it is not dependent on weather or other conditions that affect comfort heating, cooling, or ventilating loads to various degrees.

The second market analyzed indicates that consumption is minimal when only space heating is required. On the other hand, if both heating and cooling are provided for the facilities in an industrial park, the year around need improves the load factor. Spring and fall consumption is still quite low, however.

The remainder of the markets analyzed, in which the heating and cooling needs are influenced primarily by the weather, have substantially the same seasonal demands and impose similar yearly load factors. In some instances, the internal component of the cooling load has a noticeable impact on consumption.

The load factor has a significant impact on production cost, whether the fluid utilized to convey energy is steam, medium or high temperature water, or chilled water. It is axiomatic that the higher the load factor, the lower is the production cost. Thus, applications involving manufacturing processes with daily and year around demands for heat energy offer the lowest cost.

The plant and its initial cost

The initial investment in a solid waste incineration facility with its heat recovery and air pollution control equipment depends on several factors, including:
- Plant size, usually expressed in daily tons of solid waste capacity.
- The type of air pollution control equipment.
- The type of solid waste preparation and handling systems.
- The type and quality of heat energy produced together with the type of fluids used for energy transfer from the plant to points of use.
- The area of the country (that is, Northeast, Southeast, West, etc.) and the economic climate prevailing.

Some of these data may be incorporated into curves such as are shown in Fig. 35-1.

These curve data apply only to the Southeast area of the country. Investment cost data are depicted by Curve A, which covers the following:
- Heat recovery equipment, consisting of water wall furnace, main heat exchanger, superheater, and economizer section.
- A moving grate assembly that provides "turnover" and mixing to obtain more complete combustion.
- Auxiliary oil burners.
- Forced and induced draft fan assemblies.
- Stack gas pollution control equipment for both particulate and gaseous pollutants.
- A building of standard industrial type construction with solid waste storage pits sized for at least three full days of operation.
- Piping, wiring, controls, onsite labor, and site preparation.
- Design, legal, and financing fees, and the cost of money during the construction period.

Incinerator facility details

A section through such an incinerator facility is shown in Fig. 35-2. The solid waste from various transfer stations is dumped into the storage pits shown at the right. The storage capacity of these refuse pits is equal to three days' operation. Crane operators mix the residential, commercial, and industrial wastes before charging the incinerator. Wastes move down the water jacketed hopper throat onto a reciprocating type grate assembly for drying, ignition, combustion, and burnout; residue and noncombustibles are discharged into the ash disposal trailer. Combustion gases pass through the superheater section, the main boiler bank, and the economizer, then through mechanical separators and the three-stage scrubbing system. The latter two reduce the particulate and gaseous pollutants to state and federal emission levels.

The scrubbing of gases having a wet bulb temperature of approximately 150° F with systems having a saturation efficiency of virtually 100 percent results in a stack gas dew point of approximately 150° F. Thus,

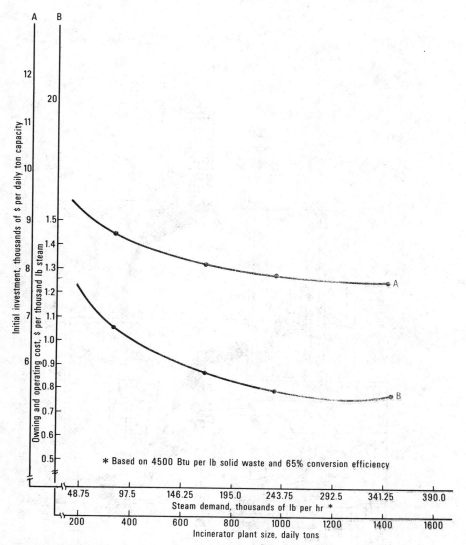

Fig. 35-1. Budget cost data for basic incinerator plant located in the Southeast, adjusted to 1975 level. Curve A indicates initial investment cost, including incinerator unit with waste heat recovery; pollution control equipment; stack and breeching; forced and induced draft fan assemblies; building with solid waste storage; piping, wiring, and controls; charging system; onsite labor; fees; supplementary fuel facilities; and financing cost during construction. Cost obtained from curve is modified by appropriate multiplier from Table 35-2 to account for chilled water plant, if required, and distribution system. Curve B, based on a 0.50 yearly load factor, indicates overall owning and operating cost for base plant, including fixed charges based on 20 year depreciation; 7 percent interest; taxes and insurance; operating personnel; electricity, supplementary fuels, water and chemicals; and routine maintenance and sevice contracts. Cost obtained from curve is modified by appropriate multiplier from Table 35-3 to account for chilled water plant, if required, and distribution system, and by appropriate multiplier from Table 35-4 to account for specific market's deviation from 0.50 load factor base.

vapor plumes can form during portions of the year when the secondary or induced air lowers the mixture temperature to the condensation point.

The induced draft fans are provided with inlet bypass so that reheated outside air can be mixed with the saturated gas from the scrubbers. This lowers the dew point of the stack gas mixture and raises its dry bulb temperature sufficiently to practically elimi-

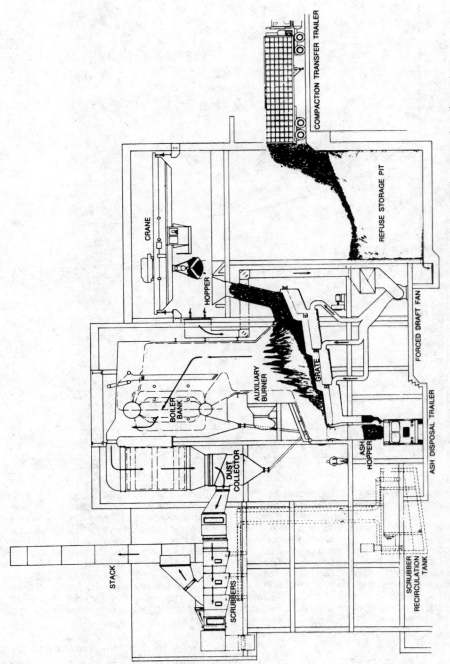

Fig. 35-2. Section through incinerator plant equipped with heat recovery and air pollution control equipment.

nate the vapor plume or "little white cloud." In addition, the stack gas is discharged at a velocity of approximately 3500 fpm so that any condensation that may form will occur at some distance from the stack and hopefully not be associated with the plant's operation.

As stated, the plant's air pollution abatement equipment provides emission levels lower than those required by state and federal regulations. The emission levels are considerably lower than those that would be produced by the many small plants that the facility replaces. It is estimated that a facility such as this will reduce annual particulate emission by a factor of two and sulfur oxide emission by a factor of four to five, even though it will burn solid waste continuously while the many small plants it replaces would burn fossil fuel only during the heating season.

Wet scrubbers are used so that gaseous as well as particulate pollutants can be removed. Bleed water from the scrubbers goes to the ash quench tank and is lost by evaporation and ash dragout. Thus, there is no net water effluent from the scrubbers and ash quench tank. Bleed from the cooling towers is used for make-up to the scrubber recirculation system. Boiler blowdown and plant cleanup water are both fed to the sanitary sewer.

Chilling plant, distribution

The cost data indicated by Curve A in Fig. 35-1 do not cover the cost of a water chilling plant if cooling is to be provided or the cost of the distribution system. These costs must be added to the initial investment for the basic plant obtained from Fig. 35-1 after determining whether steam or water is to be used for heating, the chilled water flow rate if cooling is to be provided, and the type of piping systems required.

The distribution system cost will vary depending on the fluids involved (steam or hot water for heating, chilled water for cooling if required); the length of the system (a comparatively short run for steam only to a manufacturing plant as opposed to an exten-

sive "finger" type system for an urban development project); the types of piping and insulation involved; and the amounts of excavation, backfilling, and resurfacing required for the project.

Incremental costs for chilled water plant if required, distribution system, and complementary changes in the incinerator plant are shown in Table 35-2. The cost additions are expressed as multipliers to be applied to the base cost of a conventional incinerator plant.

For example, a distribution system to provide 150 psi steam to a manufacturer will increase the base incinerator plant cost 6 percent. This assumes a close coupled arrangement with the manufacturing plant adjacent to the incinerator plant and the steam and condensate lines routed for minimum interference.

If both steam and chilled water are to be provided to several manufacturers in an industrial park, the base cost will be increased

Table 35-2. Initial investment multipliers are applied to base plant data obtained from Curve A in Fig. 35-1 to account for chilled water plant, if required, and distribution system.

Application	Initial investment multiplier
1) Manufacturing process	
• 15 psi steam	1.080
• 150 psi steam	1.060
• 265 psi steam, 100° F superheat	1.050
• 400 psi steam, 150° F superheat	1.045
2) Industrial park	
• 150 psi steam	1.060
• 150 psi steam, 41° F chilled water	1.560
3) Urban area	
• 150 psi steam, 41° F chilled water	2.050
• 290° F MTW, 41° F chilled water	2.000
4) University campus	
• 150 psi steam, 41° F chilled water	2.260
• 290° F MTW, 41° F chilled water	2.210
5) Medical center	
• 150 psi steam, 41° F chilled water	2.170
6) Airport complex	
• 290° F MTW, 41° F chilled water	2.130
7) Shopping Center	
• 290° F MTW, 41° F chilled water	2.120
8) Apartment complex	
• 290° F MTW, 41° F chilled water	2.100

56 percent. This increment accounts for a water chilling plant located within the park and two distribution systems with laterals.

Chilled water plants and distribution systems for the remaining markets analyzed, requiring either medium temperature water or steam and chilled water at approximately 41° F, will result in initial investment increments of 100 percent or more of the base plant cost. These increments include chilling plants with auxiliaries such as heat rejection equipment, low head and primary pumping equipment, piping, service valves, etc. and buildings to house the equipment. In each case the incinerator plant would be remote, and the services might be under streets, adding a premium for the distribution systems.

These data are gathered from various feasibility studies prepared during the past several years.

Owning and operating costs

The total initial investment determined by the above procedures can be used for budget purposes. It may also be used to calculate fixed charges.

Included in Fig. 35-1, however, are owning and operating cost data, given by Curve B, for a base plant operated at a load factor of 0.50. Bases for depreciation, value of money, etc., are indicated.

Owning and operating cost correction factors for various distribution systems and for plant modifications to meet the needs of various applications are presented in Table 35-3. These multipliers, when applied to data from Curve B, will provide, for budget purposes, the cost of steam production when the plant load factor is 0.50.

For example, assume that we require the owning and operating cost for an incinerator plant serving an urban complex to be a given amount. This figure must include increased fixed charges for a chilled water plant and for the distribution systems over and above those indicated by Curve B in Fig. 35-1. With steam at 150 psi for heating and chilled water at 41° F for cooling and dehumidification, the appropriate multiplier is seen to be 1.714. In other words, the owning and operating cost

Table 35-3. Owning and operating cost multipliers are applied to base plant data obtained from Curve B in Fig. 35-1 to account for chilled water plant, if required, and distribution system.

Application	Fixed charge multiplier
1) Manufacturing process	
• 15 psi steam	1.055
• 150 psi steam	1.041
• 265 psi steam, 100° F superheat	1.034
• 400 psi steam, 150° F superheat	1.034
2) Industrial park	
• 150 psi steam	1.041
• 150 psi steam, 41° F chilled water	1.381
3) Urban area	
• 150 psi steam, 41° F chilled water	1.714
• 290° F MTW, 41° F chilled water	1.680
4) University campus	
• 150 psi steam, 41° F chilled water	1.857
• 290° F MTW, 41° F chilled water	1.823
5) Medical center	
• 150 psi steam, 41° F chilled water	1.796
6) Airport complex	
• 290° F MTW, 41° F chilled water	1.768
7) Shopping center	
• 290° F MTW, 41° F chilled water	1.762
8) Apartment complex	
• 290° F MTW, 41° F chilled water	1.748

will be 71.4 percent greater than that indicated by Curve B.

Again, this is based on a load factor of 0.50. In Table 35-1, however, a load factor range of 0.60 to 0.70 is indicated for the urban complex application. Correction factors for the various markets' load factor deviations from the 0.50 base are presented in Table 35-4. This table indicates that for an urban complex, a multiplier of 0.843 will correct the 0.50 load factor on which Curve B is based to 0.65 (the average of the 0.60 to 0.70 load factor range given in Table 35-1. The net correction, therefore, is $1.714 \times 0.843 = 1.445$. Subsequent calculations will amplify on this correction procedure. The multipliers in Table 35-4 take into account the relative impact of owning as well as all operating costs.

Generally speaking, the fixed charges for an incinerator plant with heat recovery, proper pollution control for both gas and liq-

Table 35-4. Load factor multipliers are applied to base plant owning and operating cost data obtained from Curve B in Fig. 35-1 (along with fixed charge multipliers in Table 35-3) to go from 0.50 load factor on which Curve B is based to mid-points of market load factor ranges indicated in Table 35-1.

Application	Load factor multiplier
1) Manufacturing process	
• Process use only	0.816
2) Industrial park	
• Heating only	1.445
• Heating and cooling	0.936
3) Urban area	
• Heating and cooling	0.843
4) University campus	
• Heating and cooling	1.081
5) Medical center	
• Heating and cooling	0.976
6) Airport complex	
• Heating and cooling	0.883
7) Shopping center	
• Heating and cooling	1.187
8) Apartment complex	
• Heating and cooling	1.141

uid effluent, water chilling plant, distribution system, etc., will amount to approximately 65 to 68 percent of the overall owning and operating cost. Of the remaining 32 to 35 percent of the owning and operating cost, approximately 18 to 20 percent of the owning and operating cost (that is, 55 percent of the operating cost) is independent of plant output. This segment of cost includes operating personnel, management, and general administrative expense. Other operating cost items such as electricity, water, chemicals, supplementary fuel, and routine maintenance and service vary with steam production. This segment amounts to approximately 15 percent of the owning and operating cost, or approximately 45 percent of the operating cost.

As a result, the higher the load factor and the yearly production, the lower is the production cost. Expressed in another manner, if the system load factor were increased from the 0.50 base to 0.60 (an increase of 20 percent), the total cost would increase only approximately 3 percent. In addition, if there

were a 20 percent increase in gross income accompanied by only a 3 percent increase in owning and operating cost, the unit sales price could be significantly reduced in a nonprofit operation, or the profit picture markedly enhanced for an independent corporation.

An example for a typical plant

To illustrate the foregoing, let us assume that budget costs are required for initial investment, fixed charges, owning and operating cost, and production cost for a 1000 daily ton capacity plant with a water chilling plant sized for a heating to cooling demand ratio of 1.0 and serving an urban area including new and existing offices, stores, banks, hotels, etc.

Table 35-1 indicates that the load factor range for urban complexes is 0.60 to 0.70.

Curve A in Fig. 35-1 indicates an $8000 per ton initial investment for a basic plant, or $8,000,000 for the 1000 daily ton plant considered. The items included in this investment are indicated in Fig. 35-1.

Table 35-2 lists the correction factor for addition of the distribution system and chilled water plant. For an urban complex supplied with steam at 150 psi and chilled water at 41° F, the indicated multiplier is 2.050. The revised initial investment figure is thus $8,000,000 × 2.05 = $16,400,000.

Curve B in Fig. 35-1 indicates the overall owning and operating cost for a base plant operated at a load factor of 0.50. For our plant, the owning and operating cost, or production cost, is $0.81 per 1000 lb of steam.

Table 35-3 provides us with an adjustment in owning and operating cost to account for the additional expense of the distribution system and water chilling plant. The appropriate multiplier for our system is 1.714, and the production cost is therefore $0.81 × 1.714 = $1.39 per 1000 lb of steam.

The above cost is based on a load factor of 0.50. From Table 35-4, the correction factor for an urban complex load factor of 0.60 to 0.70 (use 0.65) is 0.843. Thus, the final production cost is $1.39 × 0.843 = $1.17 per 1000 lb of steam.

Results show investment return

The results of the previous exercise are as follows:

• The in-place cost of the incinerator plant, chilled water plant, and distribution system is $16.4 million.

• The production cost (overall owning and operating cost) for services to the urban complex having a load factor of 0.65 is $1.17 per 1000 lb of steam.

Information such as this may be used during preliminary discussions before a feasibility study is completed. Should there be interest in return on investment at this preliminary stage, it is approximately 5 percent based on sales at $1.75 per 1000 lb. of steam, initial investment of $16,400 per ton capacity, and conversion efficiencies that provide 6000 lb of steam per ton waste (250 lb of steam per hr per ton). If the municipality pays a nominal amount per ton for incineration in lieu of landfill (say $3 per ton), however, the return on investment will be approximately 11 percent.

In time, economical techniques for recycling certain wastes and new markets will develop to provide additional income for such an operation. Supplementary income can be obtained from sales of residue, after separation, such as paving base or building block.

In summary, we are faced with an ever-increasing amount of solid waste. Disposition will surely involve the most economical techniques consistent with air, water, and land pollution constraints. Conventional incineration and pyrolysis—processes in which energy may be recovered and put to economical and useful purposes—are disposal methods that appear to have merit.

Engineers are frequently called on to evaluate the economics of the many different solid waste disposal methods. In a given instance, the engineer recommends to the municipality, after a comprehensive study, the best solution to its disposal problem. In some sections of this country, supplies of low cost fossil fuels are not sufficient to serve the needs of industry. The heat energy in solid waste may provide a partial answer to this problem.

This study may be helpful in analyzing such problems and providing preliminary information for budget purposes. The data can be used in preliminary discussions, prior to the completion of a comprehensive study.

36

GSA's systems design and performance approach to energy conservation

The General Services Administration's method considers not only first cost but also long range system performance in its assessment of what constitutes an economical building—a new way of looking at the bottom line.

ROBERT R. RAMSEY, PE,
Vice President, Leo A. Daly

SHERI HRONEK,
Systems Information Specialist,
Leo A. Daly, Omaha, Nebraska

An untapped and long ignored energy source is available. In terms of quantity, it will have in 1990 the potential equivalent to energy available from any of the fossil fuels or from nuclear energy. This neglected source is energy conservation in the built environment.

One of the conclusions of the American Institute of Architects' Energy Conservation Steering Committee headed by Leo A. Daly, FAIA, is that "present energy policies, with their emphasis on increased supply, seriously underplay the important role of conservation in general and of conservation in the built environment in particular. This imbalance results in forfeiting major opportunities for better investment of the nation's energy resources."

Energy conservation means more than changing the thermostat. What it means is finding ways to stop wasting energy and put what is wasted to use.

Defining the problem is relatively easy.

What is difficult, is to implement solutions or to motivate individuals to strive to incorporate conservation goals into their designs for buildings. It is too easy to say our society can continue as it has and only search for new supplies. Those supplies are not inexhaustible, and the day of reckoning is on the horizon. However, attitudes relating to the gas shortages a few years ago, and other shortages around the world show that individuals will not voluntarily change habits or viewpoints concerning the energy situation.

Human nature being what it is, people need incentives. The federal government's General Services Administration offers the desired inducement to the building industry in its *Performance Specifications for Office Buildings.*

GSA's performance specifications

These specifications have been put to the test by GSA on a "real-world" experiment implemented to demonstrate the usefulness of the systems approach to designing, specifying, and constructing complex buildings. This approach is an advanced problem solving technique, and its testing ground was GSA's project for the Social Security Administration Program Centers in Philadelphia, Chicago, and Richmond, California, near San Francisco.

Because the systems approach emphasizes the interrelationships of the elements of the whole system, its use makes possible innovations that affect energy consumption. In other words, it emphasizes the interrelationships of ceiling, floor, partitions, structure, luminaires, HVAC and electrical distribution, and it demands that the designer look at these elements as part of an integrated whole.

Because system elements are interdependent and the uses of energy within a building are interrelated, the designer can find new solutions to problems of energy conservation.

Lighting: an energy glutton

Within the commercial/residential market, the largest use of energy is for space heating and cooling. And one of the villains in this energy gluttony is light. Lighting is the major consumer of electrical energy in office buildings both for the power required for luminaires and for the energy used for air conditioning, which is in part necessary because of heat from the lights.

The commercial market (stores, office buildings, schools, hospitals, and government buildings) is rapidly increasing its use of electricity with the spread of air conditioning. By 1990, this market is expected to account for 34 percent of the total electrical demand while energy consumption in general is expected to increase at more than 5 percent annually.

In low rise office buildings without elevators, lighting can consume up to 60 percent of the electrical energy supply to the building. Lighting systems in office buildings have generally constituted 25 to 60 percent of the cooling load on air conditioning equipment, according to a recently published report of the GSA, *Energy Conservation Design Guidelines for Office Buildings.*

More than 95 percent of the electrical energy used by lights is converted to infrared radiation. Or, in other words, less than 5 percent become visible light. A fluorescent lamp transforms about 20 percent of its electrical input into light. According to 1971 figures, fluorescent lamps provide about 70 percent of the country's total illumination.

Within this framework, the designer must examine the system elements for possible relationships that will conserve energy. The performance specifications compel the designer to take full advantage of all the options available and to search for new ways of coordinating the elements.

Basis-of-award calculation

The basis-of-award calculation, developed by Leo A. Daly, who with the Nolen-Swinburne Partnership, was the executive architect/engineer for the SSA Program Centers, provides the incentive to use these options.

Through the performance specifications, the government has established the method for those involved on the project. The specifications describe the performance required by the various system elements, but do not specify the hardware to be used. The designer must determine what the guidelines and goals are. It is he who must decide what hardware is to be used or who must design new hardware if existing products do not meet the requirements of the performance specifications. For example, the specifications require illumination levels, and the designer determines the luminaires needed for that level.

Within the SSA building, the approach is the same for energy conservation. The government demands that energy be conserved, but the system designer must search and research the paths to fulfill this demand. During this process, he forms his own guidelines, wrestles with his own problems, and reaches his own solutions, possibly via new and uncharted roads.

The incentive for following the laborious path comes in the form of a contract, which is presented after the basis-of-award calculation

GSA's performance specifications have been put to the test in three Social Security Administration Program Centers. From top to bottom, the buildings are located at Philadelphia; Richmond, Calif.; and Chicago. The Program Centers are designed to demonstrate the usefulness of the systems approach to designing, specifying, and constructing complex buildings.

is computed. An additional "carrot" is the possible proprietary innovation that can give the system designer an in-road in the increasing energy conservation market.

The basis-of-award calculation is basically saying to the system designer that the first cost of the building plus the cost of energy equals the building cost. Presumably, given a price limit, all the bidders' first costs will be the same: every cent within the price limit will be spent to provide a system that will save money during the life of the building. On the SSA Program Centers project, all bids came within a cost limit of $28 million, but the contract was let according to the most energy conserving system designed.

The dollar is not the evaluation; however, it is the measuring stick. The final evaluation is the performance of the system—the energy conserved.

The HVAC and electrical subsystems integrate neatly within the floor ceiling sandwich. The cylindrical component suspended in the center of the photo is a background masking sound speaker. This component, together with the ceilings and carpet, helps provide the acoustical control essential to the efficiency of an open office arrangement. Behind the speaker is the constant air induction box that maintains balanced flow of primary versus supply air. The pipes leading into the luminaires convey the chilled water used to reduce the radiant heat from each fixture's fluorescent tubes.

The measuring stick

How was the measuring stick applied? The basis-of-award calculation had five parts. First, the bidder presented the net cost of his system, the total cost of all the elements that go into the system. Next, the executive architect/engineer determined bid equalization factors based on the system designed by the bidder. These factors included the floor/ceiling sandwich thickness (which could adversely affect the cost of the building), foundation (the weight of the floor/ceiling sandwich affects the structural walls), HVAC central plant costs, conductors, fire protection, and luminaire voltage and frequency.

Under existing procedures today, the contract would be let to the bidder who had the lowest building price. But all this practice guarantees is cheaper materials. The owner and the people who use the building are the losers in the long run. A low building price could indicate inferior materials, which would have a short life and would have to be replaced. The system's safeguards include the performance specifications' requirements that

assure that quality materials are used and the basis-of-award calculation that incorporates other costs in the systems building cost. Added to the building price are life-cycle costs.

For the basis-of-award calculation only those life-cycle cost factors that could be computed were used. For example, a computer simulation was used for the HVAC subsystem. Actual weather and solar conditions for a selected average year were obtained from the U.S. Environmental Science Service Administration, and heat gains and losses were calculated for each building for each of the 8760 hr of the year.

A weather editing program was prepared by processing computer cards obtained from the United States Weather Bureau in Asheville, N.C. Using this information, an energy study was prepared to determine the number of Btu's necessary to maintain a 75° F temperature.

For the expansion of the SSA Headquarters in Baltimore, the energy life costs were based on the analysis of each proposed

A model of the system.

The finished floor subsystem is designed to quickly and economically keep pace with spatial changes. Antistatic, 18 in.-sq carpet tiles are attached to the floor with a nonsetting adhesive, permitting convenient access to flush, deck-mounted electrical and communications outlets. The outlets are self-abandoning and positioned on a 60 by 30 in. grid pattern to accommodate any interior furnishing and equipment plan.

system made by means of NECAP, a NASA computer program for energy analysis. The Leo A. Daly Co., in joint venture with Ezra D. Ehrenkrantz & Associates as the systems consultant, modified the NECAP program so that it was applicable to the GSA project.

Find common denominator

The measuring stick at this point is ineffective. A yard stick will not measure meters. On the one hand is a system price in dollars and on the other an energy consumption indication in Btu's. To compare the two, a common denominator must be found. Since that denominator is the dollar in the basis-of-

award calculation, Btu's must be converted to dollars. The cost of Btu's is based on today's prices with an inflationary factor included. While normal annual increases in labor, materials, and energy costs can be expected, energy costs are currently escalating more rapidly than labor materials. Conversion calculations for the Baltimore project are based on the expected doubling of energy costs in the next five years with a 5 percent per year increase thereafter.

Since the life of the building is considered to be 40 years for the SSA projects, the life cost was then computed.

The cost of energy used by the HVAC subsystem is now known in dollars, the common denominator. The successful bidder will have computed these costs for himself and worked from that point in a search for ways to cut the costs over the next 40 years. If he can decrease the amount of Btu's needed by the system during these 40 years, he will decrease the amount of money the system will cost and his bid will be lower. Whoever uses the least amount of energy, requires the least amount of money, and has the lowest bid.

Through this process he may discover a new system that will save energy, and the industry as a whole will benefit.

An added bonus

The system designer* for the SSA Program Centers devised a new heat recovery system that utilizes heat from luminaires to save energy and energy costs while increasing luminaire performance. This is not a new idea, but the manner in which he designed the recovery system will save money in the cost of the building.

The combination of two innovative subsystems (lighting and HVAC) is projected to result in an interior system that will use less energy than that required for conventional building interiors. The Richmond and Philadelphia buildings were dedicated in September 1975 and beneficial occupancy for the Chicago building is scheduled for October 1976.

A decision by the system designer to use luminaires that were both water and air cooled in two of the buildings led to energy and construction cost savings.

The water cooled luminaires are simple heat exchangers in the shape of a standard fluorescent fixture. Water passing through the tubes, which are an integral part of the fixture, absorbs much of the luminaire heat in the water circuit. Thus, most of the heat of light never reaches the occupied space. In winter, this excess heat will be used to preheat the cooler outside air before it is used in the ventilation system. In summer, air conditioning loads will be reduced by rejection of excess luminaire heat.

*A joint venture of Owens-Corning Fiberglas Corp. and Wolff & Munier.

Construction costs were reduced by combining the water coolant pumping systems with the sprinkler piping system. This saved the cost of an extra piping system for luminaire coolant.

The HVAC subsystem designed by the system contractor makes use of the water and air cooled luminaire system and relies on control zoned induction boxes and linear air bars to continuously maintain an even and economically balanced flow of primary versus supply air. The double induction boxes induce secondary air from the conditioned office space, the ceiling plenum area, or both, while the luminaires transfer lighting heat either to water and air, or air only for the Richmond project.

The induction box is arranged with tandem dampers to provide cooling, heating, or a combination of both. When maximum cooling is required, primary air is mixed with the induced room air, and the total mixture is supplied into the space through the air bar and air tube assembly. When less than maximum cooling is required, a combination of air induced from the room and the plenum, or all warmer plenum air is used to temper or reheat the primary air. This arrangement achieves a significant reduction in the volume of primary air required, with a resulting reduction in the size of the primary air duct system and central air handling equipment. Reheat of the primary air is accomplished through the use of rejected heat from the lights rather than additional energy consuming electrical or steam reheat.

Reduce lighting load

In addition to the HVAC costs from the computer simulation, the basis-of-award calculation included the costs of energy for luminaire operation. The use of the reheat system not only reduced HVAC energy requirements by reusing normally wasted lighting energy, but it also increased the performance output of the fixtures to the point that three tubes, rather than the normal four tubes per fixture, were ample to provide the high level of illumination performance required.

The basis-of-award calculation also in-

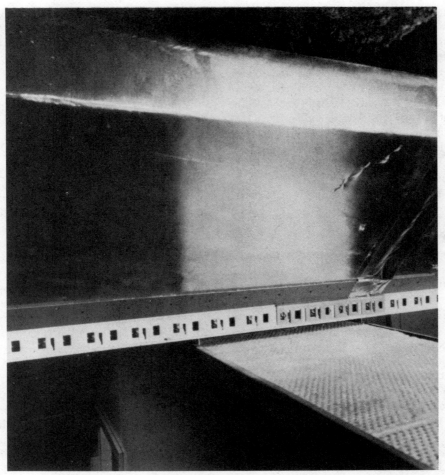

A special acoustical cap allows office partitions in the Program Centers to be arranged at any point beneath ceiling main-runners and cross-tees without interrupting the flow of conditioned air. The cap maintains a ½ in. space between ceilings and partitions, permitting unrestricted air movement while perserving acoustical and visual privacy of full, floor-to-ceiling walls.

cluded a space adjustment model to determine the costs involved for maintenance and changes in space needs. A nine year optional maintenance cost was also included with the life-cycle costs.

These were the only costs that were felt could be objectively evaluated. In response to this, some real reduction in energy consumption per square foot has been experienced—as well as an increase in the quality of some of the components. This increase in quality is reflected in lower life-cycle costs and is, therefore, competitive.

But the process does not end there. The goal for excellence continues and although

55,000 Btu per sq ft per year is the limit now, it is perfectly conceivable that this goal will be reduced.

The systems approach, which emphasizes interrelationships; performance specifications, which describe the performance required; and the use of life-cycle costs, which provide an overview of costs, have all made energy conservation in the built environment a reality. The old adage claims a bird in the hand is worth two in the bush. In this case, these solutions may not be the panacea for energy problems, but we do have a bird in the hand for all the talking about the two in the bush. This answer exists now. And if used properly, it may lead to new unconceived answers to the energy dilemma.

Architect/engineers for the three Program Centers were: Philadelphia, Deeter Richey Sippel Associates; Richmond, Pereira/Bentley/Tudor; Chicago, Lester B. Knight and Associates, Inc. Leo A. Daly/Nolen-Swinburne Partnership was the executive architect/engineer and Turner Construction Co. was the construction manager for all three projects.

Photographs depicting interior design features were provided courtesy of Owens-Corning Fiberglas Corp.

37

Life-cycle costing

WILLIAM J. COAD, PE,
Vice President, Charles J. R. McClure
and Associates, St. Louis, Missouri

Life-cycle costing is an effort to quantify the total owning costs of a system subsystem, or component. Owning costs, in turn, can best be segregated into three basic components: l. first cost or investment, 2. energy costs, 3. maintenance and service costs.

For many reasons (which will not be listed due to space limitations), the trend in product development in recent years has been to minimize first cost at the expense of the other two. However, with the recently accelerating increases in energy values and consciousness of the significance of maintenance and service costs, the trend is moving toward giving due consideration to these two components.

A recent experience in life-cycle costing was most interesting and revealing. In an existing plant, two water chillers were to be replaced with two new centrifugal chillers. The preliminary designs were completed, including development of flow diagrams and complete detailed load and energy studies. At this point, it was decided to request proposals for the chillers, and specifications were written for life-cycle proposals.

Energy cost impact features

To account for energy costs, the following energy cost impact features of the machines were considered:

● *Power impact*—From the refrigeration part load analysis, it was determined how many months a year both chillers would operate, how many months one chiller would operate, and how many months no chillers

would operate. A formula was then developed to provide a single number multiplier taking into account these demands established, the local utility demand rate, and present day value of future dollars based on the owner's economic variables. The bidder would then simply multiply the full load compressor power (kW) requirement of the proposed unit by this compressor power multiplier to establish the power cost impact for the unit.

● *Compressor energy impact*—From the combination of the refrigeration energy analysis and the part load analysis, a formula was developed, linearizing the part load performance of the units, which provided a single number multiplier to be applied to the compressor power requirement at a single stated point of performance at less than full load. This formula took into account the calculated number of operating hours of each part load condition per year, the graduated commodity rate for the local utility company integrated with the other facility loads, and the present day cost of future monies as described above.

● *Water flow resistance impacts*—Similar single number multipliers were developed relating the pressure drops through the chillers and condensers, respectively which considered the various parameters described, to be applied to the head loss for the chiller and product of head loss and flow rate for the condenser.

With calculation methodology currently available, the energy and part load requirements can be quantified quite accurately. Anticipated maintenance and service requirements are, however, much more difficult (if not impossible) to quantify. Thus, it was decided that the only way to legitimately include this ingredient in the life-cycle proposal was to have each proposal include a full maintenance and service contract for an extended time by the manufacturer. The terms of the contract were specifically enumerated in the specification.

Each manufacturer was then requested to submit a proposal on a stated series or model of machine to comply with specific performance requirements. The information given was full load system capacity, piping configuration, single machine capacity, chilled water flow rate, entering and leaving chilled water temperatures, entering condenser water temperature, and anticipated total machine hours of operation per year. Each manufacturer was invited to submit additional proposals on any alternative units in its product line that could meet the performance specifications.

Two sections in bid form

The bid form included two sections; the first was product cost and energy impact. For any unit proposed, the bidder stated the cost of the machines and cost of startup services; then entered the compressor full load kW, multiplied by the power impact factor; and entered the product as the power life-cycle value. Similarly, the compressor life-cycle energy value, chiller pressure drop life-cycle energy value, and condenser value were entered. These six figures were then added, giving the total life-cycle cost of the respective unit.

The extended maintenance and service contract proposal constituted the second section of the proposal form. This cost, by previously determined formulation, was then converted to present day investment dollars.

Six proposals were received. The results of which are shown in Fig. 37-1. The ordinate represents the amounts and the abscissa the

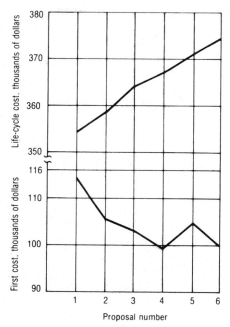

Fig. 37-1 Life-cycle and first cost versus bids received.

proposals, starting with proposal 1, the lowest life-cycle cost and extending to the sixth proposal, the highest life-cycle cost. The top curve is the life-cycle costs and the lower one is the first cost, including start-up service. (The first cost was adjusted as needed to account for difference in installation cost between various machines, such as insulation, field piping of ancillary devices, etc.) The impressive feature of the bids as revealed by the curves is the near perfect divergence or inverse relation between first and life-cycle costs. It illustrates clearly that industry efforts at investment cost reductions have been at the expense of energy consumption and to a lesser extent maintenance.

This methodology of purchasing major components, if not complete selection of systems, will go a long way toward quantifying decisions that were previously often made on the basis of judgment, but it is one more example of an ever increasing maturity in the HVAC industry.

SECTION IV

Control and system adjustments for energy conservation

The correct application and adjustment of the controls in heating, ventilating, and air conditioning systems is a very important element in the conservation of energy as discussed in these three chapters:

- Understanding the functions of controls.
- Installing correct controls at proper locations.
- Readjusting and balancing systems.
- Tuning the system for best operation.

38

Application and adjustment of controls for energy economy

Accurate instruments, optimum control cycles, and a basic understanding of cause and effect applied to system functions can save energy.

ROBERT W. FLANAGAN, PE,
Environmental Engineers, Inc.
Des Moines, Iowa

Over the past several years, an interesting series of articles dealing with "Joe's Body" has appeared in *Reader's Digest*. Each article dealt with a specific organ of the body, its importance to the body's functions, what kind of abuse it can take, and what controls it. The important factor to remember about these body parts is that eventually all of them are controlled by the nervous system. Without signals sent back and forth between the organs and the central nervous system and controller, the body cannot function. In fact, it either becomes lifeless in a sense, malfunctions to a great degree, or dies. In some cases, if an organ is destroyed, the body will continue to live; but in no case, can the central nervous system die and still sustain life in the body.

Every building has to a greater or lesser degree a very close relationship to Joe's body. Automatic controls, no matter how simple or complex, are the nervous system of a building. If a "building body" is to remain healthy, its

This chapter is based on a paper presented by Mr. Flanagan at a special series of two-day seminars sponsored and developed by *Heating/Piping/Air Conditioning* on How to Save Energy in Existing Buildings.

control system must be kept healthy; so it can sense malfunctioning correctly and transmit these data to a central nervous system panel without delay and with great accuracy. The central nervous system processor must be able to interpret the data and issue the correct logic and information to adjust the system to keep the building body functioning properly.

Just as the human body can suffer from high and low pressures in its circulatory systems, such as chills, fevers, malfunctioning of pumping devices, etc., so can the building body. Each time you look at a building complex, I hope you visualize putting on a white smock, stethoscope, and take the necessary instruments and information samples, call in outside expertise when required, and research the various systems being studied to initiate the correct program for applying and adjusting controls for minimum energy use.

Do you know what you are doing?

The proper application and adjustment of controls requires an overall perspective of a building, the various components in it, and how each is interrelated. One simply cannot turn a room thermostat down or up and come away with the conclusion that energy has been saved. Conservation of energy in an existing building requires a considerable amount of time spent examining a building's various components and control systems, and it will require in all probability a larger capital investment in automatic controls. It is possible to save from 20 to 50 percent in operation costs by simply upgrading systems where required and applying engineering techniques and automatic controls presently available.

Before the ills of a building body can be corrected, a total case history must be taken. A doctor cannot take a patient's temperature and say, "take two aspirins and go home." An engineer tends to do this in building operation and maintenance, but it really is not the right approach. An engineer must have an overall perspective of what he is doing, and what he is trying to accomplish. He must know that when he adjusts something out on the end of a duct run, he may be readjusting three other things back near the main apparatus casing. When he inserts a controller into a duct on a diagram, schematically it may look perfect, but unless he knows what he is doing, it will not perform as planned. In the end, he will blame the controls.

An engineer may have to do many things to correct a problem. This requires an overall understanding of what it is all about. Temperature controls have been blamed for many problems for which they are not responsible. It is the fault of engineers and owners because we do not spend enough money or time on automation of a building and controlling its various systems.

In all probability, many building engineers do not know what their present system is doing. He really does not know whether there is leakage through return air and outside air dampers; whether preheats are kicking on when a system is cooling with outside air; whether excess static pressures in duct systems is causing excessive leakage; or whether the mixed air controller is really giving the optimum relationship between outside and return air for mixed air purposes. He really does not know whether the chilled water temperature from the refrigeration machine is right to

handle air conditioning loads, and whether the supply air temperature is too cold, causing the reheat coils to penalize the system energy consumption by reheating back.

System components

Let us start with the room condition itself and its controlling instrument, the room thermostat. It is more economical to raise the temperature in a building than it is to add humidity and depress the temperature to achieve the same comfort level in a space. Steam humidification costs more to effect the same comfort than raising the dry bulb temperature. Claims that adding humidity to a building and depressing the thermostat 2 deg saves energy are made by individuals who do not understand the relationship between dry bulb heating and humidity in energy consumption.

A thermostat must sense the room condition. Many may think this is so trite it does not need further discussion; however, a room thermostat often senses heat output from a table lamp, an overhead spotlight, office equipment, solar gain, etc. Often the room thermostat is mounted on an exterior wall with no insulation sub-base. In a cold climate, a false signal is imposed, causing increased energy consumption. Also, in many cases, draperies are hung over a thermostat, or it is placed behind a door where air circulation is improper. A room thermostat must sense the average room condition to function properly.

In many buildings it is necessary to operate with multilevel temperatures between interior and perimeter zones to effect savings. If it is 68° F on the perimeter and 78° F in the interior, take advantage of interior to exterior heat transfer by natural temperature differential. It will not always save energy to raise the temperature in a space to 78° F during a reheat cycle. Therefore, an engineer must be very analytical in setting a room thermostat. No hard and fast statements exist because for each system (and there are an infinite number in each design concept), an engineer must look at each and say, "If I raise the thermostat setting does it add energy, or if I depress it does it save energy." He must know his system.

Another area that should be explored is

changeover instrumentation or night setback. It may be possible to conserve energy by changing all thermostats by readjusting supply air pressure to the instruments during unoccupied hours and raising them back the next morning to the desired setting. This will not be done manually. Either the system is shut down or it will continue to run and over-extend itself, in operating hours.

All thermostats placed on a changeover system must be compatible to this type of operation. It may be necessary to separate out existing room thermostats and switch them to changeover type operation.

The changeover period should be accomplished during unoccupied periods.

Occupants entering an extremely cold room will accept the temperature psychologically better if they have not been acclimated to it than if the temperature drops suddenly.

Accept temperature swings

Another method of conserving energy is taking a swing in temperature in a space before a building is fully unoccupied. Start shutting down systems and/or rescheduling the temperature before occupants leave. This must be done gradually. Most occupants will sense a $2°F$ change in temperature in a space, but a 5 to $10°F$ change can occur gradually before they complain.

One of the most unacceptable practices is to put a thermometer in the cover of a room thermostat. Many of these thermostat thermometers are in error and slide up and down in the cover, and/or they are out of calibration if they are bimetallic. An occupant notices the temperature of the wall thermometer, looks at the room temperature on his desk mounted thermometer, and concludes that the desk one is correct and the one on the wall is wrong, when in actuality, both may be wrong. It is a psychological problem that must be solved in the future to take full advantage of temperature swings and varying room conditions.

Setting of controls save

Another area of adjusting and applying controls to conserve energy is the proportional type instrument and throttling range.

All proportional instruments have an inherent drift characteristic. When temperature control people or the maintenance staff set the calibration, they should set the drift toward the side that saves energy.

Throttling ranges can also provide some energy savings. For example, consider a four-pipe heating and cooling system. The same room thermostat sequences the heating valve and then the cooling valve through variances in air pressures and spring ranges on the control valves. The spring range and control pressure should be selected so that the valves actually operate in a "no man's land" during a change from cooling to heating. Both valves should actually be closed 100 percent at this point. This is the type of detail that must be checked periodically as valve springs deteriorate.

If both the heating and cooling coil valves open simultaneously, there is a distinct loss in energy for which no value is gained. Therefore, select a spring range and pressure differential that allows an intermittent area in which to operate. Of course, this same problem occurs in a dual system, such as perimeter radiation with all-air.

One of the greatest culprits in the control business is inaccurate selection of a sensing point of the media to be controlled. For instance, in a mixed air plenum where outside and return air streams converge, it is very possible and very likely that the sensing point for a good reading of the resultant mixture will shift as the ratio of the outside to return air varies. Certainly, averaging type instruments supposedly correct this, but again, often these are erroneously placed. The real answer may be installation of anti-stratification devices, such as turning vanes and similar products.

Often a discharge controlling instrument is located on one side of a duct on a two-wheel fan system. It, of course, is very likely to measure a stratified air condition. It is not uncommon to have 20 to $40°F$ stratification across a 10 ft long steam heating coil. The secret to a mixed air plenum operating properly is creating fairly high turbulance in it. Of course, this costs in fan horsepower, but in the long run, it is cheaper than using excess energy. Not only that, the system will perform as designed.

Specific systems explained

Figure 38-1 shows a method of resetting supply air temperature for a straight cooling cycle from zone or room thermostats. Input from the various zone and/or room thermostats is fed through a high signal selector that reschedules chilled water to the cooling coil so that the exit supply air temperature from the coil is raised to the highest possible level to satisfy the most extreme zone calling for cooling. A humidity transmitter is located in the return air duct system to override the inputs from the zone thermostat should the relative humidity in the return air become too high.

Artificially depressed chilled water temperatures are very costly back at the refrigeration plant. Figure 38-1 deals with a constant volume system and shows a mixed air controller taking advantage of outside and return air through an enthalpy logic controller. Operate the supply air temperature at the highest possible level and still maintain acceptable room conditions.

Figures 38-2, 38-3, and 38-4 show a recom-

mended control sequence for conserving energy in a single duct variable air volume system with no reheat. With this type of system, it is recommended for interior zones that the space temperature be set at 76°F or higher during summer and winter for maximum energy conservation. This reduces the cooling energy load and also the fan horsepower load.

Resetting the supply air temperature is also recommended. It should be no lower than required to satisfy the zone calling for the most cooling with a terminal unit in that zone calling for maximum flow.

The installation of a static pressure probe tip in the supply duct and controlling inlet guide vanes at the fan would also reduce both leakage through the terminal unit and the amount of fan brake horsepower needed to operate the system. Static pressure in the duct system should be at the lowest possible level to maintain minimum static pressure to operate the most extreme or critical terminal box. Any additional pressure generated in the duct system must be chewed up at the other various terminal boxes and at the most extreme box itself.

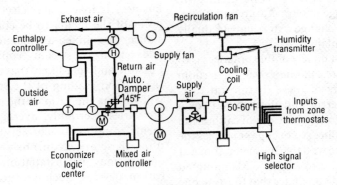

Fig. 38-1. Temperature reset from zone demand.

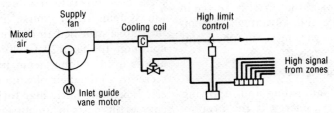

Fig. 38-2. Single duct variable air volume temperature control.

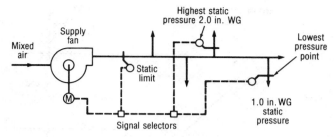

Fig. 38-3. A recommended control sequence for a single duct VAV system with no reheat.

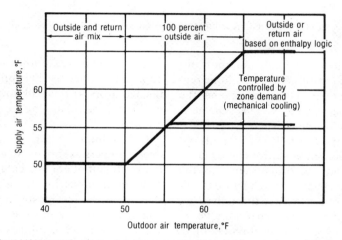

Fig. 38-4. Single duct VAV supply air temperature reset with outside or return air based on enthalpy logic.

The sensing tip must not be at the fan discharge but at the lowest pressure point in the system. A high limit static pressure controller should be installed at the fan discharge to prevent excessive duct pressure should fire dampers accidentally close or due to malfunctioning controls. The use of an enthalpy economizer mixed air package controller is definitely recommended.

One factor to remember with this type of system is that mechanical cooling required for lower supply air temperatures generally costs more than fan horsepower savings. Therefore, for mechanical cooling, deliver air at the maximum temperature possible. When the system is on all outside air cooling, deliver air at the lowest possible temperature and take advantage of fan horsepower savings due to reduced air flow at the terminal unit.

If a single duct variable air volume system with reheat is utilized, it is important to reduce all boxes to minimum air flow before reheat is energized. Also, the minimum air flow stop point on the box should be reduced if codes permit a lower minimum air flow.

A word of caution on electric reheat coils is that a minimum air flow is required before the reheat coil will operate due to safety devices. For rooms with intermittent occupancy, a control scheme should be set up to adjust the box to full close-off if possible and eliminate the reheat coil altogether. The rule to follow is to deliver system supply air at the highest possible temperature and still maintain room conditions at full flow out of the box. The space temperature should be set lower in winter and higher in summer for top savings.

Figure 38-5 shows an arrangement for a dual duct constant volume system to conserve energy by reducing static pressure in the system to a minimum acceptable level to operate the most extreme terminal unit. In the past, it has been assumed that a mechanical constant volume controller supplied with dual duct

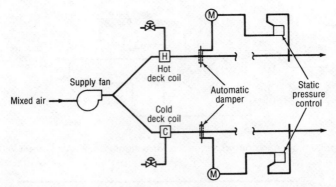

Fig. 38-5. A dual duct constant volume system arrangement to reduce static pressure.

boxes made this item seemingly unimportant. Keep in mind that these constant volume regulators deteriorate over the years, and the damper valves begin to leak. This scheme is applicable to multizone units.

Discharge dampers installed in hot and cold ducts should be sensed by static probes placed far away from the fan to maintain only the minimum pressure necessary to operate terminal units. This reduces leakage through either cold or hot inlet dampers at the terminal box. Generally speaking, a 50 percent reduction in pressure differential between hot and cold ducts reduces closed valve leakage to about 25 percent of its original value.

Study dual duct system

With a dual duct system, it is imperative that tight box close-off occurs when a terminal unit is calling for either full cooling or heating. If leakage occurs at a terminal unit, then perhaps it is necessary to adjust the linkage serving the dampers. Another item that can be applied is an increase in supply air pressure to the actuators. If this still does not correct the problems, then changing the actuators may be required.

If it is impossible to install hot and cold deck dampers as shown, an alternate solution is adjustment of hot and cold deck controllers to an optimum setting. It is also recommended that the room thermostat be set lower in the winter and higher in summer. Although not shown in Fig. 38-5, full advantage should be taken of outside air for cooling purposes, and the same type of logic should be employed as in Fig. 38-1 for mixed air control.

Figure 38-6 shows a recommended scheme for conserving energy in a constant volume reheat system. They are very fine systems and produce almost the ultimate in room control, but they require a lot of energy. Under this scheme, the room thermostats controlling individual reheat coils are fed through accu-

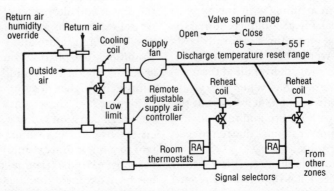

Fig. 38-6. Supply air reset control scheme for a constant volume reheat system.

Table 38-1.   Summer and winter reheat system savings through cooling coil temperature reset. Based on 100 cfm, 5 day work week, 1° F reset.

City	Summer			Winter		
	Cooling coil savings, $	Reheat savings, $	Total, $	Humidity savings, $	Reheat savings, $	Total, $
Atlanta	3.50	4.00	7.50	1.55	2.00	3.55
Chicago	2.55	2.80	5.35	1.45	1.95	3.40
Cleveland	2.70	3.05	5.75	1.60	2.15	3.75
Denver	2.30	2.60	4.90	1.70	2.30	4.00
Detroit	2.60	2.95	5.55	1.55	2.25	3.80
Indianapolis	2.75	3.10	5.85	1.50	2.05	3.55
Los Angeles	5.40	6.15	11.55	1.00	1.30	2.30
Milwaukee	5.00	5.50	10.50	0.70	0.95	1.65
Minneapolis	2.40	2.70	5.10	1.30	1.70	3.00
Portland	2.75	3.10	5.85	2.85	3.75	6.60
Sacramento	3.35	3.80	7.15	2.15	2.80	4.95

Courtesy of Johnson Controls, Inc.
The following formulas were used in calculating the above figures:
Cooling savings   = Δh x lbs per hr x hours of occurrence/Btu per ton·hr x $0.0225 per ton
Reheat savings   = Δh x lbs per hr x hours of occurrence x $2.25 per 1000 lb steam/Btu per lb
Humidification
    savings   = ΔW x lbs per hr x hours of occurrence x $2.25 per 1000 lb steam

mulators, and signal selectors reset the supply fan discharge temperature upward to the highest possible level to satisfy the most extreme room condition at any given time. This reduces the total amount of reheat needed in a system for the other areas not needing as cold an air temperature to satisfy particular zones. Again, this scheme incorporates a return air humidity override instrument in case the humidity becomes unbalanced. The mixed air control logic would be the same as shown in Fig. 38-1.

Table 38-1 lists some possible savings through summer and winter reheat system reset through cooling coils.

Use enthalpy controller

Figure 38-7 shows the psychrometrics of an enthalpy logic mixed air controller and changeover cycle. This instrument, of course, senses the total heat content of the return and outside air and mixes both streams in an optimum manner to cut the total energy load on the refrigeration plant. Any time the outside air condition is colder than the supply air temperature required in a system, outside and return air dampers mix air to maintain the proper desired supply air temperature to a space. The enthalpy logic controller maintains 100 percent outside air any time its temperature is higher than the required discharge temperature from the cooling coil, when it has

a lower enthalpy than the return air, and when its dry bulb temperature is lower than the return air temperature. Any time the outside air's total heat content is higher than that of the return air, and the outside air temperature is lower than that of the return air, the enthalpy controller mixes outside and return air to give an optimum mixed air temperature. Minimum outside air and maximum return air are utilized any time the outside temperature is higher than the return air temperature.

Table 38-2 shows possible energy savings resulting through enthalpy switchover versus conventional switchover systems per thousand cfm of system capacity at a 65° F changeover point.

What about outside air?

Another area that requires urgent review in system operation is the minimum amount of outside air intake to satisfy CO_2 and odor dilution. Historically, $\frac{1}{4}$ cfm per sq ft throughout office building spaces has been utilized simply for good odor dilution. This air quantity was determined based on a certain number of occupants smoking in a given space. The actual outside air requirements needed on a minimum basis have long been neglected. It has been fairly well established that approximately 5 cfm per person is required for good

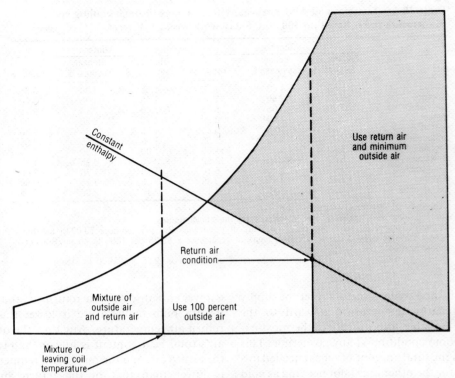

Fig. 38-7. Psychrometrics of an enthalpy logic mixed air control and changeover cycle.

Table 38-2. Energy savings achieved with various switchover systems. Savings per 1000 cfm of system capacity per year.

City	Dry bulb controller only for mixed air. 65 F switchover Per 100 cfm reduction in minimum outside air, Btu	Enthalpy controlled switchover Per 1000 cfm of system capacity per year, Btu x 10	Per 100 cfm of total system reduction in minimum outside air, Btu x 10
Atlanta	4,193,100	2,250,000	4,685,580
Chicago	1,758,870	2,090,000	2,109,510
Cleveland	1,246,290	2,330,000	1,770,320
Dallas	3,627,190	2,310,000	5,193,630
Denver	1,280,670	3,970,000	1,754,010
Detroit	1,292,220	2,560,000	1,754,010
Houston	9,163,910	1,880,000	9,529,040
Indianapolis	2,112,410	2,170,000	2,485,910
Los Angeles	11,190	2,470,000	849,150
Miami	12,835,570	2,050,000	1,315,874
Milwaukee	1,242,540	2,020,000	1,569,060
Minneapolis	1,094,490	2,320,000	1,559,812
New York	1,959,970	2,170,000	2,355,187
Phoenix	1,176,200	4,200,000	2,181,195
Portland	175,500	2,950,000	277,087
Sacramento	763,500	4,240,000	1,679,040
San Francisco	382,720	3,520,000	100,080

Courtesy of Johnson Controls, Inc.

CO_2 dilution. Any amount in excess of 5 cfm per person is used strictly for odor dilution. Thus if the average office space on the interior zones requires about 1 cfm per sq ft for supply air, then only 5 percent of the total air circulated need be outside air to satisfy CO_2 dilution.

Careful analyses should be made of the total exhaust air requirements for a building and infiltration rates into a structure. In all probability, instead of 25 percent outside air to maintain a good balance in a building, it may be that 10 to 15 percent minimum outside air is sufficient. Again, city, state, and federal codes must be reviewed.

Table 38-2 shows possible savings on a per year basis per thousand cfm of system capacity per 100 cfm reduction in minimum outside air requirements. This may require new controllers, new damper blades, etc., to reach lower requirements.

Consider damper losses

This logically leads us to our next area of concern, damper losses and leakage in outside air and return air dampers. Also, as previously mentioned, do not forget about multizone dampers on a hot and cold deck and/or the dampers in terminal units for dual duct systems where leakage can occur.

Tables 38-3, 38-4, and 38-5 show the effects of leakage in hot and cold deck dampers and in outside and return air dampers when in full closed position. Table 38-3 shows that a damper leakage of 10 percent, produces energy waste in the area of 25 percent in the hot and cold decks of a multizone unit or any dual duct system totaling 10 percent leakage in the

Table 38-4. Energy consumed by 10 percent leakage of an outside air damper during the night heating cycle, Btu per year per 1000 cfm of system capacity. Building is 60° F base.

City	Energy, Btu	Fuel oil, gal	Cost, $
Atlanta	1,368,020	8.88	2.46
Chicago	3,522,960	22.87	6.34
Cleveland	3,080,870	20.00	5.54
Dallas	1,145,470	7.44	2.06
Denver	3,193,080	20.73	5.74
Detroit	3,290,160	21.36	5.91
Houston	609,930	3.96	1.10
Indianapolis	3,369,530	21.88	4.97
Los Angeles	546,210	3.55	0.98
Miami	87,000	0.06	0.13
Milwaukee	3,770,790	24.48	6.78
Minneapolis	4,298,530	27.91	7.73
New York	2,416,560	15.69	4.35
Phoenix	713,880	4.64	1.29
Portland	2,254,700	14.64	4.06
Sacramento	1,328,800	8.63	2.39
San Francisco	1,301,260	8.45	2.34

Table 38-5. The effect of return air damper leakage in consumption. Based on per 1000 cfm outside air, 5 days per week, 12 hr per day, operating on cooling only.

City	Energy consumption, millions of Btu, at various percent leakage			
	6%	8%	10%	12%
Atlanta	1.07	1.42	1.78	2.12
Chicago	0.90	1.19	1.78	2.12
Cleveland	0.96	1.28	1.60	1.92
Dallas	0.96	1.28	1.60	1.92
Denver	1.27	1.70	2.13	2.55
Detroit	0.98	1.32	1.64	1.97
Houston	0.92	1.23	1.54	2.21
Indianapolis	0.91	1.21	1.51	2.18
Los Angeles	2.61	3.48	4.34	5.21
Miami	0.58	0.78	0.97	1.16
Milwaukee	0.93	1.23	1.54	1.63
Minneapolis	0.93	1.24	1.55	1.85
New York	1.02	1.36	1.71	2.05
Phoenix	1.41	1.89	2.37	2.84
Portland	1.45	1.93	2.41	2.90
Sacramento	1.69	2.26	2.83	3.39
San Francisco	2.10	2.80	3.50	4.20

accumulative value of all its terminal units. Past experience indicates that the average damper blade leaks 10 percent minimum unless specifically designed for 100 percent shutoff against fairly high static pressure differentials.

Table 38-4 is a summary of losses resulting from a 10 percent leakage in an outside air

Table 38-3. The effects of damper leakage based on 85° F heating and 65° F cooling.

Damper leakage, %	Hot deck temperature, F	Cold deck temperature, F	Energy waste, %
0	85	65	0
1.1	85.2	64.8	2.0
5.0	86.1	63.9	11.0
10.0	87.5	62.5	25.0

Courtesy of Johnson Controls, Inc.

damper when the unit is supposedly on 100 percent recirculating air during unoccupied hours. It is based on Btu per year per thousand cfm system capacity.

When balancing a system out airwise, it is important to take advantage of this percent leakage through the damper blades and incorporate it as part of the overall minimum portion of outside air intakes. If the design calls for 15 percent outside air at minimum and the dampers have a leakage of 10 percent at closed position, it is obvious that it is necessary to make up 5 percent more air to reach the 15 percent minimum requirement when operating in a normal fashion on minimum outside air.

Leakage through a return air damper can be costly also. We tend to think of return air as not imposing any load on a system. However, that certainly is not the case when trying to mix outside and return air and/or operate on a 100 percent outside air cycle through a mixed air controller. This is especially true when the total heat content of return air is greater than that of outside air. Table 38-5 gives some estimated data on percent leakage through return air dampers and the resultant energy consumption during a cooling cycle.

The overall effects of damper leakage can be reduced by putting all fan operations utilized to maintain building conditions during unoccupied hours on intermittent operation. Periodic checks of damper blades should be made to determine what percent leakage is occurring. This can best be accomplished in extreme weather conditions, to establish a wide temperature differential between inside and outside conditions to make testing easier. Adding neoprene seals to existing blades may help solve this problem. Checking damper leakage should be done periodically to determine that the damper is returning to a full closed position when required by controls. If a damper leaks more than 10 percent, it is probably less costly to replace the entire damper assembly with 100 percent positive closing damper blades than to use excess energy.

Systems using excess energy

Many systems are designed so that excess energy use is mandatory. I am referring to pre-

heat coils in minimum outside air intakes. Where minimum outside air requirements are less than 25 to 30 percent of the total air supplied to a space, in all probability there is absolutely no need for a preheat coil. Many times we find preheat coils with a discharge controller on the minimum outside air set at 65° F for a constant leaving temperature from the preheat coil. This becomes somewhat ludicrous when trying to maintain a mixed air temperature of 55° F in the plenum and the preheat coil controller comes on, heats up the minimum outside air, and adds to the total load of a building. In many cases, controls should be blocked completely out on a preheat coil; even better yet, take the coil completely out of the air circuit in the minimum outside air portion.

It is amazing how many people still use preheat coils when there is absolutely no need for them. They are afraid of some emergency that never occurs. Cost-wise, spending additional money to design anti-stratification chambers to properly mix the minimum outside air with the return air during severe winter conditions would be better.

In the past, rooftop systems have had very poor quality controls and outside and return air dampers. It is estimated that most rooftop mounted units can be upgraded to the point where 30 to 75 percent energy savings are possible by replacement of damper sections with higher quality ones and replacement of electrical-mechanical controls with solid state ones. Replacement of hot and cold deck zone dampers on a multizone unit should be considered if these leak excessively.

Perimeter heating systems

Now, let us turn our attention to perimeter heating systems. If a perimeter radiation system uses hot water as the heating medium, the hot water should be rescheduled as a function of the outdoor temperature. If installed by zones, it should be rescheduled through the action of solar compensators for each zone. The secret is to use the lowest possible output from the radiation. In fact, we would recommend that radiation be rescheduled to fall approximately 20 percent short in maintaining the heat balance in a space and that the net deficiency be made up by solar gain and internal loads. The ultimate in actual energy conservation with regard to perimeter radiation is individual room controls and automatic valves on each piece of radiation. The larger the zone handled by any one controlling element, the greater the likelihood of increased energy consumption.

If the control of perimeter radiation is in conjunction with other controlling elements, such as reheat coils, zone dampers, etc., a sequence should be set up so that the radiation is the last possible element called upon for heating purposes and conversely, the first element to disappear when space heating requirements are met. The best time to establish schedules for radiation is when the heating cycle is at its most extreme with regard to outside weather conditions and at night when the building is unoccupied and practically all lights are off. All other systems should be down when establishing a rescheduling sequence for perimeter radiation where water is used as the heating medium.

The refrigeration plant

Figure 38-8 shows the relationship between leaving chilled water from a refrigeration machine and leaving condenser water temperature for a 1200 ton centrifugal unit. Basically, this same set of curves applies to any centrifugal refrigeration unit. The only item that would change is the kilowatt input to the compressor at varying conditions.

Previously, the value of rescheduling supply air temperature from supply fans and raising chilled water temperature from the coil to the highest possible limit to prevent excess energy consumption at the refrigeration plant was discussed.

Figure 38-8 shows how savings occur. The top curve, based on a constant 96° F leaving condenser water temperature from a refrigeration machine, indicates that if the leaving chilled water temperature is raised from 42 to 45° F (only a 3° F spread), over a 5 percent savings in kilowatt input to a refrigeration plant can be effected. If the leaving chilled water is constant at 42° F and the capability of varying the leaving condenser water downward through some artificial means exists, depressing the leaving condensing water temperature from 96 to 93° F at a constant leaving chilled water temperature of 42° F, yields an expected savings of over 6½ percent in compressor input. Now, if elevating the leaving chilled water temperature and depressing the leaving condenser water temperature any time

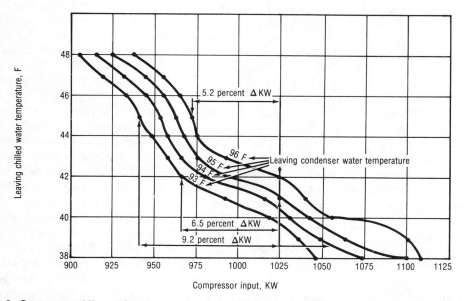

Fig. 38-8. Compressor kilowatt input versus leaving chilled water temperature and leaving condenser water temperature for a 1200 ton centrifugal refrigeration machine at 3600 gpm of condenser water.

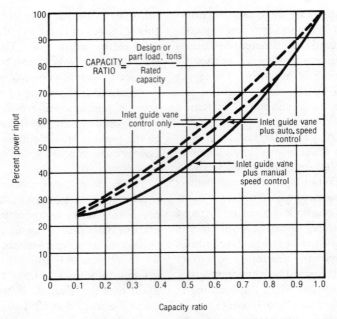

Fig. 38-9. Centrifugal refrigeration machine performance at part load operation based on different types of controls.

both are practical through the same 3°F spread on both the chiller and the condenser, it indicates that 9 percent energy savings in kilowatt input to the compressor can easily be picked up.

It is almost impossible to take full advantage of this type of savings manually. It requires the best instruments. Any wide error in this area has a significant effect savings because the temperature spread is so small. If an instrument has a 1°F spread, it has a significant impact on the output of a chiller. Generally stated, a 1½ percent savings in energy consumption can be expected any time the leaving chilled water is raised one deg. An additional 1½ percent energy savings can be expected anytime the condenser water temperature is depressed 1 deg in the ranges mentioned. Condenser water temperature must be kept high enough to make a condenser operate properly without causing refrigeration machine surging.

Figure 38-9 shows part load performance operation of a centrifugal refrigeration machine based on different types of controls. It is most efficient in its 30 to 80 percent loading ratio than at 100 percent as might be errone-

ously concluded. This curve shows that inlet guide vane control with manual speed reduction of the centrifugal wheel produces a more efficient operation than an inlet guide vane with automatic speed control. The automatic control must have a built-in safety margin to prevent the unit from surging, whereas with manual speed control, an operator with experience at handling the machine at partial load and low speed can bring the speed down until he hears surging take place.

One of the questions that naturally arises with regard to reducing condenser water temperature from the refrigeration machine is, how do you do it with a fixed system? One possibility, of course, is to increase the air flow through the existing cooling tower. Another answer might be the installation of actual additional cooling tower capacity. Or if the cooling tower is about to be replaced, select a larger one. An engineer must be careful in this area not to chew up the savings with added fan brake horsepower on the tower.

Table 38-6 shows some relationships between cooling tower capacity versus entering and leaving temperature conditions, fan brake horsepower, and erection cost. It is interesting

Table 38-6. Cooling tower capacity versus tower fan brake horsepower and erected cost.

No. cells	Water Temperature, °F		Fan, bhp	Erected cost, $*
	Entering	Leaving		
1	96	86	39	20,830
1	95	85	57	21,230
1	94	85	57	23,640
2	93	83	40	31,740

*Cost is based on labor rates in Des Moines, Iowa and includes railing, inlet streams, ships ladder, and totally enclosed motors. It does not include grillage or starters.

that to depress the leaving condensing water downward 3° F both in and off the tower and to keep the fan brake horsepower constant with a normal selection costs approximately 50 percent more in first cost of a tower installation. Since the tower uses no more energy than usual in terms of fan brake horsepower, the centrifugal unit picks up the energy savings at a rate of 5 percent throughout its life. Therefore this can be a good investment.

Figure 38-10 shows the relationship of input versus rated fan volume for a variable volume system. Many engineers are considering changing a constant volume system to a variable volume one and/or changing older fans or upgrading with different types of volume controllers. It is important to understand controls and the effects they have on the fan curve.

Figure 38-11 illustrates the need for keeping resistance in a system to a minimum. This might be through reduction in pressure drops across filters and/or selecting terminal boxes in variable volume systems or dual duct systems with the lowest possible minimum static pressure requirements. In many cases, it may be advisable to take one box with a large air quantity and a high pressure drop and substitute two smaller boxes with lower pressure drop needs to do the same job.

The curve in Fig. 38-11 is a typical curve for a backward curved double-width double-inlet fan operating in the low to medium pressure

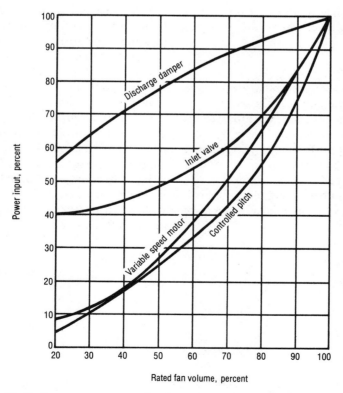

Fig. 38-10. Fan power consumption for various types of part load controls.

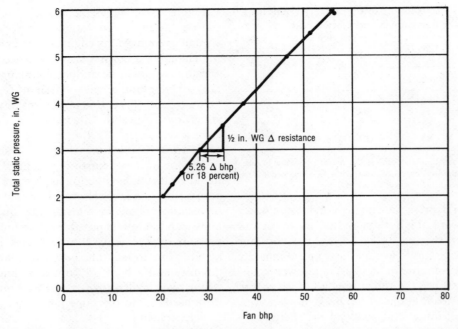

Fig. 38-11. A typical performance curve for a backward curved double-inlet double-width fan operating in the low to medium pressure range, providing 50,000 cfm.

range. A reduction in system static pressure of $\frac{1}{2}$ in. WG indicates an energy savings of about 18 percent.

Summary

The following is a list of items that are going to be demanded of all engineers in the future:
- An analytical attitude toward systems and a thorough knowledge of them.
- A knowledge of system performance curves, not only for fans but for the refrigeration machine and other equipment.
- A greater number of and more sophisticated controls with a higher degree of accuracy must be installed.

- An analysis of the various areas of a building to determine whether raising or depressing the temperature saves energy.
- A building analysis to determine the minimum outside air requirement possible.
- Initiation of greater maintenance programs.
- Examination of life cycle costing of equipment, controls, etc. If inhouse expertise does not exist, then it is recommended that a competent person be hired to do it.

Like Joe's body, we should understand the building body and proceed to correct its ills with the greatest expertise and techniques that are available to us.

39

Readjusting and rebalancing air and water systems for energy savings

A program of testing, balancing, and adjusting existing systems for maximum efficiency.

GUY W. GUPTON, JR., PE,
Gupton Engineering Associates, Inc.
Consulting Engineers, Atlanta, Georgia

If we, the American people are to continue to enjoy the highest standard of living in the world, we must evolve and implement an energy management program to combat and minimize the basic energy shortages that face us. According to reliable sources, the current energy consumption in the United States runs about 41 percent for manufacturing and processing, 25 percent for transportation, and 34 percent for people. In the last category, the largest part of that energy is for residential usage with about 15 percent of the total usage in commercial buildings. This 15 percent is the portion that we have to work with.

While ASHRAE Standard 90-75 defines the parameters for energy conservation in most new construction, there is nothing definitive available for programs of energy conservation in existing buildings. The purpose of this chapter is to help you evolve a plan of action for one part of the multifaceted area of energy conservation in existing systems. This chapter specifically deals with readjusting and rebalancing air and water systems for energy savings with the main purpose of stimulating your thinking along the lines of how to decrease energy consumption in existing systems while maintaining comfort conditions through a program of testing, balancing, and adjusting of building environmental systems.*
Let us consider how to use this tool of testing, balancing, and adjusting in an energy conservation program. But first, let's just call it testing and balancing, or T & B for short. A word of warning: the *A* for Adjust has been dropped, and that A may well be the most critical point of the whole testing and balancing procedure.

What is testing and balancing?

Testing and balancing work is known by many terms: TBA; test, balance, adjust; TAB;

This chapter is based on a paper presented by Mr. Gupton at a special series of two-day seminars sponsored and developed by *Heating/Piping/Air Conditioning* on How to Save Energy in Existing Buildings.

*A paper I presented to the University of Florida air conditioning conference entitled "Modification of Existing Systems to Conserve Energy" covers more than what is covered in this chapter and puts my remarks into context with my overall philosophy of energy conservation in existing buildings.

and T & B.† Placing these terms in a logical sequence, we have the following definitions:

• Testing, the quantitative determination of conditions within the environmental system's boundaries, including fluid flow rates, electrical energy flow rates, temperatures, humidities, and pressures.

• Balancing, the distribution of fluids within a system and to terminals by design or by manipulation of dampers and valves.

• Adjusting, the regulation of devices in a system to give desired flow rates, air flow patterns, control sequences, control device throttling ranges, and control instrument settings.

Why test and balance? It is an important procedure simply because "you can't tell the players without a program." Few complex mechanisms are as inscrutable as a building environmental system. One cannot look at a pipe or duct and tell in which direction how much of what fluid is flowing. These fluids, air and water, are the lifeblood of a building environmental system. Even after measurements of flow and temperature are made, system analysis is difficult without a plan of action.

In order to evolve a successful plan of action for energy conservation in building environmental systems, it is highly desirable (if not mandatory) that a three point determination be made on environmental systems covering the following points:

• *Point 1, the past or the designer's intent*—Information abstracted from contract documents, shop drawings, maintenance data, and the designer's files should be gathered and examined. These data can best be assembled by the original system designer, or if this is not possible, by a consulting engineer.

• *Point 2, the present*—A field test program of fluid flow rates and conditions of temperature, humidity, and pressure in the system should be carried out. These data can best be obtained by a test and balance organization.

† An excellent source of information on this subject is the *ASHRAE 1973 Systems Handbook*, Chapter 40, Testing, Adjusting, and Balancing.

• *Point 3, the future*—What conditions of temperature, humidity, and pressure are desired, along with what flow rates of air and water to effect energy conservation? This information can best be prepared by the system designer in collaboration with a test and balance organization.

Thus, as complete a review as possible of past intentions in light of present conditions will point out possible system modifications for energy conservation in the future. Testing and balancing is the tool for establishing present conditions and for providing the effectiveness of system modifications.

Which systems and when to T & B?

In considering why one should test and balance, we reviewed the three conditions of systems: past, present, and future. Using the data derived for Point 1, the past, and Point 2, the present, we can decide which systems present the best opportunities for energy conservation. This decision is a principal determinant in a plan of action.

When is the best time to test and balance: summer, winter, spring, or fall? It would be desirable to make readings at peak cooling and heating times, but time is of the essence in energy conservation. One cannot wait three months for one peak, then six months for another peak, only to find that the current year is not typical of weather patterns. Basic distribution systems must be measured, and their energy inputs predicted on hours of usage per year. The Btu processing systems energy inputs at current outside conditions can be measured, and work can proceed from that basis to estimate peak loads. Checking utility bills in comparison to degree-day reports can give a good prediction of annual energy consumption.

Who should perform T & B work?

After the scope of work has been defined, the next question is, who is to do the actual field work of testing and balancing? If the design engineer is equipped to do the work, he is the ideal party for the task. The fact is, however, that very few design engineers have the required instruments or the skilled technicians to operate them. There is, however, a highly specialized segment of the HVAC industry that provides testing and balancing services of a level of accuracy suited to these purposes. Such organizations are called test and balance organizations. They may be independent agencies having no connection with contracting organizations but operating under procedures set forth by a national organization, such as the Associated Air Balance Council; or they may be a part of a subsidiary to an installing contractor but operating under the procedures set forth by a certifying organization, such as the National Environmental Balancing Bureau.

In any event, the results of the entire energy conservation program depend heavily on the accuracy of the testing and balancing work. Do not skimp here; get the best.

How is T & B to be done?

Although the procedures for testing and balancing are complex and highly technical, there is a body of literature available that allows specifications to be written on the actual testing and balancing work desired for a specified program. The following are sources for this information:
• Associated Air Balance Council (AABC), the certifying body of independent agencies.
• National Environmental Balancing Bureau (NEBB), sponsored jointly by the Mechanical Contractors Association of America and the Sheet Metal and Air Conditioning Contractors National Association as the certifying body of the installing contractors' subsidiaries.
• Construction Specifications Institute (CSI), which offers a specification series that includes a guide specification Document 15050 entitled "Testing and Balancing of Environmental Systems." Reprints of this paper are available. It explains factors to be considered in using the guide specification for project specifications.

A plan of action

Now that we are familiar with testing and balancing, a plan of action for a project can be prepared. This plan of action consists of five steps.

The first step is to determine what the building systems were intended to do. In practice, testing and balancing data would be abstracted, as in Point 1, for study as to the relative volume of flow and basic energy consumptions for these flows. At this time, a review or reconstruction of the design criteria originally given by the system designer is necessary. Such a review should establish:

• The expected occupancy, as to type of usage and number of occupants.
• The outside air requirements for ventilation.
• The hours of operation for the systems.
• The lighting loads for the areas served.
• The fenestration and solar load control devices or methods provided.
• Technical data such as air flow, cfm; water flow, gpm; temperatures entering and leaving heat transfer elements; and design conditions for inside and outside in summer and winter.

Thus, in this step, the system design intent is determined. Now, we know how much of what should go where, when and why. We begin to have a feel for where the system energy is going, and what measures may be possible to reduce and optimize that energy consumption. This allows the formulation of the next step in the plan of action.

This second step is to determine which systems with which to work. This can be done by establishing a priority system, with high priority given to large energy users and lower priorities assigned on the combined bases of energy consumption and potential for reduction. On this priority basis, field tests of the selected systems are conducted for significant performance parameters, such as flow rates; temperature differentials; and values of temperature, humidity, and pressure at various points in the system.

Study of these results in comparison to Point 1 findings often provides major differences between design intent and actual operating conditions. Such differences can usually be attributed to three major causes.

• Lack of testing and balancing after original installation.

• Degradation of performance from equipment deterioration or inadequate maintenance.

• Improper operation of equipment and systems, either by operating staff or for a lack of operating staff. Working from the findings of this study, it is possible to formulate the next step in the plan of action: to plan for future operature of the system.

Planning is most difficult step

The third step is the planning of actions to be taken with the building systems as to adjustment, modification, or change in operating methods. This planning is the most difficult step. Decisions made here determine to a large degree the success of an energy conservation program. Planning may cover:

• *Flow reduction for air and water*—There is good potential for savings here. In a given system with a centrifugal fan or pump, a reduction to 90 percent of original flow results in 81 percent of original frictional losses and 65 percent of original energy requirements. Remember that a flow reduction of 1/10 yields an energy reduction of $\frac{1}{3}$, but only if the basic fan or pump is changed. Change belt drives for fan speed reduction of the same percentage as the desired flow reduction, and trim pump impeller for pump capacity change as determined from the manufacturer's characteristic curves. In each of these procedures, adequate provision must be made for system changes over a time, such as filter loading and changing cycles, or piping system friction increases due to interior pipe roughening and strainer blockage. To minimize required head allowances, investigate the possibility of changing filters to a type with increased dirt holding capacity and lower dirty pressure drop, or of changing strainer media to the largest openings that will protect pumps, valves, heat exchangers, and spray nozzles from damage or blockage.

• *Ventilation air reduction*—When basic air flow rates are changed, outside air volume often is reduced in the same proportion. Consider how much outside air is required and the exhaust rates required for toilet rest rooms.

Code requirements, though intended to establish absolute minimums for public health and safety, often result in excess outside air over that needed for makeup of oxygen and dilution of carbon dioxide and odors. Some cities have adopted ASHRAE Standard 62-73, which establishes a basic 5 cfm per person outside air requirement. A word of caution should be applied here: It is necessary to obtain approval of the code enforcement agency at the place of the building to reduce outside air below code requirements, and OSHA enforces codes, too.

The 2 cfm per sq ft rule in common use as a toilet exhaust rate has its basis in very old plumbing codes. There is no justification for this rate. Restudy exhaust requirement in light of ASHRAE Standard 62-73. If the code body will not give easement, to say 1 cfm per sq ft, it may be possible to reduce statutory requirements by close measuring of floor space just for toilet rooms, omitting vestibules and powder rooms.

• *Lighting load reduction*—With lighting loads often accounting for 25 to 40 percent of total space loads, load reduction by relamping or lamp removal can be significant. Some cautions to follow are: watch for changes in heat-of-light systems, and other systems that offset roof heat losses in winter with recessed lighting heat. Also, watch for high latent heat areas. The system may depend on the heat of lights to avoid need for reheat for humidity control.

• *Solar load reduction*—Most buildings offer no possibility for reduction in the glass areas on the skin. Solar load reduction must come from shading or reflective techniques. Among those reasonably available are:

• Venetian blinds or draperies—Light colored blinds and opaque draperies with metallic thread linings are most effective. Existing blinds often can be improved by cleaning and repair of operating devices. Existing draperies can be lined with material similar to the astronauts' sunsuits.

• Outside shading—Sunshade screens have a good cost-effectiveness ratio. Low shadow angle screens can be used on southern exposures to cure that paradoxical condition where maximum cooling loads occur on December 21st. Architectural shading devices can be equally effective, but they are much more expensive. In Point 1, determine whether new buildings around the structure are providing more shade or whether old buildings have been removed, giving more solar load exposure.

• Reflective films—A number of easily applied, semipermanent, silvered plastic films are available that are very effective heat reflectors, but they may be objectionable from a purely aesthetic point of view.

• Change of glazing—In some cases, a change to reflective coated glass or to double glazing in existing or new frames may be justifiable.

• *Transmission and infiltration reduction*—The geometry of the building is established. The orientation of the areas of exposure and the stack effect due to height cannot be changed, but transmission and infiltration loads can be reduced. Available techniques include:

• Walls—Insulation is the answer and dead air is an insulation. Thus, furring inside existing solid masonry walls with foil-faced gypsum boards can be effective. Brick veneer walls can have insulation blown in. Steel buildings can have a combination of sprayed-on and batt-type insulations. Light colored paints can reduce the solar heating of exposed walls to reduce cooling loads.

• Roofs and ceilings—Once again, insulation is an answer, but reduction of summer temperature differentials is another useful tool. Insulation can be most readily applied over lay-in ceilings but may also be blown in over solid ceilings. Caution: do not insulate over backs of recessed lighting fixtures where reduction in room load has been taken due to heat transfer to the ceiling cavity. Where a *good* roof exists, spray cooling of a roof by evaporation of water is a practical approach. Once again, light colored coatings can reduce solar heating to reduce cooling loads. Consideration may be given to converting ceiling cavities to return air plenums in fireproof construction with insulation on the roof. The ceiling return will take up to 50 percent of the heat of lights onto the return air stream, thus reducing the supply air flow requirements to the

space plus giving some decrease in roof cooling and heating loads.

• Infiltration at doors and windows—Sealing cracks through weatherstripping and caulking is a simple and effective remedy for low buildings. In higher structures, crack sealing at all levels is needed as well as pressure differential reduction through addition of vestibules at entrance doors, closing doors and windows in stair towers, changing smoke relief openings in stair towers to automatic opening types, and addition of fan equipment for pressure control in return air shafts.

• *Equipment operation optimization*—A little tender loving care may be very effective in reducing the energy input to a piece of central station equipment, such as a boiler, chiller, or cooling tower. At least once a year, preferably just before entering the next season, a factory trained technician with current operating instructions should be retained to inspect, check out, and adjust the equipment for optimum performance. Work done in an inspection might include:

• Boilers—Adjust burners for maximum efficiency at firing rate other than maximum to suit actual system load, adjust air and draft set controls for proper cut-in and cut-out points, and check flame safety controls.

• Chiller—Clean tubes, charge refrigerant to proper level, trim concentration of absorption refrigerants, and set controls for desired temperatures.

• Cooling towers—Adjust fan blade pitch or fan speed for minimum power input, set condenser water bypass controls, set winterization control set points, and clean heat transfer surfaces.

• *Control system checkout*—More energy is wasted through automatic control systems than through almost any other means. Controls are often just installed and left to do their best. In some cases, the control system installer may set overlapping cooling and heating functions to give smooth and tight control without a dead band. This is good control but poor economics.

For those companies trying to meet the 68 to 78°F conditions of the 1973 presidential order, a widened throttling range can give low/high limits at 68 and 78°F with cooling in the mid-band at light load but swinging to the high limit on maximum load.

Reset schedules bear review and rejustification on outside temperature reset for hot deck and hot water circulating temperatures. Water temperatures in reheat systems should be experimented with to find the lowest temperature that will satisfy the space.

• *Operating hours reduction*—After the system cooling-heating loads are reduced, the fluid flow rates reduced to satisfy those loads, the central system performance optimized, and the control system adjusted, the next step is to reduce the actual hours of operation. Techniques used may include the following and an infinite number of combinations of them:

• Time clock or program control—These inexpensive devices may be had in 24 hour, day-night astronomic dial, and seven day versions with spring carryover for power failures of several hours duration. One clock per system is justifiable.

• Manual interval timers—These should be used to overcall the off cycle of program timers and allow the system to run for the time interval manually set before automatic shutdown.

• Night limit cycles—To avoid excessive space temperature drops in winter, inside space thermostats start up the heating system when the temperature reaches about 50° F. The heating system runs until about a 5° F rise is achieved without outside air or interlocked fan operation.

• Warm up cycle—Delay opening outside air dampers and starting exhaust fans on cold startup until space temperature is within a few degrees of normal operating temperature.

• Outside temperature control of hot water circulators—While many systems may be set up for full-time operation of hot water circulators below 40° F outside temperature, it may be possible to change the cut-on point to 30° F with safety.

Execute the plan

The fourth step is to execute the plan. We can cut and try, or do it all at once. The system designer should oversee the execution of the work in order that unexpected changes are not made by overzealous craftsmen that would nullify the energy conserving aspects of the changes.

By this time, an additional party should have been brought into the program; he should be the utility representative. He may be one man or a team of several, but he can help by providing for the temporary use of metering equipment in the final step of the plan.

The proof of the pudding is in the eating. Your efforts in the earlier steps of the plan of action bear fruit now in the fifth and final step. After completion of the planning work, another set of test and balance observations must be made. By comparison of these readings with the readings of Point 2, an indication of the effectiveness of the work can be obtained. By reading temporary meters set up on specific electric loads, usage patterns can be established. After a month's operation under the energy conservation program, compare the meter readings with previous year's readings. If the operating results are not as predicted, find the reason, and make corrections.

Combine the month-by-month review and comparisons of energy consumption for the first year, making any adjustments to the program that are indicated in a month's operating results. At the end of the year, study the data in terms of energy consumption, before and after. Extend the "before" period consumption using the "after" period rates, and compare the actual "after" billing.

This is the proof of a program's effectiveness. This is the bottom line!

40

System tuning

WILLIAM J. COAD, PE,
Vice President, Charles J. R. McClure
and Associates, St. Louis, Missouri

In the study of energy economics relating to energy conversion systems, the first step is to identify the two fundamental components of the analysis, product energy and process energy.

In building systems, *product energy* directly satisfies space environmental needs, in the form needed. Some examples are: the heat introduced into a room at a rate identically equal to the rate at which heat leaves the room to the surroundings, and heat extracted from the room at a rate exactly equal to the rate at which it enters the room from the surroundings, or relevant internal sources, such as occupants, lighting devices, and appliances.

Process energy is consumed by the "system" to satisfy the product need. Examples are: energy consuming subsystems, such as pumps, fans, psychrometric control, conditioning excess ventilation air, and all forms of energy losses.

The total energy required, say annually, by a building is then the sum of the product energy and the process energy.

Experience in analyzing numerous actual buildings and building systems has revealed that the process energy components far exceed the product components in the vast majority of cases—especially if the product energy for refrigeration is based on ideal Carnot relationships.

This chapter is directed at the need for careful initial and ongoing adjustments of the various subsystems and components to achieve minimization of the process energy.

It should be noted that all process energy is considered energy loss.

Background of design changes

The field of building systems technology has undergone significant technological advances throughout the past 30 years. The pressures motivating these advances were such that the primary design parameter (and evaluation of success of the design) was the system performance—that is, did it satisfy the product energy requirement?

As a result, very little attention was directed to process energy requirements, and the systems designed were most forgiving of improper adjustments. Some examples are:

• Oversized primary conversion capacity negated the need for fine tuning the apparatus. Even when poorly adjusted, the oversized equipment or subsystem produced the output needed.

• Low design temperature ranges in fluid transportation or conveying systems enabled most heat transfer elements of the system to perform with flow rates far short of the intended design. For example, the flow rate versus capacity curve in the ASHRAE Systems Volume illustrates that for a 20°F temperature drop coil, 50 percent design flow provides 90 percent design heat transfer.

• Multiple zoning of high velocity dual duct air systems, when designed for the sum of the peak flow rates, were capable of maintaining space conditions regardless of system imbalance—a control that resulted from the correct proportionate mixture of high and low temperature air streams.

In the past decade, however, two significant changes have occurred—the realization of the limited supply of energy in the form needed for building systems and increased costs of investment monies. These, plus simultaneous advances in computer hardware and software technology, have resulted in the recognition of other design parameters that have been recognized as an equal to that of ultimate performance. This recognition has resulted in the following:

• Extremely exacting analysis of the product requirement in the form of building design loads, daily time integrated loads to carefully identify not only peak individual space or room loads but maximum coincident system loads and annual load energy requirements.

• The selection and design of equipment and subsystems to match the calculated diversified loads.

• A reduction in the dependency of simultaneous heating and cooling in an effort to minimize psychrometric control process burdens.

• Maximizing temperature ranges in fluid transportation systems for the purpose of both reducing pipe sizes and pump energy.

• Variable flow fluid systems (both transport fluids and conditioned air) to reduce annual energy burden.

The systems thus designed are much less forgiving of poor adjustment or balance. As a result, the performance has not been as successful in cases wherein the adjustments were not achieved.

Also in the earlier systems, in virtually all of the maladjustments that went unnoticed because of the ability of the system to adjust, such inherent system adjustments were at the expense of increased process energy consumption.

Consider adjustment, balancing

There are two fundamental requirements in the concern of system adjustment and balancing.

• Initial adjustment—this must be performed before putting a new system into operation. In the case of existing systems either being modified or subjected to a newly established energy management program, the initial adjustments must be completed before the management program commences.

• On-going monitoring—tests and adjustments that must be made as a part of the continued ownership responsibility in carrying out the preventive or planned maintenance program.

Initial adjustments are the responsibility of the system designer. Unless the designer recognizes these requirements at each and every phase of both design development and contruction monitoring, there is little possibility of successful accomplishment. Suggested guidelines for the designer to follow in addressing this responsibility are:

• Develop flow diagrams during each phase of design development for all thermal fluid subsystems, such as air, water, steam, ahd refrigerants, prior to the actual layout, and modify and update them as the layout proceeds. The flow diagram is the only tool that presents a visual indication of the overall system. Subsequently, all balancing, testing, and adjustment points and devices should be identified on the flow diagrams and be included in the construction documents.

• Develop system schematics for psychrometric and other energy conversion subsystems. These are analogous to the flow diagrams for the fluid systems and should be developed for such subsystems as air handling units, chillers, and converters. The schematic should include such devices as control dampers, control valves, and temperature control devices. These, like the flow diagrams, should include heat transfer data, flow data, points of measurement for adjustment, and adjustment devices, and be included in the construction documents.

• Design a "balanceable system." With the help of the diagrams, the hydraulic and thermal dynamic interrelationships of the system components upon one another at both full and reduced load conditions should be identified. It is at this phase of design that "designed-in" problems are minimized. Such potential problems as unanticipated series pumping phenomena, reverse flow paths, fluid overdraw, etc., become evident when the dynamic interrelationships are carefully considered.

- Assure adequate metering characteristics at measuring points. At all points where velocities are to be used to measure volumetric flow rates, for instance, adequate velocities must be assured and designed in. Likewise for pressure differential measurements, adequate drops must be provided.
- Construction plans and details should be carefully coordinated with the flow diagrams to assure compatibility and to assure adequate access to measurement and adjusting points.
- A logistics mechanism for effecting the testing and adjustment program must be achieved. This mechanism is dependent on the availabilty of technical expertise in the locale of the building. The technical expertise may be in the form of independent testing and balancing firms; construction contractor in-house capabilities; or in some cases, this service may have to be performed by the design engineer himself.

Following recognition of the available technical expertise, the logistics mechanism must include recognition of the extent of the work to be done when establishing the construction budget; a clear definition of the work and identification of the acceptable technical expertise in the construction specifications; a mechanism for achieving coordination between the design engineer, the installing contractor, and the testing and adjusting team; and a method for verifying that the work has been successfully accomplished.

System monitoring is necessary

The on-going monitoring of the system adjustments is, needless to say, the responsibility of the building owner. However, this phase of preventive or planned maintenance is a totally new concept to most owners, if not to many building systems design practitioners. Thus, a good deal of education is needed.

This education must proceed as a dissemination of information among the engineering community through engineering societies, technical publications, and seminars. Simultaneously, the informed design professionals should initiate the task of informing their clients, the building owners and managers, during the embryo stages of project description and economic analysis.

The flow diagrams and subsystem diagrams discussed previously will prove a useful tool in the continued monitoring and adjustments of a system. Design firms with the capability should consider the possibility of providing the services of maintenance management consultation to building owners, and construction contractors with the capabilities should consider the possibility of providing contracted system maintenance that should include on-going testing and adjusting.

The preceding discussion was directed toward procedures relevant to new building projects. However, existing building systems, which have been constructed without the benefit of this planning, should be addressed. These very likely consume more process energy by inherent design than the future buildings will. As a result, there is an ever-increasing demand for energy analysis and resulting steps at reducing the total energy consumed by such systems.

The first step in this analysis is to develop a mathematical model of the building energy systems that will accurately identify the sources of energy consumption—both product and process. This initial step (the results of which can be verified by historical energy consumption records) requires, in most cases, precise testing of the present operating adjustments and maladjustments. Once the operations modes have been verified by field testing, the mathematical modeling currently available through computer programs should be acceptably accurate. If deviations from acceptable norms of accuracy are not achieved, additional field testing will be necessary to reveal the cause.

When this iterative process of calculation and testing achieves compatibility with the operations records, many of the areas of excess energy use will have already been identified quantitatively.

The next step in the retrofit process is to use the mathematical model (the accuracy of which has now been confirmed) to determine the effects of changes on the product requirements (such as alternative fenestration systems) and process requirement (such as reductions in total air flow or variable water circulation, and adjustments) on the annual energy consumption. After an acceptable program for modifications in both apparatus and operations techniques has been established, the fundamental requirements of initial adjustment and on-going monitoring discussed previously should be effected.

Needless to say, any program that does not continually monitor the results to confirm the success of the initial or on-going adjustments will not have confirmed either the benefits of the effort itself or the degree of success of the responsible agent.

SECTION V

Operation and maintenance of systems for energy conservation

If proper operation and maintenance of heating, ventilating, and air conditioning systems is not performed, much energy will be lost. These 4 chapters discuss good preventive maintenance procedures as follows:

- Planned vs. breakdown maintenance.
- Scheduling and optimizing system and building use.
- Maintaining and servicing equipment efficiently.

41

Planned versus breakdown maintenance

WILLIAM J. COAD, PE,
Vice President, Charles J. R. McClure
and Associates, St. Louis, Missouri

A building systems design engineer is now expected to provide considerably more input to a life cycle cost study of a building venture than he was in the recent past. The current dynamic nature of technology and the state of the art in the building industry have invalidated many of the statistical norms widely used as recently as the last decade. In a life cycle cost analysis, the three major ingredients controlled primarily by a system designer are: first cost amortization, energy costs, and maintenance/service/operations costs. The latter will be discussed with emphasis on maintenance and service costs.

Maintenance and service costs (M/S) should be grouped into a single entry, simply because well managed maintenance efforts initiated during the environmental system startup or debugging phase will reduce service costs during a building's life to a minimum. Conversely, without well managed maintenance efforts, service costs can be a serious financial burden.

Before attempting to quantify M/S needs and subsequent costs, two questions must be answered:

- How long does the owner intend to own and manage the property?
- Are special skills that might be needed for maintenance and service of complex apparatus available in the area, and at what cost?

The answer to the first question dictates whether at time zero a planned maintenance program should be put into effect. If the owner-developer intends to retain the property for a long time, the only intelligent course of action is to instigate such a program. Examples of this type of building are institutional buildings or commercial buildings planned for long time ownership by major real estate holding firms or major tenants.

The other alternative, short term ownership (less than 15 years), might dictate the breakdown maintenance or total dependency on service approach. Whether by plan or not, this latter approach is employed in the *majority* of today's buildings. Examples of projects where this approach is intended are the so-called blue shoe commercial buildings where the owner-developer intends to retain ownership long enough to achieve a financially beneficial crossover point between equity growth and depression tax shelter benefits.

Annual M/S cost versus years after startup for long term ownership is shown in Fig. 41-1. If a planned maintenance program is followed, system components will be in as good a condition after a projected 30 years as at the beginning. The only deterrent to permanency is obsolescence.

Figure 41-2 represents a typical M/S cost curve for short term ownership. Generally, the crossover point occurs between 5 and 12 years.

The answer to the first question not only

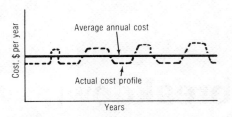

Fig. 41-1. Annual M/S costs of a planned maintenance program.

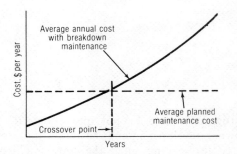

Fig. 41-2. Annual M/S costs of a breakdown maintenance program.

affects a designer's projection of life cycle cost, but it will also seriously affect the selection of machinery and components. The absence of the answer to this question has in numerous cases resulted in system designers being severely criticized for misdirected decisions. The results of which are that long term owners are stuck with a property planned otherwise, or rebound owners are not financially prepared for high M/S costs.

Unless there exists a pre-programmed written maintenance program and procedure, it is *most* likely that breakdown maintenance is the operating procedure.

The second question is best addressed by a few brief examples from personal experience.

Of the many so called total energy systems (systems which convert available fossil fuel into all the necessary end use energy forms for the building system) installed, several have been removed and replaced with conventional forms of available energy. Considering that the concept is valid, once installed, the investment monies spent, and unless extremely ill conceived in energy and load balance, the *only* reason these plants have been removed is the unavailability of necessary skills to maintain and service the machines.

An absorption refrigeration system installed in a remote area is another example. A lack of readily available maintenance and service skills led to its removal and replacement by a vapor compression system.

After addressing these two questions, a designer can establish the philosophy underlying the M/S aspects of a system design and eliminate certain classes of machines or subsystems. The next step is to assign M/S values to the selection of all system components or subsystems. This is perhaps the most difficult parameter of a system to quantify.

42

Scheduling and optimizing equipment operation and building use

Five building modifications reduced energy cost 60 percent in nine months.

CHARLES F. SEPSY,
Professor and Research Supervisor,
Mechanical Engineering Dept., and

ROBERT H. FULLER,
Research Associate, Ohio State
University, Columbus, Ohio

The Ohio State University is actively engaged in an energy conservation program, the results of which are surpassing the energy savings anticipated when the program was proposed. The program goal stated in the original proposal was to obtain a 20 percent average energy reduction in buildings programmed. The estimated dollar savings on over 20 buildings now studied range from 30 to 60 percent. Actual metered results on several buildings already modified show that the estimated savings are conservative.

The following information is presented to give some of the background and results of this project at the University.

This chapter is based on a paper originally presented by Professor Sepsy at a special series of two-day seminars on "How to Save Energy in Existing Buildings" that were sponsored and developed by *Heating/Piping/Air Conditioning* magazine.

Background details of project

In early 1973, a proposal for the study of energy conservation in campus buildings was presented to Dr. Edward Moulton, vice president of Business Administration. At that time, the University utility bill was approximately $5,000,000 per year and had been increasing at an average rate of 15 percent per year, due to increased demand from new and existing building construction and remodeling. While the past rate of increase was determined during a period of stable utility rates and before the occurrence of oil embargoes and reductions in gas allocations, it was evident that increased utility rates would occur. The University saw that the energy situation demanded action to conserve energy on the campus. Therefore, project funding was obtained for a two year period, and an energy conservation coordinating committee was created to oversee the campus activities.

The organization and funding of the project are unique. The coordinating committee consists of members from the faculty and physical facilities. The committee develops and coordinates conservation efforts throughout the campus. Some typical conservation programs with which the committee has been involved are:

- Summer closing of classrooms to reduce air conditioning.
- Advertisements in campus newspaper to solicit the help of faculty, staff, and students, in energy conservation and inform them of progress made on the campus.
- Scheduling building chiller spring start-up and fall shutdown dates.
- Establishing room thermostat set point temperatures for summer and winter operation.
- Developing guidelines for lighting levels by which building lighting is reduced.
- Shifting summer office hours one-half hour earlier to allow cooling systems to be shut down earlier in the day.
- Moving fall quarter ending date two weeks earlier to allow buildings to be shut down for a longer period during the break at the end of the quarter, thus reducing gas use during the winter when allocation reductions are expected to be most severe.
- Establishing a central campus control center to monitor and control building systems.
- Conducting classes for maintenance personnel to make them aware of how to operate systems and conserve energy.

Although the committee is involved in many projects, the area that is of primary importance to them is the study of campus buildings. This project is directed by Charles Sepsy of the Mechanical Engineering Dept. and Dallas Sullivan from Physical Facilities. It is a joint effort between Mechanical Engineering and Physical Facilities and is treated as a research project of the Mechanical Engineering Department with Physical Facilities as the sponsor. Funds are provided to support the staff of faculty, engineers, and students. In this way, money is used not only to reduce energy costs but provide financial support for students and also involve them in an educational research project.

The objective of the program is to study energy use in campus buildings and identify areas where systems can be modified to reduce consumption. One point should be emphasized. The operational function of a building is not changed to suit the needs of energy conservation. The environmental needs of the operations performed in this building are evaluated, and systems modifications are proposed within the limits set by those requirements.

The results of the building study are presented to the coordinating committee in a formal report. It reviews the report and recommends modifications for implementation based on the estimated payback period.

Building studies

To demonstrate how the project works, it will be helpful to discuss a specific building.

The second building study performed by the project was on McCampbell Hall, which is a four story building consisting of offices and conference rooms and is used by the Nisonger Center for the Mentally Retarded. It also contains six small apartments used by those receiving the services of the Center. Gross floor area is about 109,000 sq ft.

The cooling system is a 385 ton centrifugal chiller (sized for future expansion) with a cooling tower in the penthouse. Heating hot water is provided by two 150 hp gas fired hot water boilers, which are also located in the penthouse. Steam for humidification and for equiment located in a small kitchen used for teaching purposes is provided by a 60 hp steam generator located in the penthouse.

The four main building air handling units are high velocity, dual duct; each unit originally provided approximately 25,000 cfm, which breaks down to 0.95 cfm per sq ft of floor area, with approximately 20 percent outside air.

The building control system is pneumatic with a central control console located in the penthouse.

This study was conducted in three stages. In the first, contract documents were reviewed, and operations were confirmed by tests and observations of the existing system. The tests established actual air flow rates and motor amperage readings, while a room by room survey showed individual space usage, lighting wattage, and occupancy.

The second stage was the arrangement and analysis of data that allowed the energy consumption of McCampbell Hall to be broken down into categories. First the electrical consumption was divided in the following classifications: ventilation, lights, cooling auxiliaries, elevators, heating auxiliaries, and miscellaneous.

Manual calculations were used

Manual calculations were used to establish the percentage of use of each classification. The results are shown in Fig. 42-1A.

Then, manual calculations were made to divide the gas consumption into winter, summer, and mixed seasons heating, and miscellaneous gas categories. The percent breakdown is shown in Fig. 42-1B.

Figure 42-1C shows the cost of energy consumed by McCampbell Hall using actual metered data from the fiscal year 1972-73 and average gas and electrical cost to the University in September 1974. The utility costs used in the calculations shown in this chapter are average costs to the University for the month stated. They do not reflect the actual costs that would be incurred if the building were separately metered with a demand meter and billed by the utility. That is, all electrical costs and savings are calculated at an average rate; thus, lights that are on during a peak demand period are assumed to cost as much to operate per kilowatt-hour as lights that are on during evening hours when actual cost per kilowatt-hour is lower.

In the same manner, gas costs and savings are calculated using average gas costs to the University on a per mcf (thouands of cubic feet) basis. This is not completely accurate, since in the winter, reductions in natural gas allocations have forced the firing of oil in many boilers at approximately 2.5 times the cost of gas.

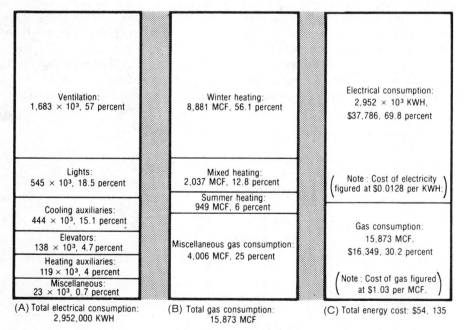

(A) Total electrical consumption: 2,952,000 KWH

Ventilation:
1,683 × 10³, 57 percent

Lights:
545 × 10³, 18.5 percent

Cooling auxiliaries:
444 × 10³, 15.1 percent

Elevators:
138 × 10³, 4.7 percent

Heating auxiliaries:
119 × 10³, 4 percent

Miscellaneous:
23 × 10³, 0.7 percent

(B) Total gas consumption: 15,873 MCF

Winter heating:
8,881 MCF, 56.1 percent

Mixed heating:
2,037 MCF, 12.8 percent

Summer heating:
949 MCF, 6 percent

Miscellaneous gas consumption:
4,006 MCF, 25 percent

(C) Total energy cost: $54, 135

Electrical consumption:
2,952 × 10³ KWH,
$37,786, 69.8 percent

(Note: Cost of electricity figured at $0.0128 per KWH.)

Gas consumption:
15,873 MCF,
$16,349, 30.2 percent

(Note: Cost of gas figured at $1.03 per MCF.)

Fig. 42-1. Bar charts show the results of the study made on McCampbell Hall. 42-1A shows the electrical energy distribution; 42-1B shows the distribution of gas consumption; and 42-1C shows the costs of electricity and gas for the building. All data collected from July 1, 1972 to June 30, 1973.

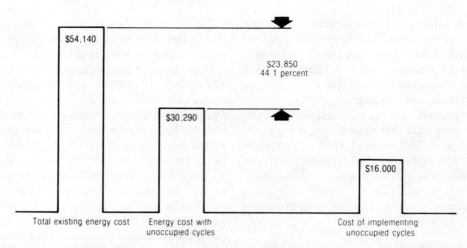

Fig. 42-2. The effect on energy consumption in McCampbell Hall by Modification No. 1, unoccupied cycles.

Energy savings methods analyzed

With this information, the study group was able to establish five energy savings methods that were analyzed in the third stage of the study. Calculations were made using the same manual methods previously used to determine the gas and electric usage breakdown.

The first energy modification studied was the initiation of unoccupied shutdown of air handling systems. A review of occupancy showed that, with the exception of the six apartments, operation of the fan systems was needed only ten hours a day, five days a week. Thus, for 14 hr per day plus weekends and holidays, 190 hp of electric motors could be

shut down. Therefore, during these periods, 19,000 cfm of outside air would not require heating or cooling. Figure 42-2 shows a saving of $23,850 per year in gas and electricity, or a total reduction of 44 percent.

The second modification analyzed was the reduction of outside air quantities. The survey showed that 3000 cfm of outside air could be eliminated by shutting down several unnecessary exhaust fans. The savings realized by reduced exhaust during occupied hours is shown as $120 per year in Fig. 42-3. This economy was calculated assuming that the unoccupied modification would be implemented; therefore, savings for night and

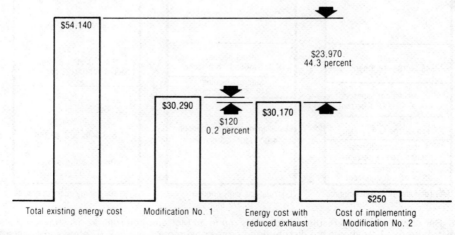

Fig. 42-3. The effect on energy consumption due to Modification No. 2, reduced exhaust in McCampbell Hall.

weekend heating and cooling outside air are not included.

The third energy conservation modification analyzed was removal of excess lights. Room surveys showed that lighting levels were above those required for the tasks being performed.

Save energy by reducing lighting

Lighting reduction was accomplished by use of a guide to lighting level adjustment, which was developed during the energy conservation project. The guide contains some points of value to those involved in building lighting, and an abstract of the document is shown in the chapter.

Review of the building lighting, using these guidelines, reduced lighting and cooling loads and allowed the supply air volume to be lowered; as an added benefit, air noise that had been a problem was eliminated.

Figure 42-4 shows an estimated savings for reduced lighting and air volume of $4980 or 9.2 percent of the total utility costs. Again, the unoccupied modification was assumed to be implemented so that savings for air volume reduction were calculated only for the 10 hr per day, five days per week operation period.

Convert to variable volume system

The fourth energy saving modification study was the conversion of the constant volume system to a variable volume system. This analysis was made using the approach of changing the existing constant volume boxes to provide variable volume in interior spaces where large cooling load variations were anticipated due to widely fluctuating occupancies. Figure 42-5 shows the estimated yearly energy savings for this modification as $7110.

The conversion to variable volume was rather simple for this system since the fans were equipped with inlet vanes. Fan speeds were reduced to accommodate the air requirement of 0.62 cfm per sq ft versus the original 0.95 cfm per sq ft. The fan system required only the proper controls to modulate the inlet vanes to change volume according to the demand of the terminal units.

The existing constant volume terminal units were converted to variable volume units by installation of a sheet metal blank-off plate in the hot duct connection to the boxes. Figure 42-6 shows operation of the converted box. With flow through the hot duct, the mixing damper acts as a volume damper on the cold duct.

The fifth modification includes enthalpy control and economizer reset. The original system controls were set so that at outdoor temperatures above 55° F, the chiller comes on, and 100 percent outside air is drawn in until the outside temperature reaches 70° F. Above 70° F, only minimum outside air is

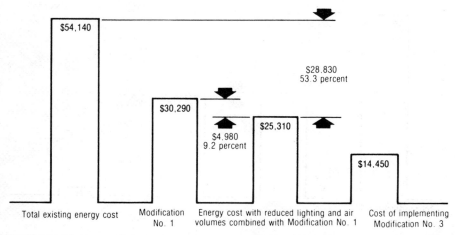

Fig. 42-4. The effect on energy consumption due to Modification No. 3, reduced lighting levels and air volumes combined with unoccupied cycles.

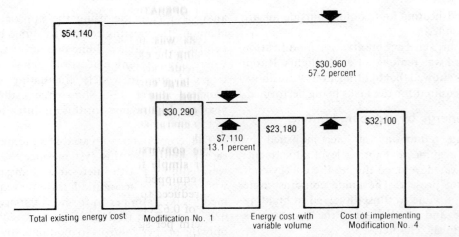

Fig. 42-5. The effect on energy consumption due to Modification No. 4, variable volume with unoccupied cycles.

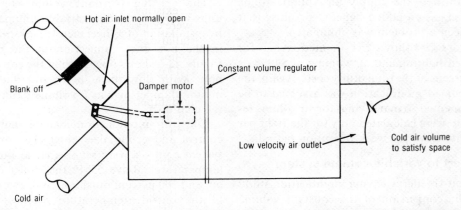

Fig. 42-6. Conversion of high velocity dual duct, constant volume mixing box to variable volume cooling box.

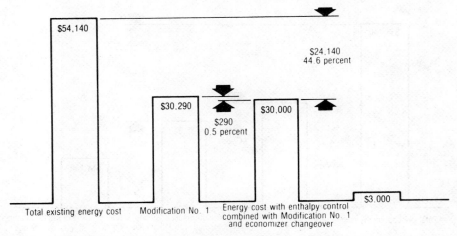

Fig. 42-7. Effect on energy consumption due to Modification No. 5, enthalpy control and economizer changeover combined with unoccupied cycles.

taken in. The 70° F setpoint for changeover to minimum outside air is a very good choice for Columbus, Ohio. Therefore, the added savings for enthalpy control shown in Fig. 42-7 of $290 per year are very low.

Since it was felt that all five of these modifications should be incorporated into McCampbell Hall, and it was realized that the savings calculated for each indivdual modification were not additive, it was necessary to perform a calculation estimating savings realized when all five modifications were incorporated into the building. As shown in Fig. 42-8 the yearly estimated cost saving was $34,580 or 63.9 percent.

It should be noted that these predicted energy savings were accomplished without computer analysis, using information currently available from ASHRAE and the National Oceanic and Atmospheric Administration.

Another noteworthy point is the great advantage inherent in working with an existing building. A survey can be made to determine exact space usage, and then the environmental system can be adapted more precisely to the actual requirements of the space, often resulting in improved comfort and reduced energy usage.

Estimates on energy savings are an academic exercise. The actual results make a program successful. For McCampbell Hall, the results are impressive for the first nine months

of 1975. As shown in Fig. 42-9, when compared to the same period of the 1972–73 fiscal year, the gas and electric usage are both 60 percent lower. This resulted in a 60 percent reduction of utility costs for the first nine months of 1975 when calculated at average rates for September, 1975, which were $0.0180 per kWh and $1.32 per mcf as compared to the rates used for the original report of $0.0128 per kWh and $1.03 per mcf.

Results show big savings

Table 42-1 shows results for McCampbell Hall and for three other buildings now in the process of being modified.

The savings achieved for the individual buildings along with the savings from the other programs initiated by the coordinating committee have reduced overall campus energy consumption. A report prepared by Dallas Sullivan for the fiscal year 1974–75 credits energy conservation efforts for reducing the use of heat energy (natural gas and oil) by 13 percent and the consumption of electricity by 8 percent. These were accomplished in spite of the fact that over 500,000 sq ft of new buildings have been added to the campus energy requirements since the base period used for the comparison, fiscal year 1972–73. The total dollar savings attributed to energy conservation for the 1974–75 fiscal year is $1,230,900.

The savings reported in this chapter have

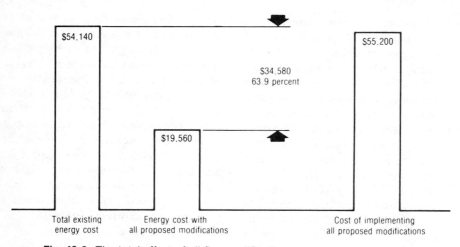

Fig. 42-8. The total effect of all five modifications on energy consumption.

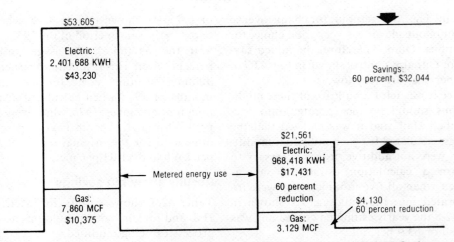

Fig. 42-9. Estimated energy savings for the period January 1, 1975 to September 30, 1975. Savings were calculated using September 1975 costs of $1.32 per mcf (thousands of cubic feet) and $0.018 per kWh.

Table 42-1. Metered results of buildings being retrofitted for energy conservation for the period January 1, 1975 to September 30, 1975, Ohio State University.

Building	Estimated implementation cost, $	Calculated savings from meter readings at Sept. 1975 rates, $	Metered savings, %	
			Electricity	Gas
Allied Medical	30,000	39,188	40	75
Howlett Hall	31,740	10,943	25	57
Biological Sciences	37,156	29,520	12	47
McCampbell Hall	55,200	34,580	60	60

not been adjusted to compensate for differing weather or occupancy conditions. They are the actual difference in the meter readings for the two periods.

Study each building

In closing, it is necessary to emphasize the need for a professional evaluation of each building and its system. There is no simple list of rules that can be universally applied to saving energy in buildings. Each system must be evaluated on an individual basis in order to achieve maximum operating efficiencies.

One unique example of what results can be achieved by a proper examination has been demonstrated in the Legal Aids Building on the Ohio State University campus. Research projects, funded by ASHRAE and the National Science Foundation and directed by Professor Sepsy, have been conducted on the energy use in this building since 1970. One of

the research goals is to evaluate energy saving modifications that could be made to the HVAC system.

The building's HVAC distribution is by means of a dual duct system. All of the supply air is passed through the cooling coil. To control space temperature, the hot deck heating coil operates year around. This design, while closely controlling space humidity, is a high energy use system that essentially has the same characteristics as a terminal reheat system.

Cooling is provided by four staged direct expansion cooling coils and a 60 ton air cooled compressor. A hot water heating coil in the hot deck provides the winter space heating and provided summer reheat energy. One energy conservation modification being studied is the use of a refrigerant condensing coil to provide the air conditioning reheat. This permits shutting down the gas fired boiler during the cooling season.

The refrigerant piping and control were the primary concerns in installing this modification. The final arrangement installed in the building was to cut into the hot gas discharge line off the compressor, install the hot deck condensing coil in series with the rooftop air cooled condensing unit, and install a refrigerant pressure controller to modulate dampers on the air cooled condenser to maintain adequate system pressure. Face and by-pass dampers were installed at the hot deck condensing coil to prevent excessive temperatures. Savings for this modification are derived from two sources:

- Reduction of yearly gas consumption by the boiler.
- Reduced refrigerant condensing temperatures at design conditions.

The system has been operating successfully and showing significant energy savings. Past gas meter records show a cooling season consumption in excess of 200 mcf per month for the 25,000 sq ft building. Virtually all of this gas is being saved.

Savings modifications of this nature cannot be accomplished without the knowledge of those individuals who are qualified as engineers, experienced in their field.

Training program is essential

Finally, successful implementation of an energy conservation program involves the cooperation of people. At a facility the size of Ohio State University, many people are involved.

A great deal of reliance had to be placed on the maintenance supervisors and their staff. Therefore, it was important to communicate early in the program what the goals were and how they could be achieved.

One hour classroom sessions were held weekly over a 20 week period, during normal working hours, for all involved maintenance personnel.

The various systems that would be affected by the program were explained in detail—how they are intended to operate and the reasons why they do not always operate as intended. For example, the participants learned that there was more to analyzing why an air system does not maintain desired space temperature than checking a thermostat.

Pertaining to air side considerations, most of the systems employed in the campus are dual duct and terminal reheat. Since conversions of some dual duct systems to VAV operation was contemplated, the energy saving characteristics of VAV operation were explained.

Hydronic systems and some aspects of control systems and refrigeration were also covered in the training program.

The classroom sessions helped to demonstrate what the program was intended to accomplish, how it could be done, and emphasized the important role that the maintenance staff would play.

The results have been most gratifying. The excellent results achieved to date stem from the teamwork of all involved personnel. In turn, these results reinforce the desire of all to cooperate fully. Everyone can readily see that they are a part of an important and successful program that is helped considerably by adequate communication and training.

A guide to lighting level adjustment in existing buildings

First, it should be noted that there is no evidence that low lighting levels cause any permanent eye damage. Reaction such as eye strain or headache may result from too high or too low lighting levels or from glare as a result of improper lighting; but once the lighting problem is corrected, the adverse reactions will be eliminated.

The following steps are intended to be used primarily for lighting level adjustment in existing buildings, but many of the points noted should be considered when designing lighting for new buildings.

1 Space survey

Consult with occupants to determine actual work areas in a given space and the type of tasks performed in the work areas. In areas within a given space where specific tasks requiring good visibility are performed, the lighting levels should generally be higher than in areas used for aisles, storage, etc. That is, desk tops, laboratory tables, reading tables, and similar work areas will generally require higher lighting levels than the aisles between work areas.

It should be realized that the overhead lighting is dependent on the arrangement of desks, tables, and other work areas within the room and that if the arrangement is changed, the lighting may also have to be rearranged.

Once specific task areas are determined, the lighting should be adjusted to provide the levels given in Section 2.

2   Recommended lighting levels

Area	Approximate footcandles
• Storage areas.	10
• Corridors, hallways, aisles, and circulation areas within a given space where tasks requiring good illumination are not performed.	20 to 30
• Office areas where tasks such as reading, writing, and typing are performed, and occupants are over 45 or have poor vision.	75 to 100
• Classrooms, offices, libraries, study rooms, and general work areas in which reading, writing, and similar tasks are performed by occupants under 45.	40 to 60
• Laboratories and shop areas.	40 to 60
• Laboratories, offices, shop areas, and work areas where critical or very fine work is performed for extended periods.	100 to 120

While this list can be used as a general guide, exceptions will be found and thus individual judgment will be required.

3   Measuring lighting levels

A light meter should be used to determine actual lighting levels obtained before and after lighting revision. Readings should be taken at the actual task level by placing the meter on the desk top or similar surface.

4   Position of lighting source

Where it is practical, small area lamps should be used to increase lighting levels in areas where tasks are performed. For instance, desk lamps or drafting table lights can be used to increase illumination instead of using ceiling lights.

When ceiling lights are removed, fixtures in front of the point where the task is performed should be removed first. If the lighting level is still too high, then lighting directly above the task area should be reduced. The light from behind and to either side of the task area is the best source and should provide the largest percentage of illumination. Following this procedure will reduce glare. In addition, where tubes are removed from three and four tube fixtures, it is preferred to remove tubes from only one ballast so that some illumination is provided by each fixture rather than have light and dark spots in the ceiling.

5   Lighting contrast

The eye can adjust to differences in illumination rather quickly as long as the differences are no larger than 10 to 1. Thus, the lighting levels listed in Section 2 must be modified in certain areas to accommodate the adjustment capabilities of the eye.

When room occupants look from one area of the room to another or move from one space to another, the lighting level differences should be within the ratio of 10:1.

6   Fluorescent tube removal

At the same time tubes are removed, the appropriate ballast fuses should be removed. If ballasts do not have fuses, the hot wire to the ballast should be disconnected and the bare end taped. If ballasts are left energized, they often use as much energy as when the tube is in place and no saving is accomplished.

After the tubes are removed and ballasts are disconnected, the fixtures should be identified with the date of the tube removal so that tubes will not be replaced at some unnecessary future date.

7 Lamp replacement

The practice of replacing lamps with lamps of smaller wattage or lamps of greater efficiency should be encouraged. Use of higher efficiency lamps will provide the required lighting levels with lower wattage lamps. The relative efficiency of various lamp types is indicated by the following listing, starting with the most efficient first: high pressure sodium vapor, metal halide, fluorescent, mercury, and incandescent.

In addition to saving energy, the higher efficiency lamps will reduce maintenance because of longer lamp life.

8 Light switches

Provide switches for lighting so that areas of intermittent use can be switched off. In large rooms, provide switching so that small portions of the room can be lit; thus, providing the capability of adjusting lighting to actual use.

9 Summary

The primary consideration here is for the individuals using the spaces. While reduction of energy usage is an important goal, it is not the intent of these guidelines to reduce the safety, comfort, or efficiency of the building occupants. Thus, adjusting lighting levels must be tempered with the judgment and experience of the individual removing the lighting. For instance, the lighting contrast ratio of 10:1 as outlined must be considered when illumination is reduced. Otherwise, a condition conducive to accidents may exist in dimly lit areas adjacent to areas of high illumination.

43

Equipment maintenance for energy conservation

The proper maintenance of mechanical equipment cannot be ignored in an energy conservation program. Here is a detailed account of what factors to include in a maintenance program.

KENNETH LeBRUN,
Sales Engineer, RMC, Inc.,
Chicago, Illinois

In this chapter I will offer an analysis of energy conservation as it applies to heating, air conditioning, and ventilation economics.

Before tackling such a broad term as *economics*, perhaps it should be defined. *Webster's Dictionary* defines it as "The careful management of wealth, resources, etc. Avoidance of waste by careful planning and use; thrift, or thrifty use."

Through good management, energy consumption can be analyzed and controlled, and the tight spots in a schedule can be handled by planning. Utility costs can be reduced without cutting back on production, or production can be expanded without increasing utility consumption. Good management is the key.

Combustion equipment

Let us consider the various energy consuming devices on a heating and air conditioning system. This should begin at the heart of a system: boilers and refrigeration compressors. Combustion equipment is the first item to consider when examining heating. A leading group of combustion experts recently discovered, while conducting field tests, that for most installations without provisions for flue gas analysis, proper adjustment of combustion equiment could yield fuel savings of 5 to 10 percent. In extreme cases, savings ran as high as 30 percent. Combustion equipment should be adjusted through the entire range of firing, and it should be checked at regular intervals throughout the heating season.

Table 43-1 gives the basic compositions of natural gas and fuel oil. These are made up of elements common to both: carbon, hydrogen, sulfur, oxygen, and nitrogen. The actual percentage of each component can vary from that listed in the table.

Table 43-2 illustrates the chemical reactions of combustion for the elements shown in Table 43-1. One molecule of carbon combined with two oxygen molecules results in the formation of one molecule of carbon dioxide, CO_2, and the heat released from this reaction is 14,600 Btu per pound of carbon burned.

Carbon monoxide, CO, is formed during incomplete combustion. The heat released during this reaction is only 4440 Btuh per pound of carbon burned, compared to 14,600 Btuh per pound when complete combustion is obtained. The balance of the carbon's heat content is lost.

Table 43-1. Comparative analysis of typical fuels.

Fuel	Carbon	Hydrogen	Sulfur	Oxygen	Nitrogen	Heat value Btu per lb	Specific weight, lb per cu ft	Air combustion
	Percent by weight							
Natural gas	76.1	23	—	—	0.9	22,800	0.044	9.7 cu ft per cu ft
Mfd. gas	43	10.4	—	41.6	5	11,850	0.048	4.3 cu ft per cu ft
Fuel oil	85.6	12	0.35	0.6	0.5	18,500	55.6	1365 cu ft per gal.

Table 43-2. The chemical reactions of combustion for the elements shown in Table 43-1.

Combustible material	Chemical reaction	Oxygen, lb	Nitrogen, lb	Air, lb	Flue products, lb	Heat value Btu per lb
		Weight per lb of combustible				
Carbon (C)	$C + O_2 \longrightarrow CO_2$	2.67	8.85	11.62	12.62	14,600
Carbon (C)	$2C + O_2 \longrightarrow 2CO$	1.33	4.43	5.76	6.76	4,440
Carbon monoxide (CO)	$2CO + O_2 \longrightarrow 2CO_2$	0.57	1.90	2.47	3.47	10,160
Hydrogen (H)	$2H_2 + O_2 \longrightarrow 2H_2O$	8	26.56	34.56	35.56	62,000
Sulfur (S)	$S + O_2 \longrightarrow SO_2$	1	3.32	4.32	5.32	4,060

It is important to obtain the proper amount of combustion air as indicated in Fig. 43-1. Too much air increases sensible flue gas losses, reducing the efficiency of the combustion process. Too little air causes incomplete combustion, which wastes fuel, creates poisonous CO, and can result in reduced heat transfer efficiency due to deposition of soot (unburned carbon) on heat transfer surfaces within the boiler or furnace.

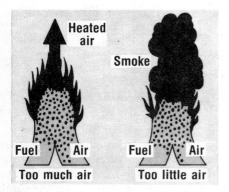

Fig. 43-1. Too much combustion air increases sensible flue gas losses; too little causes incomplete combustion.

Figure 43-2 illustrates the range of heat loss in the combustion process with varying air-fuel ratios. The lower horizontal line shows the heat lost through radiation and transmission losses from the boiler shell at a given firing rate. The *incomplete combustion loss curve*, which starts at the left, shows that high losses occur with incomplete combustion and decrease as the fuel-air ratio approaches the chemically correct mixture for "perfect" combustion. The *flue gas loss curve* increases as the quantity of air is increased. Combining these three curves results in the *total losses curve* that theoretically would have its minimum value at the point of the chemically correct air-fuel ratio but which in practice is to the right along the upper horizontal line since in actuality some excess air is required to obtain complete combustion.

Figure 43-3 shows the relationship between the CO_2 content of flue gas and the amount of excess air typically found with natural gas and fuel oil when the burners are clean and properly adjusted. The higher the CO_2 content of the flue gas, the smaller the amount of excess air being used in the combustion pro-

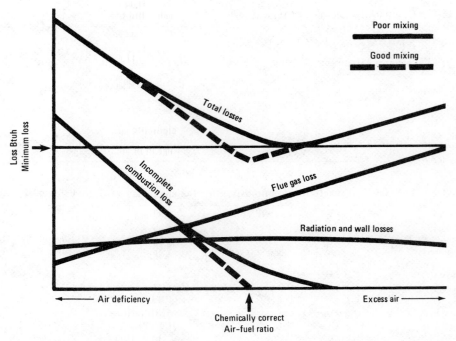

Fig. 43-2. The range of heat loss in the combustion process with varying fuel air ratios.

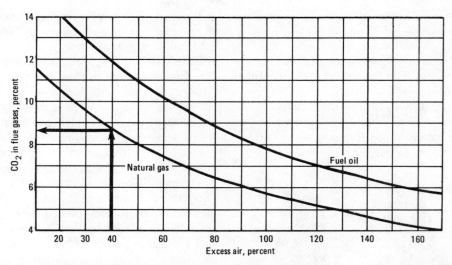

Fig. 43-3. The relationship between CO_2 contents in flue gas and the amount of excess typically found with natural gas and fuel oil when burners are clean and well adjusted.

cess. A typical situation for a space heating installation using natural gas is illustrated— 8.60 percent CO_2 with 30 percent excess air. Reducing the amount of excess air increases combustion efficiency with most equipment. The degree to which combustion air volume may be reduced depends on the following: 1) whether or not sufficient combustion air is available (adequate provisions for combustion air exist); 2) whether or not the venting system has sufficient draft. Too high a draft wastes fuel; too little promotes inefficient burning, particularly with atmospheric burners; 3) the type of burner installed—atmospheric or powered.

With proper burner adjustment, adequate combustion air, and draft, operation with as little as 10 percent excess air without carbon monoxide formation is generally possible.

Periodic checks of flue gas content and adjustment of burners accordingly result in higher efficiency. Flue gas analyzers are available to determine CO_2 content.

It is just as important to maintain clean heat transfer surfaces on the water side as on the fire side of a hot water boiler or steam generator (Fig. 43-4). Improper water treatment or no water treatment often result in a buildup of scale on the heat transfer surfaces. Sediment or sludge may settle out in the boiler. These occurrences create an insulating effect on heat transfer surfaces, reducing the efficiency, and if uncorrected may severely damage the boiler.

Foaming water in the boiler is an indication of oil or dirt in the system. This is characterized by a rapidly fluctuating water level in the boiler sight glass. Oil has to be removed by a cleaning process. Sludge may be removed by blowing down the boiler, but a water treatment expert should be consulted to determine what the proper treatment should be to minimize scaling and sludge formation, and to determine a blowdown schedule that is not excessive since heat energy is thrown away in this operation.

External boiler factors

When efficient boiler operation has been assured, it is time to review components external to the boiler that affect its operation. A chimney should be checked for two conditions: blockage and overdraft. A blocked chimney restricts combustion air, resulting in a high fuel-to-air ratio, poor combustion efficiency, and a potentially dangerous condition, especially with atmospheric type burners. Overdraft causes a high air-to-fuel ratio that also lowers boiler efficiency.

A large open exhaust fan in a building, a restricted combustion air louver, or adverse wind conditions can cause a negative pressure in the boiler room that also reduces the combustion air to the combustion equipment, resulting in a high fuel-to-air ratio.

Another consideration that affects a boiler room is outdoor air temperature. As outdoor air temperature drops, air density increases. The stack effect in the chimney also increases, and if the combustion air is untreated, the burner fan moves a greater weight of air. Both conditions result in a high air-to-fuel ratio and

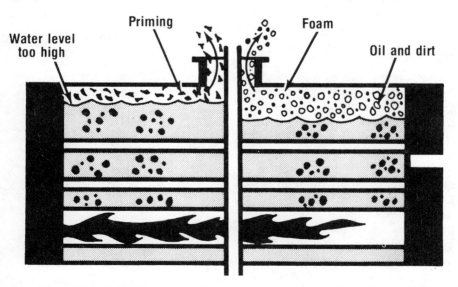

Fig. 43-4. It is important to maintain clean heat transfer surfaces on both the water and fire side of a hot water boiler or steam generator.

may require readjustment of the burner controls. This is of greater consequence with atmospheric type burners.

Operating efficiency

The boiler water temperature or steam pressure should be set to operate at the lowest temperature that will satisfy load demands. This results in minimal heat loss through the boiler shell and the supply and return piping.

Many systems have never been balanced. If one or more zones receive an inadequate supply of water or steam, the problem is normally resolved by higher operating temperatures and pressures or longer operating hours. These measures may overheat other spaces. I recommend consulting a certified test and balance company to solve this problem.

A leaky system that is losing water or steam requires additional heat to treat the entering cold water and additional water treatment to treat the added impurities.

An undersized expansion tank on a water system or a defective relief valve cause unnecessary steam or water loss through the relief valve.

Improper air venting in a water or stream system can cause many serious problems affecting efficiency and reliability of a heating system. This problem also applies to steam traps. These items are important and should not be overlooked in a maintenance program.

Air conditioning systems

The air conditioning side of a system must also be examined. The heart of an air conditioning system is the compressor. Three major types are associated with the air conditioning industry: reciprocating, centrifugal, and absorption.

The single greatest cause of excessive energy consumption or failure in an air conditioning unit is dirt or fouling of its surfaces. If any of these three types of machines have dirty condensers or evaporators, capacity drops and energy consumption rises. Extraction of heat from a dirty evaporator requires a lower than normal suction pressure, and heat discharge through a dirty condenser requires a higher than normal head pressure.

Figure 43-5, a cutaway of a compressor cylinder, helps to illustrate how a reciprocating compressor loses capacity. Every compressor must have a small amount of clearance between the valve assembly and the face of the piston at the top of the stroke. This prevents the piston from striking the valve assembly. This clearance space is left full of high pressure gas at the end of every compression stroke. The high pressure gas re-expands when the piston travels through its downward suction stroke and limits the amount of

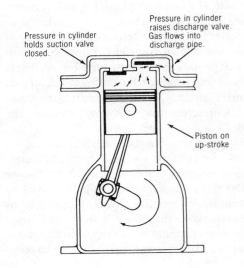

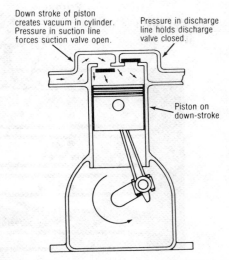

Fig. 43-5. Cutaway of compressor cylinder illustrates how a reciprocating compressor loses capacity.

suction gas that can enter the cylinder. The higher the discharge pressure, the greater the density and pressure of the residual gas left in the cylinder as the piston begins its suction stroke. This residual gas reduces the amount by weight of refrigerant that can be drawn into the cylinder during the suction stroke.

Also, as the suction pressure is lowered, the gas density decreases, and less refrigerant by weight enters the cylinder on the suction stroke. Although the reciprocating compressor is a constant volume device, changes in the head and suction pressures change the mass flow of refrigerant, which is the true determinant of compressor capacity.

Now, let's examine change in capacity of a water cooled condensing unit with a direct expansion evaporator coil. Table 43-3 contains data on a compressor rated at 15 tons and operating on R-22. Normal operating conditions are 105° F condensing temperature and 45° F suction temperature. At these conditions, net capacity is 17 tons, 15.9 bhp. If condensing temperature rises to 115° F, as it would with a dirty condenser, cooling tower, or poor tower water circulation, while suction temperature remains at 45° F, the compressor capacity would decrease to 15.6 tons and brake horsepower would increase to 17.5 bhp.

On the evaporator side, if suction temperature drops to 35° F due to dirty filters, dirty fan wheel, or slipping belts, while maintaining

105°F, compressor capacity would drop to 13.8 tons at 15.3 bhp. If the two problems are combined and a compressor is selected to operate at 35° F suction temperature and 115° F condensing temperature, compressor tonnage drops to 12.7 tons and the brake horsepower rises to 16.4 bhp. Table 43-4 summarizes how various operating conditions affect capacity for this same 15 ton compressor.

It is not unusual to find a reciprocating compressor operating at these conditions, or worse. The brake horsepower per ton is exorbitant and wasteful. In addition, high compression ratios and short cycling seriously reduce the reliability of a system. The data listed in Table 43-4 are based on an industrial quality compressor with a relatively high volumetric efficiency. The effects of dirt on window air conditioners and smaller commercial air conditioners are much more pronounced. These data apply not only to reciprocating compressors but to centrifugal and absorption types as well.

Centrifugal and absorption chillers are affected adversely by dirt and improper maintenance in a similar way. The following tables are based on chilled water and tower water temperatures rather than upon the refrigerant pressures directly. There is a direct relationship between these water temperatures and the refrigerant pressures. Leaving chilled water temperature is a function of the suction

Table 43-3. Effects of various suction and condensing temperatures on a 15 ton compressor operating on R-22.

Suction		Condensing temperature and pressure							
		85°F, 155.7 psig		95°F, 181.8 psig		105°F, 210 psig		115°F, 242.7 psig	
Temperature, °F	Guage pressure, psig	Tons	Bhp	Tons	Bhp	Tons	Bhp	Tons	Bhp
0	24.1	6.8	10.3	6.1	9	5.6	10.9	—	—
5	28.3	7.5	10.8	7.1	11.3	6.3	11.7	5.7	12.1
10	32.9	8.9	11.2	8.1	11.9	7.3	12.4	6.6	13
15	38	10.1	11.7	9.3	12.4	8.4	13.2	7.6	13.8
20	43.3	11.4	12.1	10.5	13	9.6	14	8.7	14.5
25	49	12.9	12.4	11.9	13.4	11.0	14.5	9.9	15.3
30	55.2	14.4	12.5	13.3	13.8	12.4	14.9	11.3	15.8
35	61.9	16.1	12.5	14.9	14	13.8	15.3	12.7	16.4
40	69	17.7	12.6	16.6	14.1	15.4	15.6	14.2	16.9
45	76.6	19.6	12.7	18.3	14.3	17	15.9	15.6	17.5
50	84.7	21.5	12.7	20	14.4	18.6	15.9	17.3	17.7

Table 43-4. The effects of various conditions on a 15 ton compressor operating on R-22.

Condition	Suction temperature, °F	Condensing temperature, °F	Tons	Brake horsepower	Brake horsepower per ton	Percent
Normal	45	105	17	15.9	0.93	—
Dirty condenser	45	115	15.6	17.5	1.12	20
Dirty evaporator	35	105	13.8	15.3	1.10	18
Dirty condenser and evaporator	35	115	12.7	16.4	1.29	39

Table 43-5. The effects of various combinations of leaving chilled water and condenser water temperatures on tonnage of a 555 ton capacity centrifugal chiller.

Leaving chilled water temperature, °F	Leaving condenser water temperature, °F						
	85	87.5	90	92.5	95	97.5	100
	Tons	Tons	Tons	Tons	Tons	Tons	Tons
40	566	564	535	528	490	440	—
42	578	574	557	546	522	490	—
44	591	585	580	565	555	542	507
45	597	597	590	575	562	549	519
46	603	603	600	585	570	557	531
48	616	616	616	606	585	573	555

Table 43-6. Effects of various operating conditions on a centrifugal chiller, 555 ton capacity.

Condition	Chilled water temperature, °F	Condenser water temperature, °F	Tons	Kilowatt per ton	Percent
Normal	44	95	555	0.82	—
Dirty condenser	44	100	507	0.86	5
Dirty evaporator	40	95	490	0.84	2.5
Dirty condenser and evaporator	40	100	N.A.*	N.A.*	N.A.*

*Not available.

pressure, and the leaving condenser water temperature is a function of the condensing pressure. Table 43-5 shows how leaving chilled water and condenser water temperatures affect tonnage on a 555 ton capacity centrifugal chiller. Table 43-6 shows how various operating conditions affect capacity for this unit. Tables 43-7 and 43-8 give similar data for an absorption chiller, 520 ton capacity. Data presented in these tables are extrapolated from a manufacturer's catalog ratings.

Water versus electricity

In some instances, such as water cooled air conditioners operating on city water without a cooling tower, there are two utility sources that must be evaluated: water and electricity. Increasing the flow rate of city water through a condenser decreases the head pressure with a resultant increase in total capacity and decrease in horsepower per ton. A decrease in flow rate causes the opposite condition. At an approximate aver-

Table 43-7. The effects of various combinations of entering cooling water temperatures, steam pressure, and leaving chilled water temperatures on tonnage of a 520 ton absorption chiller.

Entering cooling water temperature, F	Steam pressure, psig	Leaving chilled water temperature, F						
		40 Tons	42 Tons	44 Tons	45 Tons	46 Tons	48 Tons	50 Tons
80	12	528	551	572	585	595	616	635
80	10	510	530	551	562	578	590	614
80	8	483	502	523	535	546	569	588
80	6	455	473	493	505	519	546	577
80	4	422	438	458	473	484	515	554
80	2	387	405	426	439	452	484	525
85	14	—	—	—	546	562	583	604
85	12	468	494	520	533	546	567	588
85	10	447	473	494	510	521	546	570
85	8	421	443	468	481	494	519	544
85	6	391	411	432	447	462	489	520
85	4	356	377	400	412	426	453	489
85	2	322	338	361	374	389	420	452
90	14	421	447	473	489	504	530	551
90	12	396	426	457	469	483	504	525
90	10	376	405	434	447	459	463	504
90	8	348	377	406	420	432	456	478
90	6	317	345	374	385	399	422	445
90	4	286	313	341	353	354	389	411
90	2	255	283	302	315	328	353	377

Table 43-8. Effects of various operating conditions on a 520 ton capacity absorption chiller.

Condition	Chilled water temperature, F	Tower water temperature, F	Tons	Pounds of steam per ton	Percent
Normal	44	85	520	18.7	—
Dirty condenser	44	90	457	19.3	3
Dirty evaporator	40	85	468	19.2	2.5
Dirty evaporator and condenser	40	90	396	20.1	7.5

age of 0.75 kw per ton compressor power consumption and 1 gpm per ton city water consumption, it may be worth the effort to analyze the load and utility requirements on any system operating on city water.

Heat exchangers need attention

The first type of air conditioning heat exchanger to consider is the refrigerant-to-water type. The formula for determining the size and performance of a heat exchanger is:

$$Q = U \times A \times TD_m$$

where

Q = the total heat transfer, Btuh
U = the overall coefficient of heat transfer
A = the net area of heat transfer
TD_m = the log mean temperature differential between the refrigerant and the water, °F

A dirty or scaled heat exchanger on either the evaporator or condenser surface changes the coefficient of heat transfer, or the U factor, drastically. If an attempt is made to simultaneously maintain constant heat rejection, Q, and the area, A, remains essentially constant, the temperature differential, TD, between the refrigerant and water must be drastically increased. The end results of this higher temperature differential are higher condensing temperatures, lower suction temperatures, and longer operating hours.

A second factor that significantly affects the U-factor of either an air-to-refrigerant or water-to-refrigerant type is the flow rate of water or air through the heat exchanger. Both air and water have a thin insulating film on the heat exchanger surface. Higher velocities reduce the insulating effects of this film and improve the efficiency of the heat exchanger. Low velocities have the opposite effect of decreasing the heat exchanger's U-factor or heat transfer efficiency. This problem arises with dirty air filters or reduced chilled water or condenser water flow as a result of clogged strainers.

The construction characteristics of heat exchangers are also important. A very important factor in selection of chilled water or refrigerant coils for an air handling system is the construction quality of the copper tubes and aluminum fins. The manufacturer's available selections of these should be carefully considered. The industry tends toward configuration fin design, high fin spacing, and thin tube walls and fins.

A configurated fin design provides high turbulence. This breaks down the insulating film on any surface and signficantly increases any heat exchanger's heat transfer efficiency. This additional efficiency means smaller and less expensive coils, but it also means a higher air side coil pressure drop requiring additional fan horsepower to achieve equivalent air flows. It becomes a trade-off of first cost versus operating cost.

High fin spacing significantly increases the net heat transfer area and also the heat transfer capability for a heat exchanger. For instance, a six row refrigerant or chilled water coil in an air handler with 12 fins per in. has a higher capacity and lower cost than

an 8 row coil with 8 fins per in. under equal operating conditions. A 12 fin per in. coil will create more service headaches since it is more susceptible to moisture carryover problems and dirt collection than an 8 fin per in. coil. The old story about a coil is a coil is a coil does not apply in this business. There are significant differences in copper tube wall thickness, fin thickness, and bonding between fins and tubes.

Thin tube wall design means shorter coil life expectancy, and thin fins and poor mechanical bonding between fins and tubes means a rapidly deteriorating heat transfer efficiency. Because of the difference in the coefficients of expansion between copper and aluminum, poorly bonded fins will in time develop an air gap between the tube and the fins. Air is a good insulator and reduces the efficiency of a coil.

A direct expansion coil in an air handler is a more critical application than a chilled water coil. Although both types of coils require uniform air distribution and loading for maximum efficiency, they have different results when misapplied.

Direct expansion vs. chilled water

Uneven loading on a chilled water coil produces less capacity and probably creates an uncomfortable space condition. This is normally compensated for by lowering the chilled water temperature until the condition is satisfied, with resultant inefficiency.

If a direct expansion coil is unevenly loaded because of misapplied roll filters, air bypassing filters, or poor duct design, such as sharp turns entering the direct expansion coil, the refrigerant coil circuits become unbalanced. On most large air handlers with direct expansion coils, the coils have thermal expansion valves with temperature and pressure sensing devices at the coil outlet. These meter refrigerant flow to the coil to maintain a preset superheat condition leaving the coil. If any one circuit is starved for air, the refrigerant on that circuit fails to evaporate and passes through to the superheat control as a saturated rather than a superheated gas. The thermal expansion valve

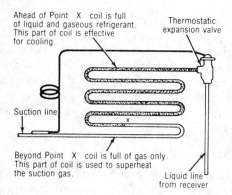

Ahead of Point X coil is full of liquid and gaseous refrigerant. This part of coil is effective for cooling.

Thermostatic expansion valve

Suction line

Beyond Point X coil is full of gas only. This part of coil is used to superheat the suction gas.

Liquid line from receiver

Fig. 43-6. The effects of selecting the proper superheat setting on a direct expansion coil.

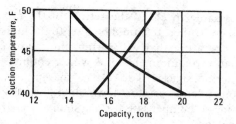

Fig. 43-7. Curves showing evaporator compressor balance.

then reacts to throttle flow of refrigerant to all circuits with the net result of starving the coil due to an unbalanced circuit.

One of the purposes of a thermal expansion valve is to guarantee a constant superheat leaving the coil. Lowering the superheat setting increases efficiency of the coil and direct expansion system because of a better quality refrigerant in the coil and the compressor. Too low a superheat setting can cause the coil to pass liquid refrigerant and result in slugging of the compressor (see Fig. 43-6).

Control and parallel flow

In a typical refrigerant or water coil installation, it is important to consider the advantages of counter flow versus parallel flow performance. Counter flow provides greater heat transfer than does parallel flow because of the greater average temperature differential between the fluid doing the cooling and the fluid being cooled. Counter flow effect is accomplished in a different manner for refrigerant and chilled water coils.

In a chilled water coil, the air entering the coil passes first over the tubes with the warm leaving water and leaves the coil passing over the tubes with the coldest entering water. In a direct expansion coil, coil temperature is highest where the refrigerant enters and decreases as the refrigerant flows through the coil. This means that a counter flow effect with chilled water coils is accomplished by true counter flow of the two fluids while

it is accomplished with parallel fluid flow with direct expansion systems.

It is important to understand the effects of pressure drop in a suction line between the evaporator and compressor. Most R-22 systems are designed for a 3 psi pressure drop between the evaporator and compressor. This normally allows for good oil return and minimum penalty to the performance of the equipment. In unusually long suction runs, oil return becomes a critical design consideration, and excessive pressure drop penalizes the performance of a system by increasing the average evaporator temperature while simultaneously lowering the suction pressure at the compressor. Figure 43-7 shows balance between evaporator and compressor.

Proper air system balance

Let us look at the air side of a system to explore what proper maintenance will do to conserve energy.

A properly balanced air handling system uses a minimum amount of outdoor air—just enough to meet building ventilation and code requirements. It provides mild pressurization of a building, except in those instances when industrial processes require negative areas. A negative building pressure is generally undesirable. It introduces large quantities of unfiltered outdoor air through the building perimeter rather than through the air conditioning system. This causes draft problems in the winter, humidity problems in the summer, and excessive consumption of utilities year around.

If mixed air dampers are included in an economizer system with a reheat coil, it is advisable to check the setting of the mixed air controller. It should be reset to the highest temperature that will permit minimum ventilation requirements and winter cooling for interior spaces. Too often, the mixed air controller is set too low; sometimes this is done to compensate for a system air balance problem.

The mixed air controller is also frequently set too low in packaged multizone systems when hot and cold deck controls are reset to meet the demands of two extreme zones. If these two zones are not properly balanced, the entire system will operate inefficiently and operating costs will rise.

Select proper insulation

Another important consideration in the economical performance of any air condditioning system is the proper selection of insulation. Uninsulated pipe and ductwork are excellent heat exchangers. This can be a serious problem for supply air ducts and water pipes, especially in a ceiling plenum or an untreated space. If a system is designed for 55°F air leaving the air handler and it arrives in the space at 65°F, the system becomes unbalanced because some areas are being overcooled while others are being undercooled. The net results are excessive equipment operation and uncomfortable space temperatures.

Understand the system

I have tried to present some insights into the fundamental criteria for efficient operation of heating, air conditioning, and ventilating systems. Much was omitted for the sake of brevity. One last thought that I would like to leave you with is: When you do make adjustments to a system for purposes of improving the operating efficiency, make sure that you understand the full system before you start adjusting individual controls.

All illustrations and Tables 43-1, 43-2, 43-3, 43-5, and 43-7 are courtesy of The Trane Co.

44

Maintenance and service for HVAC and energy conservation

The best energy conservation program cannot be effective without proper maintenance.

E. R. AMBROSE, PE,
Mount Dora, Florida

Reductions in energy usage attributable to better design and application practices can be quickly nullified by improper operation and care of heating and air conditioning systems. Owners, tenants, bulding managers, supervisors, or other such personnel charged with the responsibility for equipment are not always aware of the disastrous consequences of abusing and neglecting equipment.

Consequently, planned maintenance and service at predetermined intervals by competent and qualified personnel are strongly recommended for residential installations as well as the larger commercial systems. Proper care and attention can pay handsome dividends by extending equipment life, by savings in operating costs, and by general overall satisfaction.

Service Agreements

Service agreements are frequently sponsored by the original air conditioning manufacturer in cooperation with local service organizations. This is particularly true for unitary equipment usually used for residences as well as for small and medium size commercial structures. Larger buildings, having central

plant equipment, can usually justify stationing qualified service personnel on the premises.

Service agreements, in addition to repair and replacement of defective parts, should provide for a specified number of preventive inspections each year, depending on system size and complexity. These inspections, made to assure satisfactory and efficient operation of equipment, should check:

• For correct distribution of air handling apparatus.

• For proper operation and correct positioning of volume dampers, particularly outdoor supply and exhaust dampers.

• To verify that equipment operation agrees with the predetermined occupancy schedule for the building. Usually, all ventilation supply and exhaust fans can be completely shut down, with dampers in closed position during all unoccupied periods. Similarly, cooling equipment can be inoperative, and heating equipment operated intermittently (at lower temperatures) during unoccupied periods.

• Pressure, temperature, and safety devices for proper operation at correct settings. Clean contacts and make necessary adjustments and replacements.

- Filters—both air and water—should be examined, measured for excessive pressure drop, and cleaned or replaced when necessary.
- All heat transfer surfaces for removal of scale, foreign matter, and other impurities that may impede heat transfer.
- Heating equipment—partial and full loads—flue gas analysis for proper adjustment of combustion equipment and for suitable draft. Blockage causes burning inefficiencies and overdraft causes wasted heat.
- Cooling equipment—partial and full load—head pressure, suction pressure, performance and electrical demand of all motors.
- Adjustment of all system components, motors, starters, drives, and accessories. Clean and lubricate components per manufacturer's instructions and/or as conditions warrant.
- General competence of building operator.
- Record of all operating expenses. Make sure system is operating as intended.

Evaluate minimum energy needs

The objectives of reducing a structure's heating and cooling loads and/or designing equipment for higher thermal efficiencies are usually more easily obtainable in a new building for several reasons. First, an evaluation of minimum energy requirements can be made prior to choosing the system design and selecting equipment. Secondly, ASHRAE Standard 90-75, Energy Conservation in New Building Design, along with federal, state, and local codes and ordinances can be used as a guide for selecting the exterior envelope construction of new buildings to comply with allowable heat flow requirements.

In contrast, there are usually many restrictions tied to selecting the proper system or equipment for an existing building. Even so, considerable possibilities exist for reducing heating and cooling loads. However, practical and acceptable methods of increasing thermal efficiency are limited and face some difficulties. Nevertheless, existing buildings taken as a whole offer the best chance for

materially reducing energy usage in the immediate future.

The eight important factors contributing to lower energy usage are listed in Table 44-1. These are fully applicable to new buildings. Their practical application to existing buildings, however, is subject to the following limitations.

- Thermal insulation, double glazing, reduced ventilation air quantities, and lower lighting levels usually offer practical means for reducing heating and cooling loads without disturbing the basic system design.
- Heat reclaiming cycles are also frequently feasible as an auxiliary system without major changes in the basic system.
- Controls must be dependable and accurate to improve overall performance by taking advantage of the inherent operating characteristics of the equipment. They should receive primary consideration in any upgrading program. Older equipment has a tendency to deteriorate (if not properly cared for), get out of adjustment, or be bypassed because of defective parts, carelessness, or from lack of understanding by the operating supervisor or serviceman.
- Thermal efficiency improvement usually requires extensive changes in heating and cooling equipment. These changes are normally not economically justifiable if only the efficiency of the equipment, at standard conditions, is involved. If, however, full capacity modulation is intended in response to outdoor temperature fluctuations, the resultant increase in seasonal efficiency should be evaluated.
- Maintenence and service to assure repair and replacement of defective parts and inspection of the system on a regular and competent basis are mandatory to keep a system at maximum efficiency. It is essential to have properly written agreements and to make sure they are understood and followed by the service organization as well as the owner.

Reduce heating and cooling loads

Practices and procedures for reducing the heating and cooling load (the first six items in Table 44-1) usually can be appraised and

Table 44-1.   Contributing factors to lower energy usage in existing buildings.

1 INSULATION
- Sidewall: Not feasible in most instances.
- Roof/ceiling: Practical in many residential structures but difficult to justify in commercial and industrial buildings.

2 FENESTRATIONS
- Insulating glass of double glazing: Can generally be justified in colder climates.
- Awnings/shading devices: Depends on exterior construction. Usually justifiable when subjected to full sun exposure.

3 VENTILATION AIR
- Reduction and purification: Minimum outdoor air quantities combined with purification equipment has considerable merit. Updated federal and state codes needed.

4 LIGHTING
- Reduce levels and redesign system: Directly related to HVAC equipment capacity and operating cost. New Illumination Engineering Society standards being developed.

5 CONTROLS
- Revision, replacement, addition, and adjustments: Up to date, reliable system mandatory to assure maximum efficiency under all operating conditions.

6 HEAT RECLAIMING
- Double circuit condenser, domestic hot water, heat storage, air-to-air heat exchanger, and rotary wheel or drum: Such auxiliary equipment can frequently be incorporated in a system to recover excess heat from one area for use in another or to supplement the total output at extreme temperatures—thereby improving the overall efficiency.

7 THERMAL EFFICIENCY
- Coefficient of performance, seasonal performance factor, equipment capacity modulation: Complete capacity modulation to balance load caused by outdoor temperature variation offers an attractive means of energy input reduction.

8 MAINTENANCE AND SERVICE
- Planned contract, qualified personnel, repair and replacement of parts, and equipment operating properly: Proper care and attention can pay handsome dividends by extending life of equipment, savings in operating cost, and general overall satisfaction.

Also see "133 Ways to Save Energy in Existing Buildings," Section X, Summary, following Chapter 79.

the corresponding savings in energy usage determined quite easily by a competent engineer or contractor. Determining possible improvements in overall seasonal efficiency (the seventh item), however, requires a more thorough evaluation of the design and operating cycles. Perhaps the most promising means of improving seasonal efficiency or seasonal performance factor is to take advantage of variable loads occurring throughout the heating and cooling seasons due to changes in outdoor temperature and internal load conditions. It may be that full capacity modulation equipment is not readily available for residential and small commercial installations. This is a phase of the subject requiring the attention of manufacturers and no doubt extensive research and development are needed.

In this regard, fuel fired heating equipment for existing residential and small commercial structures also needs considerable research and field investigation to determine factual seasonal efficiencies that can be used in energy conservation investigations. The relative low seasonal efficiencies frequently cited must either be confirmed or disproved. In this category, there are literally millions of existing residential installations in the 10 to 30 year age bracket that have been converted from coal or changed from one type of system to another without regard for overall seasonal thermal efficiency. Many of these installations are handicapped by oversized furnaces and/or chimneys, short cycling and standby losses, improper air-fuel ratio adjustments, and low combustion efficiencies.

The big question is, How can the necessary higher standards of design and installation be established and enforced to bring about the necessary reduction in annual energy consumption in heating and air conditioning systems. It is a tedious and time consuming task to reach an agreement between technical societies, trade associations, equipment manufacturers, federal and state building codes, and other such groups. This must be done, however, and a universal standard developed before full compliance can be expected. ASHRAE Standard 90-75 is a step in this direction, but such standards must be mandatory and enforceable.

Conservation education needed

In the educating process, the professional engineer, architect, and contractor should become energy conscious as well as cost conscious. The general public must be made aware of the importance of energy conservation. One form of education can result from sales promotion efforts. Suppliers of material and products for reducing heating and cooling loads of a structure, such as thermal insulation, window glass, controls, heat reclaiming equipment, etc., have and will continue to actively and effectively promote the advantages and possibilities of energy conservation for all types of structures. These efforts should continue to be successful as long as economic justification can be established.

Economic justification is perhaps the dominant reason for improving seasonal performance of the existing systems in residential and commercial buildings. If outstanding success is to be realized in reducing the energy usage of these systems to the degree predicted by government agencies and other groups, additional incentives must be created whether they be patriotic, forceful compliance, or some type of monetary rebate. acceptable incentive is the basis of the entire program.

Finally, proper leadership must be established to bring all factors together in an united effort. Otherwise, existing installations, with their exhorbitant waste of energy, will be with us for years to come.

SECTION VI

Replacement, modernization, and upgrading systems for energy conservation

To maximize energy conservation in many existing installations, it may be most feasible to replace, modernize, and upgrade the heating, ventilating, and air conditioning systems. These 5 chapters outline the necessary factors:

- Equipment and system changes in commercial, industrial, and institutional buildings.
- Energy saving techniques as applied to university buildings.
- Retrofitting and replacing roof top systems.

45

Commercial and institutional buildings: replacement, modernization, upgrading

When it comes to saving energy, the author suggests that we should not be intimidated because the building exists, but rather look at the problem as if the building were in the design stage.

MAURICE G. GAMZE, PE,
Gamze-Korobkin-Caloger, Inc.
Chicago, Illinois

Energy consumption in existing buildings depends upon the following: building skin, rate of air infiltration, building height, operating characteristics of mechanical and electrical systems and their general maintenance, type of building service, heating and cooling space terminals, outside air requirements, internal loads, and climate. All of these characteristics would be considered if the building were new and in the design stage. Therefore, one method of approaching conservation in an existing building is to approach it as if it were in the design stage.

Since the methods of conserving or recycling energy in institutional and commercial buildings are varied and numerous, some categorization is necessary to logically pursue them. For purposes of this discussion,

This chapter is based on a paper presented by Mr. Gamze at a special series of two-day seminars sponsored and developed by *Heating/Piping/Air Conditioning* on How to Save Energy in Existing Buildings.

the examples presented are limited to distinct applications.

Architectural considerations

The U-value, infiltration rate, and solar penetration characteristics of the building's skin should be evaluated first. For example, the majority of the thermal energy consumed in hospital patient areas and hotel guest rooms is related to these items. Usually, very few basic mechanical changes can be made in these spaces other than installation of temperature controls to prevent overheating and overcooling.

Hospital patient areas and hotel guest rooms have other common traits, such as relatively small internal lighting loads and basically, 24 hr per day operation. In such buildings, the most profitable change, depending on climate and feasibility, would be to reduce the amount of glass, and if the glass is single glazed, replace it with an insulated glazing system and tighten or caulk the building to reduce air infiltration.

Particular care should also be taken with ventilation systems since these are often related to code requirements, and codes can often be interpreted in such a way as to reduce energy consumption.

Motel energy needs studied

The benefits derived from approaching an existing building as if it were being designed for the first time were demonstrated by a study made of an existing motel in the metropolitan New York area. The motel had approximately 200,000 sq ft of guest room area and about 40,000 sq ft of restaurant and public facility area. The outer walls were concrete black with masonry facing, and single glazed windows constitute slightly under 40 percent of the gross wall area. The window frames were of a relatively low cost aluminum with a substantial amount of air leakage. The heating and cooling system consisted of fan-coil units beneath the windows and wall mounted thermostats. A two-pipe system served these units.

The public spaces were served by a four-pipe system consisting of many air handlers, both single and multizone, serving the multitude of spaces at the base of the building.

The first obtainable data came from studying fuel bills. The guest room areas consumed approximately 7800×10^6 Btu per year for heating and about 378,000 ton-hr per year for cooling.

For economic reasons, the rehabilitation work was divided between the guest room areas and the public spaces to allow an orderly expenditure of money on an annual basis.

The windows were so bad that the owners considered replacing them, but we recommended equipping the windows with a relatively tight fitting aluminum frame with vinyl covering on the aluminum and double glazing. This was done, and heating consumption dropped from 7800×10^6 to about 4218×10^6 Btu per year. The new windows did not have much effect on cooling consumption, however.

A study of the effects of partially blocking up the windows was also performed. The calculated heating consumption went down again quite noticeably, and the cooling consumption also went down. However, implementation of this plan would have caused the original architect to drive off the Brooklyn Bridge. These studies and later construction indicated the value of exploring methods of modifying a building's architecture before starting detailed studies of the HVAC systems.

The additional cost of the vinyl covering and insulating glazing was approximately $64,000. The fuel saved was approximately worth $12,000 annually, based on a cost of 35 cents per gal of oil.

If a conservative estimate of 25 years is assigned to the remaining useful life of the building, the rate of return on $64,000 investment (discontinued cost flow method) is 18.7 percent as shown in the calculations below.

PW @

$$15\% = -64,000 + 12,000 \ (pwf - 15\% \\ -25 \ \text{yr})$$

$$= -64,000 + 12,000 \ (6.464)$$

$$= -64,000 + 77,568$$

$$= +\$13,568$$

PW @

$$20\% = -64,000 + 12,000 \ (pwf - 20\% \\ -25 \ \text{yr})$$

$$= -64,000 + 12,000 \ (4.948)$$

$$= -64,000 + 59,376$$

$$= -\$4,624$$

By interpolation, the rate of return before taxes is 18.7. Subsequent rates of return shown are also before taxes.

It must be pointed out that if the window leakage had not been so high, the investment for double glazing alone might not have been economically viable.

Buildings constructed in the last 10 years, typically, have curtain wall construction. Over a period of time, these walls generally

Table 45-1.   Monthly and annual heating load data on a motel in the metropolitan New York area.

| Motel | Heating consumption, MBtu | | | | | | | | | | | | |
	Jan	Feb	Mar	Apr	May	Jun	Jul	Aug	Sep	Oct	Nov	Dec	Annual
Building 1: single glazed, standard ventilation	1,318,090	1,284,083	916,691	599,202	159,088	55,255	9,023	25,992	93,380	386,316	619,464	1,275,082	6,741,666
Building 1a: double glazed	1,070,338	1,045,113	733,412	469,375	120,219	39,682	5,768	18,427	67,517	296,747	487,330	1,035,348	5,389,276
Building 1b: double glazed, variable ventilation	422,880	421,221	249,754	130,952	27,219	5,914	0	1,228	9,198	76,503	150,872	412,043	1,907,784
Building 1c: double glazed, tinted, variable ventilation	442,588	443,460	266,449	141,923	27,998	5,951	0	1,229	9,371	79,135	156,659	427,258	2,002,020
Building 2: single glazed, standard ventilation	1,531,128	1,470,523	1,023,296	658,509	189,104	72,233	15,261	37,755	119,896	450,193	726,439	1,508,996	7,803,332
Building 2a: double glazed, standard ventilation	882,128	853,195	546,279	319,962	83,312	25,728	2,464	10,669	41,836	211,377	368,403	872,698	4,218,051
Building 2b: double glazed, variable ventilation	625,967	608,424	369,561	200,355	49,848	13,215	311	3,918	21,020	131,375	241,665	622,020	2,881,680
Building 2c: reduced glass, double glazed, variable ventilation	408,291	103,786	228,728	111,268	22,705	4,222	0	574	6,362	67,870	138,803	402,545	1,795,153

become as leaky as a sieve, and from a practical standpoint this is very difficult to correct. This means that large amounts of infiltration and makeup air enter and leave the building in a highly uncontrolled fashion. It is possible to reface and recaulk a building's face, but the economics involved are often questionable, with the possible exception of residential buildings.

Exhaust and makeup air

For this same motel, it was also recommended that an exhaust system regulated by use be installed for all guest room toilets. The 20 story hotel had about 350 toilet exhausts running 24 hr per day at an exhaust rate of 100 cfm per toilet (Fig. 45-1). Using data accumulated in the Chicago area, it was

estimated that the actual toilet room usage followed very closely the data shown in the *ASHRAE Handbook and Product Directory* for domestic water usage in hotels. This indicated that the average toilet room usage is approximately 20 percent at maximum and perhaps 5 to 10 percent on a diurnal basis. This waste was reduced by installing motorized boxes in the exhaust registers, opening up the base of the risers to allow for static pressure regulation, and putting inlet vane controls on existing fans. This cost about \$35 per toilet or a total cost of \$12,000.

This reduction in air quantity decreased energy consumption an additional 1400×10^6 Btu per year and energy costs of \$14,700 annually for a rate of return of nearly 40 percent on investment. The cooling con-

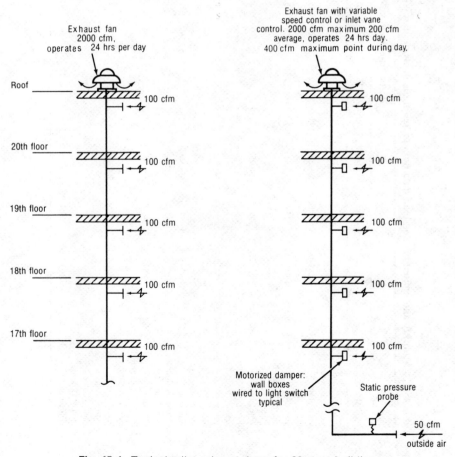

Fig. 45-1. Typical toilet exhaust risers for 20 story building.

sumption was affected very little, however, since hotel and patient rooms have loads, such as televisions and lights that run fairly constantly.

Another area worked on in this motel was the large amount of outside air heating required for makeup air to the various kitchen exhaust hood sytems, which amounted to approximately 40,000 cfm. Hotel restaurant and hospital kitchens run for extended periods of time, generally speaking, anywhere from 15 to 24 hrs per day. Since employees cook at random in the sense that kitchen equipment is turned on and off in a random manner, the general practice is to run the air supply systems and air exhaust systems whenever the kitchen is in service.

Many restaurants take conditioned air from the eating facilities and use it either totally or partially for hood exhaust. The better designed facilities have completely independent makeup air systems for kitchens. However, in both cases, a generous supply of 120 to 150°F air is exhausted outward from the kitchen hoods.

Since kitchen makeup air and hood exhaust are about equal, with the exception of the quantity exhausted for dishwashers and some other minor equipment, it would seem profitable to extract the heat from the kitchen exhaust stream and channel it into the makeup air stream for winter heating. A configuration as shown in Fig. 45-2, which allows for the coils to be cleaned, results in a complete run-around cycle of operation where kitchen exhaust heat is usually more than adequate for tempering winter air at practically any condition for use in the kitchen makeup system. Obviously, the recovery system must be turned off. This type of system costs approximately $0.40 to $1 per cfm, based on Chicago area construction costs. Using a fuel rate of about $3.50 per million Btu, the system generally will pay for itself in about one to three years.

The maintenance of this system is not as bad as it seems because by separating the coil sections and keeping coils in the horizontal pass, it is relatively simple to steam clean coils and flood them with detergents two to four times a year to keep coil surfaces clean. If the exhaust stream is clean, the system will operate more efficiently. Several systems of this type have been operating for up to six years in the Chicago area.

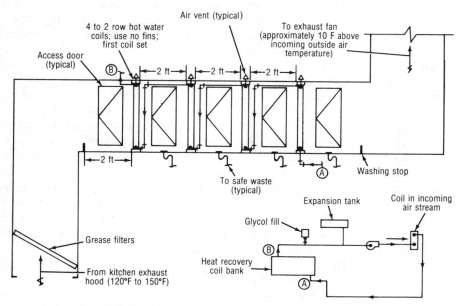

Fig. 45-2. Heat recovery system from kitchen hood exhaust.

Using range hood exhaust, the actual maintenance has been two washdowns per year plus steam cleaning once per month.

If space is available, it may be advantageous to install enthalpic heat exchange wheels instead of run-around coils. However, in many existing facilities, ductwork is so configured that this is unfeasible due to the downtime required for duct remodeling to install these wheels. The advantage of the run-around coil system in installation time is minimal; the disadvantages are its inactivity during the cooling season, and it is really only applicable when high temperature differences exist to obtain good heat exchange.

In hospitals and public facilities, such as restaurants, hotels, auditoriums, etc., large amounts of outside air are used basically to satisfy codes, and in these cases, various air-to-air heat recovery cycles can be useful. Only after looking at these systems and points of consumption should further investigation be made as to additional energy reducing systems.

Heating and cooling systems

The starting point to reduce consumption further is to examine the amount of air side fan horsepower required to drive the air systems and the characteristics of the air and primary energy systems at part load operation. Since most facilities operate at part load the vast majority of the time, system auxiliaries, basic piping configuration, air systems, and control systems should be examined.

The typical layout of most chilled water plants under approximately 3000 tons has chillers in parallel with an evenly split flow, and the chillers are operated to maintain a header temperature to satisfy the largest cooling load. Part load operation can be noticeably improved by modifying this system slightly in two ways. First, water from an operating chiller should not be mixed with water from an unoperating one. Mixing can be prevented simply by providing automatic valves in the chilled water piping to hydraulically isolate the unoperating chiller. Since this may cause hydraulic difficulties in parts of the system, the hydraulics of the entire system should be carefully examined.

Select correct pump

Secondly, by judicious pump selection or refitting the pump impeller, it is possible to get about 70 percent of the design chiller flow (second chiller) with only one chiller and without eroding the tubes in the operating chiller. A marginal portion of water can be bypassed around the chiller to assure that tube erosion does not occur. The system is now operating off a fixed chilled water point and the auxiliaries to operate the second chiller; basically, cooling tower fans and condenser pumps are not running. In most buildings, the output of the second chiller is usually limited to less than 30 to 40 percent of total cooling season hours. Therefore, significant electrical cost reductions can be achieved by this type of operation.

When a constant flow chilled water system is changed to a modified variable flow system, piping modifications may be required at the cooling coils to assure continued satisfactory operation. Coils should be examined to see how closely they match the actual load. If they are marginal, a blending pump can be installed to assure the coils receive the desired chilled water flow (Fig. 45-3). If the coils have sufficient surface, this can be omitted.

It should be apparent that if a chilled water system serves a small number of coils, this approach is more valuable, but if a system serves a large number of coils, it may not be economically viable. Before recommending this approach to an owner, the original system must be examined to determine whether there is any excess coil capacity. If the cooling coils were computer selected, additional care must be exercised because on most installations, if the cooling loads were "honest," there will be little excess cooling surface.

Consider boiler hook-up

Another area that should be looked at if a system has hot water boilers is the piping of boilers in series rather than in parallel. A

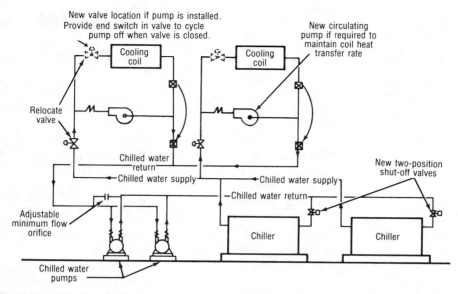

Fig. 45-3. Chilled water system conversion from constant volume flow to semi-variable flow.

typical parallel boiler installation is shown in Fig. 45-4. It is common in many existing installations. Theoretically, at reduced loads, one boiler can handle the space heating and process loads. With one boiler off, however, heated and unheated water are mixed. This causes the supply water temperature to fluctuate, which in turn affects the performance of the heating terminals and the process loads as well.

To eliminate this mixing problem, both boilers are often operated together over the entire load range. Better control over leaving water temperature is obtained, but effi-

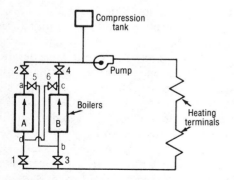

Fig. 45-4. Conversion from parallel to series boiler operation. Piping for a dual boiler utilizing series flow is also shown. Boiler isolation valves should be installed above each boiler.

ciency is sacrificed. All the auxiliaries for each boiler—burner motors, controls, forced or induced draft fans—operate whenever boiler operation is required.

Furthermore, boiler heat transfer efficiency diminishes at part load, particularly below about 30 to 40 percent load. Fuel is wasted because full space heating loads exist only a few hours per year.

The problem is compounded further from oversized boilers. Typically, with a two boiler installation, each boiler is sized to handle 75 percent of the design load, while three boiler installations are sized so each can handle 50 percent of the load. This is inefficient since all boilers may be fired throughout the heating season, but no boiler is ever 100 percent loaded.

Figure 45-4 also shows how a dual boiler may be piped for series flow. With this arrangement, good temperature control of leaving water is easier to obtain. Also, one boiler operating at higher load levels much of the time increases fuel utilization efficiency. Finally, shutting off the other boiler(s) and removing them from the flow path by closing the boiler isolation valves when not required to handle a portion of the load reduces electricity usage since the attendant auxiliaries are also off. The closing of the isolation valves also reduces the boiler stand-by losses.

Reduce pump horsepower

Pumping horsepower may also be reduced with a series operation. Supply water temperature may be increased with corresponding reduction in water flow. Many older heating installations have a 20°F design temperature differential. Doubling this halves the water flow rate. Existing heating coils designed for a 20°F Δt generally will perform satisfactorily with up to a 40°F Δt and 50 percent of the original flow. Temperature differences across radiation devices, such as hot water baseboard, convectors, etc., often can be increased from a design 20°F Δt to 30 to 35°F Δt.

Water flow reductions yield large pump horsepower savings. Cutting flow in half reduces pump horsepower by 66 percent or more. Either pump impellers should be replaced or new pumps installed to obtain the most efficient pump performance with the new operating conditions.

Finally, temperature and pressure ratings of the boilers must be checked when contemplating converting to larger temperature rises through the boilers to meet code requirements.

Series versus parallel flow

A brief discussion of series versus parallel flow of chillers is also in order. Series flow through two chillers permits greater Δts and reduced flow and pumping energy as does series flow through boilers, usually with somewhat lower horsepower per ton requirements. The effects on system hydraulics and cooling coil capacities must be examined if this is contemplated. Piping changes to convert from parallel to series flow generally are more involved because of interference from condenser water piping. However, they may be feasible when major work, such as enlarging an existing chilled water plant, is done.

Figure 45-5 shows typical operating conditions with chillers piped in parallel and in series. The double cross hatched area at the

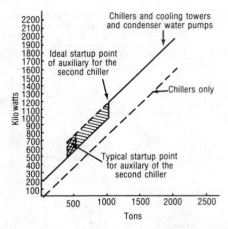

Fig. 45-5. Typical operating conditions with chiller pipes in parallel and in series.

left indicates normal startup range of the second chiller in a parallel configuration. The single and double cross hatched areas denote typical range of operation for both chillers with series flow. Typical startup of second unit is shown by the right hand boundary of the cross hatched area. Often, there are many operating hours at reduced cooling loads. This is particularly true when a plant operates on a 24 hr per day basis as at hospitals, campuses, etc. Considerable savings may be achieved by reducing operating hours of the chiller as well as by shutting down auxiliaries. Series flow is another way to achieve this.

Look at the air-side

The air-side of a system is a fertile field for applying conservation methods, but again, the original building design must be examined to determine the most rational approaches. For example, if a system has very high duct pressure drops, consideration should be given to adding auxiliary duct risers in order to relieve pressure drop and obviously, the power requirements. This generally is feasible only if a major building renovation is to be performed. If the system is of a more recent design, generally, the pressure drop within coils and filters is so high that running additional risers is of questionable value. Unfortunately, this is common in modern design. In these buildings, the duct pressure distribution is really only a small part of the total system pressure drop, and there is very little that can be done to reduce electricity consumption except to tear them out and build something different, which generally is impractical.

From a thermal standpoint, there are many things that should be looked at. If the system is multizone or double duct, obviously, the problem of hot deck discharge temperature versus mixed air control should be examined (Figure 45-6). This will vary with the hours of operation and the total profitability of operating with variable outside air quantities (economizer cycle). However, there may be energy savings available from studying this situation. For example, by changing the economizer control to a fixed percentage control, there is a penalty for cooling 20 percent of the return air from 72 to 55°F, but a large amount of energy is saved by heating 80 percent of the air 23°F instead of 50°F.

Since conditions vary continuously, replacement of the typical economizer control with enthalpic mixed air control coupled with averaging air flow controls—meaning the hot deck versus the cold deck flow*— and resetting of mixed air control can be used to advantage when zone loads are mildly divergent or subject to high solar loadings. A great amount of energy can usually be found in the problem of heating cool air streams or cooling hot air streams. For example, a very typical case is the use of perimeter outside radiation with either a fixed or a variable volume air system (Fig. 45-7). In this case the outdoor radiation acts as a reheater forcing more cool air into the space. The result is similar to a dog chasing its tail. Consequently, the variable air volume system often will use, in a true analysis, as much or more energy than an old fashioned, low pressure, single zone system. Only by slaving the air terminal with the radiation terminal, is it possible to completely overcome this problem. Here

*For a further discussion of enthalpy control see Chapter 64, *Enthalpy Control Systems: Increased Energy Conservation*, Dr. Gideon Shavit.

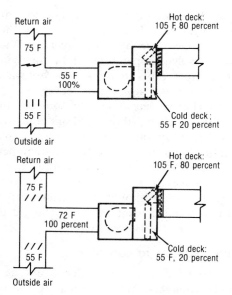

Fig. 45-6. The top drawing shows a typical multizone system with economizer cycle active. With DX systems, a load balancer or hot gas bypass is needed to prevent cycling and eventual destruction of the compressor. The bottom drawing shows a typical multizone system with the economizer cycle inactive. Similar psychrometric problems occur with double duct systems and induction air systems.

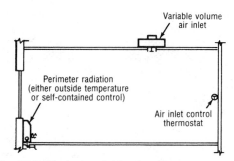

Fig. 45-7. Typical variable air volume system with perimeter radiation. Three options are available: zone perimeter radiation, install time control on perimeter outdoor temperature control to prevent operating during working hours unless outdoor temperature is below approximately 35 to 40°F, or slave perimeter radiation control to air valve control.

again the solution is simple, but the costs are significant, running around $200 per space.

Zone perimeter radiation

Zoning the perimeter radiation and only operating this radiation when absolutely required by the outside temperatures are less expensive solutions. Varying the hot water temperature by zones in accordance with outside temperature and employing solar compensators to adjust zone hot water temperatures will save further energy.

This chapter has described some of the ways in which existing buildings and systems may be upgraded to reduce energy consumption and provide better system performance. Often, the largest energy savings can be achieved by architectural modifications, and these should be examined first. Then, determine what may be achieved by alterations to the HVAC systems.

Some cautions to consider

The following discussion offers some cautionary thoughts. When estimating the effects of building or system modifications on energy consumption, degree-day methods may give very erroneous results. For example, with the reductions in infiltration and transmission loads achieved by installing double glazing and better quality windows in the hotel project detailed, a degree-day analysis would not take into account the effects of solar gain in heating the structure, and thus, it would give false (high) fuel savings. Solar gain can be considerable, and an hour-by-hour simulation program that accounts for all major variables often provides more accurate results, which can vary significantly from those achieved by simple manual methods.

The pitfalls of the degree-day method are that it may give either too low or too high a value for fuel saved. This can result in making uneconomical investment decisions either by not implementing useful energy conservation methods, or in wasting money in implementing others that a more thorough type of analysis would have rejected.

Not only is what you study in attempting to save energy in existing and new buildings important, but also how it is studied.

46

Replacement, modernization, and upgrading existing industrial plants

When money is tight, it is difficult to justify the replacement, modernization, or upgrading of existing equipment—until you look at the economies available from improved efficiency.

O. F. SMITH, JR., PE,
Senior Engineer, Western Electric Co.,
Reading, Pennsylvania

When the words replacement, modernization, or upgrading are used to refer to industrial equipment, they connote the necessity of spending money—and that is correct. Part of today's energy troubles are due to our obsession with first cost. The procedure of getting three quotations and buying the lowest price has too often been the practice in purchasing industrial equipment.

Economics of energy use

This emphasis on first cost has led to a systematic development of inefficient energy use at the point of consumption. For instance, a 4000 Btuh window unit air conditioner with an energy efficiency ratio (Btuh to watts) of 4 costs $100. The same unit with an energy efficiency ratio of 7 has a price tag of $135, and it is overwhelmingly outsold by its lower first cost competitor. Yet, at the end of an expected 10 year life, the buyer who paid the higher first cost would be $100 ahead when the total cost, including energy costs, are determined. Life cycle cost (the discounted present worth of total owning and operating costs over the expected period of use) provides the justification for efficiency.

One important caution enters into this discussion. Higher energy rates should be anticipated in calculations of future energy costs. Continued increases may run as high as 4 to 7 percent annually. This represents a 1.5 to 2 multiplier for energy costs in ten years. Others have estimated a multiplier of 3 for energy rates as soon as 1981. What can be used as a benchmark?

ASHRAE Standard 90-75 has been adopted for application in new buildings to conserve energy. It can serve as a benchmark to compare with an existing system. A 20 percent increase in energy consumption efficiency should be a reasonable goal for an existing building. All right, how are such savings to be realized? In existing buildings,

Table 46-1.   Energy reductions possible from applying various glass shading schemes to one 6 ft high by 1 ft wide window.

| | Estimated reduction in annual ton-hrs | | | | | |
| | New York City | | | Dallas | | |
Shading scheme	East	South	West	East	South	West
Single plate, no shading	15.6	18.0	27.0	33.6	36.0	58.2
Single plate, heat absorbent	8.6	11.1	16.5	18.6	22.2	35.4
Single plate, inside venetian blinds	8.4	10.5	14.7	18.6	21.0	31.5
Single plate, reflecting	9.1	8.7	13.5	19.6	17.4	29.0
Single plate, 23 bar louvered screen	4.5	6.4	9.5	9.8	12.8	20.5
Single plate, outside vertical, automatic louvers	3.3	5.1	7.5	7.2	10.2*	16.2

*Vertical fins and fixed louvers or horizontal manual louvers, outside.
Source: Jaros, Baum, & Bolles, Consulting Engineers, New York, N.Y.

increased efficiency will result from attention to details—from concentrating on a large number of small problems.

The building itself serves as a starting point. In an air conditioned office building or factory, as much as 120 ton-hr annually can be saved by properly shading one 6 by 3 ft window. Table 46-1 shows how. In one building, the installation of louvered screens on 2½ ft high strip windows on two levels reduced the refrigeration peak by 53 tons and paid out in six years at energy rates current in the 1960s.

Specific items for saving

Replacement or modernization of mechanical equipment in a plant can increase efficiency. Pumps are a good example. Too often, pumps are taken for granted. All pumps wear internally; excessive wear can be checked by a simple shutoff pressure test. If the discharge pressure of a centrifugal pump when shut off is below the expected performance curve, action to overcome the waste should be taken. Replacement of worn parts would be the normal maintenance action. However, the pump's performance curve should be checked for efficiency before it is repaired.

Pump efficiencies of 30 percent are not uncommon; yet efficiencies exceeding 50 percent are available. A higher efficiency pump can often be obtained by looking at the catalog carefully. For example, from one manufacturer's catalog, it is possible to pick four pumps for the same duty (20 gpm at 350 ft head) that require brake horsepowers of 6, 7¼, 11 and 12½. Many pumps in service

today were selected in happier times before the energy crunch.

If cooling water is circulated to process equipment, it is essential to obtain the maximum temperature rise to tolerable to each piece of equipment. Maximizing heat transfer reduces the circulation flow rate, pumping head, and energy consumption. Recirculating cooling water systems exist that operate on 2, 3, or 4°F temperature rises because of unregulated flow. Such a system can be modernized by adding self-actuated temperature regulating valves at specific equipment, or at least, by adding control with plug disk valves with dial settings.

Rinse waters respond to control also. Conductivity probes can be used with throttling or shutoff valves to limit or stop rinse water when conductivity drops to clean water level.

Applying metering to utilities

Meters are perhaps one of the most potent tools for controlling plant utilities. The following example is from the municipal field, but the principle applies. In the 1950s, the city of Allentown, Pa. faced a water demand approaching the water works' pumping capacity. The first, and most obvious, solution was to add one or more pumps. (Adequate supplies were available.) The consulting engineer recommended installation of a single unit that would increase pumping capacity by 25 percent. The preliminary report pointed out, however, that essentially the same effect could be obtained by a program of meter installations and by changing from a flat monthly rate based on the size of the service pipe to billings based on actual usage. In other words, for the same population, metered usage would be approximately 25 percent less! The reasons for this are all too familiar—the old human reasons.

How can metering be applied to plant utilities, such as water, soft water, hot water, steam, air conditioning, electricity, city gas, process gases, compressed air, sewage disposal, industrial waste, water disposal, and solid waste disposal? The principle is universal. First, the total annual cost for the particular utility must be established. Table 46-2 shows an example of the total yearly expense for soft water service. Then, the unit of allocation or base on which this expense will be distributed should be determined. For utility services, this is very often, but not always, the annual consumption in weight or volume units.

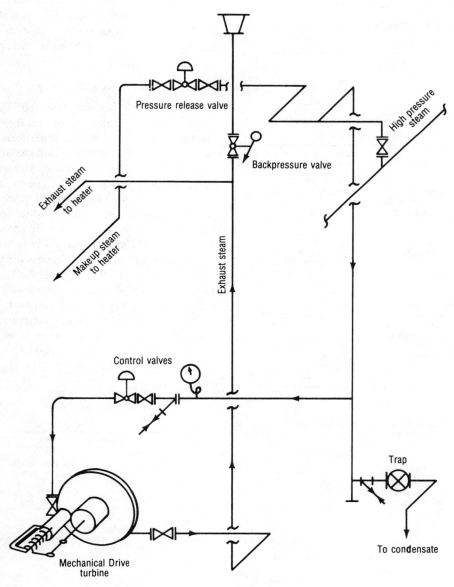

Fig. 46-1. Schematic of turbine drive steam piping.

Table 46-2. Analysis of utility expense for 1973 soft water service.

Depreciation	$ 1,000
Repairs and changes	13,000
Purchased supplies	15,000
Salaries and wages	2,000
City water	16,000
Waste water disposal	14,000
Electricity	9,000
Building rent charge	7,000
Other	1,000
Ttoal yearly expense	78,000

Total yearly consumption: 46,000,000
gal (base)

Soft water 1973 rate =
Yearly expense/yearly consumption
or
$78,000/46,000 thousand gal
or
$1.696 per thousand gal

Table 46-3. Plant utilities subject to cost distribution and control.

Utility	Base
Water	
Soft water	Volume, gallons or
Hot water	thousand gallons
Demineralized water	
Recirculated cooling water	
Steam	Weight, lbs or thousand lbs
Air conditioning	Floor area or total cooling coil load
Electricity	KVA or equipment time on
City gas	Volume, cu ft or thousand cu ft
Process gases	
Compressed air	Volume, cu ft or gal
Sewage disposal	
Industrial waste water disposal	Volume or chemical usage
Solid wastes disposal	Weight (preferred) or volume

The total yearly consumption of soft water was predicted as 46,000,000 gal. The yearly cost was divided by the yearly consumption (the unit basis of allocation). This established the usage rate per thousand gallons. The installation of submeters is relatively simple for soft water. At the end of each month (fiscal period), each soft water user receives a charge calculated as the product of the actual thousands of gallons used and the $1.696 per thousand gallons rate. The human tendency to walk away and leave the soft water running is now inhibited somewhat. The typical bases for distribution as shown in Table 46-3 indicate the relative possibilities of metering or other accounting basis.

Check combustion efficiency

A combustion efficiency test will indicate the need for modernization or replacement of any size boiler. In addition, regardless of the type of equipment, periodic flue gas analyses for CO_2 and stack temperature can guide firing adjustments to save 10 to 15 percent of the fuel used. Table 46-4 lists efficiency goals that boiler and burner equipment should reach. If tests show that boilers are significantly short of these values, modernization is justified.

The goals for percent of excess air tell another story. If more than one boiler is installed, the load should be placed on the one boiler that is likely to be closest to full load whenever possible. If a boiler generates more than 50,000 lb per hr or more, it might be feasible to utilize air heaters. Boiler efficiency can rise 2 percent for each 100° F rise in air temperature.

Table 46-4. Combustion efficiency goals.

			Power plant boilers	
Firing	Small heating boilers, %	Large heating boilers, %	No air heaters, %	With air heaters, %
Coal (stoker)	70	74	78	84
No. 2 fuel oil	70	—	—	—
No. 6 fuel oil	72	76	78	82
Gas	75	76	77	79

Firing rate, %	Percent excess air
10 to 12½	35
20	25
25	17½
50	15
100	12½

Table 46-5. An electric motor, 3600 rpm and a steam turbine, 3600 rpm compared for boiler room service.

Horsepower	Electric motor Cost, $	KVA	Steam turbine Cost, $	Lbs steam per hour
100	1500	93.5	3200	5100
70	1100	68.5	2400	3600
40	600	38.8	1900	2480
14	250	14.0	1550	1148
6	165	7.3	1300	660

Table 46-6. Typical power factors for high speed motors.

Horsepower	Load	Approximate power factor
10 and above	Full	0.87
10 and above	¾	0.83
10 and above	½	0.72
1 to 10 hp	Polyphase	0.75 to 0.90
1 to 10 hp	Split phase	0.75 to 0.85
1 to 10 hp	Low speed	0.70 to 0.85

Electric motors vs. turbines

Modernization can sometimes cause us to look backward. For instance, in a boiler room, there are a number of opportunities to replace electric motor drives with steam turbines and effect economies in two ways. Turbines can replace induction motors that have lagging power factors, which add to an electric bill both in kilowatts and lower power factor (Table 46-5). Also, backpressure turbines can exhaust to deaerating heaters, fuel oil heaters, water heaters, absorption refrigeration, and to any low pressure steam main. A steam turbine can easily be applied to a boiler feed pump, condensate return pump, forced and induced draft fans, compressors, and centrifugal refrigeration machines.

A lagging power factor imposes an energy cost penalty for oversized, small, and low speed motors. An oversized motor should be replaced; a high speed motor should be used wherever possible; and capacitors and synchronous motors should be used to correct power factors (see Table 46-6).

Finally, it is worth repeating that energy savings in existing industrial buildings will result from a number of people concentrating on and solving a large number of small problems. This is, of course, one definition of hard work, but it can be satisfying.

47

Energy saving techniques for existing buildings

Five techniques for saving energy in existing buildings are detailed and evaluated by common sense hand calculations.

PAUL K. SCHOENBERGER, PE,
formerly, Energy Conservation Engineer,
Ohio State University, presently,
Energy Management Consultant,
Columbus, Ohio

Increasing utility costs are encouraging the search for ways to conserve energy. There is a widespread belief among engineers that utility costs will rise at about 30 percent annually until at least 1980. At this inflationary rate, costs in 1980 would be almost triple present costs. The energy usage of a facility may become the single most important consideration in determining the feasibility of construction. In fact, the availability of energy sources for new construction is already a problem in many areas of the nation.

It is imperative that a common sense approach be utilized toward energy conservation. Systems cannot be operated to prevent normal utilization of a facility, however, they should provide the minimum conditions necessary for occupancy and no more. Conservation measures should be aimed at eliminating wastes. The energy usage of many existing buildings can be cut by more than 50 percent and still maintain adequate comfort. In some instances, the implementation of conservation measures can actually improve the environment within a facility.

Evaluating the effects of proposed conservation measures on energy consumption can be extremely difficult. This can be done by fragmenting a facility's usage into known components and examining the effects of proposed measures on each component. This method is readily adaptable to computerized analysis, however, it is not economically feasible to program all systems and all possible conservation measures. The results obtained from some computerized analysis programs are not any more accurate than can be obtained from some simplified, common sense hand calculations. Simplified calculations with ballpark results are all that is necessary to evaluate many conservation measures.

Five energy conserving techniques

The following discussion outlines five basic techniques for conserving energy in existing buildings.

• *Programmed start and stop of HVAC equipment*—Turning equipment on and off in accordance with occupancy schedules can save energy. Restart should be delayed until the latest possible time that will allow space conditions to be under control prior to occupancy. Motor operation should be sequenced to prevent increasing a facility's electrical demand.

Total shutdown of all systems may be feasible during unoccupied hours of the cooling season. If perimeter heating radiation is available, shutdown of supply and exhaust air systems may be possible during unoccupied hours of the heating season.

Building temperatures should be reduced during unoccupied hours of the heating season. Ventilation air dampers should remain closed when the air supply systems are restarted to permit rapid return to normal temperatures prior to occupancy if outside air must be heated or cooled. When outside conditions permit, 100 percent outside air can be used to purge the building. In many cases, opening outside air dampers and starting exhaust systems can be delayed for a period of time after occupancy and then closed and shut down a period of time before workers leave.

In many instances, minimum ventilation requirements can be met by closing the outside air dampers a part of every hour. When they are opened, a ventilation purge cycle is initiated.

Facilities with low pressure air distribution systems can use night setback thermostats to cycle units on 100 percent return air and maintain lower space temperatures during unoccupied hours of the heating season. These should have wide deadbands to prevent short cycling of supply air units. Units should operate with outside air dampers closed to overheat areas and warm interior furniture and equipment for a short time prior to occupancy.

Facilities with all-air high velocity distribution systems require operational tests to determine the best operating methods during unoccupied hours. Shutdown of all possible exhaust systems, reductions in outside air intake to systems, and reduced temperatures should be initiated during unoccupied hours of the heating season in all instances. When exhaust systems can be totally shut down, it may be possible to maintain acceptable temperatures by running the return air fans

with the supply fans off and the outside and relief air dampers closed. As long as the return air fans are not overloading, they may force enough air through the systems to maintain acceptable unoccupied temperatures. The constant volume feature of terminal high velocity boxes is lost when operating in this mode, but the box regulators remain wide open at reduced supply pressures and permit air to pass through the units. As long as no area is totally starved for air, the return fans alone may supply adequate air for the majority of unoccupied times. In some cases, the return fans alone will not be adequate during extremely cold weather. Cycling of the supply fans may be necessary then.

Cross-connecting supply air plenums of adjacent air handling units permit shutdown of some of the units during low load and unoccupied times. This may be adequate for a considerable portion of the time. It is imperative that close-off dampers be used to assure proper air flows and that fans and fan drives are adequate for this mode.

Return to normal operating conditions should be accomplished as described for low pressure air distribution systems.

• *Minimizing reheat with supply air reset*—Supply air temperatures should be kept at maximum levels to reduce heating and energy waste. Many air supply systems utilize mixed air control for free cooling during the winter. Outside and return air are mixed to obtain a fixed design supply air temperature, which is often 55° F. This 55° F air is then used directly through the unit cold decks to cool spaces with heat gains, and it is either reheated through the hot decks or at the spaces for areas with heating loads. Elevating the mixed air temperature to slightly overheat spaces with cooling loads reduces the amount of energy used in reheating. This basically overheats interior areas and areas with solar heat gains, but it can add comfort as well as reduced energy consumption. It may be possible to use a mixed air temperature as high as 63° F to maintain 76° F in areas with cooling loads. This is higher than recommended by Federal guidelines, but it conserves energy since a greater percentage of return air is recirculated. Some bleedoff of heat also occurs between interior and perimeter areas maintained at 68° F, which reduces the amount of heating necessary to keep perimeter spaces at the reduced temperature.

During the cooling season, elevated cold deck temperatures can be used when loads and humidity conditions permit. Increasing the cooling supply air temperature also permits increasing the chilled water supply temperature utilized at the cooling coils, which can increase chiller efficiency and reduce energy consumption.

• *Reduced lighting levels and air qualities*—General lighting levels can be reduced in many instances, and task lighting employed for special needs. This reduces electrical demand and consumption and also permits reduction in air quantities supplied to spaces. Reduced air quantities allow decreased fan power inputs and reduces false loading of many systems—fan horsepower varies as the cube of the fan wheel speed, rpm.

Some guidelines for reducing lighting levels are:

• The eye can adjust to differences in illumination rather quickly as long as they are no larger than 10 to 1. When occupants look from one area to another, the differences should be within this ratio.

• Specific tasks requiring lighting levels higher than overall levels can be handled by desk lamps, drafting table lights, or elevated area lighting.

• Where four tube fluorescent fixtures are used, removal of two tubes and associated ballast fuses from each fixture is preferable to removing alternate fixtures. This provides reduced lighting from each fixture without alternate light and dark spots on the ceiling.

• When ceiling lights are removed, fixtures in front of the task area should be removed first. If the lighting level is still too high, remove lighting directly above. The lighting to either side of the task area as well as from behind is the best source and provides the largest percentage of illumination while reducing glare.

It is imperative that maximum air quantities be adjusted to offset peak loads with no mixing or false loading at this time. Many

systems will not operate properly with excessive amounts of supply air. In most multizone and dual duct systems, the air utilized through the hot decks is a mixture of return and outside air. When outside air temperatures and humidities are high, increased amounts of air needed through the hot decks increase humidity levels and reduces comfort. In terminal reheat systems, excess air must be falsely loaded by reheating to prevent subcooling of the spaces. Excess air in single duct, constant volume systems without mixing or reheat, can make proper dehumidification at the cooling coils impossible.

• *Electrical demand load limiting*—Electrical savings can be realized in many instances by reducing demand peaks and maintaining more uniform usage. Electrical usage can be continuously monitored, and interruptable loads can be sequenced off for periods of time to reduce a facility's maximum demand.

• *Conversion to variable volume of supply air*—Varying the amount of air supplied to a space in accordance with the load rather than mixing hot and cold air or reheating at the space can reduce energy consumption. Diversity in space loads can be used to reduce air quantities and fan power inputs. Air quantities should be cut to minimum amounts before reheating or mixing is permitted for final temperature control. Many constant volume systems can be modified to operate with a variable volume of supply air. Often, a constant volume single duct reheat or double duct constant volume box can be converted to variable volume operation. Also, variable guide inlet vanes can be added sometimes to an existing fan installation. Interior areas can be converted to variable volume cooling only, while perimeter areas may require heating as well as cooling.

Use utility and weather records

Actual utility cost records and data published by the United States Weather Bureau can be used to establish an energy use history of a facility. If utility records are available for the past several years, the severity of weather conditions can be compared to the energy usage to determine if usage varies significantly with changes in weather. False loading of many systems causes their energy usage to remain fairly constant while weather conditions may vary considerably. Variances in operations and usage can also affect energy requirements. The advantage of analyzing existing facilities is that an actual history of energy usage as well as operation can be established.

The following examples illustrate some simplified methods of predicting reduction in energy consumption for the five techniques detailed in the first half of this chapter. The examples cover two types of systems: a constant volume, high velocity reheat system and a constant volume, high velocity double duct system.

Office building systems studied

The hypothetical facilities served by each system are office buildings. Each is utilized 48 hr per week and unoccupied the remaining 120 hr per week. Both are located in Columbus, Ohio.

The building served by the reheat system has two central air handling units that provide 55°F air to terminal reheat boxes the year around. The building with the double duct system also has two central air handling units, but they supply hot and cold air to terminal mixing boxes.

Double duct systems can operate anywhere between these two extremes during the cooling season:

• Minimum amount of reheat energy with the return air providing most of the hot deck heating capacity. This may result in poor humidity control under some load conditions.

• Maximum amount of reheat energy, which results in energy consumption equivalent to a terminal reheat system. This provides excellent humidity control and comfort under all load conditions.

The second system characteristic is assumed for the double duct system example.

Perimeter heating radiation is not used in either facility, and the systems run continuously year around. Both office buildings are 70,000 sq ft in area, and the total air

delivery of the central fan systems in both examples is 70,000 cfm.

The building with the reheat system has a gas fired boiler plant that supplies media to absorption water chillers for cooling and coils for heating.

The facility with the double duct system has a gas fired boiler plant that supplies media to heating coils in the air handling units, and an electrically driven centrifugal water chiller that supplies media to cooling coils in the air handlers.

The energy use history of the reheat facility indicates that 2,200,000 kWh of electricity and 300,000 therms of gas are consumed annually. The history of the double duct facility indicates that 2,800,000 kWh of electricity and 140,000 therms of gas are consumed annually.

The savings initially shown in each operation are calculated on the basis of each operation being performed independently of the others unless specifically noted. Plus and minus signs denote increases and decreases, respectively.

Terminal reheat system example

The first example considered is the constant volume, high velocity terminal reheat system.

A. Unoccupied cycles of operation.

1. During the heating season, supply and exhaust air fans are shut off for 120 hr per week. Return air fans are run on 100 percent return air to maintain unoccupied winter temperatures.

2. During the cooling season, systems are totally shut down during unoccupied hours.

3. Supply air fans draw 94 amps each; return air fans, 48 amps each; and exhaust fans draw 20 amps at 480 V on three-phase electrical service.

4. The cooling season is 17 weeks long and the heating season is 35 weeks long.

5. *Electrical savings during heating season:*

Supply fans: 94 + 94 = 188 amps
Exhaust fans: 20 amps
———————
208 amps

The total kilowatts used are calculated:

$$I \times E \times 1.73 \times \cos \theta / 1000 \text{ W per kW}$$
$$= \text{kW}$$

Terminal reheat system

- Office building.
- 70,000 sq ft in area.
- Columbus, Ohio.
- Occupied 48 hr per week.
- Unoccupied 120 hr per week.
- Cooling season lasts 17 weeks.
- Heating season lasts 35 weeks.
- Two central air handling units provide 70,000 cfm.
- Gas is used for heating and cooling.
- Gas fired boiler plant uses gas at 1000 Btu per cu ft (1 therm = 100,000 Btu) at 75 percent efficiency.
- Absorption chiller uses 18.5 lb of steam per ton-hr.
- One ton of cooling = 12,000 Btu per ton-hr.

$$208 \times 480 \times 1.73 \times 0.85 = 146.8 \text{ kW}$$

The total for the heating season is:

$$146.8 \times 120 \times 35/1000 = -616,000 \text{ kWh}$$

6. *Electrical savings during the cooling season:*

Supply fans:	188 amps
Exhaust fans:	20 amps
Return fans: 48 + 48 =	96 amps
Cooling auxiliaries:	60 amps
	364 amps

The total number of kilowatts used is:

$$364 \times 480 \times 1.73 \times 0.85 = 256.9 \text{ kW}$$

The total saved per cooling season is:

$$256.9 \times 120 \times 17 = -524,000 \text{ kWh}$$

7. *Gas savings during heating season.* Mixed air is controlled at 55°F. Without any space loads, the entire air supply must be reheated to space conditions of 73°F before offset of heat losses can be effected. The reduction per heating season is:

$$70,000 \times 1.08 \times 18 \times 35 \times 120$$
$$= -5720 \times 10^6 \text{ Btu}$$

The gas fired boiler plant uses gas at 1000 Btu per cu ft at an efficiency of 75 percent. Considering these factors, the reduction over the heating season is:

$$5720 \times 10^6/(10^5 \times 0.75) = -76,300 \text{ therms}$$

8. *Gas savings during the cooling season.* Supply air is controlled at 55°F. With low space loads, it must be reheated to maintain thermostat settings.

Since the majority of the unoccupied hours occur at night when there are reduced cooling loads, it is assumed that the average summer reheat during unoccupied periods is 13°F, and the average cooling load for these periods requires a 10°F drop through the cooling coils. Total air delivery is 70,000 cfm.

The reduction in energy use per cooling season for reheat during unoccupied hours is:

$$70,000 \times 1.08 \times 13 \times 17 \times 120$$
$$= -2000 \times 10^6 \text{ Btu}$$

Using gas at 1000 Btu per cu ft (1 therm = 100,000 Btu) and a boiler efficiency of 75 percent, the gas savings for reheat during unoccupied hours of the cooling season is:

$$2000 \times 10^6/(10^5 \times 0.75) = 26,670 \text{ therms}$$

The number of Btu's saved per cooling season during unoccupied loads is:

$$70,000 \times 1.08 \times 10 \times 17 \times 120$$
$$= -1540 \times 10^6 \text{ Btu}$$

With an absorption chiller using an average of 18.5 lb of steam per ton-hr and boilers using gas at 1000 Btu per cu ft (1 therm = 100,000 Btu) at 75 percent efficiency, the gas savings for the cooling season during unoccupied hours is:

$$1540 \times 10^6/(10^5 \times 0.75) \times 18,500/12,000$$
$$= -31,660 \text{ therms}$$

9. *Total electrical savings for unoccupied cycles of operation is as follows:*

Heating season:	−616,600 kWh
Cooling season:	−524,000 kWh
	−1,140,600 kWh

This means that unoccupied operation of the systems as described reduces electrical consumption by approximately 50 percent.

10. *Total gas reduction in therms for unoccupied cycles of operation is as follows:*

Heating season: −76,300 therms
Cooling season
   for reheat: −26,670 therms
Cooling season
   chiller loads: −31,660 therms
                   −134,630 therms

Unoccupied shutdown of the system as described reduces energy consumption by about 45 percent.

Minimize reheat

B. Minimizing reheat with supply air reset.

1. During the heating season, mixed air is elevated from 55 to 63°F during occupied hours. No savings are realized during unoccupied hours with the systems shut off.

2. During the cooling season, cooling supply air temperature is elevated when humidity conditions permit. It is assumed that its temperature can be increased from 55 to 59°F as an average condition during unoccupied hours of the cooling season.

3. *Gas savings during the heating season.* The mixed air requires 8°F less reheat during occupied hours. The reduction in Btu's per heating season is:

$$70,000 \times 1.08 \times 8 \times 35 \times 48$$
$$= -1020 \times 10^6 \text{ Btu}$$

Assuming the same rate of consumption and efficiency as previously, the gas savings per heating season in therms are:

$$1020 \times 10^6/(10^5 \times 0.75) = -13,500 \text{ therms}$$

4. *Gas savings during the cooling season.* Cooling supply air requires 4°F less reheat for occupied hours.

The reduction in Btu's per cooling season is:

$$70,000 \times 1.08 \times 4 \times 17 \times 48$$
$$= -247 \times 10^6 \text{ Btu}$$

The reduction per cooling season for reheat during occupied hours is:

$$247 \times 10^6/(10^5 \times 0.75) = -3290 \text{ therms}$$

The absorption chiller uses an average of 18.5 lb of steam per ton-hr, and the boilers use gas at 1000 Btu per cu ft (1 therm = 100,000 Btu) at an efficiency of 75 percent. The reduction in chiller loads for cooling during occupied hours is:

$$247 \times 10^6/(10^5 \times 0.75) \times 18,500/12,000$$
$$= -5080 \text{ therms}$$

5. *Total gas reduction for minimizing reheat with supply air reset during the occupied hours is:*

Heating season: −13,500 therms
Cooling season reheat: −3,290 therms
Chiller loads: −5,080 therms
                   −21,870 therms

This means that increasing the supply air temperatures during the occupied hours as described would reduce gas consumption by approximately 7.3 percent.

Reduce lighting, air quantities

C. Reduced lighting levels and air quantities.

1. During the heating season, it is assumed that the lighting intensities can be reduced by one watt per sq ft. during occupied hours. The area of the building is 70,000 sq ft. This results in the following reduction:

$$70,000 \times 1 \text{ W per sq ft} \times 1 \text{ kW}/1000 \text{ W}$$
$$= -70 \text{ kW}$$

It is also assumed that the systems can be shut off during unoccupied hours.

During the cooling season, air quantities can be adjusted to handle the reduced cooling loads during occupied hours. It is assumed that systems can be shut off during unoccupied hours.

2. *Electrical savings during cooling season.* The reduced lighting load as calculated in Point 1 is 70 kW. This permits the following decrease in air delivery quantities. (A reduction of 1 W per sq ft means a load reduction of 3.4 Btu per sq ft.)

$$3.4/(1.08 \times 20° \text{F}) = -0.16 \text{ cfm per sq ft}$$

This is a 16 percent reduction compared to the original 1 cfm per sq ft. From the fan laws and electric motor performance curves, it is determined that a 16 percent cut in air quantities permits an approximate 40 percent decrease in fan power inputs necessary to supply the reduced air circulation, thus:

$$\begin{array}{ll} \text{Supply fans:} & \text{188 amps} \\ \text{Return fans:} & \underline{\text{96 amps}} \\ & \text{284 amps} \end{array}$$

Hence, applying the reduction, 40 percent of 284 is 113.6 amps. The reduction in fan loads is:

$$113.6 \times 480 \times 1.73 \times 0.85 = -80.2 \text{ kW}$$

The total reduction in electrical demand for reduced lighting and air circulation is:

$$70 + 80.2 = -150.2 \text{ kW}$$

The reduction in electricity usage for the cooling season is:

$$150.2 \times 17 \times 48 = -122,600 \text{ kWh}$$

3. *Electrical savings during the heating season.* The electrical demand in the heating season is reduced by the same amount as in the cooling season. The reduction in electricity usage in the heating season is (note that the heating season is 35 weeks long as opposed to 17 weeks of cooling):

$$150.2 \times 35 \times 48 = -252,300 \text{ kWh}$$

4. *Gas savings during cooling season.* The reduction in cooling loads due to reduced lighting load is:

$$70 \text{ kW} \times 3412 \text{ Btu per kWh} \times 17 \times 48$$
$$= -195 \times 10^6 \text{ Btu}$$

The reduction in the annual cooling load due to reduced fan power inputs is:

$$80.2 \times 3412 \times 17 \times 48$$
$$= -223 \times 10^6 \text{ Btu}$$

Now, considering the consumption of the absorption chiller and the efficiency and gas usage of the boilers, the reduction in chiller loads due to reduced lighting and fan energy is:

$$418 \times 10^6/(10^5 \times 0.75) \times 18,500/12,000$$
$$= -8600 \text{ therms}$$

5. *Increase in gas consumption during heating season.* The reduced lighting levels increase the amount of heating required in spaces with heat losses. It is assumed than an average of 60 percent of the areas are heated during the heating season, perimeter spaces and areas under the roof. It is also assumed that the reduced heat obtained from lighting is offset with increased amounts of reheat at the spaces. Since 60 percent of 70 kW is 42 kW, the additional heat energy required to offset reduced lighting levels during the heating season is:

$$42 \times 3412 \times 35 \times 48 = +241 \times 10^6 \text{ Btu}$$

When the boiler conversion factor is added to this, the increased consumption due to reduced lighting levels is:

$$241 \times 10^6/(10^5 \times 0.75) = +3210 \text{ therms}$$

6. *Total electrical savings due to reduced lighting levels:*

$$\begin{array}{ll} \text{Cooling season:} & -122,600 \text{ kWh} \\ \text{Heating season:} & \underline{-252,300 \text{ kWh}} \\ & -374,900 \text{ kWh} \end{array}$$

The reduced lighting levels and air quantities decreased the facility's electrical demand by 150 kW and electrical consumption by about 17 percent.

7. *Total gas savings for reduced lighting levels and air quantities:*

$$\begin{array}{ll} \text{Cooling season:} & -8600 \text{ therms} \\ \text{Heating season:} & \underline{+3210 \text{ therms}} \\ & -5390 \text{ therms} \end{array}$$

Gas consumption is decreased by approximately 1.8 percent.

Consider demand load limiting

D. Electrical demand load limiting.

1. The electrical utility cost records should be examined to determine if demand control would be economically feasible. It may be possible to shut portions of lighting or other equipment off during peak demand periods to reduce the cost of electricity. The exact effect of this measure can be easily analyzed once the interruptable loads are determined. In most instances, this measure would not

affect the electrical consumption more than a few percent, but it may have a very short payback on implementation costs due to significant demand charge reductions.

Convert to VAV system

E. Conversion to variable volume of supply air (VAV).

1. Diversities in occupancies and solar loads can be compensated for by varying the air supply in accordance with the loads rather than supplying constant quantities of air to the spaces and reheating for final temperature control. Diversities in loads may allow operation on variable volume with only 60 percent of the air circulation needed with constant volume systems; this reduces fan draws as well as false loading. The actual reduction can be approximated by comparing the block building load with the sum of the peak zone loads. For purposes of this analysis, it is assumed that air quantities can be cut by 30 percent when variable volume of supply air is instituted.

2. During the heating season, air quantities are reduced to minimum amounts before reheat is permitted in spaces with heating loads. No savings are realized during unoccupied hours when systems are shut off.

3. During the cooling season, total air quantities are reduced by 30 percent due to swing of solar loads and use of average occupancy loads. There are no reductions during unoccupied hours when systems are shut off.

4. *Electrical savings for the year.* It is assumed that the electrical demand for fan inputs is reduced by about 65 percent when the total circulation is cut 30 percent. The original full load fan power draw is:

$$\begin{aligned} \text{Supply fans: } & 180 \text{ amps} \\ \text{Return fans: } & \underline{96 \text{ amps}} \\ & 284 \text{ amps} \end{aligned}$$

Cos θ: 0.85

$$284 \times 480 \times 1.73 \times 0.85/1000 = 200 \text{ kW}$$

The new fan power draw at 34 percent of the full load motor horsepower, which is 45 percent of the full load amps, or: $0.45 \times 284 = 127.8$. The power factor cos θ, decreases to

0.65, and the new fan power draw is:

$$127.8 \times 480 \times 1.73 \times 0.65/1000 = 69 \text{ kW}$$

This represents a 65.5 percent reduction since (200 kW − 69 kW) 100/200 kW = 65.5 percent. Fan power reduction is 131 kW.

The reduction in electricity usage due to reduced fan loads over the year is:

$$\begin{aligned} 131 \times 35 \times 48 &= -220,100 \text{ kWh (Heating)} \\ 131 \times 17 \times 48 &= \underline{-106,900 \text{ kWh (Cooling)}} \\ & -327,000 \text{ kWh} \end{aligned}$$

The conversion to variable volume of supply air reduced electrical consumption by approximately 14.8 percent.

5. *Gas savings during the heating season.* Reductions of the total air circulation by 30 percent reduced the reheat load necessary to elevate this excess air from 55 to 73° F. This reduction in reheat load for the heating season due to conversion to VAV is:

$$\begin{aligned} 0.3 \times 70,000 \times 1.08 & \times 18 \times 35 \times 48 \\ &= -686 \times 10^6 \text{ Btu} \end{aligned}$$

Adding the boiler efficiency consideration, the reduced consumption is:

$$686 \times 10^6/(10^5 \times 0.75) = -9150 \text{ therms}$$

6. *Gas savings during the cooling season.* Reductions of the total air circulation by 30 percent reduces the reheat load necessary to increase this excess air as well as reduce the cooling load of the facility. It is assumed that the average temperature rise for reheat over the cooling season is 6° F. The cooling load is cut by the same amount as the reheat is reduced. The reduction in reheat load due to conversion to variable volume is:

$$\begin{aligned} 0.3 \times 70,000 \times 1.08 & \times 6 \times 17 \times 48 \\ &= -111 \times 10^6 \text{ Btu} \end{aligned}$$

Again, adding the boiler efficiency factor to this results in the following reductions:

$$111 \times 10^6/(10^5 \times 0.75) = -1480 \text{ therms}$$

The reduction of fan power inputs during the cooling season reduced the annual cooling load by:

$$\begin{aligned} 106,900 \text{ kWh} & \times 3412 \text{ Btu per kWh} \\ &= -365 \times 10^6 \text{ Btu} \end{aligned}$$

The reduction in false loading and in fan power reduces the cooling load by an equal amount. The reduction in chiller loads due to conversion to VAV is:

$$476 \times 10^6/(10^5 \times 0.75) \times 18,000/12,000$$

$$= -9780 \text{ therms}$$

7. *Total gas savings for conversion to VAV is:*

Heating season:	−9150 therms
Cooling season reheat:	−1480 therms
Chiller loads:	−9780 therms
	−20,410 therms

This reduced gas consumption by about 6.8 percent.

Summarizing total savings

F. The total effect.

The total effect on energy consumption for implementing all of the measures discussed would be somewhat greater than the sum of the measures figured individually. When the system is converted to VAV, the effect of supply air reset is diminished. However, implementing the other items reduces the maximum air quantity even further. Reducing the lighting load 16 percent reduces the supply air volume from 70,000 cfm to 58,800 cfm. If the system is also converted from a constant volume to variable air volume operation, diversity in space loads reduces the maximum delivered air volume even further. The 30 percent reduction assumed in the example yields a maximum supply air volume of 41,200 cfm, and heating, reheat, cooling and fan loads are further reduced, offsetting the diminished savings from supply air reset when the air quantity is reduced. Whether this 40 percent reduction (70,000 to 41,200 cfm) from the original design would adversely affect performance must be determined prior to implementation.

The reduction in gas and electric consumptions for *each* individual measure is shown in Table 47-1.

Double duct system example

The effects of the five basic techniques for reducing energy consumption will now be assessed for the facility with a constant volume, high velocity double duct system.

Table 47-1.   Reduction in electric and gas consumption for the terminal reheat system example.*

Technique	Electricity		Gas	
	KWH, x 10⁻³	Percent	Therms	Percent
A. Unoccupied cycles of operation	−1140.6	−51.8	−134,630	−44.8
B. Minimizing reheat with supply air reset.	—	—	−21,970	−7.3
C. Reduced lighting levels and air quantities.	−374.9	−17.0	−5,390	−1.8
D. Electrical demand load limiting (estimated)	−44	−2.0	—	—
E. Conversion to VAV	−327	−14.8	−20,410	−6.8

*Quantities and percentage shown are for each item treated independently. Cummulative savings would be somewhat greater than the total of the individual conservation steps.
In this system, gas is used for heating and cooling (steam absorption).
Initial consumption: 2,200,000 KWH per year and 300,000 therms per year.

A. Unoccupied cycles of operation.

The same conditions exist for the double duct system as were outlined in Section A, Points 1–4 of the first example.

The electrical savings during the heating season are also the same: 616,000 kWh. The reduction in electricity usage during the cooling season varies somewhat, however, so the analysis shall begin there.

1. *Electrical savings during cooling season:*

Supply fans: 94 + 94 =	188 amps
Exhaust fans:	20 amps
Return fans:	96 amps
Cooling auxiliaries:	40 amps
	344 amps

Double duct system

- Office building.
- 70,000 sq ft in area.
- Columbus, Ohio.
- Occupied 48 hr per week.
- Unoccupied 120 hr per week.
- Cooling season lasts 17 weeks.
- Heating season lasts 35 weeks.
- Two central air handling units provide 70,000 cfm.
- Gas is used for heating only.
- Gas fired boiler plant uses gas at 1000 Btu per cu ft (1 therm = 100,000 Btu) at 75 percent efficiency.
- Electrically driven centrifugal chiller uses 0.8 kW per ton of cooling.
- One ton of cooling = 12,000 Btu per ton-hr.

The total number of kilowatts used is:

344 × 480 × 1.73 × 0.85/1000

$$= 242.8 \text{ kW}$$

The reduction for fans and auxiliaries during the cooling season is:

242.8 × 120 × 17 = −495,000 kWh

In addition to this, the chiller would be turned off during the unoccupied hours. Since the majority of unoccupied hours occur at night when cooling loads are low, it is assumed that the average unoccupied cooling load is 100 tons. The chiller load reduction for the 17 week cooling season is:

100 × 0.8 × 120 × 17 = −163,000 kWh

2. *Gas savings during heating season.* Mixed air is controlled at 55°F. There are no space loads, and the entire air supply must be reheated to space conditions before offset of heat losses can be affected. The reduction over the 35 week heating season is:

70,000 × 1.08 × 18 × 35 × 120 =

$$-5720 \times 10^6 \text{Btu}$$

Using gas at 1000 Btu per cu ft (1 therm = 100,000 Btu) and a boiler efficiency of 75 percent, the gas conserved over the heating season is:

$5720 \times 10^6/(10^5 \times 0.75)$ = −76,300 therms

3. *Gas savings during the cooling season.* Mixed air is reheated to 95°F through the hot decks on the air handling units. With no space loads, 50 percent hot air at 95°F is mixed with 50 percent cold air at 55°F to maintain a neutral temperature of 75°F in the spaces. It is assumed that this condition only occurs during nighttime for 70 hours per week. The remaining 50 unoccupied hours impose a load due to the heat storage effect of the structure. The air is reheated from 75 to 95°F. The reduction per cooling season is:

35,000 × 1.08 × 20°F × 17 × 70

$$= -900 \times 10^6 \text{Btu}$$

Using gas at 1000 Btu per cu ft and a boiler efficiency of 75 percent, the reduction for the

cooling season is:

$900 \times 10^6/(10^5 \times 0.75)$ = 12,000 therms

4. *Total electrical savings.* Total reduction in electrical consumption for the unoccupied cycles of operation is as follows:

Heating season: −616,000 kWh
Cooling season: −658,000 kWh

−1,274,600 kWh

The unoccupied operation as described reduced electrical consumption by about 45 percent.

5. *Total gas savings.* The total reduction in gas usage is as follows:

Heating season: −76,300 therms
Cooling season: −12,000 therms

−88,300 therms

Shutdown of the systems during unoccupied hours reduced gas consumption by about 63 percent.

Reset supply air to save reheat

B. Minimizing reheat with supply air reset.

The same conditions exist as outlined in Section B, Points 1 and 2 of the reheat analysis.

The reduction in gas consumption for both the heating and cooling season is also the same as for the reheat system, however, the double duct system has an electrically driven chiller, yielding a higher percentage of savings for this example:

Heating season: −13,500 therms
Cooling season: − 3,290 therms

−16,790 therms

Elevating supply air temperatures during the occupied hours reduced gas consumption by about 12 percent.

Lighting levels and air quantities

C. Reduced lighting levels and air quantities.

Points 1 and 2 from Section C of the reheat analysis are also applicable here.

1. *Electrical savings.* The reduction in electricity over the cooling season due to reduced lighting levels and fan power inputs is

the same as for the reheat system: −122,600 kWh and −418 × 10⁶ Btu. However, the double duct system has a centrifugal chiller that uses an average of 0.8 kW per ton for cooling.

The total reduction in chiller output for the cooling season due to reduced fan horsepower and lighting levels is:

$$418 \times 10^6 / 12,000 \times 0.8 = -27,800 \text{ kWh}$$

The total reduction for the cooling season is:

Load reduction from lights and fan horsepower:	−122,600 kWh
Chiller load reduction:	− 27,800 kWh
	−150,400 kWh

2. *Electrical savings during the heating season.* This is the same reduction as achieved by the reheat system: −252,300 kWh.

3. *Increase in gas consumption during heating season.* As in the reheat system, gas consumption is increased +3210 therms per heating season due to the reduced lighting levels. The loss of the heat from lighting must be offset by increased amounts of reheat at the spaces. This results in an increase in gas consumption by 2.3 percent in the double duct system.

4. *Total electrical savings.* The total reduction due to reduced lighting levels and air quantities for the year is:

Cooling season:	−150,400 kWh
Heating season:	−252,300 kWh
	−402,700 kWh

Lower lighting levels and air quantities reduced the facility's electrical demand by at least 150 kW and the electrical consumption by approximately 14 percent.

Demand load limiting

D. Electrical demand load limiting.

The same remarks as for Section D, Point 1 of the reheat system analysis are applicable here. It may be possible to shut off portions of the electrical equipment during peak demand periods to reduce the cost of electricity, but in most cases, this does not affect consumption more than a few percent. How-

ever, it may have a very short payback period on implementation costs.

Variable air volume supply

E. Conversion to variable air supply.

Points 1, 2, 3, of Section E of the reheat system are applicable here.

1. *Electrical savings for the year.* As in the reheat example, when the total circulation is cut by about 30 percent, the electrical demand for fan horsepower is reduced by about 65 percent. Hence, fan horsepower draw drops from 200 kW to 69 kW. This amounts to a reduction of −327,000 kWh for the year. Chiller input is reduced due to decreased reheat: $111 \times 10^6 / 12,000 \times 0.8 = -7,400$ kWh or a total of −334,900 kWh. Conversion to variable volume air supply reduced electrical consumption by 11.9 percent.

2. *Gas savings for the year.* As in the first example, reduction of the total air circulation by 30 percent reduces the reheat load necessary to elevate this excess air from 55 to 73°F. This reduces the load during the heating season by −686 × 10⁶ Btu, or when all factors are considered by −9150 therms.

During the cooling season, the reduction in reheat load amounts to −111 × 10⁶, when the chiller and boiler factors are added, −1480 therms. This amounts to a total reduction of 10,630 therms during the year for the double duct system. This is a reduction in gas consumption of about 7.6 percent.

Total savings summarized

F. The total effect.

The total effect on energy consumption for implementing all of the measures discussed would be somewhat greater than the sum of the measures figured individually for the same reasons as given for the reheat example. The same cautions for converting to VAV apply.

The reduction in electric and gas consumption for the double duct system for *each* individual conservation measure are summarized in Table 47-2.

Summary of studies show big savings

Many systems have been operating as described in these examples. The energy con-

Table 47-2. Reduction in electric and gas consumption for the double duct system example.*

Technique	Electricity		Gas	
	KWH, x 10⁻³	Percent	Therms	Percent
A. Unoccupied cycles of operation.	−1274.6	−45.5	−88,200	−63
B. Minimizing reheat with supply air reset.	—	—	−16,789	−11.9
C. Reduced lighting levels and air quantities.	−370.8	−13.2	+3,120	+2.3
D. Electrical demand load limiting (estimated)	−56	−2	—	—
E. Conversion to VAV	−334	−11.9	−10,630	−7.8

*Quantities and percentages shown are for each item treated independently. Cumulative savings would be somewhat greater than the total of the individual items.
In this system gas is used for heating and electricity for cooling.
Initial consumption: 2,800,000 KWH per year and 140,000 therms per year.

sumption of these facilities could be reduced by more than 50 percent, and the feasibility of instituting any of the proposed measures is easily determined with ballpark calculations.

Most HVAC systems do not automatically adjust to varying loads to reduce energy usage in accordance with the reduced loads. The systems have been designed for maximum possible loads in all areas, but this situation rarely occurs. Most systems do not automatically compensate for reduced occupancy and swing of solar loads to reduce energy usage; they falsely load themselves to maintain high energy usage at reduced loads.

Common sense can be used by operating personnel to reduce energy usage through manual manipulation of the system controls; however, justification of expenditures to implement changes in operation or equipment require a ballpark prediction of their effects on energy consumption. This chapter was prepared to assist owners in evaluating the economics of proposed conservation measures.

Following is an example of the data necessary for evaluating proposed conservation measures. A terminal reheat system is the basis of the example.

A. Present utility costs (obtained from utility records). For this example, it is as-

sumed that gas costs $0.10 per therm, and electricity costs 1.5 cents per kWh, including the demand charge.

1. Gas costs:

300,000 therms per yr × $0.10 per therm

$$= \$30,000 \text{ per year}$$

2. Electric costs:

2,200,000 kWh per year × $0.015 per kWh

$$= \$33,000 \text{ per year}$$

The combined energy cost is $63,000 per year.

B. Savings in utility costs due to unoccupied cycles of operation (conservation measures calculated in Section A of the reheat example).

Reduction in gas costs:

$$45\% \times 30,000 = 13,500$$

Reduction in electrical costs:

$$50\% \times 33,000 = 16,500$$

The total amount saved for one year is $30,000.

In an actual analysis, the incremental costs of the energy sources have to be applied, *i.e.*, the proper rate step must be used since most utility rates have decreasing charges as consumption increases.

If the programming equipment and controls necessary to automate the unoccupied cycles is $5,000, the return on initial investment would be almost six times per year. Obviously, the accuracy of these ballpark calculations is unimportant since the energy reductions for unoccupied cycles would be economically sound with 20 percent of these savings.

This same procedure can be used to evaluate any of the proposed conservation measures. It should be noted that implementation of some of the measures would reduce energy usage enough during the first year to pay for implementing further measures the next year.

48

Upgrading system performance in existing installations

What can be done to improve efficiency.

E. R. AMBROSE, PE,
Mount Dora, Florida

This chapter considers possible improvements in efficiency by replacing equipment and by altering the basic system design. Such changes are usually classified as retrofitting and generally include everything from installing different types and sizes of heating and cooling units to drastic changes in control systems.

Heating and cooling distribution

Table 48-1 lists some of the more acceptable heating and cooling distribution systems found in today's structures. Brief comments are made as to the upgrading potential of each.

Some of these are not always readily adaptable to improvements in efficiency. This category includes high pressure, high velocity air systems and terminal reheat systems. However, there are examples where: dual duct systems have been converted to single duct VAV systems[1]; perimeter induction systems have been converted to VAV[2]; and terminal reheat systems have been optimized by readjusting winter and summer supply air temperatures.[3] The first two ex-

amples cited retained the existing double duct mixing boxes and induction units.

A number of existing systems can be easily upgraded at reasonable costs. This group includes low pressure, low velocity air systems, and fan-coil installations. Frequently, the unitary, decentralized, and self-contained terminal installations also offer attractive possibilities for overall efficiency improvement.

It must be pointed out that each system must be analyzed separately on its own merits. The many variables involved make it difficult, if not impossible, to judge existing installations without an inspection. Also, owning and operating cost evaluations must be made by a competent engineer to assure that the proposals are economically justifiable.[4]

Lighting levels important

A structure's lighting layout is not the responsibility of the HVAC engineer. However, he must be familiar with the required illumination levels and work closely with the illumination engineer. It is the HVAC engineer's responsibility to assure that the selected heating-cooling-lighting combination is practical and efficient.

Considerable attention is now being given

---

[1]Superscript numerals refer to references at the end of the chapter.

Table 48-1.   Heating and cooling systems likely to be found in existing buildings.

System	Comments
Low Pressure, low velocity air	Usually has an upper pressure limit of 3.75 in. WG. Outlet velocity approximately 1800 fpm.
High pressure, high velocity air • dual duct • single duct • induction units	Has total pressure range of 6.75 to 9.75 in. WG. Outlet velocity is approximately 5000 fpm. Fan brake horsepower varies directly with pressure. Has low overall efficiency. Sometimes feasible to convert to other systems.
Variable air volume (VAV)	Regulates the amount of air to a given area to match the load. Provides for reduction in fan motor load by fan inlet guide vanes with higher efficiencies.
Terminal reheat	Energy waste because of falsely loading the equipment at partial loads. Energy waste may be reduced by resetting of temperatures.
Unitary, decentralized, and two package • floor • ceiling • perimeter • rooftop	Conditioner-compressor sections can be located jointly or remotely. Exceedingly flexible system, as to location and operation. Low first cost and easy maintenance are attributes. Compressor-condenser section is usually located outside conditioned space. Cooling can be provided for specific areas outside of normal working hours without the need to operate all air handling systems and a central cooling plant.
Self-contained, terminals • air cooled condenser • water cooled condenser	Through-the-wall air conditioning. Probably the most flexible with the lowest first cost. Equipment noise can be a problem. Cooling can be provided for specific areas outside of normal working hours without the need to operate all air handling systems and a central cooling plant.
Fan-coil units • two-pipe • three-pipe • four-pipe	Usually requires minimum floor space. Duct work throughout the building unnecessary. Noise levels generally acceptable. Simultaneous heating and cooling possible. Fan-coil controls can be interlocked with overhead supply air systems so that the two systems do not buck one another.

to the development of new lighting standards by the Illumination Engineering Society (IES) with assistance from electric utilities and lighting equipment manufacturers. The importance of a practical lighting layout is discussed and recommended illumination levels for various types of spaces are included in ASHRAE Standard 90-75 Energy Conservation in New Building Design.[5] Generally, these recommendations are the same as listed in the latest *Illumination Engineering Society Handbook*.

Before changes are made in an existing building's lighting system, a survey is necessary to check the illumination levels against standards covering the tasks performed in various areas and, equally important, to learn how the lighting system interrelates with the air conditioning system.

Frequently in modern commercial buildings, lighting is 25 to 30 percent or more of the air conditioning cooling load. Therefore, changes in existing lighting levels can affect heating and cooling loads and the required capacity of the HVAC equipment. Heat from lights may help to heat a building in winter, and this must be considered if a reduction in lighting is being contemplated. In fact, additional heating capacity may be necessary. Also, the cooling system may require extensive alterations to obtain any benefit from the proposed lighting reduction. Consequently, in some cases when all factors are considered, it may be more economical and practical to keep the existing lighting system.

Controls may need replacing

A dependable control system that furnishes the desired heating and cooling when and where needed is essential to obtain maximum efficiency under all operating conditions. Several manufacturers have developed control cycles for energy conservation that are computerized for maximum reliability and accuracy. A monitoring device that sounds an alarm or shuts down a system when efficiency falls below a certain level can also be included in a control system. In fact, the performance and satisfaction level of an older air conditioning installation can often be materially improved by merely replacing defective control equipment and by upgrading the operating cycle.

Determine seasonal efficiency

The thermal efficiency of an air conditioning system is influenced by various design and operating conditions. These include fluctuations in outdoor temperatures with corresponding changes in heating and cooling loads, together with the inherent application features of a particular installation.

Consequently, the estimation of energy savings for a given system must be based on an evaluation made over a reasonably long time period, such as an entire season. The terms seasonal efficiency and seasonal performance factor describe such an evaluation. The first relates to the heating system, and the second applies to the cooling system.

• *Heating*—Seasonal heating efficiency (in percent) is defined as the useful energy output (at the point of use) for an entire heating season divided by the total energy input. The energy input should include the heat equivalent of the work input to fans, pumps, controls, and all other such components.

Reliable seasonal efficiencies of fossil fuel heating equiment, particularly, are difficult to obtain because of the many factors involved that are more directly related to the respective installation characteristics than to the design of the equipment. For instance, heating equiment operates at both full and partial loads, and stack losses (due to wind and thermal head), short cycling losses, and losses due to variable combustion, and makeup air requirements will occur.

Equipment manufacturers should be able to supply efficiencies for boilers and furnaces at full and partial loads under controlled, steady state conditions, but actual field tests are necessary to determine the effects of these other factors. The rather meager data published so far for residential and small commercial installations indicate probable seasonal efficiencies from 35 to 50 percent, or lower, for fossil fuel equipment. However, considerably more data are needed before dependable seasonal efficiencies can be established.

Electric resistance and electric heat pump systems are not burdened with the application disadvantages of fossil fuel equipment. Consequently, considerably more actual operating data are available as a guide in establishing realistic seasonal efficiencies, but additional data are needed for these installations also.

Until such data are established, equipment manufacturers cannot publish meaningful efficiency information needed to correctly evaluate possible improvements and corresponding savings in energy usage resulting from retrofitting existing installations.

• *Cooling*—Comparable to seasonal efficiency for heating is the coefficient of performance for cooling equipment. By definition, COP is the ratio of the rate of net heat removal (at the point of use) to the rate of total energy input, expressed in equivalent units. Similarly, seasonal performance factor (SPF) is the total net heat removal for the entire cooling season divided by the respective heat equivalent of the total energy input to produce the effect. COP and SPF should include the input to compressors, fans, blowers, pumps, and other such auxiliaries.

The term, energy efficiency ratio (EER), is defined as the net capacity at a given rating point divided by the power input in watts. (This input includes the condenser fan but not the evaporating fan). EER is not applicable to seasonal performance evaluations but can be used to compare unitary equipment at a predetermined operating condition. About 85 percent of the so-called room air conditioners now on the market have an EER of 5.0 to 8.0—an average of 6.3.

ASHRAE Standard 90-75 states the improvement in COP or EER that can logically be expected in the near future.[5] For instance, larger, entirely electric equipment (65,000 Btuh and over) is expected to have a minimum COP (EER) at a standard rating condition of 2.0 (6.9) beginning January 1, 1977 and of 2.2 (7.5) by January 1, 1980. Similarly, the minimum COP for heat operated cooling equipment for the same two periods is 0.5 and 0.55. These projected increases represent an improvement of about 10 percent in three years.

It is apparent, therefore, that the hope for conserving energy will not materialize from design improvements in the COP or EER of equipment.

Capacity modulation benefits

A much more realistic approach to reducing energy usage is to take full advantage of variations in heating and cooling loads due to changes in outdoor temperatures and internal load conditions throughout the respective seasons.

A structure's heating and cooling demands and consequently, equipment capacity are based on 100 percent capacity at preselected design temperatures. However, the design conditions actually occur for a small number of hours and represent a rather insignificant portion of the total energy consumed during a season. In fact, moderate outdoor temperatures requiring less than full capacity prevail a majority of the time.

For instance, the percentage of total capacity related to total heating and cooling hours in four geographical areas is given in Table 48-2. The need for 100 percent heating capacity in Area 3, for example, excluding solar heat effect and internal heat gain, is about 3 percent of the total heating hours.[6] In contrast, 75, 50, and 25 percent of the heating capacity are needed 14, 36, and 47 percent of the total heating hours, respectively. Therefore, outdoor temperatures in this area require 50 percent or less of the total heating capacity 83 percent of the season.

Similarly, the need for 100 percent cooling capacity in the same location occurs 10 percent of the total hours. Likewise, 75, 50, and 25 percent of the total capacity are needed 32, 46, and 12 percent or less of the total cooling hours, respectively. Therefore, 50 percent or less of the total cooling capacity is required approximately 58 percent of the cooling season.

Compressor/equipment balance

The occurrence of moderate temperatures for a majority of hours during both the heating and cooling seasons provides an attractive means of using capacity modulation to improve equipment and system performance.

Cylinder unloading is sometimes used to modulate compressor capacity. This is not very acceptable because of the potential 30 percent drop in COP between 100 and 50 percent load.[7] Two-speed compressors are more attractive. For example, a vapor compression, reciprocating compressor can be reduced to half-speed by switching from a

Table 48-2. Percentages of total capacity related to total heating and cooling hours in four geographical areas.

% of total capacity needed	Percent of total heating hours				Percent of total cooling hours			
	Area 1, 6034 hr	Area 2, 6229 hr	Area 3, 6417 hr	Area 4, 6234 hr	Area 1, 2600 hr	Area 2, 2175 hr	Area 3, 2125 hr	Area 4, 2050 hr
25	53	52	47	31	12	12	12	12
50	39	35	36	53	43	46	46	49
75	7	12	14	15	32	32	32	27
100	1	1	3	1	13	10	10	12

two-pole 3600 rpm to a four-pole 1800 rpm motor, resulting in a 3.3 percent increase in COP. Rotary compressors also have all the desirable operating characteristics and offer attractive possibilities as an acceptable, efficient, multi-speed compressor for full capacity modulation. Certainly, more attention and effort should be devoted to producing larger units with practical means of full capacity modulation.

Such compressor designs would be a big improvement because, as indicated by Table 48-2, 50 percent or less of the cooling capacity is needed 55 to 61 percent of the time. If a compressor's capacity could thus be modulated to 50 percent or less during these periods, the heat transfer surfaces and other equipment, which are sized for design loads, could be so balanced to have a much smaller temperature gradient between the refrigerant and heat sink medium, as well as the heating and cooling medium, with corresponding reductions in energy input to fans, pumps, and other auxiliaries—all conducive to a higher seasonal performance factor.

Similar reductions in energy usage can often be realized in heating systems. It may not be as feasible, however, to effectively modulate the capacity of a single fuel fired boiler or furnace such as normally used for residences or small commercial structures. One possibility would be to use several small boilers in place of one larger unit, which could be operated in sequence to match the load. In the case of electric resistance heating and heat pumps, full capacity modulation can be achieved to balance the load in a manner similar to that described for cooling systems.

Maintenance and service

Maintenance and service is highly recommended to keep heating and cooling systems in good condition and operating at maximum efficiency.

References

1. Sepsy, C. F. and Fuller, R. H., "Scheduling and Optimizing Equipment Operation and Building Use," Chapter 42.
2. Grant, S. M., "Variable Volume Induction Systems," *Heating/Piping/Air Conditioning*, March 1976.
3. Achenbach, P. R., "Government Regulations and Activities Directed Toward Energy Conservation For Buildings," Chapter 79.
4. Ambrose, E. R., "Architectural Aspects of Energy Conservation in HVAC," Chapter 15.
5. ASHRAE Standard 90-75. "Energy Conservation in New Building Design," The American Society of Heating, Refrigerating and Air-Conditioning Engineers, New York, N.Y.
6. Ambrose, E. R., "The Heat Pump: Performance Factor, Possible Improvements," *Heating/Piping/Air Conditioning*, May 1974, p. 77.
7. Versagi, F., "2 Speed Compressor," *Air Conditioning, Heating and Refrigeration News*, June 25, 1973.

49

Retrofitting and replacing energy intensive rooftops

As fuel costs rise, many existing rooftop air conditioning units are becoming burdensome to their owners. Here are some housekeeping, retrofit, and replacement measures that can be applied to reduce their energy intensity and operating costs.

ROBERT T. KORTE,
Editor, Heating/Piping/Air Conditioning, Chicago, Illinois

During the 1960s and early 1970s, packaged rooftop air conditioning units were installed by the thousands on top of schools, factories, shopping centers, and low rise offices throughout this nation. The avid acceptance of this "black box" approach came from a variety of reasons, not all of them bad. Mostly, they involved attractive prices for equipment and installation labor—low initial cost. And why not. Energy costs were a tremendous bargain, and few were looking to future days when this might change. Available at low cost and requiring no in-building space, rooftops became a natural for the many speculative plants, stores, and offices being built, as well as for the extensive classroom space that beleaguered school boards were attempting to provide within restrictive budgets.

Now the energy crunch is upon us, and while the first cost philosophy will be a long time a-dying at the hands of the life cycle costers, some progress in this direction is being made, hopefully to pick up momentum soon. Nevertheless, rooftop units, many

(but not all) of them greatly improved to be sure, continue to be installed in large numbers, indicating that while a host of arguments and charges have been brought against them, they are meeting a definite demand and have a valid role to play.

The rooftop concept, then, is certainly alive, but thousands of the units that have been installed are not at all well. Premature mortality is common, as is substandard performance. But the straw that is breaking owners' backs, at least in parts of the country where energy shortages and cost increments have hit hardest, is high operating cost.

This then is the premise of this chapter. We have in this country a large stock of rooftop air conditioning equipment that is five years old or more and energy and maintenance intensive. The burdens they impose on their owners may currently range from irksome to onerous, but they will intensify as energy costs continue to soar. Clearly something, in terms of retrofit or replacement, must be done about those units that are operating as energy hogs.

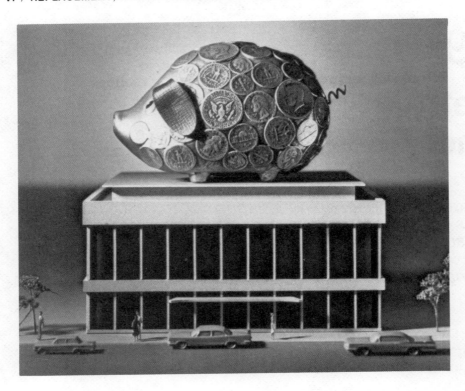

Why so energy intensive?

There are few surprises in store for someone investigating why rooftop equipment is energy intensive in many applications. The reasons relate primarily to equipment location, misapplication, and equipment shortcomings.

The ambient conditions under which rooftop equipment is called upon to perform are the worst imaginable. The cooling cycle labors under the highest ambient temperature, the heating cycle under the lowest. Regardless of how well the equipment is insulated, these extremes take their toll. Similarly, wind losses can be high on both cycles. Airborne dust and debris take their toll as well, progressing deep into the units and lodging on coil and control surfaces. In addition, sun loading on condenser coils plays havoc with equipment performance. Water is always a problem for rooftop equipment. It leaks into units and corrodes them, and it freezes in the worst places in cold weather, causing all sorts of problems. And unless curbings and flashings are installed with the utmost care, water will cause roof damage through freezing and thawing and ultimately leak into the building.

Also related to equipment location is the old out-of-sight/out-of-mind syndrome, with the result that maintenance is frequently performed on a breakdown rather than a preventive basis. It is easy for buiding personnel not expert in things mechanical to regard that black box up there as part of the roof, like the chimney. And even for those who know their way around mechanical equipment, it is easy to say why carry these heavy tools up that ladder to work on a cold, windy, icy (or hot or wet) roof with 480 volts to worry about.

Misapplication problems

Also blackening the reputation of rooftops are the numerous instances of misapplication that can be found. Rooftop air conditioning equipment offers pronounced advantages, among them low first cost, freeing up of valuable building space, ease of installation, and complete factory assembly, wiring, and testing. Unfortunately, these advantages proved so attractive to the building community, especially architects and speculative builders, that the rooftop concept became a panacea. Rooftop units were frequently applied without consideration of the unique characteristics of the building and its occupancy, without any cost or performance comparisons against alternate mechanical system approaches, and, indeed, sometimes without the services of a consulting engineer completely. In some cases, rooftops were right for the job; in others, problems were soon encountered.

Nor can consulting engineers and mechanical contractors escape the blame for misapplications. Some went along with the architects and builders in too blindly endorsing the advantages of rooftops, enjoying the easy engineering and installation they offered. Others, despite what they had to say about rooftops in engineering circles, happily accepted their commissions and contracts, acquiescing in the architect's decision to go rooftop and then perhaps botching the job because their hearts were not in it. And a few, just a few, insisted on detailed engineering analyses, providing showcase installations when rooftops were right and arguing forcibly and with fortitude for alternative approaches when they were not.

What about the equipment?

Which brings us to rooftop manufacturers and equipment shortcomings. Here we are talking specifically about older units now operating in energy intensive applications, not necessarily (though perhaps) about modern products available. Perhaps the earliest shortcomings of rooftops resulted from inadequate field testing prior to marketing, and hence an underestimation of the severity of rooftop operating conditions and a corresponding overestimation of performance and longevity.

Cabinetry in particular reflected this. Very light gauge metal was employed, poorly treated for corrosion resistance and fabricated with little thought to air and moisture leakage. Internal insulation was exposed and skimpy, perhaps $\frac{3}{4}$ in. of 1 lb density, which would be minimal for a heating application in Orlando, let alone Chicago. Access panels were completely removable, held in place by sheet metal screws that inevitably became lost, and poorly gasketed if at all. Dampers were designed not to bind rather than not to leak, and components were arranged for unit compactness rather than to invite cleaning and maintenance.

Other complaints heard from engineers, contractors, and owners include poor compressor selection, inadequate condenser and evaporator surface, poor matching of components, cheap controls, inadequate fan bearing supports, and the like.

Perhaps the overriding influence on rooftop shortcomings was the attitude that if these units were a low first cost buy, they better be low in first cost. Engineers and contractors complained that rooftops were designed down to a price instead of up to a quality and put together for the manufacturer's convenience rather than the owner's benefit. Manufacturers on the other hand said they would love to market more rugged equipment but the buying influences would simply not pay the added cost. And they could point their fingers at suppliers who had taken the high road and were no longer in business.

What's to be done?

Once an owner has determined that he must do something about the high energy costs his rooftop equipment is burdening him with, he has basically two options, upgrading the existing units or replacing them. A lot of factors come into play in this decision, and engineering expertise is called for. In essence, only a redesign of the system together with life cycle costing of the various alternatives will assure the owner he is making the right decision. Certainly he does not want to go the replacement route if a few simple housekeeping measures and the addition of some inexpensive controls will provide the savings he is looking for. And if he decides that replacement is necessary, he must be very careful not to perpetuate the problems of the existing installation by simply throwing new black boxes on the roof.

Consider these factors

Following are just a few of the factors that must be taken into account in making the retrofit-replacement decision:

- Is the comfort performance of the existing equipment adequate, or would it be if the units were operating up to capacity?
- Has the facility been expanded or have changes been made in space usage or occupancy to the extent that even rated capacity would be inadequate? Are any such changes being considered for the near future?
- How severe are the maintenance requirements and breakdown history of the units?
- What is the age and physical condition of the units? Obviously, little can be done with a unit if the housing is literally falling apart, and any unit that has been on the line for 10 years or more may well be a candidate for replacement regardless of its past history.
- What is the size of the equipment? You can not afford to spend much on repairing a

3 or 5 ton unit when a new one would cost only $1200 to $1500. Moreover, the smaller rooftops are extremely compact and rarely lend themselves to much in the way of retrofit. They have always been basically black boxes. A 30 or 40 ton unit, on the other hand, may have had all kinds of accessories hung on it—return air fans, humidifiers, economizers, multiple compressors and in many cases compressor unloading—and more retrofitting options, as well as dollars, are available.

- If the problem seems to be basically one of control, are the units in fact controllable?

First step: housekeeping

If a thorough analysis shows that desired performance can be achieved at reasonable cost via various upgrading techniques, the place to start is with basic housekeeping measures. This is extremely important, for the savings that can often be achieved through a few simple no or low cost efforts are absolutely amazing.

Perhaps the greatest single cause of excessive energy consumption or failure in a rooftop unit, or any air conditioning unit,

---

Economizer case study

The following data indicate the savings achieved through the application of an enthalpy control economizer cycle in a fast food restaurant, as shown by comparison of performance with identical equipment at an identical facility in the same location for a one-month period from mid-September to mid-October 1975.

- Compressor operating time

	Restaurant without economizer	Restaurant with economizer
1st stage compressor	175.7	31.7
2nd stage compressor	146.1	21.8
Totals	321.8 hr	53.5 hr

The difference is 268.3 hr, or an 83.3 percent saving in compressor operating time credited to the economizer. Unit efficiency was determined as 27.02 kW/12.5 tons = 2.16 kW/ton. At a $0.04/kWh electric rate, the dollar savings are:

$$268.3 \times 2.16 \times 0.04 \times 12.5 = \$289.75$$

Since each restaurant has two rooftops units, the savings amount to $580 for a one-month period.

Data courtesy Honeywell

is dirt or fouling of its surfaces. If a unit has a dirty condenser or evaporator, capacity drops and energy consumption rises dramatically. If condensing temperature rises, as it does with a dirty condenser or a slipping belt on the condenser fan drive, compressor capacity decreases and brake horsepower increases accordingly. The same result may occur if the unit is disoriented to the prevailing winds. In Chicago, for example, hot weather usually comes with southwest winds; if the condenser fan blows in that direction, the hot air recirculates and capacity drops way down.

On the evaporator side, as suction temperature drops, as it does with dirty surface or reduced air flow caused by dirty filters, dirty fan wheel, or slipping fan belt, the same results—decreased capacity and increased brake horsepower—occur.

Combining these effects, it is not uncommon to find reductions of 25 percent and more in capacity together with increases of 40 percent and more in brake horsepower per ton, simply because of dirt. This is wasteful and exorbitant, and there's just no reason not to keep heat transfer surfaces and air filtering and moving equipment clean: free of scale, airborne dirt, and corrosive contaminants. Moreover, high compression ratios and short cycling seriously reduce equipment reliability.

Table 49-1 provides an indication of the effects of fouled heat transfer surfaces.

Another housekeeping measure is thorough inspection of the unit followed by appropriate repairs. Special attention should be paid to the cabinetry. It is not unheard of to find access panels flapping in the breeze, because half the sheet metal screws have been lost, or even left off completely. The result can be a unit operating with excessive outside air or loss of excessive amounts of conditioned air. All screws should be replaced or, better yet, marine latches provided. Weatherstripping on access panels can be replaced or in the case of some products installed for the first time. Loose seams in the cabinetry can be welded or caulked. Internal insulation that has torn away can be replaced, and perhaps an additional layer added throughout.

Dampers should also be checked. The integrity of their seals, hopefully provided, should be assured, and damper settings should be examined. It is apparent that in many rooftop installations, damper settings were based on the notion that a minimum outside air requirement of say 10 percent would be met with a minimum damper opening of 10 percent. This is not a linear relation, and the result is probably more like 30 to 40 percent outside air.

Obviously, a 10 percent outside air requirement cannot be met if the dampers leak 15–20 percent when fully closed. Unit configuration may make it difficult or impossible, but consideration should be given to replacing the dampers with new ones of the low leakage type. Several manufacturers

Table 49-1.   The effects of poor maintenance on the efficiency of a 5-ton capacity reciprocating compressor.

	Suction		Discharge						
Condition	Temp., °F	Gauge pressure, psig[1]	Temp., °F	Gauge pressure, psig	Compression ratio[2]	Compressor capacity, Btuh	Power input, watts[3]	EER[4] Btu per watt	Percent capacity reduction
Normal	40	69.02	120	262.6	3.31	62,000	6500	9.54	
Dirty evaporator	37	64.69	120	262.2	3.49	58,500	6300	9.28	5.6
Dirty condenser	42	71.99	130	299.3	3.62	59,000	7100	8.31	4.8
	44	75.04	140	341.3	3.96	55,000	7600	7.24	11.3
Dirty evaporator and condenser	35	66.89	130	299.3	3.84	50,000	6500	7.69	19.4

[1]Refrigerant 22
[2]Ratio of absolute discharge and suction pressures.
[3]Three phase power
[4]Energy efficiency ratio

stock these in standard sizes and guarantee them for 1 percent leakage or less.

The savings that can be achieved through reductions in outside air, whether through sealing off leaks or replacing or resetting dampers, are fairly easy to quantify. For example, say we have a minimum outside air requirement of 10 percent, we find we are actually getting 35 percent, and we take whatever measures are needed to get down to 10 percent, thus obtaining a 71 percent reduction in the outdoor air load. On the heating side, if we are providing 6000 cfm around the clock (not uncommon) and working with a return air temperature of 73°F, an average outdoor temperature of 40°F, and a combustion efficiency of 75 percent, we obtain a saving of:

$$1.085 \ (6000 \times 0.25) \ (73 - 40)/0.75$$
$$= 71{,}610 \text{ Btuh}$$

This is the equivalent of 72 cu ft per hr of gas. For the heating season of 4940 hr (7 months × 30 days × 24 hr), our saving, at a gas rate of $1.80 per 1000 cu ft, is:

$$72 \times 4940 \times 1.80/1000 = \$640$$

(Obviously, even greater savings could be achieved if we added nighttime and weekend setback and reduced outside air to zero during unoccupied periods.)

On the cooling side (neglecting the latent load), if our return air temperature is 73°F and the average outdoor temperature is 80°F, we can save:

$$1.085 \ (6000 \times 0.25) \ (80 - 73)$$
$$= 11{,}400 \text{ Btuh or 1 ton}$$

If electricity costs us $0.035 per kWh and our efficiency factor is 1.7 kW per ton, the saving for a 2400 hr cooling season (5 months × 30 days × 16 hr) is:

$$2400 \times 1.70 \times 0.035 \times 1 = \$143$$

This amounts to $885 a year in what is really a conservative example.

One-shot expenditures

There are a number of additional steps that one can take to upgrade performance that go somewhat beyond housekeeping but fall short of unit retrofit. They involve one-shot expenditures that cumulatively can provide significant savings. For example, we have talked about tightening up cabinetry. In some cases, it might be wise to consider as an alternative or adjunct the complete enclosure of the unit in a well insulated penthouse.

The condenser coil should not be exposed directly to sunlight, at least not more than 4 percent of it. A good shade can be erected rather inexpensively.

Frequently, the drain connections of the evaporator section are not provided with seals, and you'll find that they are blowing out water and cold air constantly. This may amount to only 3 percent of the total air flow, but it is a total waste and 3 percent here, 3 percent there which, of course, can add up quickly. It is easy and inexpensive to put in a couple of 90 deg ells and a nipple and provide a water trap.

Another step is to go through the conditioned space taking temperature gradients to determine whether the thermostat is adequately sampling room conditions. Move it to the right spot and then take whatever provisions are necessary to keep anyone from readjusting it. Tinkering, which people inevitably do if they can, is very wasteful and can have an adverse effect on the equipment. For example, consider a restaurant or fast food service where an adjustable stat is located too near the supply outlet. Workers may turn it down to 50°F to get the 70°F they want. At night, when the system is off, the stat is trying to pull the building down to 50°F; the expansion valve slugs liquid over and tears up the compressor. So locate the thermostat in a representative position and if possible make it nonadjustable. A shield is often inadequate.

Institute PM

Once these or other steps have been taken to optimize unit performance and energy consumption, institute a sound and rigorous program of preventive maintenance. Do not skimp in either its conception or assuring its execution.

If there is even the slightest doubt that

building personnel will be able to carry out the necessary inspections and the routine or seasonal service operations, then execute a service contract with a reputable and knowledgeable mechanical contractor.

Controls are very important

It is probably safe to say that if you have an energy consumption problem in a rooftop installation, you often have a control problem. Inadequate control is undoubtedly the most common problem in older vintage units; happily, many existing rooftops are quite readily adaptable to the upgrading and retrofitting of controls.

The most significant improvements in rooftop technology have been in controls, specifically in the use of solid state electronic control systems. These provide a higher degree of flexibility than electromechanical systems and thus are able to maximize energy savings. A number of controls manufacturers have been quite active in applying this technology to the development of modern, integrated control packages for the retrofit market. Information on such options has been widely distributed and is readily available.

Perhaps the commonest type of controls that can be applied for energy conservation are automatic time clocks to cut off or reduce heating or cooling in whole buildings or specific areas during unoccupied periods. The savings that can be achieved from the setback of heating or shutdown of cooling overnight and on weekends are truly dramatic. Careful analysis is required in setting these devices for shutdown and startup, both to take maximum advantage of any coasting prior to shutdown or setback and to assure adequate morning pickup. Also, all necessary interlocks should be added to prevent unnecessary operation of fans, pumps, pilots, etc. Controls and interlocks should prevent unnecessary consumption of energy to maintain standby heating during the cooling season or to maintain cooling during the heating season.

Apply economizer cycle

Another effective tool for energy conservation is the economizer cycle to allow the use of outside air for "free cooling" when weather conditions permit. Economizer cycles should use enthalpy controllers and be integrated or interlocked with the cooling cycle. If the unit is already equipped with an economizer, of an older type, it is well to analyze its operation, for it may be requiring far more energy on the heating side than it is saving on the cooling side. The savings that an economizer cycle can provide are illustrated by the calculations for a case study involving a fast food service, shown in the accompanying item entitled "Economizer case study."

Other control areas to investigate include new gas trains including electric ignition in lieu of pilot lights, replacing on-off heating control with modulating control to prevent heating overruns, staging compressors, and hot gas bypass or cylinder unloading.

Following are some summary guidelines for reducing energy consumption and cost through control measures:

1. Equipment must be controllable.
2. Control outside air judiciously:

• An economizer cycle will permit the use of outdoor air for "free cooling" when weather conditions permit.

• The economizer cycle should employ an enthalpy controller to determine when outdoor air is suitable for space cooling.

• The economizer cycle should be integrated or interlocked with the cooling cycle.

• All outdoor air dampers should be closed in the unoccupied mode or during the morning warmup cycle in the heating season and the pulldown in the cooling season.

• All dampers should leak less than 3 percent when closed and at rated velocity and static pressure.

• Dampers should be properly adjusted to meet ventilation requirements and no more.

• System air balance must be optimum. This may require relief or barometric dampers with an economizer cycle.

• Electronic air filtration should be considered to permit reduction of outside ventilation air quantities.

3. Reduce the temperature differential

between hot and cold decks on multizone units.

4. De-energize equipment when not needed:
- De-energize motors and blowers during unoccupied periods.
- Cycle the blower on heating during unoccupied periods of nighttime setback periods.

5. Maintain proper temperature setpoints:
- Reduce temperature levels during unoccupied periods.
- Maintain heating at 68°F and cooling at 78°F except when terminal reheat is employed.
- Replace adjustable controls in accessible places with controls having locked settings to prevent unauthorized changes.

6. Utilize system supervisory controls to indicate malfunctions, dirty filters, etc.

Playing with components

Once the owner has taken care of housekeeping and controls, he has probably gone as far as he can go in taming the beast on his roof. And the odds are that if he has diligently seen to these, he will have achieved the energy and cost reductions required. Only rarely does one have the opportunity to go further and juggle components within a rooftop unit to obtain a better mix from the standpoint of energy conservation. This is simply due to the nature of a rooftop as a very compact factory assembled device.

Such opportunities are virtually nonexistent with smaller units and limited even with larger units. In cases where such measures are considered, detailed engineering analysis and close coordination with the manufacturer are required. Some possibilities in this area include:
- Conversion of rooftop systems to variable air volume. There would appear to be few obstacles to creating a dump type VAV system where single zone rooftops have been applied to single duct terminal reheat systems. Providing true VAV at the fan would be less likely because of problems in having suitable fan characteristics, applying inlet guide vanes, etc.

- Adding heat recovery. The difficulty would be in getting the device into the rooftop unit, and it would probably have to be installed remote to the unit.
- Conversion of rooftop units to heat pump operation.

The above are offered only for reflection and without endorsement or precedent, except in the last case. The details on one conversion of this type are told in the accompanying item entitled "Rooftop unit converted to heat pump."

Rooftop replacement

If the decision is made that existing equipment cannot economically be rehabilitated to provide the performance and operating cost reduction sought, replacement is the answer. But it is important at this point that one broaden his perspective and totally re-evaluate building requirements, for it may be possible to go to another type of system if this is desirable. In effect, the engineer should assume that it is a new building, albeit with more design restrictions than most, and start from scratch in analyzing building requirements and loads. These may have changed over the years as occupancy swelled and office or production equipment was added.

In some applications, particularly industrial, space may be allocated inside the building to house a new central system. Or perhaps the analysis of current and future needs will indicate a need for facility expansion, and this then can be planned to encompass the mechanical services.

It may be possible to erect a penthouse on the roof to accommodate a built-up system. Or at least at the larger sizes, it may be possible to apply a penthouse type rooftop unit which is factory assembled and tested but offers the designer the capability of selecting all components and auxiliaries to meet the specific requirements of the building. Rooftop air handling units that can accommodate various heating and cooling options (hot water, steam, gas or oil fired heat; chilled water or DX cooling) and/or heat recovery options installed in conjunction with a remote heating-cooling plant may be the answer.

Comparison of all practical approaches on the basis of life cycle costs will point the way to the right solution.

But perhaps we find that there is no building space for an alternate approach, or there's no power available over in the corner where we might steal a little, or there is no ceiling space for an air handler, and we really have little choice but to replace in kind. If rooftop manufacturers are designing more efficient products, then there is certainly no reason why the replacement system shouldn't be rooftop.

Rooftop improvements

Much has been heard about the improvements in rooftop units of the third, or fourth, or whatever generation, and the tacit assumption in this chapter has been that this is true. Certainly it is with respect to controls, available accessories, and in most cases cabinetry. But what about the proper matching of components, the provision of adequate coil surfaces, etc.?

In some cases, units built years ago were superior in this respect to later models. A contractor tracing the history of rooftops he has worked with described their energy efficiency with a curve starting at a relatively high point in the early years, dropping off rapidly, and only now beginning to come back up.

This is a reasonable view, for advances in compressor technology—changing from four-pole to two-pole, going from 1800 to 3600 rpm, improving compressor windings and winding insulations, upgrading of valves—had the ironic effect of permitting satisfactory operation at higher compression ratios and encouraging reductions in condenser surface.

Quite frankly, when the Arabs turned off the oil, some manufacturers were caught with their EERs down. And given the lead time required for product development, the necessary design changes may or may not

have been made at this point. Perhaps they came on the '76 units; maybe they are on the '77 units. Thus, it is extremely important that the person selecting replacement rooftops carefully examine product data and EERs to obtain the best unit possible and not perpetuate mistakes made with the earlier installation.

The above remarks are intended only to encourage caution, not to undermine rooftops, for there are some truly fine products available today. A brief listing of some of their features, augmented by the objections and shortcomings previously detailed, will serve as a guide as to what should be sought in new rooftops for either existing or new buildings.

As we have said, marked improvements have come in controls. New solid state controls, embracing enthalpy controlled economizers, facilitate a much closer control of building comfort while minimizing operating time and hence energy and costs. Other improvements include the addition of higher quality insulation, installation of low leakage outside and exhaust air dampers, and a general overall improvement in the construction of products.

Dampers have less leakage

Not untypical of the construction of equipment now being sold are dampers having a maximum leakage rate of 0.5 to 1 percent of rated air flow and packages with insulation values equivalent to a K factor of 0.12. Also included in construction improvements are higher quality gaskets around all removable panels, the provision of hinged panels with marine latches, better insulation fastening methods to eliminate its loss, cabinet finishes that are more weather resistant, and general improvement in the leakproof characteristics of the cabinetry, coupled in some cases with means to remove moisture if it does enter. Improvements have also come in mechanical components utilized such as condenser fan motors, compressors, fans and fan motors, and, very important, all damper operators. And increases have come in the

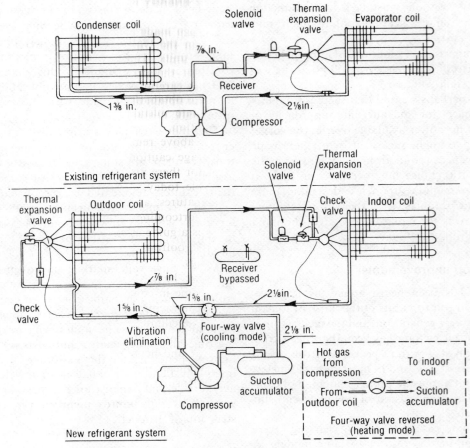

Information courtesy Mammoth Div., Lear Siegler, Inc.

Scope of work

Mechanical

• Remove and save refrigerant charge from four refrigerant circuits.

• Remove eight condenser coils.

• Remove intermediate condenser coil supports.

• Disconnect four receivers and valves from liquid and discharge lines (receivers left in place but bypassed). Cap connections to prevent water or dirt accumulation.

• Prepare and seal entire condenser floor with mastic sealer.

• Fabricate and install sheet metal deflectors along periphery of condenser section.

• Install four full height outdoor coils on condensate deflectors and attach to unit frame. (Coils supplied by manufacturer with distributors and headers attached.)

• Mount four four-way valves in condenser section.

• Mount four horizontal suction line accumulators in condenser section.

• Provide refrigerant piping as shown in the diagram for each circuit.

Controls

• Mount four time delay relays.

• Remove existing heating and cooling sequencers.

• Mount logic panel.

• Mount sequencers.

• Mount defrost timer.

• Mount two changeover relays and rewire ambient changeover control.

• Mount eight defrost relays.

• Mount four pressure switch initiate defrost controls, set at 50 psig.

• Mount four temperature stat defrost terminate controls, set at 80 F; strap bulb on line to outdoor coil discharge header.

• Mount four head pressure fan switch controls.

• Rewire existing controls and wire new controls as required.

Rooftop unit converted to heat pump

Early in 1975, a rooftop heating-cooling unit was converted to heat pump operation. The unit was one of 24 shipped in 1969 and installed on the West Andover Assembly Plant of the Raytheon Co., Andover, Massachusetts. The 24 units were of nearly identical design, specially modified penthouse units with 140 tons cooling capacity, 300 kW of electric heat, and integral condenser sections.

The plant consisted of nearly 15 acres under one roof, completely heated and air conditioned. When the plant was planned and constructed, the owner was assured of a plentiful and inexpensive supply of electric power. From an original average of 0.7¢ per kWh, however, electricity had risen to nearly $3\frac{1}{2}$¢ per kWh in 1974. At this point, the manufacturer was asked to assist in the conversion of one of the rooftop units to heat pump operation to determine whether successful operation could be achieved.

Heat pumps produce between two and three times as much heat per kilowatt-hour of electric energy as conventional electric resistance heating coils. This advantage convinced the owner that the trouble and expense of converting the trial unit would be recovered through reduced electric bills, which had reached a staggering $1 million in 1974.

The basic difference between a heat pump and an air conditioner, of course, is the ability of the heat pump to reverse the flow of refrigerant so that the indoor coil, an evaporator when cooling, becomes a condenser and heats the indoor air when heating is required. Meanwhile, the outdoor or condenser coils, which release heat to the atmosphere when indoor air is being cooled, become evaporator coils in winter operation and cool the outdoor air.

It was necessary to replace the condenser section coils and to change nearly all controls. Most of the engineering was done by the manufacturer, who also provided the hardware and controls in kit form. The accompanying schematic and specifications indicate the scope of the work.

Subsequent operation in the heat pump mode has been satisfactory, averaging a COP of approximately 2.2 allowing for blower and defrost power. Conversion of additional units is being evaluated.

condenser surface provided. Noise and vibration characteristics have also been improved.

There has been a trend away from a multiplicity of small packaged units toward larger single units of the penthouse type. There has also been a move away from the traditional types of rooftop equipment such as multizone units, particularly the draw-through type.

An innovation in the rooftop industry is the introduction of packaged heat pump equipment, with the heat pump category including both the water-to-air type as well as the more traditional air-to-air. Variable air volume rooftop units are available, as are units with heat reclaim capabilities of various types.

Major changes in equipment design have been directed toward improving the overall energy consumption characteristics. In conjunction with changes in total system design, there have been changes such as larger overall equipment sizes, larger fans requiring less power input, better coil circuiting and compressor loading design to allow greater utilization of compressor efficiency at part load conditions, and, again, much better control systems.

These are only some of the product changes in broad outline. More are coming, and the sophisticated equipment of today and tomorrow promises the owner improved performance and longevity together with minimal operating costs in rooftop installations.

While many persons provided help and input for this chapter, the author would like to especially thank the following: HPAC Board of Consulting and Contributing Editors members: F. L. Brown, Mechanical Contractors' Chicago Association; Maurice G. Gamze, Gamze-Korobkin-Caloger, Inc.; and James M. Whalen, RMC, Inc.; also Robert A. Kegel, Beling Engineering Consultants. The following companies also contributed: Honeywell; Penn Div., Johnson Controls, Inc.; Ranco; Alton Mfg. Co.; Lennox Industries Inc.; Mammoth Div., Lear Siegler, Inc.; McQuay Group, McQuay-Perfex Inc.; and York-Shipley, Inc., Jackson & Church Div.

SECTION VII

Proven ways to save energy

The 10 chapters in this section explain in detail how energy has been saved by applying sound engineering approaches to the design, installation, and operation of heating, ventilating, piping, and air conditioning systems. Among the subjects discussed are:

- HVAC design guidelines.
- Pump energy savings for hydronic systems.
- Energy saving methods for commercial and institutional buildings.
- Saving air and energy in industrial plants.
- Estimating air infiltration.

50

HVAC design guidelines prove conservation of energy

These design guidelines show how energy can be saved by applying basic concepts that have been proven in many installations over the years.

PAUL K. SCHOENBERGER, PE,
Energy Management Consultant
Columbus, Ohio

Heating, ventilating, air conditioning systems can and should be designed to consume minimum amounts of energy and still maintain the controlled environment inside buildings. Reducing the size of HVAC equipment will make it possible for energy to be minimized if the building envelope is designed for low heat transmission and low solar effects. Also, lower lighting levels and task lighting should be applied to further reduce the size of the HVAC and electrical systems. However, to be sure that the mechanical and electrical systems will not waste energy, the environmental control systems must respond to actual weather conditions and occupancy levels. Unfortunately, most existing facilities operate most of the time with partial heating, cooling, and electrical requirements, while maximum loads may occur only a small portion of the time.

Energy standards will help

New Federal Guidelines and ASHRAE Standard 90-75 will permit the use of energy efficient environmental control systems that most engineers would not have considered acceptable a few years ago. The recommendations of these standards are to maintain occupied interior temperatures as low as 66 to 68°F during the wintertime and as high as 78 to 80°F with 60 percent RH during the summertime. These changes will permit the use of an energy efficient system that will maintain comfort conditions better than these minimums for the majority of the time, but these extremes may only be reached during just a few days of the heating and cooling seasons.

It is imperative that heat generated inside a building be used to offset the heat losses if minimum energy usage is the goal. The maximum heat generated inside a building constructed with minimum glass areas and well insulated walls and roof areas can be greater than the maximum heat losses. The occupied building may generate several times the heat required to offset the losses at average wintertime conditions. Any energy conserving system must make use of the internal heat gains available from occupancy by moving this heat from where it is generated to where it is needed. Spaces that generate more heat due to occupancy than is required to offset their heat losses should be maintained at 76° F rather than 68° F to improve comfort and fully utilize the internal heat gains.

Design energy saving perimeter heating

Perimeter wall area heat losses can be offset by installing small low-pressure heating fan-coil units that pull air from the warm interior of the building and distribute this air to the perimeter wall areas. The perimeter units can pull air through the ceiling spaces to warm this air to above room conditions with the heat from the lighting fixtures. This warm air from the building interior can be discharged to blanket the wall areas and neutralize the cold transmission effects of the wall and glass areas. Separate units must be installed for different exposures to compensate for solar effects. The perimeter heating units should operate only when the exposure has heating requirements and should be shut off at all other times. During unoccupied periods, the perimeter units can be operated to maintain reduced temperatures in the building if desired.

The perimeter heating units should be controlled to maintain their exposure temperatures between 67 and 73° F by cycling the unit fans between these limits. The six degree temperature differential will prevent short cycling of the units and should also permit occupation of the facility much of the time without the units operating. When the temperature at the exposure drops to 66° F, the heating coil valve should modulate to the open position to allow hot water to flow through the coil to maintain the temperature of 66° F with continuous fan operation. Control of the units to maintain this minimum temperature at the exposed wall areas of the building is in accordance with the Federal Guideline minimum temperature when outside temperatures are at their lowest. Under these conditions, only internal heat is being used for control of perimeter temperatures between 67 and 73° F, and energy is not being wasted within this range. With interior areas

maintained at 76° F, the ceiling plenums will be a few degrees warmer than this temperature due to stack effects and the heat from lighting. A 12 to 14 degree temperature difference may be available from the inlet air to the fan-coil units for heating at the perimeter when the units are maintaining the 66° F minimums, and maximum utilization of the interior heat is being accomplished.

Cycle units with full water flow

The perimeter heating units should be controlled to maintain temperatures between 55 and 60° F during unoccupied periods. The units should be cycled on and off at these limits while full hot water flow is maintained through the heating coils. A warm-up cycle can be used to elevate building temperatures prior to occupancy by operating the unit fans continuously while full water flows through the heating coils. Upon completion of the warm-up cycle, normal occupied control of the units should be resumed.

Use heaters at entrances

Cabinet unit heaters should be installed in entrance vestibules and in stairwells. Opening and closing of exterior doors can cause wide variations in heating requirements. These areas should be controlled between 66 and 68° F during occupied times and between 55 and 60° F when the building is unoccupied.

When electricity is used as the primary heating energy source, the unoccupied heating load may require the same energy input as the lighting system. Since this heat is not required when the lighting is in use, an off-peak electrical heating system is possible without increasing the size of the electrical power distribution system for simultaneous use of the lighting and heating systems.

Provide water tank for storage

A water storage tank can be provided as a heat sink to store sufficient heat for warm-up and operation during the occupied periods. A multi-stage electric heater can be provided for the tank with a load limiting device to decrease the number of stages that are permitted to operate as the lighting load increases. It should not be necessary to apply a safety factor to the maximum heater size to allow for water temperature pick-up. When the unoccupied cycle is initiated and lights are turned off, the well-insulated building will have a heat sink effect, and it should take several hours before temperatures drop to 55°F. The heat sink effect of the building construction will permit a regain in water temperature before additional heat is required for maintaining the reduced unoccupied temperatures. An auxiliary heat exchanger can also be added to the tank to heat domestic hot water. With operation of the system in this manner, the size of the electrical power distribution system is minimized and the maximum electrical demand of the facility is reduced. Special electrical rates may also be available for off-peak usage of the electricity during the unoccupied hours of the heating season.

Consider solar assisted heating

Solar assisted heating systems should also be considered for maintaining unoccupied building temperatures at the mimimum levels. Considerable heat is available from water heated through solar collectors to temperatures between 120 and 200°F when this heated water is circulated through coils with reduced air inlet temperatures of between 55 and 60°F. See Chapters 68, 69, 70, and 71 for additional information on solar heating design details. Care must be taken to prevent auxiliary heating systems from operating until the maximum usable heat has been extracted from the solar system.

How to save cooling energy

Isolation of the perimeter wall area heating loads permits the installation of a cooling system for the interior of the building. Such a system must neutralize the excess heat generated from occupancy and solar gains during wintertime, and at the same time it will cool the building to acceptable levels for occupancy during summertime. All areas, including rooms with perimeter exposures, must be supplied with adequate cooling capacities to offset their maximum heat gains. An energy

conserving system, such as a variable air volume system, must compensate for diversities in occupancy levels and swings in solar loads without false loading by reheating or mixing of warm and cool air. Space temperatures can be maintained by varying the quantity of cooling supply air in direct proportion to the actual space cooling requirements. The variable air volume system must be capable of varying the supply air quantities between 100 percent and zero percent of the maximum values. Tight shutoff is imperative if false loading is to be eliminated. In this way, cool supply air cannot be supplied to spaces that are totally unoccupied and have no cooling requirements. Room air supply outlets used for variable air volume systems should be of the induction type to provide draftless variable air volume performance. The air outlets must be capable of supplying reduced air quantities to the spaces without dumping this air and causing drafts. Most areas rarely operate at maximum loads, and the supply air outlets will deliver reduced volumes of air most of the time if properly selected.

Minimum outside air ventilation rates should be supplied through the variable air volume cooling system. When lights are turned on in an area, cool air is required to offset this heat gain and in most cases, ventilation air is automatically supplied. "Free" cooling with outside air should be utilized whenever outside conditions permit by mixing it with return air. Extreme care must be taken to prevent the "free" cooling economizer from providing excess ventilation air when outside temperatures are low. Mixed air should be reset from 60°F at an outside temperature of 60°F, to 66°F at an outside temperature of zero °F. Maximum air delivery quantities to the interior areas should be purposely oversized to permit the use of this 66°F air for cooling of the spaces. With zero °F outside temperature and 76°F return air temperature, 6.6 parts of return air are mixed with one part of outside air to obtain an air mixture at 66°F. This means that the system will be operating using 13.2 percent outside air at an outside air temperature of zero °F.

With the volume of the variable volume

cooling system throttled to about 40 percent, the actual minimum ventilation rate is 5.3 percent of the maximum air delivery rate, which is more in line with true minimum ventilation requirements. It is essential that minimum air delivery rates of the variable volume cooling system be prevented so that the fans will not become unstable because of the low air flow rates. Reset of the mixed air will help to prevent this condition. Non-overloading centrifugal or airfoil fans with inlet vane controls should be used to reduce the fan motor loads at reduced air delivery rates. Between 60 and 80°F outside air temperatures, enthalpy control of the economizer should be utilized.

Enthalpy control permits flexibility

Enthalpy control allows the system to select either 100 percent outside air or maximum return air, whichever contains the least amount of total heat. It may be necessary to operate the mechanical refrigeration when outside temperatures are above 55 or 60°F; however, the use of 100 percent outside air containing less total heat than the return air reduces the load on the mechanical refrigeration system and saves energy. Above outside temperatures of 80°F, minimum outside air should be used in the cooling system. The enthalpy controlled economizer uses only true minimum outside air ventilation rates when the outside air has more total heat than the return air or if the outside air temperature is greater than 80°F. During all other times, minimum ventilation rates are exceeded because the building is purging excess heat from the occupancy load, or the mechanical refrigeration load is being minimized through the use of 100 percent outside air. When the outside air temperature is 30°F, approximately 2.5 parts of return air are mixed with one part of outside air to obtain a mixture at 63°F, and the system is using 28 percent outside air. For further details regarding enthalpy control, see Chapter 64.

Operate VAV system to conserve energy

The variable air volume cooling system should be turned off during all unoccupied hours to conserve energy. Tight low-leakage dampers should be used for the damper sections on the outside air side, return air, and relief-air sides to prevent infiltration through the dampers when the unit is turned off. A return-relief fan should be used to provide positive control of return and relief air. This fan should also have an inlet vane control to decrease the fan capacity and motor loads at reduced air delivery rates.

Size cooling system for maximum load

The maximum calculated cooling load should be used for sizing the air flow rate in the cooling system. In some cases, diversities in usage and occupancy levels will permit reductions in the maximum air delivery and the load on the cooling system. Caution must be exercised when using a variable air volume system to assure that no areas become starved for adequate air. Maximum air delivery quantities must be provided to offset maximum occupancy and solar loads for the different exposures. The maximum air delivery rates for interior conditions should be purposely oversized to permit reset of mixed air during wintertime. If some areas become overloaded at design conditions and no diversity exists in the other areas, the total air delivery of the system will become inadequate and some areas will become starved for proper air circulation. Where diversities in usage exist, spaces calling for reduced air quantities will make excess air available for other areas.

The maximum calculated cooling load may be exceeded for short periods, but oversizing the system to allow for this condition is not necessary. Lighting levels in some areas can be reduced during these extreme conditions to reduce the interior heat gain. Minimizing of the cooling system operation by requiring some reduced lighting levels for a few hours each summer is not an unreasonable imposition on the occupants. These maximum electrical loads must be able to be supplied by the power plants at the hottest summertime conditions. To provide for these peak loads, most power plants operate auxiliary peaking units, but the use of these peaking units often becomes the most expensive power that is generated. Trimming of summertime peak loads will not only save costly energy, but it will also enable the power plants to accommodate additional loads for other purposes.

Use outside air to purge heat buildup

During summertime operation, heat absorbed in the insulated walls and roof during the daytime will be dissipated into the building during the evening hours. This heat buildup can be purged by operating the cooling system fans using 100 percent outside air during the cool early morning hours without using the mechanical refrigeration system. Restart of the variable air volume cooling system during the summertime should permit immediate use of the enthalpy controlled economizer. However, restart of the mechanical refrigeration system should be delayed as long as possible. The variable air volume system should be restarted with the fan inlet vanes at their minimum positions, and the outside and relief air dampers fully closed with the return air dampers fully open.

Normal control of the economizer and inlet vanes should be gradually activated over a three minute period to reduce the sensitivity of the controls at start-up. The variable air volume devices used for individual rooms should be normally open and should modulate to the closed position in response to the reduced cooling requirements. During wintertime operation, restart of the variable air volume cooling system should be delayed until completion of the perimeter units' warm-up cycle. The variable air volume system should again be restarted with the fan inlet vanes at their minimum positions, and outside and relief air dampers should be fully closed with the return air dampers fully open. Opening of the minimum outside air damper section and return to normal economizer control should be further delayed until the building begins to overheat due to occupancy.

Provide for unstable fan operation

Provisions should be made for preventing unstable variable air volume fan operation. Inlet vane travel should be limited to the stable fan ranges. Stops should be provided to limit maximum closure of the fan inlet vanes, and when air delivery reaches this minimum, the cooling supply air temperature should be allowed to rise if overpressurization of the air supply system is possible at this minimum setting.

Control chiller capacity in stages

When mechanical refrigeration is in use, cooling supply air temperature should be maintained at the maximum temperature possible to satisfy the cooling requirements and maintain acceptable humidity levels. Increased efficiency can be obtained from chillers when operating at elevated chilled water temperatures, and this may be more advantageous than operating the fan system at reduced air volumes.

Several stages of capacity control should be provided to reduce chiller energy requirements at partial building cooling loads. Selection of the chilled water temperatures and the rise in water temperatures should also be optimized to provide minimum energy consumption. It is common practice to select cooling coils and chillers using a water temperature difference between inlet and outlet of $10°$ F from 45 to $55°$ F. Pipe sizes and pump power requirements can be reduced by using larger temperature differentials; however, this will also increase the depth of the cooling coil and add resistance to the air supply system. Lowering the entering water temperature can be used to increase the coil capacity at the expense of increasing the size of the chiller required to produce this colder water.

When to shut down chiller

The mechanical refrigeration system should be shut off during all unoccupied periods. In some situations it may be necessary to operate the system during extended unoccupied periods to prevent excessive interior temperatures or humidities. Shutdown of the chiller should be introduced 15 to 30 minutes prior to shutdown of the chilled water pump and cooling air supply system. Chilled water temperatures should be allowed to rise to about 70° F before the pump and air handler are stopped.

The chiller should not be allowed to start unless the cooling air system is in operation, the outside air temperature is above 55 or 60° F, and the chilled water and condenser water pumps are in operation. Load limiting devices should also be used to start the chiller at reduced power inputs. The chiller motor should have inherent motor overload protection and an automatic pumpdown cycle where applicable.

51

Pump energy savings for hydronic systems

A detailed analysis of the methods applied to reduce flow and head loss and, consequently, pumping power requirement.

G. F. CARLSON,
Director of Technical Services, ITT
Bell & Gossett, Morton Grove, Illinois

The selection of pumps for hydronic systems has been oriented toward safety factor considerations for many years. Systems have been overpumped to overcome potential flow balancing problems; to assure more than adequate terminal flow for heat transfer reasons; and in many cases, to surmount known deficiencies in oversimplified selection procedures. All of the above have led to a high order of power wastage.

Impending power shortages and projected cost increases for electricity require that we use available design and operational tools to decrease pumping power costs. This, of course, requires that more time be spent on design and also on fine tuning new and existing systems for the lowest operating cost.

Establish pump brake horsepower

The pump shaft brake horsepower requirement is established by dividing the amount of power a pump puts into the water (hp_w) by the pump's efficiency (E_p). The correlation is as follows:

$$bhp = hp_w/E_P$$
$$= H \times Q \times \rho \times hp/33,000 \times 1/E_P$$

where

bhp = pump brake horsepower
hp_w = water horsepower
E_P = pump efficiency stated as a decimal
H = head, ft
Q = flow rate, gpm
ρ = density, lb per gal.
hp = horsepower, ft lb per minute

Most pump curve to brake horsepower relations are stated to about 85°F water density (8.33 lb per gal.). For this condition, the following relationship obtains:

$$bhp = H \times Q/3960 E_P$$

Engineers are interested in efficiency. One of the problems with the above relationship, however, is that we may become mesmerized by pump efficiency and neglect the part of the formula where true energy savings can be found ($H \times Q$). For example, a pump efficiency of 70 percent correlates to a system pumping efficiency of 35 percent if the pump is selected and operated at twice the required pump head. If the flow requirement were similarly specified, system pumping efficiency would decrease to about 17.5 percent.

In an unbalanced system, it would actually drop to about 12 percent.

It can truly be said that overall system pump operating efficiency (operating cost) is established more by those concerned with system design, installation, balance, and operation than pump engineers.

Pump curves usually illustrate pump brake horsepower requirements for specific flow and head conditions, and efficiency is also indicated. The larger the pump, the higher the efficiency, is a generality, but again, we should not become hypnotized by it, since it is also usually true, that a smaller pump matched to a system operates at lowest costs—even though it is less efficient.

Pumping costs can be determined

Pumping costs as paid to a utility can be assessed from Table 51-1, which is based on 85 percent motor efficiency over various operating time periods.

Even at present prices, generalized as approximately $0.015 per kWh, pumping costs are high, and projected increases of approximately twice the current price, require thought and work by design engineers and mechanical contractors if operating costs are to be held to reasonable levels.

The pumping power requirement for any system is fundamentally determined by flow need and head loss. The basic methods applied to reduce these factors are:
- Increased operating temperature difference.
- Utilization of appropriate piping pressure drop data.
- Avoidance of excessive pump head safety factors.
- Pump impeller trim after proportional balance.
- Zone pumping.
- Primary-secondary pumping.
- Pumping in parallel and in series.
- Two-way control valves with variable volume pumping: variable speed, pump banks, etc.

The following discussion examines each of these basic methods in further detail.

Use temperature difference in design

It is common practice to select a temperature difference as a design tool to determine the flow rate. The long standing, conventional practice has been to use a 10°F rise for chilled water (1 gpm per 5000 Btuh), and a 20°F drop for heating (1 gpm per 10,000 Btuh). These traditional parameters establish a high safety factor against flow balance problems, but they result in wasteful systems that are generally overpumped when compared to those designed with other temperature differences.

Many systems designed today use higher operating temperature differences to reduce flow rates, for example, 15°F Δ T chilled 1 gpm per 7500 Btuh) and 40°F Δ T heating (1 gpm per 20,000 Btuh). In the future, a more sophisticated design approach will achieve minimum flow rate needs. Such a method is used today in some coil selection procedures.

In this approach, it is realized that terminal units require flow as a function of size relative to load and supply water temperature;

Table 51-1. Pumping cost per 1 bhp based on 85 percent motor efficiency.

Operating time	$ per kilowatt hour						
	0.01	0.015	0.02	0.025	0.03	0.035	0.04
1 hr	0.00879	0.0132	0.0176	0.022	0.0264	0.0308	0.0351
12 hr	0.105	0.158	0.212	0.263	0.316	0.368	0.42
24 hr	0.21	0.317	0.423	0.526	0.632	0.735	0.84
30 days (1 month)	6.32	9.48	12.60	15.80	19.00	22.10	25.20
6 months	38.60	58.00	77.00	95.20	116.00	134.00	154.00
9 months	58.00	87.50	105.00	143.00	175.00	202.00	232.00
1 yr	77.20	116.00	154.00	193.00	232.00	270.00	308.00

individual terminal flow needs are assessed on this basis. Hence, the final temperature difference becomes the result of, rather than the basis for, a design. Major decreases in overall flow need are possible since for any fixed load, the larger the terminal surface area, the lower the flow need. However, decreased flow need per Btu transfer establishes more stringent flow balance requirements.

Flow rates determine pumping power

The following example illustrates how oversized terminals set lower flow rate needs to reduce pumping power needs in existing systems.

Baseboard in an existing hot water system was selected for 200° F supply and 20° F Δ T. The system operates at 210° F supply water, and a 10 hp pump is used for circulation. In any hot water system, an increase in supply water temperature will, in effect, oversize the terminal and allow a decrease in water flow because heat transfer drive is determined by mean water-to-air temperature difference. Changes in the internal coefficient affect overall heat transfer only slightly compared to drive influence. For a given heat transfer rate, the mean water temperature is set by that of the supply water and the flow rate as in the following examples: 210° F supply − 190° F return = 20° F Δ T at 100 percent flow; 220° F supply − 180° F return = 40° F Δ T at 50 percent flow; 230° F supply − 170° F return = 60° F Δ T at 33 percent flow. The mean water temperature through the terminal is 200° F in all cases, and 100 percent heat transfer is available because mean water-to-air temperature difference is maintained.

In a moderately well balanced 20° F Δ T base design system, a 10° F supply temperature increase allows a 50 percent flow decrease. A 20° F increase (200 to 220° F) assures a 50 percent flow reduction without comprehensive flow rebalance except in extreme instances. Flow in existing systems should *not* be reduced more than 50 percent because of potential air purge distribution problems (flow velocity). Flow can be reduced 50 percent in the example system if

the supply water temperature is increased from 210 to 230° F.

A 50 percent flow reduction in existing piping establishes a head loss of 25 percent of that originally needed. Power need is reduced to about $\frac{1}{8}$ of the initial requirement since 50 percent flow times 25 percent head equals 0.125 power. By trimming the existing 10 hp pump impeller or by substituting another pump to match the new flow head, pump brake horsepower is less than 2 hp— a net savings of 8 bhp. A $700 per year pump power savings would be realized for 9 months operation at $0.015 per kWh. Derivitives of this example, particularly when applied to hot water operation of dual temperature systems, will also establish substantial pump power savings.

Appropriate data

System head loss should be determined by using piping head loss values that are most suitable for a closed loop heating/cooling system as outlined by: ASHRAE, Hydraulic Institute, or equivalent data. Tables stated to a William and Hazen C factor of 100 should not be used because they are apparently intended for open city water distribution type systems. These tables establish twice the true head loss, and the consequent selection of a matching pump increases power consumption.

Excessive safety factors

Pumps are also often safety factor overheaded.* A 10 to 100 percent safety factor head is added to assure terminal flow rates, but those in excess of design return very little in terms of increased terminal heat transfer and the relative increase in pumping cost is very high.

A 100 percent safety factor head causes about 140 percent required flow. The additional 40 percent flow increases hot water terminal capacity only about 3 percent, and

---

*Design engineers must provide sufficient pump head to overcome the highest head loss possibility for installed equipment. Excess pump head will occur when lower head loss equipment is actually installed.

that of typical chilled water only about 7 percent. Power requirement is 2.5 times (actual value dependent on pump curve shape: flat or steep) than otherwise needed. If the actual needed pump power draw were 10 hp, a 100 percent safety factor increases it to 25 bhp—a wastage of 15 hp or $1750 per year at $0.015 per kWh, as an example.

Impeller trim, proportional balance

Proportional balance techniques and impeller trim can eliminate unnecessary power consumption by overheaded pumps. Proportional balance can be achieved by either preset or special field read and set procedures. However, a simple balance specification *does not* necessarily assure minimum power usage. This can be best illustrated with a simple example.

Figure 51-1 depicts the basic system. Actual pump head requirement is 60 ft; the pump, however, is selected for 100 ft. Pump flow need is 2000 gpm, and each circuit requires 1000 gpm.

Pumping power requirements, circuit flow rates, and head losses for five different balance modes are shown in Figs. 51-2 through

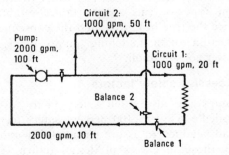

Fig. 51-1. Basic scheme for balance and pump power relationship example.

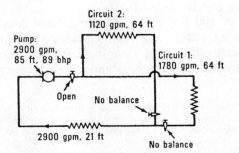

Fig. 51-2. System without balance results in a maximum draw of 89 bhp.

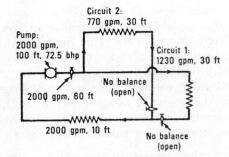

Fig. 51-3. Balance at pump only.

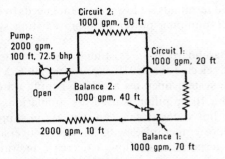

Fig. 51-4. Balance at terminal only.

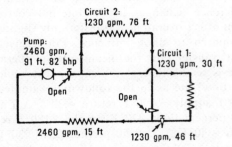

Fig. 51-5. Proportional balance only.

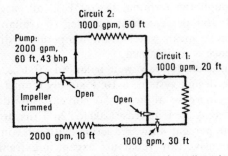

Fig. 51-6. Proportional balance, impeller trim.

51-6, and their operating characteristics plotted in Fig. 51-7. For simplicity, pump efficiency is assumed as 70 percent at all operating points. Circuit flow rates are determined by analysis methods.

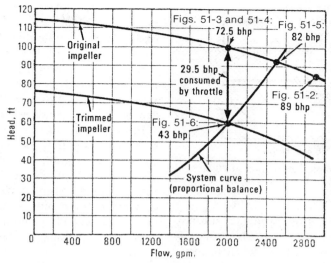

Fig. 51-7. Pump curve operating points for the various balance schemes shown in Figs. 51-2 through 51-6.

When flow balance is not used (Fig. 51-2), power draw is at a maximum 89 bhp. Balance at the pump discharge (Fig. 51-3) reduces brake horsepower draw, but it can introduce circuit flow balance problems. Balance at the terminals only (Fig. 51-4) reduces pump power draw to 72.5 bhp. However, sharply set terminal balance valves absorb pump excess head and waste about 40 percent (30 bhp) of the applied pump power.

The proportional flow balance method first sets each terminal to receive its percentage share of the total available flow (Fig. 51-5). In this case, each terminal is overflowing 23 percent. Establishing a system curve (Fig. 51-7) determines the new impeller* diameter that will provide design flow to each terminal with minimal pump power consumption (Fig. 51-6), resulting in a power savings of more than 50 percent. Power draw is cut to 43 bhp. This represents an annual savings of about $5000 at $0.015 per kWh.

Zone pumping

Sharply set balance valves consume power. The power loss through a sharply set throttle

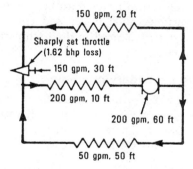

Fig. 51-8. A single pump uses 4.35 bhp at 70 percent pump efficiency.

valve is determined by the same relationship as power input by a pump. Valve power consumption is obtained from the following:

$$\text{bhp} = Q \times H_L / 3960 E_P$$

where

$$H_L = \text{head loss, ft}$$

A great deal of power waste can be avoided by zone pumping. High flow, low head circuits and low flow, high head circuits should be individually pumped. This is more economical than a single pump basically selected for a high head, low flow circuit and that will need excessive throttle on its high flow, low head circuit. Figures 51-8 and 51-9 illustrate the arrangements and savings possible in brake horsepower.

---

*Impeller trimming cost should be specified as part of the initial pump cost and performed by the pump supplier rather than by the mechanical contractor.

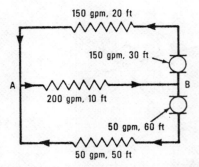

Fig. 51-9. Zone pumps use 2.7 bhp at 70 percent pump efficiency. A-B indicates variation in head loss.

Zone pumping not only provides substantial energy savings but also tends to simplify the overall balance problem when properly applied. The head loss common to all zone pumps must be considered because as it varies so does the flow. Head loss between supply and return heads must be maintained at a constant value: boiler and chiller circuits must be balanced in dual temperature systems, etc.

Primary-secondary pumping

Isolation type primary-secondary pumping should be used where feasible to reduce operating power. There are three basic applications for this.

Primary-secondary pumping increases primary loop temperature difference and decreases primary flow rate. Variations of the circuitry shown in Fig. 51-10 are used to reduce primary flow need (primary pumping power need) for both hot and chilled water systems. Primary flow reduction is dependent on secondary terminal size.

Primary-secondary pumping of high pressure, low-drop flow circuits is used to decrease primary pump power need by decreasing its head requirement. This is similar to zone pumping except that with the elimination of common piping pressure drop, primary and secondary pumps can operate in complete isolation—neither interferes with the other.

In Fig. 51-11, the pump head requirement is set by the high head loss encountered in Circuit 3. Circuit balance valves must be set hard, and pump power requirements are high.

In Fig. 51-12, the high head encountered in Circuit 3 is eliminated for the primary pump by applying separate secondary pumping. Further power reduction is established by primary-secondary pumping of Circuit 4. Overall pumping power needs are reduced to about 90 bhp, and system control is increased.

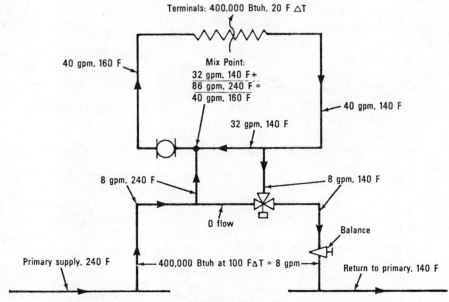

Fig. 51-10. An 8 gpm primary flow serves a 40 gpm secondary circuit.

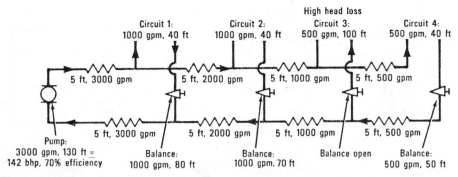

Fig. 51-11. No primary-secondary pumping. Primary pump at 142 bhp with balance valves set hard.

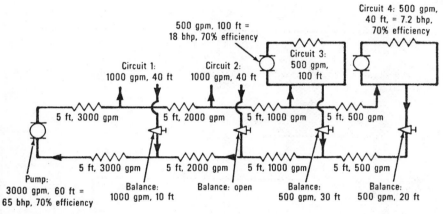

Fig. 51-12. Primary-secondary pumping application saves approximately 52 bhp, reduces balance valve setting, and improves overall system control.

If all secondary circuits were isolation secondary pumped, the primary pump would only need to overcome primary circuit head loss—3000 gpm, 40 ft—and total pumping power would drop to 98 bhp if each pump were 70 percent efficient. Control would be enhanced and balance problems eased.

Primary-secondary pumping will not always decrease power needs; it is *not* a power saving panacea. It is, however, one of the working tools available.

When dual temperature systems are serviced by a single pump, the chiller head loss is continually applied against that pump; even though the chiller is used only about 4 to 6 months in a year in northern areas (Fig. 51-13). A separately pumped chiller consumes energy only when operating and returns other side benefits such as constant chiller flow; simple, reliable changeover from heating to cooling; and zone pumping

troubles resulting from changing equipment room head loss are eliminated.

A separately pumped chiller (Fig. 51-14) would save about $550 per year based on six months winter operation at $0.015 per kWh, as an example.

A more sophisticated arrangement that utilizes zone pumps for the chiller and boiler is shown in Fig. 51-15. The changeover control permits heat-to-cool and cool-to-heat changeover without boiler shock and/or chiller charge blowing. This control is usually automatic, but it can be manual. In this illustration, control is established on inlet chiller or boiler temperature by varying main system flow rate during changeover.

Parallel and series pumping

The fact that flow can be reduced in a dual temperature system during the winter estab-

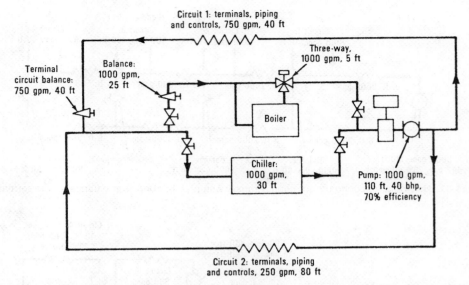

Fig. 51-13. A single pump uses 40 bhp with boiler circuit balanced against chiller. The location of the balance valve avoids backpressure on boiler with possible relief valve discharge.

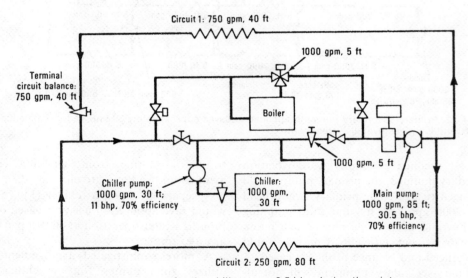

Fig. 51-14. A separate pump for the chiller saves 9.5 bhp during the winter season.

lishes one reason for introducing the energy saving possibilities inherent in pumping in parallel or in series. A system's head loss characteristics essentially determine which arrangement should be used. Systems with lower heads, 60 to 70 ft, should use a parallel arrangement, above that a series arrangement. While this generality can often be denied in specific cases, it is still a good overall principle.

Figure 51-15 shows a low circuit (Circuit 1: 750 gpm, 40 ft) that is adaptable to parallel pumping. If pumps of equal capacity are applied, each must provide 375 gpm at 40 ft for the total cooling flow of 750 gpm, 40 ft.

Figure 51-16 illustrates the operational characteristics applied to Circuit 1. Both pumps combined provide total design flow, and total power draw is 10.8 bhp. A standby flow of 625 gpm is available. Single pump

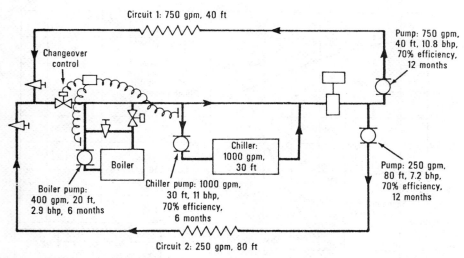

Fig. 51-15. Modified pumping arrangement saves about 15 bhp per year ($1750, $0.015 per kWh) and permits simple and controlled changeover.

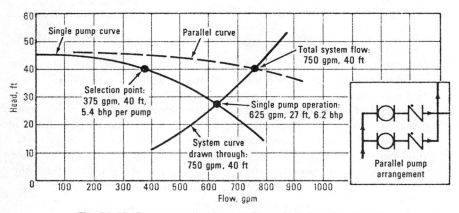

Fig. 51-16. Pumps applied in parallel to Circuit 1, Fig. 51-15.

power draw increases from 5.4 to 6.2 bhp when both pumps operate. *Pump motor size must be adapted to single pump operation.*

During the six month winter operating season, single pump operation will save 4.5 bhp or about $250 per year at $0.015 per kWh, while maintaining more than adequate terminal flow rates.

Circuit 2 in Fig. 51-15 is adaptable to a series application since it has a relatively high head, 80 ft, 250 gpm. Two pumps of equal capacity would each provide 50 percent of the total head at the required system flow rate (40 ft, 250 gpm).

Figure 51-17 shows the operating characteristics of pumps applied in series to Cir-

cuit 2, Fig. 51-15. Design flow is provided when both pumps operate. Adequate standby flow of 185 gpm is available during the cooling season and this is more than adequate during the heating season. For six months, 2.9 bhp will be used rather than 7.2 bhp applied during cooling—a savings of 4.3 bhp per six months, or an annual saving of $240 at $0.015 per kWh.

In summation, the well balanced system initially shown in Fig. 51-13 used 40 bhp per yr at a cost of $4650 per year (an overhead pump of 150 rather than 110 ft with an unbalanced system would increase power draw to about 68 bhp or $7900 per year at $0.015 per kWh). The various modifica-

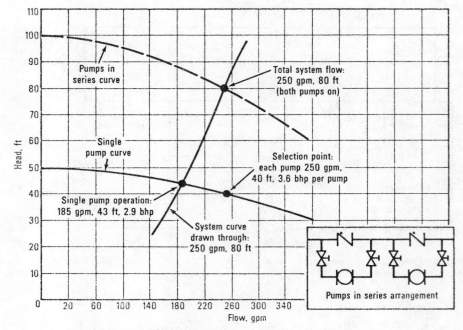

Fig. 51-17. Pumps applied in series to Circuit 2, Fig. 51-15.

tions such as chiller pumping, zone pumping, pumping in parallel, or in series reduced operating power draws to 33.5 bhp during the cooling season and to 11.2 bhp during the heating season, reducing operating costs to about $2000 per year at $0.015 kWh. Escalating costs will increase proposed savings.

Variable volume pumping

High energy costs will result in abandonment of three-way valve control (constant flow) in favor of two-way valve control (variable flow). A variable flow system lends itself to variable volume pumping, which

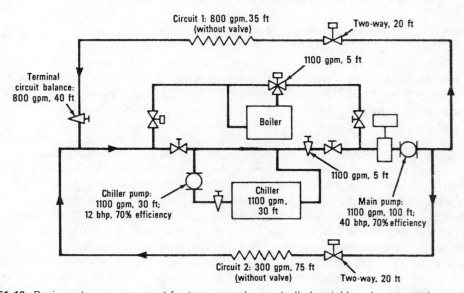

Fig. 51-18. Basic system arrangement for two-way valve controlled variable volume pumping example.

results in substantial pumping power savings and increases two-way valve controllability. Variable volume scheme can be either variable speed pumps or staged constant speed pump sets in parallel or in series. The advantages of this system can be best described by reference to a simple example shown in Fig. 51-18.

Pumping costs decrease even with a constant speed pump when two-way variable volume flow is used, but other problems occur. Flow reduction means head loss in the equipment room, distribution terminals, etc. This lost head reappears as increased head loss (power waste) across control valves and may cause control problems. The overall problem is illustrated in Fig. 51-19, which also shows a constant speed pump curve characteristic: pump brake horsepower decreases at shutoff to about 50 percent at best efficiency.

Figure 51-19 shows that at 50 percent flow, pumping power is reduced from 40 to 30 bhp, and that 24.5 bhp is absorbed by the control valves at this point.

Figure 51-20 illustrates application of a variable speed pump to the system. A differential control placed out in the system attempts to maintain constant valve pressure drop. As the differential increases, pump speed decreases. At 50 percent flow, pump power requirement decreases to about 8 bhp in comparison with 30 bhp needed for the constant speed pump for the same flow condition—an 80 percent pump shaft power need reduction.

Utility cost will not be reduced 80 percent, however, because most variable speed

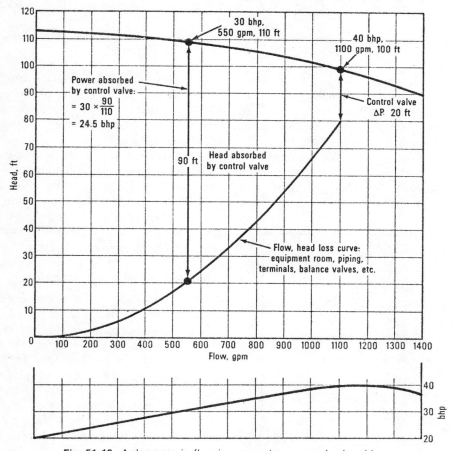

Fig. 51-19. A decrease in flow increases two-way valve head loss.

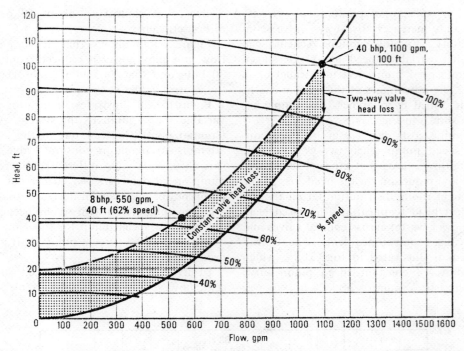

Fig. 51-20. Variable speed pump maintains constant valve head loss: reduced pump brake horsepower at reduced flow.

pump drives use "slip" as a means of reducing speed. Slip drives, whether they be interrupted voltage SCR, wound rotor with variable resistance, fluid drive, or slip clutch, all exhibit a powerful decrease in drive efficiency at reduced speeds. Variable frequency SCR and other nonslip drives maintain much higher efficiency, but they are costly and generally unavailable.

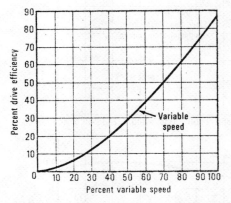

Fig. 51-21. Slip type (SCR) drive efficiency. The percent of full load slip speed is being considered, not the percent of synchronous speed.

A typical illustration of decreased efficiency for an SCR drive is shown in Fig. 51-21. Wound rotor with variable resistance exhibits higher efficiency; some other slip drives yield less.

At 62 percent speed, drive efficiency of about 40 percent establishes that the variable speed motor drive must draw 20 hp (15 kW) to provide 8 bhp at the pump shaft. At the same flow rate, a constant speed pump requires 30 bhp, and its motor draws 36 hp (26.8 kW) at 85 percent efficiency.

Constant speed pump staging

Low variable speed drive efficiency provides an entry for constant speed pump staging. For this example (Fig. 51-18), two pumps in series could be provided. At design flow, one pump provides 35 percent design head and the other 65 percent. A small low head, low flow pump provides for light load operation. The pumps are staged in and out depending on flow need. Fig. 51-22 illustrates operating characteristics.

The shaded areas in Fig. 51-22 represent

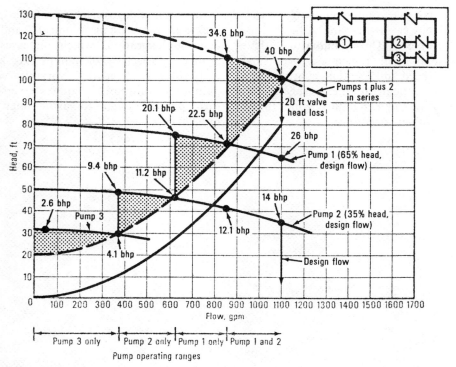

Fig. 51-22. Staged pumps reduce pumping power and retain tolerable valve head pressure drop characteristics.

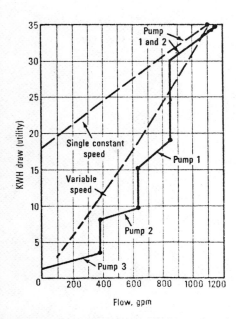

Fig. 51-23. Comparative power draws for single constant speed, variable speed, and staged constant speed applied to the basic system shown in Fig. 51-18.

valve differential head loss increase above that established by the variable speed pump. The variable speed pump retains a valve control advantage, but the staged pumps operate within acceptable valve control limits. (Valve differential head increase to twice the initial head is acceptable; over four times is considered unacceptable.) Pump choke valves eliminate valve differential head problems and can be used if needed. Pump brake horsepower requirements are shown for the various staged operating points.

Figure 51-23 shows comparative kilowatt per hour draws for single constant speed, variable speed, and staged constant speed pumps. The staged pumps propose a power saving advantage at less than 75 percent system flow demand. Other staged pump arrangements illustrate even greater power saving potential and introduce the possibility of using similar arrangements for fans on variable air volume systems.

Zoned variable volume pumping increases potential power savings and reduces pump control problems. Those familiar with three-pipe systems will remember that single variable speed differential control point location is difficult at times; especially for systems with high building zone (north, south, etc.) load variations.

In summation, variable volume pumping requires design, control, and balance sophistication, which increases with a diverse selection of pumps. As a general rule, a diversified variable volume pumping arrangement requires reversed return piping design to avoid high load unbalance.

52

Proven ways to save energy in commercial buildings

A look at what ails many existing commercial buildings and energy saving remedies.

THOMAS IMPERATORE,
Senior Vice President, Cushman & Wakefield, New York, New York

It is unfortunate that it took a move by the oil exporting nations to wake up to the fact that non-renewable energy resources must be conserved and used sparingly. The human brain is considered a great instrument, and it is unfortunate that more brains had not considered the need for energy conservation before the need became shockingly clear.

With few exceptions, we have not used our talents in the building industry to conserve energy. Most systems designed in the last 50 years show little ingenuity as far as energy conservation is concerned.

Architects were in control

Equipment has been applied just to create the effects of heating, cooling, lighting, etc., and architects have dictated in most instances that aesthetics rather than function dominate the mechanical and electrical systems design. The architect was more concerned with how to aesthetically display a radiator than with its efficiency. Similar emphasis has been placed on how to hide piping, ductwork, and appurtenances of HVAC systems, sometimes at the expense of efficient performance.

This chapter was originally presented by Mr. Imperatore at a special series of two-day seminars on How to Save Energy in Existing Buildings sponsored and developed by *Heating/Piping/Air Conditioning* magazine.

Equipment rooms have been considered as wasted space. The mechanical designer was allocated this waste space to fit equipment, and then, he had to try to fit a distribution system into the remainder of the building. An examination of most older buildings—and some newer ones—will verify this.

The problem of saving energy in existing buildings has been further compounded by the fact that much of the equipment was installed on the basis of price rather than efficiency.

Today, there is a great opportunity for engineers, architects, contractors, and others in the construction industry in the area of energy conservation. While they are solving this problem, they will be simultaneously relieving the building owners of the overall dilemma of controlling mounting operating costs. Building owners would welcome propositions from the design and construction industry in the area of energy conservation that would update some of their prestigious building structures by:

- Offering the amenities tenants have grown accustomed to expect in modern buildings.
- Providing better heating, cooling, lighting, life safety, etc., with less energy consumption.

• Providing a better financial return for the owner.

Older existing buildings

In order to embark on a meaningful energy conservation program for a modern commercial building, it is most essential that management develop a Feasibility Program that will, while it solves for energy, improve the overall economy of the building operation.

The program should, therefore, focus on modernization and the improvement of income by:

• Applying as many modern and efficient concepts as applied to lighting systems, air conditioning systems, elevator systems and security systems.

• Revitalizing unused areas such as storage areas, obsolete mechanical service areas and ventilating shafts that will produce usable space and thereby generate income.

The sizable increase in energy costs to date and expected future increases should activate management to examine the building with the intent of saving energy and reducing operating costs. Architectural, structural, mechanical, and electrical changes—along with any others—that will contribute to improved operating efficiency and building utilization should be considered. Properly conceived and implemented, the cost of these modifications should be more than offset by operating cost reductions. In many cases, rental income can be increased as well.

An older existing building is defined in this chapter as being 25 years or older. Such buildings generally have several of the following faulty features:

• Excessive air infiltration due to leaky fenestration.

• Obsolete HVAC systems.

• Lighting systems are inefficient and provide poor quality illumination.

• Inefficient elevator systems.

• A poor core to gross area ratio (a high percentage of wasted space).

These buildings produce relatively low incomes, have a high maintenance requirement, and use or waste an undue amount of energy. Operating costs are often double those of well designed and efficiently operated modern buildings.

These buildings tend to be bulky, and they often have courtyards. Outdoor exposures for all office space was required since most were built without any mechanical air conditioning or with mechanical cooling provided for only selected areas. Air conditioning systems were added later and fitted in as best as possible. It is not unusual to find a multitude of systems and a hodgepodge of equipment scattered throughout the building. Efficient operation and maintenance are difficult to achieve under these circumstances.

Consider all building aspects

To take a bold cut at conserving energy, it is necessary to raise one's sights beyond the mechanical and electrical systems and consider all aspects of the existing building.

Coordinated planning is necessary to create a program to cure all of the defects either in one phase or in an orderly succession of projects within the scope of the income that is expected as a result of the improvements. Income resulting from remodeling can be sizable. In addition to saving money by cutting energy use, remodeling often creates more rentable tenant area.

New buildings, if designed for efficient use of space, have a core to gross area ratio of about 15 percent; some have a ratio as low as 10 percent. Older buildings have a much higher ratio. Often as much as 50 percent of a floor is taken up by core services, such as the following:

• Many corridors.

• Large low pressure duct shafts (originally designed to ventilate or exhaust the corridors) that may be as large as an elevator shaft.

• Stairwells that may be fire hazards by today's standards.

• A multitude of slow speed elevators that can handle as little as one-half the number of people that a modern, high speed elevator can move.

It pays to devote considerable thought to

modifying the core area. If the core area represents 50 percent of the gross area, it is often possible through various alterations to increase usable floor area by 10 to 20 percent. This is achieved by eliminating some corridors; reducing corridor widths; eliminating some elevator shafts by installing modern, high speed elevators; and replacing existing ventilation shafts with smaller, higher velocity duct risers.

Current rent levels in major cities can pay for a large portion of these physical changes. Additional usable space will generate more income, and upgraded existing space commands higher rents.

In many instances, a comprehensive and economically feasible modernization program can be developed to correct most or all unsatisfactory conditions. The program will usually suggest replacement of the major components with the exceptions of the basic structure and the outer shell.

Installation of modern, efficient mechanical, electrical, and elevator systems will improve comfort and working conditions while reducing energy usage.

Changes for efficiency

The following five examples describe problems typically found in older buildings along with solutions to these problems and the benefits of the solutions.

1. *Obsolete heating, ventilating, and air conditioning systems*

Generally, the heating and (if provided) air conditioning systems in older buildings suffer from one or more of the following drawbacks:

- Equipment is deteriorated and can no longer be kept in working order at a reasonable cost.
- A large percentage of building volume is taken up by large, space consuming equipment, such as coal fired (or converted from coal) boilers, old coal bins, flues, hot wells, and storage tanks.
- Ventilating systems are ineffective, occupy considerable space, and sometimes do not recirculate any air and hence, consume excessive energy.

- Inefficient heat transfer equipment and piping systems have little or no operating flexibility and are not adaptable to efficient control.

Replacing the antiquated system with a modern efficient one will provide more usable space; the new system may occupy as little as one-third the space of the old one. Maintenance can also drop to as low as one-third that previously required, and energy for HVAC purposes can be lowered by one-third to one-half.

2. *Poor lighting systems*

A building originally designed for a power density of 3 to 4 watts per sq ft has often been sporadically altered in an unorganized manner to have a power density as high as 15 to 20 watts per sq ft. The additional power requirement is a result of lighting and air conditioning systems that have been added over the years. It is not unusual to find a mixture of incandescent and fluorescent lighting in an older building.

A well designed modern lighting system with efficient luminaries supplied from a properly engineered distribution system will provide improved lighting, often with fewer fixtures and energy reductions as high as 50 percent.

3. *Antiquated elevator systems*

Often, slow speed, manually controlled elevators are found in older buildings, or such elevators have been converted to automatic operation. Converting to computer controlled, high speed cars permits about one-third fewer cars to be used while dramatically improving elevator service.

4. *Leaky fenestration*

Most older buildings were designed with large operable windows, such as hung sash, hopper, etc. Window frames were made of wood or iron and have badly deteriorated over the years. Caulking and weather stripping systems for such windows have been difficult to maintain.

If it is not feasible to upgrade these windows with modern caulking and weather stripping systems, replacement should be considered. In many cases, if properly conceived, replacement can be achieved at a

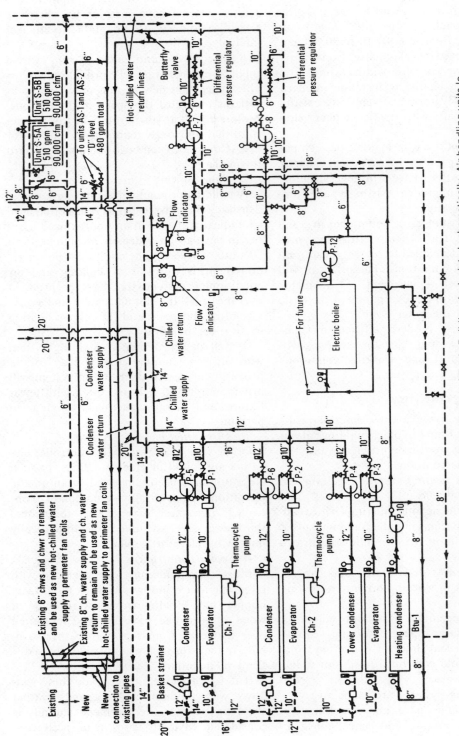

New heating plant on basement level D in Chicago Board of Trade Building, is depicted schematically. Air handling units located on the 23rd floor to serve floors 24 through 43. Another electric boiler is installed on level B are shown at upper right.

nominal cost and energy consumed for heating and cooling will be reduced.

5. Core to gross area ratios

Core areas of older buildings often equal 50 percent of the gross area compared to 10 to 15 percent in a well planned modern building. If the space requirements for building services within the core, such as HVAC duct shafts, power and piping classes, elevator shafts, etc., can be reduced, then additional usable space will be freed.

Each of these five problems affects the use of building space in some way. Implementation of the improvements suggested in the first four problems automatically reduced core area requirements, thereby decreasing the core to gross area ratio.

A well planned and properly executed program covering these five problems can reduce energy use by as much as 50 percent. The effective use of the building may improve to the extent that the cost is more than justified.

Buildings that have adopted this type of building improvement program have reduced operating costs and increased income while reducing energy consumption include: 20 Exchange Place, New York City, 0.8 million sq ft, 60 stories; 40 Wall St., New York City, 1 million sq ft, 70 stories; Board of Trade, Chicago, 1 million sq ft, 50 stories. The gross areas listed here are approximate.

Highlights of a case history

The Board of Trade Building in Chicago is an outstanding example of what can be accomplished through a comprehensive renovation program. Since the upgrading of the HVAC system was previously discussed in the October 1972 issue of *Heating/Piping/Air Conditioning* in the article entitled, "Totally Renovated Chicago Landmark goes All-Electric," only the highlights and additional information on planning the job are included here.

The building, which was completed in 1929, has 43 floors above grade and four levels below. It encompasses nearly 1 million sq ft of floor space; originally 590,000 sq ft was rentable.

Four coal designed, gas converted boilers provided steam to cast iron radiators. Year

New equipment included two electric space heating boilers (one shown here) and heat reclamation chiller.

around air conditioning was not provided originally. Over the years, eleven refrigeration machines were installed throughout the building. A total capacity of 2550 tons was provided by a mix of electric, both recipro-

The Chicago Board of Trade

cating and centrifugal, and steam absorption machinery.

Air was supplied by low velocity, single duct systems. Both central fan rooms serving several floors and multiple suspended air handlers with two or three on a floor were in use. A two-pipe system supplied chilled and hot water during summer and winter, respectively.

During the renovation, most of the existing HVAC equipment was scrapped. A central refrigeration plant incorporating an internal source water-to-water heat pump was installed in the area formerly occupied by the boiler plant. Three, 1000 ton centrifugal water chillers now supply all air handling units. One chiller is equipped with a double bundle condenser enabling the redistribution of heat generated within the building for space heating. A major proportion of the annual heating requirements are provided in this manner. Two electric space heating boilers totaling 3640 kW capacity were installed to augment the heat pump.

Two new fan rooms were built on setbacks at the 24th floor; others were completely rebuilt with new fans, coils, etc.

All cast iron radiators were ripped out, and two-pipe perimeter fan coils were installed. The fan coils have 750 w resistance coils to give heating/cooling flexibility during spring and fall weather. A single duct, high velocity terminal reheat system supplies ventilation air and controls interior space loads.

Lighting was upgraded to modern standards—illumination levels range from 100 to 150 fc. Six lamp fluorescent fixtures were installed. Four tubes are used during occupied hours, two during the cleanup period. Lighting circuits are clock controlled to minimize energy use.

The previous lighting system consisted of a mixture of incandescent and fluorescent lamps; illumination levels were 30 to 50 fc. Standard fluorescent lamps provide about 60 percent higher lighting output than incandescent lamps (based on initial lamp lumens). Therefore, when converting to an efficient fluorescent lighting system design, lighting power does not increase in direct proportion to higher light output.

All floor spaces were gutted, and new partitions, ceiling systems, floor coverings, plumbing fixtures, etc., were installed. Open stairwells were enclosed. The net result of all changes was to bring the building into compliance with the existing provisions of the city of Chicago Building Code.

• *Benefits*—Rentable space was increased from about 590,000 sq ft to 700,000 sq ft. Basement areas formerly occupied by the obsolete heating system were turned into 35,000 sq ft of modern office space. (It should be noted that the installation of a modern gas fired heating plan would also have freed considerable space for other uses.) The balance of the usable space increase came from reducing corridor widths, eliminating unessential workshops, storerooms, etc.

It is a matter of pride to all concerned that work was accomplished while maintaining virtually a full occupancy rate. Tenant cooperation was outstanding. When a floor was renovated, tenants from another area were moved in, and the vacated floor was then rebuilt.

Over the past five years, income has increased from about $3 million to $4.7 million. Operating costs have declined from $2.35 to $2.25 per sq ft.

A substantial reason for the increase in income is the 18.6 percent increase in rentable space. Also, upgrading existing office space to modern standards permitted management to increase rents.

The decrease in operating costs has been achieved in the face of inflationary pressures driving up labor, material, and energy costs.

An important point is that creation of more usable space permits a higher occupancy rate, which is the purpose of an office building. A more efficient building plus increased occupancy reduces per capita energy consumption.

Design team worked as experts

In order to accomplish a renovation the size of the Board of Trade Building, a team of experts worked as a group to develop the program. The team consisted of an architect, engineer, structural engineer, and a space programmer. One firm acted as project director.

Each party was asked to contribute ideas on what could be done to upgrade the building in every respect. From this input, a program emerged as a feasible approach to handle the renovation. It included:

- The estimated cost of construction.
- The effects on the building operating cost.
- The effects on building income.

The plan was attractive to management, and construction expenditures totaling about $10 million were approved.

The modernization program and the increased income potential it creates have made the Board of Trade Building a viable structure in today's rental market. The Board of Trade is now essentially a new building that retained its original skin, and thus its life expectancy is comparable to that of a totally new structure. Moreover, a building that by virtue of its architecture, history, and location is considered a Chicago landmark has been given a new lease on life.

Newer existing buildings

Buildings built within the last 25 years also offer many opportunities for energy conservation, but the feasibility study required to embark on this program is quite different in that there is no major construction required to change the modern existing building architecturally since the core configuration is generally more efficient. This precludes the necessity to cause major improvements in space utilization, however, generally there is a major requirement to effect mechanical and electrical changes. Although these structures were built with year around air conditioning systems that can be classed as modern, in most instances they are inefficient. If the feasibility program is properly established, the impact of the mechanical and electrical changes on the operating cost should more than offset the cost of these improvements within a relatively short period of time.

However, major improvements can be made in the operations of the mechanical systems in these buildings. Most buildings constructed during this period have poor control systems. Even if the control system was basically good, the designer's intent has been lost through poor operating practices, changes in operating personnel, lack of understanding of equipment, etc.

In one of the largest air conditioned buildings ever built in New York City, the only way the outdoor temperature could be obtained was to send a building engineer up to the roof with a thermometer. There was no feedback of space conditions. Sending a man around with a clip board did not help. When he returned an hour or so later, it was too late to make corrections. Outdoor conditions and the temperature had changed, and or the sun had passed behind another building, or was obscured by clouds, etc.

In the last ten years, computerized building control has provided management with a tool that when properly applied can dramatically cut the energy used for heating, cooling, lighting, and power.

Data can be logged continuously and an operator can get instantaneous readouts of outdoor and space temperatures, hot and chilled water temperatures, power inputs, and other essential information. Data can be placed into memory, where it can later be retrieved and studied. Programs can be written so that the computer can monitor system performance and "make" operating decisions.

Computerized controls save

A brief case history of a building computer system shows what can be accomplished. Approximately four years ago, such a system was installed in the Four New York Plaza Building. During the first year, the computer itself was not operating, but considerable data were gathered automatically. These data were logged, and it was possible to determine what was happening within the conditioned spaces and to monitor the performance of compressors, converters, etc. Outdoor air quantities were controlled, and chilled water and supply air temperatures could be reset. The data assisted in operating the building manually. During that year, 165 million lb of steam were used for space cooling and heating and domestic water heating.

During the second year, the computer was in operation, and the previous year's history

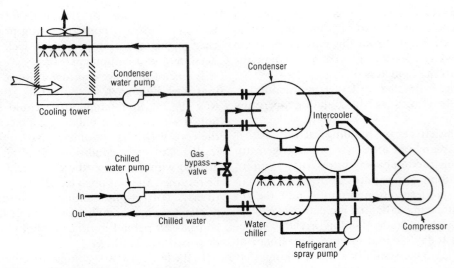

Courtesy Thermocycle, Inc.

Fig. 52-1. Off-season free cooling system is incorporated into centrifugal chillers. System provides chilled water during periods when outdoor temperature is lower than that necessitating refrigeration plant operation without having to operate compressors. Refrigeration evaporates, cooling the chiller water. It then migrates to the condenser because of the lower pressure created by the cool tower water, condenses, and recycles to the chiller.

served as an aid in decision making. Adjustments in operation of the air side systems were made; chilled water temperatures for various operating conditions were determined; and the steam consumption of the two refrigeration machines at part load conditions—one unit versus two—were checked. During this second year, steam consumption dropped to 145 million lb.

In the third year, computer programs based on the previous two years of experience were written and applied to operating the building. The computer in effect was now running the building, and annual steam consumption dropped to 105 million lb. This is a reduction of 36 percent.

Building automation systems may be installed in existing buildings as well as new ones. In the example just given, the building was new, but the computer system actually was installed after the building was finished in order not to delay construction.

It takes time to become familiar with a computer system, but there is no necessity, in fact it is impossible, to operate it as soon as it is installed. Data must be collated and evaluated to determine how the computer can

control energy use most effectively. However, it is essential to get away from relying on the old pneumatic controls and bells and alarms; the latter just got people excited.

Other ways to save energy

Further discussion of the heat pump concept is in order. The following idea may be practical under some circumstances. If an existing building has two or more chillers, one unit can be selected for heat pump duty. With piping modifications to tie the condenser into a building heating loop, the chiller can operate as an internal source heat pump, redistributing heat from interior spaces to offset perimeter heat losses. This conserves energy by reducing the output of the main heating plant.

Prior to running heating system water through the condenser, the condenser must be isolated, drained, and thoroughly flushed with fresh water. The draining, flushing, and refilling operation introduces oxygen into the heating loop, and proper water treatment to inhibit corrosion is thus a factor to consider.

Many existing heating systems are designed to operate at higher water temperatures than are available with air conditioning condenser water. Therefore, greater coil surfaces may be required. This in turn affects air flow resistance and fan horsepower.

Engineers have demonstrated creativity and ability to innovate in many ways; perhaps the approach just outlined will prove feasible for selected applications.

Utilizing existing refrigeration machinery to provide free cooling is not an idea but a proven accomplishment. Figure 52-1 shows a cycle that is adaptable to most existing chillers. During mild weather when cooling requirements are greatly reduced (and there is no need to dehumidify), condenser water leaving a cooling tower is considerably colder than it is at design conditions due to the greatly reduced wet bulb temperature. A refrigerant spray pump wets the tubes in the chiller, the refrigerant vaporizes, cooling the water in the chiller tubes (Fig. 52-1). This vapor then passes through the gas bypass valve into the condenser, which is at a lower pressure created by the cold tower water. There, it condenses and then flows back to the refrigerant spray pump.

Existing buildings: a challenge

I would like to see the engineering community spend much more time with improving existing buildings. We are all enthused by what is new. Offer an engineer or architect an assignment on a new building, and he immediately is interested. It is very difficult to get many professionals interested or involved in existing buildings; although there are signs that this is changing. I believe that the time will come when the situation will be reversed, and the design professional will be turned on by what is old.

From the descriptions presented in this chapter, the reader can readily ascertain that a considerable amount of energy can be expended by implementing a major modernization project. Yet, far more energy will be used in demolishing a large building, which basically is structurally sound, and erecting a new building on the same site. Perhaps the day will come when "energy economics" are interfaced with dollar economics when important decisions are made.*

*See "Energy Economics: a Needed Science," Chapter 5.

53
Energy conservation in commercial buildings

A detailed study examines methods to prevent wasting valuable energy in commercial buildings and an approach that will get even more economy out of an economizer system.

RICHARD R. STUBBLEFIELD,
Vice President, Broyles & Broyles, Inc.,
Fort Worth, Texas

Many of the approaches taken in commercial and institutional building construction embody a needless waste of our natural resources, contributing immensely to our current fuel shortages as well as to our environmental problems. For an understanding of how this can be, we must look to some of the basic principles involved in the practice of engineering design.

Traditionally, an architectural or engineering contract establishes a fee that is expressed as a percentage of the construction contract. This method of compensation appears to be reasonable. It is lacking in one major ingredient, however, that being the incentive required to assure substantial progress above and beyond the engineering norm.

In most instances, the owner engages the architect, and the architect in turn hires the engineer. The engineer must please his client, the architect; and he can best accomplish this by satisfying the architect's client, the owner. The owner is usually satisfied if he ends up with a building that operates under all circumstances without complaints from the occupants. One of the easiest ways to achieve this goal is to make sure that all components and systems are adequately sized. For this reason, we typically see safety factor upon safety factor applied in mechanical and electrical design. Rarely is the owner adept in architecture or engineering, and thus he remains unaware of the mechanical and electrical system inefficiencies that may exist.

Use up-to-date standards

In many cases, the design standards applied in mechanical and electrical engineering are derived from standards previously established by governmental institutions, such as the U.S. Corps of Engineers, the General Services Administration, and the Veterans Administration. Many of these standards are designed to be "safe" for extreme conditions that may never occur. Many were established years previously without the benefit of modern technology. Present day technology and the art of engineering design have progressed to the point of antiquating many of these design standards.

Similarly, city, county, and state governmental authorities may incorporate into their building codes standards that are completely unfounded and contrary to logical, common sense evaluations. Certainly standards must be applied to assure public welfare, but when a code imposes needless waste, it is an injustice to the public and provides no benefit. Such codes should be revised periodically to reflect reality.

Minimize outdoor air introduction

The outside air introduced into a mechanical system for ventilation purposes requires substantial cooling in summer and substantial heating in winter. The energy expenditure can thus be significant in either case. It has long been established that outside air ventilation rates for commercial buildings should be adequate to do the following:

- Provide a minimum volume per occupant.
- Replenish the exhaust air volume.
- Pressurize the building so as to minimize infiltration.

Many local authorities require a minimum outside air supply of 0.1 to 0.2 cfm per sq ft for general office spaces. The same authorities may require 0.25 cfm per sq ft for commercial spaces. In other areas, codes may specify outside air as a percentage of total supply air volume.

Meanwhile, many engineers design air conditioning systems on the basis of a minimum air supply rate per square foot of floor area. This minimum can range from 1 to 1.5 cfm per sq ft and higher regardless of the actual load requirements of the space. When such arbitrary and wasteful design parameters are utilized, an outside air requirement expressed as a percentage of total supply volume becomes absolutely meaningless.

Table 53-1 indicates fuel consumption rates per 1000 sq ft of floor area represented by just 0.1 cfm per sq ft of excessive outside air volume for an office building located in the Washington, D.C. area. Extension of these rates to the hundreds of millions of sq ft of office space now supplied with excessive outside air volumes indicates a huge waste of energy.

Size systems carefully

There are numerous other areas of exorbitant fuel waste in building design, many of them built into the design approaches employed. Hydronic distribution systems for both heating and cooling are invariably oversized, resulting in the expenditure of excessive pumping horsepower. Such waste can be

Table 53-1. Effect of 0.1 cfm per sq ft excess outside air supply on heating and cooling energy consumption of office building in the Washington, D.C. area. Annual system efficiency of 65 percent is used for fuel burning on premises; overall system efficiency of 35 percent is used for electric power generation.

	Annual consumption per 1000 sq ft						
	Gas heating, electric cooling		Oil heating, electric cooling		Electric heating and cooling		
	Gas, cu ft	Electricity, kWh	Oil, gal	Electricity, kWh	Gas, cu ft	Oil, gal	Electricity, kWh
Consumption on premises for heating	5300		41				1050
Consumption on premises for cooling		320		320			320
Energy consumption on premises	5300	320	41	320			1370
Consumption at power plant for heating					10,000 or 75		
Consumption at power plant for cooling	3050		23			3050 or 23	
Total fuel consumption for heating and cooling	8350		64			13,050 or 98	

eliminated by replacement or trimming of the pump impeller to match the pumping head of the system. This will reduce the brake horsepower and energy consumption of the pump motor. See Chapter 51 for more details regarding pump energy savings.

Many air distribution systems are designed as high velocity, high pressure systems; and it is very common for these to operate at static pressures in excess of those required for proper performance. Obviously, fan motor brake horsepower requirements are also excessive in such cases.

During the past 20 to 25 years, many of our design practices and standards have been derived from manufacturers of equipment to be employed. Utility companies have also had an influence on design practices. As a result, large numbers of commercial, institutional, and industrial buildings are served by equipment and fixtures with capacities beyond those required for proper use of the buildings. Lighting is just one example. Very recently, we have seen in some first class office buildings a lighting level decrease to 75 fc. Many at present, however, like most in the past, are still being designed for 100 to 150 fc.

Lighting adds to cooling load

Electric lighting represents a primary cooling load in almost every commercial building. Each excess watt of electric energy provided in the form of lighting must be removed through the expenditure of energy in the air conditioning system.

Tremendous energy savings can be achieved in commercial building air conditioning, using available equipment through innovative design coupled with the exercise of intelligence and plain common sense. Savings in the area of 40 percent, as compared with "standard" practices, can be achieved without sacrifice of quality or comfort.

Let us look now at some of the many steps that can be taken to reduce the energy waste that presently occurs in most existing buildings and to minimize waste in future buildings.

Ways to save energy

There are basically two categories of mechanical installations that are of concern: those now in operation and those yet to be installed. Once installed, the mechanical system takes on a personality much as an individual does; you can alter some portions without much difficulty, but basically it will remain the same. A new system, on the other hand, can be designed to reflect a desired operating personality.

Study systems separately

In general, existing installations must be considered on an individual basis. Nevertheless, there are some common items that should be evaluated to achieve power savings in existing commercial building mechanical systems. Some of these are:

1. Reduce the outside air supply to building mechanical systems to the absolute minimum required. The effects of this can be seen in Tables 53-1, 53-2, and 53-3.

• Reduce exhaust air volumes in all areas; use chemical air purifiers for contaminant and odor control in small toilet rooms as found in hotels, apartment buildings, etc. These air purifiers have been tested extensively and are approved for use in many areas, including all those governed by the Southern Building Code. Many authorities do not allow them, however, simply because their codes require positive mechanical exhaust.

• To control infiltration, set outside air makeup to maintain a slight positive pressure differential (about 0.05 in. WG), indoors to outdoors.

2. For systems with ventilation cycle capabilities, supply 100 percent outside air to the cooling system whenever the outside air enthalpy is below that of the return air. For the immediate Baltimore metropolitan area, U.S. Weather Bureau data for 1972 indicates 1100 hr between 7 AM and 7 PM during which outside air temperatures lay between the normal 55° F supply air temperature and the average design return air wet bulb for a typical office building. These conditions indicate the following approximate annual savings in operating energy (see Fig. 53-3 for calculations):

• No. 2 oil, 0.24 gal per sq ft
• Gas, 32.5 cu ft per sq ft
• Electricity, 3.4 kW per sq ft

3. Reduce the temperature of domestic hot

Table 53-2.   Fuel requirements per cfm or outside air supplied to a building in Washington, D.C. area. Annual system efficiency of 65 percent is used for fuel burning on premises; overall system efficiency of 35 percent is used for electric power generation.

| | Annual consumption per cfm outside air | | | | | | |
| | Gas heating, electric cooling | | Oil heating, electric cooling | | Electric heating and cooling | | |
	Gas, cu ft	Electricity, kWh	Oil, gal	Electricity, kWh	Gas, cu ft	Oil, gal	Electricity, kWh
Consumption on premises for heating	164		1.24				31.9
Consumption on premises for cooling		9.85		9.85			9.85
Energy consumption on premises	164	9.85	1.24	9.85			41.75
Consumption at power plant for heating					305 or 2.3		
Consumption at power plant for cooling	94		0.70		94 or 0.7		
Total fuel consumption for heating and cooling	258		1.94		399 or 3.0		

water for public building lavatories from the conventional 140° F to 90° F.

4. Adjust air supply systems to the minimum air volumes and static pressures required for proper performance.

• Decrease supply air volumes to minimum; increase temperature differences between room design and supply air; reduce or eliminate safety factors.

• Adjust fan speeds to the minimum static pressures required for proper performance.

5. Increase, where possible, temperature rises in both hot and chilled water systems. Drop needless restrictions in hydronic sys-

Table 53-3.   Typical fuel savings that can be achieved if chemical air purifiers are substituted for mechanical toilet room exhaust. Annual system efficiency of 65 percent is used for fuel burning on premises; overall system efficiency of 35 percent is used for power generation.

| | Annual consumption per toilet room | | | | | | |
| | Gas heating, electric cooling | | Oil heating, electric cooling | | Electric heating and cooling | | |
	Gas, cu ft	Electricity, kWh	Oil, gal	Electricity, kWh	Gas, cu ft	Oil, gal	Electricity, kWh
Consumption on premises for heating	10,667		80.6				2073
Consumption on premises for cooling		640		640			640
Energy consumption on premises	10,667	640	80.6	640			2713
Consumption at power plant for heating					19,808 or 150		
Consumption at power plant for cooling	6,132		40		6,132 or 46		
Total fuel consumption for heating and cooling	16,799		120.6		25,940 or 196		

tems, and trim impellers to the actual sizes required.

6. Reduce commercial kitchen exhaust (when makeup air is heated) to the minimum required to prevent migration of odors.

7. In systems requiring reheat for zones or terminals, install devices to reduce air volumes to minimum prior to the addition of heat.

Similar considerations can be listed for future mechanical systems:

1. Apply the latest technology and equipment to maximum advantage.

• Design and arrange air systems to provide only the heating or cooling required to achieve desired results, without simultaneous application. From both the first and operating cost standpoints, the variable air volume concept can be applied economically, using dual systems. One system, with 100 percent outside air capability, can supply cooling only on a variable volume basis. The other, of constant volume/variable temperature designed and arranged to supply only building return air, can provide heating or cooling to offset building skin losses or gains. Used with return air type light fixtures capable of transferring 60 to 70 percent of lighting heat to the return air stream, this type of system can eliminate any outside air heating requirements in winter in many areas of the country. It can also enable building heating requirements during normally occupied hours to be met with lighting heat until the outside temperature falls below approximately 48°F (see Table 53-4).

• Use better quality air supply devices and reduce air circulation rates to the absolute minimum. Lower air volumes can be achieved by using greater temperature differences between space design and supply air temperatures. Approximately 1 bhp can be saved for every 900 cfm reduction (based on approximately 5 in. WG in the system).

• For water systems, use greater design temperature rises and reduce pumping horsepower (Figs. 53-1 and 53-2).

• Use ventilation cycle operation based on enthalpy control where practical.

• Require commercial kitchen exhaust hoods to be of slot type design. This type of

Table 53-4. Typical office building loads when lighting heat is picked up by return air system to offset winter ventilation load as described in text.

Energy use	Btuh per sq ft
Cooling	
Lighting at 3.2 w per sq ft (75 fc), 3.2 × 3.41 =	10.9
Miscellaneous electric at 0.3 w per sq ft, 0.3 × 3.41 =	1.0
People at 75 sq ft per person, 450 Btuh per person ÷ 75 sq ft per person =	6.0
Outside air load at 7.5 cfm per person, 7.5 × 4.5 × 11.53 Btu per lb Δh ÷ 75 sq ft per person =	5.2
Transmission and solar gains	6.7
Air moving equipment at 1 bhp per 900 cfm, 1/900 × 2545 Btu per bhp per hr ÷ 0.88 efficiency =	3.2
Refrigeration equipment at 1.2 kW per ton,[1] 1.2 × 3412 But per kWh ÷ 320 sq ft per ton =	12.6
Miscellaneous equipment	1.6
Total maximum energy consumption	47.2
Heating	
Lighting, as above	10.9
Miscellaneous electric, as above	1.0
Transmission losses[2]	6.6
Air moving equipment, as above	3.2
Miscellaneous equipment, as above	1.6
Total maximum energy consumption	23.3

[1]Including system auxiliaries.
[2]Ventilation load of 6.5 Btuh per sq ft (1.08 × 7.5 cfm per person × 60°F Δt ÷ 75 sq ft per person) offset by lighting heat picked up by return air system as discussed in text.

hood can reduce exhaust requirements by approximately 50 percent as compared to a conventional hood.

• Establish minimum coefficients of performance (COP) for both air conditioning equipment and for air conditioning systems as a whole.

• Work for modification of codes and standards to permit realistic satisfaction of actual requirements.

• Work to establish code regulations specifying that a building must be designed to consume no more than a maximum amount of energy per square foot at peak conditions.

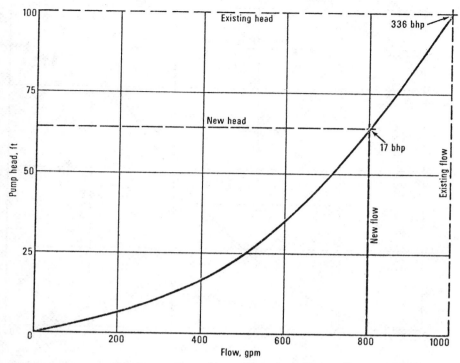

Fig. 53-1. Pumping system curve for typical chilled water, 500 tons. Initial conditions were: pump bhp, 33.6; pump flow, 1000 gpm; pump head, 100 ft. The new conditions resulting from an increase in the water temperature from 12 to 15 ° F are: pump bhp, 17; pump flow, 800 gpm; and pump head, 64 ft. The annual energy savings in electricity over 8 months of operation, 6 days per week, and 12 hr per day are: (33.6 – 17) × .746 × 8 × 4.33 × 6 × 12 × 1/0.88 = 34,900 kWh. The amount of fuel burned at the power plant is equivalent to either 332,000 cu ft of gas or 2520 gal. of No. 2 oil.

Let us now examine some of these energy saving concepts in detail.

Air side reductions

Energy savings will accrue if the delivered air volume is just sufficient to handle the space load rather than being based on arbitrary standards or code requirements. Let us illustrate this with an example.

An office building with 23,000 sq ft of floor area has an $8\frac{1}{2}$ ft ceiling height. The space sensible heat gains, totaling 488,700 Btuh, are as follows:

- Transmission and solar, 285,900 Btuh
- People, 58,000 Btuh
- Lights, 144,800 Btuh

If the supply air volume is based on a minimum air change rate of 8 per hr, the required volume is:

$$(23,000 \times 8.5 \times 8)/60 = 26,000 \text{ cfm}$$

Based on a 20° F temperature difference between space and supply air temperatures, the design air flow rate needed to control the design space load would be:

$$488,700/(1.08 \times 20) = 22,600 \text{ cfm}$$

This represents a reduction of 3400 cfm, or 13 percent.

In a new building, the fan brake horsepowers would be 29 for the air change basis and 25 for the actual load basis, assuming 1 bhp per 900 cfm. Based on an operating schedule of 12 hr per day, 6 days per week, the energy savings would amount to (29–25) × 72 × 52 × 0.746/0/88 = 12,729 kWh. (See Table 53-5.)

Reduce air volume to save

Reducing air volume in an existing building can yield even more impressive savings. A lower air volume in this case leads to signifi-

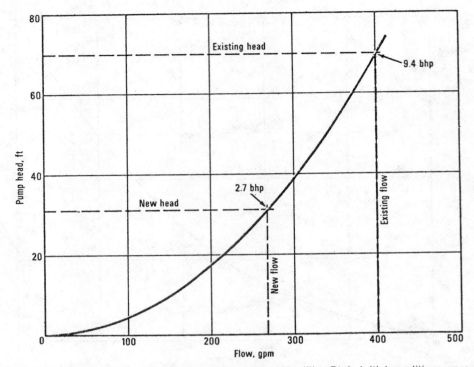

Fig. 53-2. Pumping system curve for typical heating water, 4000 million Btuh. Initial conditions were: pump bhp, 9.4; pump flow, 400 gpm; pump head, 70 ft. The new conditions resulting from an increase in the water temperature drop from 20 to 30°F are: pump bhp, 2.7; pump flow, 266 gpm pump head, 31 ft. The annual energy savings in electricity over 6 months of operation, 6 days per week, 14 hr per day are: (9.4 - 2.7) x .746 x 6 x 4.33 x 6 x 14 x 1/0.88 = 12,440 kWh. The amount of fuel burned at the power plant is equivalent to either 934,000 cu ft of gas or 7050 gal. of No. 2 oil.

cant reductions in both system pressure and brake horsepower requirements since static pressure varies as the square of the air volume and horsepower varies as the cube of air volume. Working with the same air volumes as above and assuming a static pressure external

to the fan of 5 in. WG, we find that the annual energy saving in this case is 31,824 kWh. These data are also summarized in Table 53-5.

At a power cost of $0.02 per kWh, the above energy saving translates to $636.48 annually. The present value of this amount at 8 percent

Table 53-5.   Energy savings achieved in new and existing buildings by reducing air volume to meet loading rather than arbitrary standard.

	New building			Existing building		
Air volume	Static pressure, in. WG	Fan brake horsepower	Annual power use, kWh	Static pressure, in. WG	Fan brake horsepower	Annual power use, kWh
26,000 cfm based on 8 changes per hr	5	29	92,289	5	29	92,289
22,600 cfm based on space loading	5	25	79,560	3.77*	19*	60,465
Annual energy saving with reduced volume			12,729			31,824

*From the fan laws, pressure varies as the square of air volume, horsepower as the cube of air volume.

interest for 20 years is $6250; so it is apparent that checking and adjusting of even small existing systems can pay real dividends to owners. It should be noted, however, that the carrying charges for the investment involved in oversized equipment cannot be eliminated by relatively simple operating adjustments.

Water side reductions

Figures 53-1 and 53-2 illustrate the reductions in water flow and pumping head that result from working with greater temperature differences through heat transfer surfaces. Increasing the temperature rise through cooling coils from 12 to 15° F reduces chilled water flow 20 percent and pump horsepower and power consumption approximately 50 percent (Fig. 53-1).

Increasing the temperature rise on space heating from 20 to 30° F provides a 33 percent reduction in flow and requires approximately 70 percent less horsepower (Fig. 53-2).

Outdoor air cooling

Many approaches have been used to achieve economy of operation by employing outside air for cooling purposes. In the most common of these, 100 percent outside air is taken at an outside temperature that is equal to the supply air temperature required to cool the conditioned space under maximum load. This usually occurs somewhere between 55 and 60° F. Various types of paraphernalia and hardware are employed to achieve the mechanical functions required for such installations. If properly designed, these systems can result in lower power consumption rates during periods when outside air is utilized for cooling. In some instances, however, the economics of the system in the manner it is employed do not benefit the project from a power consumption standpoint.

A few basic types of systems and design criteria requirements must be considered in the utilization of an economy cycle system. First, let us examine some of the conditions that must be considered, then the types of systems that can be employed, and finally, the methods to achieve maximum economy during ventilation cycle operation.

For each building system design, some design parameters used for computations must be established. Of these design parameters, the following affect an economy cycle: outside and return air wet bulb temperatures, outside dew point temperature, and design space dew point temperature. The supply air temperature required to establish and maintain temperature control within the conditioned space should also be considered.

Three types of systems

Three basic types of air conditioning systems are utilized to provide temperature controlled conditions within an occupied space.

1. A refrigeration system that cycles the refrigeration equipment to provide cooling as required by the space thermostat set point. These systems normally employ constant air volume at a variable supply air temperature that depends upon the operation of the refrigeration equipment and the cooling coil load. Usually, such systems are of the roof package variety.

2. A system that regulates the temperature of the supply air by modulating a refrigerant to a cooling coil or by blending two different temperature air streams is commonly used. Primarily, a constant volume of supply air at variable temperatures determined by the requirements of the conditioned space is employed.

3. The third basic system uses a constant supply air temperature from the cooling coil, and the final temperature of the air delivered to the space is regulated either by varying the volume or by a reheat coil system. Air is supplied at a predetermined temperature set point to achieve proper dehumidification and consequently, design relative humidity control within the occupied space.

A system commonly switches to an economizer cycle when outside air reaches 55 to 60° F. This means that when the outside temperature decreases to a predetermined temperature point between 55 and 60° F, a system will index from a normal cooling cycle to utilizing outside air mixed with return air as necessary to satisfy temperature demands. A properly employed system of this nature can achieve savings in energy and operation, but there are approaches that can result in even greater operational savings than those achieved from the vent cycle operation usually applied. Before such improved approaches can be applied, it is necessary to realize the importance and the effects of the basic design data previously discussed.

In most instances, it is economically feasible to use 100 percent outside air if its total heat content is less than that of the return air. The conventional economizer system changeover set point has been previously established as approximately 55° F outside temperature. At this dry bulb temperature, we are assured

that both the temperature and the humidity level within the occupied space can be maintained. However, in most building systems where the maximum load is added to the return air, it is not only practical but definitely feasible to employ outside air on a 100 percent basis at temperatures above the 55° F point. This is possible and practical if the condition of the air supplied to the space will not reduce temperature or moisture control.

In a system that varies the supply air temperature by modulating a refrigerant (the second basic system), the return air wet bulb temperature and the space design dew point temperature must be determined. Also, the outside dry bulb and wet bulb temperatures

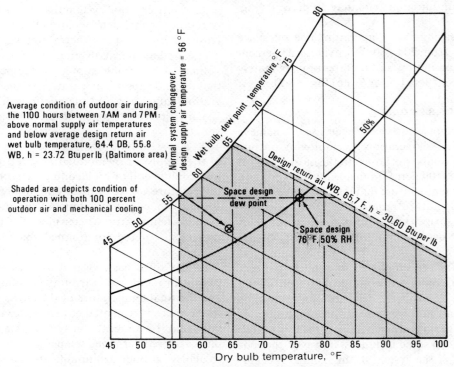

Fig. 53-3. Approximations of the effects of 100 percent outside air used in conjunction with mechanical cooling in an office building in the Baltimore and Washington, D.C. area. The following values were used in the calculations: supply air volume, 1.0 cfm per sq ft of floor area; normal system changeover temperature, 56° F DB; space design dry bulb temperature, 76° F; space design dew point temperature, 56° F; design return wet bulb temperature, 65.7° F; design return air enthalpy, 30.60; average outside wet bulb temperature below design return air wet bulb temperature and above the normal changeover dry bulb temperature (as defined on the psychrometric chart above), 55.8° F; the average outside air enthalpy is defined above, 23.72 Btu per lb; $\Delta h = 6.88$. The amount of mechanical refrigeration saved per year per sq ft in ton-hr to 4.5 x 1 cfm per sq ft x 6.88 Btu per lb/12,000 Btu per ton-hr x 1100 hr = 2.84 ton-hr per sq ft per year. The number of kWh saved per year per sq ft is equal to 2.84 ton-hr x 1.2 kW per ton = 3.4. The amount of fuel saved at the generating station when burning No. 2 oil is equal to 3.4 kWh x 3412 Btu per kWh/(135,000 Btu per gal. x 0.35) efficiency = 0.24 gal. If natural gas is burned, the amount conserved is equal to 3.4 kWh x 3412 Btu per kWh/(1020 Btu per cu ft x 0.35 efficiency) = 32.5.

must be monitored. If the outside wet bulb and dry bulb temperatures indicate a dew point temperature less than that of the design temperature and if the outside wet bulb temperature is less than that of the design temperature, a system can be indexed to 100 percent outside air with mechanical refrigeration still provided to depress the sensible heat, or the dry bulb temperature to the point required to maintain control over the space temperature. If this is done, space design conditions will still be met and economical operation will result.

In a system that uses a constant supply air temperature (the third basic system), the wet bulb temperatures of the outside and return air are the only two factors that must be evaluated (Fig. 53-3). If the wet bulb temperature of the outside air is less than that of the return air, then the total heat value of the outside air is less than that of the return air. In such cases, a logical method used to achieve energy savings is to provide mechanical refrigeration to cool the outside air in lieu of the higher wet bulb temperature of the return air. If the supply air temperature is maintained at a constant of 55 to 58°F, depending upon the design parameters, both space temperature and humidity control will be established and maintained.

Appendix

The data given in Tables 53-1, 53-2, and 53-3 are based upon experiences in the Baltimore and Washington D.C. area to obtain approximations of energy consumed in heating and cooling outside air. Operating conditions and hours may be readily changed to conform with conditions in other areas. Sophisticated computer programs may be utilized to determine the effects of different air quantities on specific design projects if conditions warrant doing so.

The data in Table 53-2 were derived in the following manner. Based on the following assumptions, one cfm excess outside air entering a building results in the following approximate fuel consumptions.

• The average summer temperature difference, Δt, between the outside temperature and the indoor temperature is equal to 60 percent of the maximum summer design condition, $\Delta t = (92-78°\,F) \times 0.60 = 8.4°\,F$.

• The average summer outside wet bulb temperature is 74°F for a four month period, and the inside design wet bulb temperature is 65°F.

• In the winter, design temperatures are 75°F DB inside and 15°F DB outside. The average winter dry bulb temperature for a four month period is 40°F. Neither heating nor cooling is required for a four month period.

The outside air enthalpy at 86.4°F DB and 74°F WB is 37.66 Btu per lb. The inside air enthalpy at 78°F DB and 65° WB is 30.06 Btu per lb.

In the cooling cycle, the amount of heat removed is $4.5 \times cfm \times \Delta h$: $4.5 \times 1.0 \times (37.66-30.06) = 34.2$ Btuh.

The kWh per cfm is obtained from: 34.2 Btuh $\times 1.2$ kW per ton / 12,000 Btu per ton-hr. This figure multiplied by 24 hr per day times 30 days per month times 4 months equals 9.85 kWh.

In the heating cycle, the amount of heat added per 1 cfm $\times \Delta t$ is equal to: $1.08 \times 1.0 \times (75-40) = 37.8$ Btuh.

The amount of fuel energy utilized in gas heating is obtained from: $37.8 \times 24 \times 30 \times 4/$ (1020 Btuh per cu ft $\times$ 0.65 efficiency) = 164 cu ft per year.

The amount of energy consumed in oil heating is obtained from: $37.8 \times 24 \times 30 \times 4/$ (135,000 Btu per gal. $\times$ 0.65 efficiency) = 1.24 gal. per year.

The amount of electricity used in resistance heating is obtained from: $37.8 \times 24 \times 30 \times 4/$ (3412 Btu per kWh $\times$ 1.0 efficiency) = 31.9 kWh.

The amount of fuel consumed to generate the electricity for air conditioning and for electric space heating was calculated similarly by using an overall generating and transmission efficiency of 35 percent.

The values listed in Tables 53-1 and 53-3 were obtained in a similar manner. Those given in Table 53-1 were adjusted by the ratio 60 hr occupied to 168 hr per week to reflect typical operating conditions in an office building.

54

Proven ways to save energy in institutional buildings

Energy conservation steps can and must be good investments if a program is to obtain top management approval. Two examples show how a program can save energy and dollars.

ROBERT A. KEGEL, PE,
formerly, Senior Project Manager,
Capital Development Board,
State of Illinois, presently with
Beling Engineering Consultants,
Joliet, Illinois

Americans are, relative to the rest of the world, very affluent. We enjoy a standard of living considerably higher than any other nation. We would not be nearly as comfortable were it not for the fact that we have been blessed with an abundance of inexpensive energy.

Our quality of life cannot improve materially or even be maintained without continued energy consumption; consumption not necessarily at the current prodigious rate but rather at some level more in harmony with what we can afford from the standpoint of preserving natural resources as well as economically. This, of course, implies conservation.

The energy problems are very real. They are real now, and will be tomorrow, and in the foreseeable future. In the future, there will be new energy sources, such as solar,

This chapter is based upon a paper presented by Mr. Kegel at a special series of two-day seminars sponsored and developed by Heating/Piping/Air Conditioning on How to Save Energy in Existing Buildings.

breeder reactors, fusion reactors, or new indefinite sources; but they are not available now. Until then, conservation is the only answer. We have an obligation to ourselves, to society, and to posterity to conserve energy.

This chapter deals with methods of saving energy in a large and important segment of society, institutional buildings.

An institutional building is different from a commercial or industrial building; it is used differently. A hospital is a 24 hr operation, but a school is typically occupied 18 to 24 percent of the time. Also, heating, ventilation, cooling, and humidification are all considered differently than they would be in an office building.

Conservation is simple!

Earlier, I stated that conservation was the current solution to the energy crisis, and I implied that this was simple! The necessary technical knowledge, equipment, material, and labor exist now; it is simply a matter of applying them.

People must be motivated to put the available resources to work, and nothing motivates people like money. When presenting a conservation program, it is important to point out not only the potential energy that will be saved, but also the corresponding dollars. In our profession, we should be energy managers. In this role, we report to the chairman of the board, president of the company, corporate managers, etc. Prior to implementing a program, we should be prepared to convince these decision makers that the program's obvious merits will also save money.

State of Illinois programs

The state, of course, is very involved in owning and operating institutions and is interested in saving energy in them. The State of Illinois is doing a great deal along these lines. There is an extensive educational program for state employees. This includes an energy conservation guideline (the first state to our knowledge to enact such a document).[1] Informative seminars on conservation (the largest in the country to date) are sponsored. The results of these seminars are available in a comprehensive textbook.[2] Also, several other seminars on specific subjects, such as power management (demand control) and improved performance of refrigeration equipment through proper maintenance and operation, have been organized.

From these efforts, the state works with architects and engineers on the design of new buildings as well as initiating conservation programs for existing structures.

The typical institution

A typical state owned and operated institution is a cluster of small buildings scattered over a wide area. As in any existing building, there are many actions that will save money (energy). For simplicity, I will illustrate only one step in this example.

This example institution has its own generating plant and central heating and cooling

plant. At the time of analysis, the plant needed extensive turbine repair work.

The problem was a rather straightforward engineering economics question: continue to generate power independently (Plan 1), or purchase power from an electric utility (Plan 2). As a modification to Plan 2, a demand limiting computer system is analyzed (Plan 2a).*

The State of Illinois already has a demand control computer system installed in several institutions. At Southern Illinois University at Carbondale this type of system has saved significant amounts of money, not only on limiting demand but also through substantial reductions in electricity.

Table 54-1 delineates energy consumption for the plant at the example institution and specifies the known cost for operating the plant under Plan 1, and the estimated cost under Plans 2 and 2a.

Table 54-1 separates the cost of electricity between respective demand and energy components. In this case, demand is not as large a percentage of the total as I have seen on other projects where the dollar savings are even more dramatic.

Gas for heating and cooling is a large component of the total annual operating cost. The next significant cost is labor. Plan 2a indicates the cost of one computer operator. Since computer input is not required on a 24 hr, seven day basis, only one man can handle the job. The savings in labor expected from such a reduction in operation are shown. One would normally expect for such a major reduction in services a greater reduction in labor impact—wouldn't one?

Table 54-1 also shows the amortized cost of the relative capital expense items, such as turbine repair at 10 percent interest, etc. Salvage value has been allowed for under alternative Plans 2 and 2a for the existing turbines and related equipment.

Summary data are delineated at the end of

---

[1] Superscript numbers refer to references at end of chapter.

*For further information on computerized demand limiting see G. E. Smetak, "Building Automation Systems," *Heating/Piping/Air Conditioning*, Jan. 1975, p. 125. The reader might also contact an electric utility or manufacturers of such equipment for information. Of particular concern should be the method by which the demand component of the electric bill is being charged.

Table 54-1 for both operating and fixed costs. Outside purchase of electricity is $17,059 more under Plan 2 than Plan 1. Normally, Plan 1 would have been favored, however the addition of a demand limiting system to Plan 2 has changed this. Plan 2a cost $30,327 less than Plan 1. Although the 1.8 percent savings (30,327/1,663,453) is relatively insignificant, I would recommend Plan 2a. Although I have stressed monetary con-

Table 54-1. Data on the feasibility of three different means of supplying power to an institution.

	OPERATING SUMMARY		
Energy	Plan 1: Power generated from own plant	Plan 2: Purchase power from local utility	Plan 2a: Purchase power with demand control
Electricity			
Maximum demand, kVA	3,560	3,417	2,289
Annual energy, kWh	21,185,000	20,337,600	17,286,960
Gas, therms			
Water	952,076	952,076	952,076
Heating	908,132	908,132	908,132
Cooling	809,251	809,251	809,251
Generation	6,519,748	—	—
Annual consumption, therms	9,189,200	2,669,459	2,669,459

	ANNUAL OPERATING COSTS, $		
Energy	Plan 1	Plan 2	Plan 2a
Electricity			
Demand	—	75,432	49,314
Energy	—	291,408	242,365
Total (exclusive of taxes)	—	366,840	291,679
Gas	815,225	530,832	530,832
Supplies	19,500	9,500	10,500
Maintenance (parts, etc.)	15,000	10,500	11,000
Labor	450,000	400,000	420,000
Annual operating cost	1,200,725	1,317,672	1,264,011

	FIXED COSTS, $		
Item	Plan 1	Plan 2	Plan 2a
Turbine repair at 10 percent interest	17,578	—	—
Excess facility charge and utility	—	24,360	24,360
Annual plant cost at 10 percent interest	346,150	—	—
Plant cost less salvage for turbo generators	—	338,480	338,480
Demand control computer system at 10 per cent interest	—	—	6,365
Total fixed cost	363,728	362,840	369,205

	SUMMARY OF COSTS, $		
Cost	Plan 1	Plan 2	Plan 2a
Operating	1,299,725	1,317,672	1,264,011
Fixed	363,728	362,840	369,205
Total	1,663,453	1,680,512	1,633,216

siderations, other priorities must also come into play.

When selecting fuels or making similar engineering decisions, dollars are not the only consideration. Most of the electricity consumed in this area (and in our example plant) is generated with coal or uranium. In fact 96 percent of electricity in Illinois is generated from these sources. Gas and oil, the remaining 4 percent, are used on a declining basis. Plan 2 or 2a allows the natural gas now being used to generate steam and subsequent electricity with a turbo-generator to be conserved and then used by more critical consumers, such as commercial and industrial consumers who must use natural gas for certain processes for which there is no alternative today. Millions of residential consumers who really do not have a viable alternative to present fuel sources might also be considered as critical.

One cannot arrive at the same conclusion for all parts of the country, but the thought process is valid. There are electric utilities that derive a substantial portion of their total energy from natural gas or oil and not from the less critical fuels, such as coal or uranium.

Before leaving this example, it may be of interest to look at the investment in demand limiting another way. The system costs $60,000. On a very simple payback basis, the investment pays for itself in 1.98 years (60,000/30,237). On a present worth basis, capital will be recovered in 2.33 years. Management generally desires to evaluate the recovery of money on capital expenses. An engineer should include this in his total proposal.

Saving energy in a high school

The physical characteristics of a recently occupied high school are listed in Table 54-2. Let's examine some of the measures to reduce energy consumption in the "new" existing building.

The building has fixed multizone, draw-through rooftop equipment. It has 15 heating and cooling units and six of the heating only configuration. It is not suggested by any means that a draw-through rooftop multizone is the most efficient use of energy. There

Table 54-2. Summary of data on high school before energy conservation techniques were applied.

Constructed	1974
Area	174,890 sq ft
Occupancy	1,400 students
Loads	
• Heating	8,000,000 Btuh transmission
• Cooling	11,680,000 Btuh ventilation
• Ventilation	397 tons transmission
	213 tons ventilation
	610 tons total
• Lighting	3.8 watts per sq ft (typical classroom)

Annual cost of operation, $		
Utility	Total	Per sq ft per yr
Gas, $0.08 per therm	42,350	0.24
Electricity, cooling	36,727	0.21
Electricity, other	59,462	0.34
Total	138,539	0.79

Based on degree-days, not actual.

are many other systems not necessarily more expensive considering life cycle or even first costs that are more available; but this is an existing building with existing equipment.

Reduce temperatures

Reducing occupied space temperature from 72 to 68° F will be examined first. This does not impose any hardship and will save both dollars and energy. (See Table 54-3). This measure is controversial and may not always be economical. Individual buildings must be considered independently. I am not suggesting that each room thermostat be adjusted accordingly, but that the entire system operating temperature be reduced.

Conversely, the space temperature in the cooling mode is increased from 75 to 78° F. This also causes little, if any, discomfort. This building already operates at a 55° F night setback temperature.

Reduce ventilation

The outside ventilation load can be reduced because the state code has been changed since the school was designed. The Illinois state code previously specified 20 percent outside air,* which is a great deal in a cooling sit-

*Efficient and Adequate Standards for the Construction of Schools, Circular Series A., No. 156. Actually the original code specifies more than one method of calculating the quantity of ventilation air. The typical requirement, however, was 20 percent of the total air suggested.

Table 54-3. Summary of effects of each of the six methods of conservation applied to high school described in Table 54-2.

Method	Investment, $	Life, Yr	Annual fixed cost, $	Annual savings, $		
				Gas	Electric cooling	Electricity
Adjust temperatures	—	—	—	2,066	7,345	—
Reduce ventilation air	—	—	—	9,521	6,743	—
Seal ducts and dampers	16,800	10	2,734	4,014	—	—
Roll filters	25,200	15	3,313	—	—	9,800
Kitchen hood exhaust	6,520	10	1,017	1,054	—	—
Lighting maintenance	—	—	—	+	-	10,034
Sub-totals			7,064	16,655	14,088	19,834

Totals: −$16,655
        −$14,088
        −$19,834
        +$ 7,064
        $43,513 or a 31 percent overall savings

uation. The code now specifies 5 cfm per person depending upon design occupancy. Classroom situations are now appearing where 1.4 to 1.8 cfm per sq ft is supplied, depending on people and light loads.

Relaxation of ventilation codes is a very positive step, and frankly, codes can probably be reduced even further. Outside air is needed only to replenish oxygen, eliminate odors, and to provide positive pressure within the space. In this school, ventilation fell from 59,000 to 28,000 cfm. This will save $9521 in gas and $6743 in electricity for cooling.

Seal ducts and dampers

The next method is to seal outside air dampers and ductwork to prevent leakage. From this project, it is estimated that 5 percent of the total air circulated can be saved, which converts to an additional $4014 per year in gas. Sealing will cost $2734 (amortized) per year, but clearly the savings justify the investment. The proper way to assess any conservation measure is to spread the cost out over the useful lifetime and compare it to the benefits gained or lost by the investment.

Many states, including Illinois, have completely gone overboard on filtering systems in general and particularly in schools. Many schools are designed with operating room level filtering systems that have very efficient but very expensive bag filters. Schools should

be designed with appropriate filtering systems, and the 45 percent efficiency bag filters should be replaced with roll filters. The effectiveness is satisfactory for a suburban school, and it will save on fan horsepower.

Roll filters replace bag filters

Roll filters replaced bag filters in this school's system, and static pressure resistance decreased a little over ½ in. WG. Bag filters are large consumers of air horsepower. When the roll filters are installed, it will cost $25,200 plus wiring, but it will save $9800 per year in electricity. The investment of $25,200 amortized will cost $3313 per year. This is another good investment for saving energy. Of course, the throw-away filters must be maintained on a regular basis to achieve these savings.

Kitchen hood exhaust

In kitchen areas, revising the hood exhaust to induction exhaust might save energy. I am not suggesting that all the air be swept out of the cafeteria and all of the air that has already been heated in the kitchen be exhausted, but that outside air be induced before this hot air circulates in the building. Exhaust just a portion of the heated air, and a great deal of money and operating dollars will be saved on exhaust systems.

While this exhaust system has been used in kitchens extensively, it is also applicable for paint and welding exhaust. The hood can be revised for $6250 or at an annual cost of $1017, but $1054 will be saved per year. This pays for itself in one year and is a good investment. In fact, all new designs should consider this type of hood.

Lighting maintenance

Lighting maintenance is another important area. This might be looked at on the basis of a regular system of lighting maintenance and replacement. The state has used a computer analysis for lighting studies. On a 36 month basis, by providing 100 percent relamping and cleaning, there will be less than a tenth of a percent interim large replacement, but that will provide 47 percent more light. Plus, it only cost $1.36 per fixture per year. Not a big investment of money, but let's look at it another way.

If a regular program of lighting maintenance and replacement had been considered at design time, the number of fixtures could have been reduced by 32 percent. If such a program had been planned and really carried out in the beginning, the number of total fixtures installed at the start could have been reduced, and the same light provided. The energy consumption is reduced by 32 percent for the lighting, and operating costs are cut by 27 percent.

Lighting should be maintained on an engineered basis and placed where it is needed. Task and direct lighting should be placed where the students or people need it. Blanket lighting is not good engineering, and it is not energy efficient.

In this particular school, 3.8 watts per sq ft, which is about 82 fc, is maintained in the classrooms. This is adequate for most classroom situations, but it is excessive in corridors, toilets, or other remote areas.

Heat reclaim not feasible

Another conservation measure not evaluated in Table 54-3 is the installation of a thermowheel or other similar heat reclaim equipment. As mentioned previously, this school has a rooftop system and units are scattered all over the roof. The complicated system of ductwork that accompanies this system

makes this a questionable measure for this school, but it should not be ruled out for other buildings. For central systems, it is a very viable consideration.

For this particular school, it would have cost $67,200 to install a heat reclaim device. This converts to an annual cost of $8835, and a little over $9348 would be saved. Even on that basis, it is hardly worth the investment even though energy would be saved. Also, many maintenance problems would be introduced.

I would like to point out that even though a certain measure won't work in a particular case, it should not be permanently ignored. Do not neglect anything that might save energy and dollars.

The total picture

Table 54-3 illustrates the total effects of each of these suggestions. The overall operating costs of this particular school were reduced 31 percent, or $43,513 was saved per year.

The first key to saving energy is to admit that a problem exists, and it is really desired to do something about it. The right attitude must include saving dollars as well as energy.

Management will have to be convinced that energy saving is important, and the program must be justifiable from an economic standpoint.

The conservation program must be a managed program; it must be planned and controlled. Plan the work, set up a program, outline specific steps, and plan how to present the program to management. Then at a later time, place the program in effect. Plan the work, and work the plan.

All conservation steps will not work in every situation, but the feasibility of implementing anything that you can think of should be examined. It is possible to conserve energy, conserve natural resources, and provide for not only a growing economy but also for tomorrow's generations.

References

1. *Energy Conservation Guide for the Construction of State Funded Buildings*, State of Illinois, April 26, 1974.
2. *Energy Crisis*, a regional conference sponsored by the Capital Development Board of the State of Illinois, Sept. 19, 20, 21, 1973.

55

Proven ways to save energy in industrial plants

When it comes to saving energy in industrial plants, the only real limitation is your imagination.

K. E. ROBINSON,
Retired, Ventilation Engineer,
General Motors Corp.
Mears, Michigan

Many people have no concept of how to start an energy conservation program for industry. Yet without one, very little will be accomplished. Effective energy conservation programs geared for industry are possible and can result in significant savings.

Figure 55-1 depicts an actual situation that occurred in an industrial plant. It shows the plant's rising utility costs from 1965 to 1969. After 1969, energy expenditures began to decline, even though more floor space was added. An energy conservation program began in 1968 and became truly active in 1969. The program's effectiveness is demonstrated by the fact that in 1970, which might be considered the first full year of the program, operating costs were reduced over $300,000. This was accomplished in spite of utility cost increases of 6.3 percent for electricity, 14.7 percent for gas, and 8 percent for coal. In addition there was also a slight increase in floor space.

This example is important also because it happened before energy conservation became

a popular item. If this degree of success was achieved in 1960 to 1970, even better results should be possible now since there have been improvements in equipment and more knowledge has been accumulated. Today such programs not only offer greater dollar savings because of higher utility costs, but in the near future they also may keep a plant running that otherwise would be shut down due to a lack of available energy.

Steps to success

The basic and most important steps that must be employed if a successful program is to be developed and carried out include the following.

- The program must have the backing of top management.
- The person in charge not only must be qualified, but also he must be a *positive* thinker.
- Goals should be determined and records developed to show the program's status. Unless it can be demonstrated that the program is progressing, it may be lowered in priority or phased out completely.
- An education program for everyone from top management to the new hourly

This chapter is based on a paper presented by Mr. Robinson at a special series of two day seminars sponsored by *Heating/Piping/Air Conditioning* on How to Save Energy in Existing Buildings.

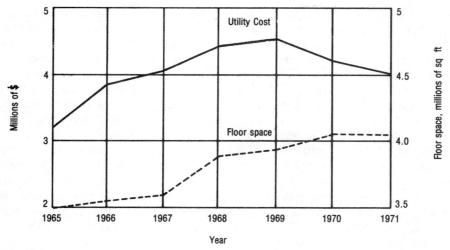

Fig. 55-1. Utility cost in a plant before and after an energy conservation program was started in 1968 and became truly active in 1969.

employee must be initiated. This is necessary to break life-long habits of wasting energy. Education must be extended to production management because in some cases, they will have to change their methods of production. Energy conservation may well take priority over production.

• Reduction in energy, as it occurs, should be reflected in each department's budget.

• There must be an accurate energy inventory. Although most people seem to ignore this point, it is basic to know what energy is being used, where it is used, and what is being thrown away. All energy cannot be lumped into one big package. When an energy inventory is placed on paper, areas of conservation often seem to jump right out.

• Metering of all energy to each department is desirable. Accurate information about energy consumption in individual departments lends authority to recommendations.

• Console operation of all energy consuming equipment is necessary if maximum conservation is to be obtained.

Each program should be looked at in terms of both immediate and long-range results. In many cases, immediate results can be achieved without studies. In every plant that I have walked through, it was possible to make a list of energy saving ideas that could be implemented rapidly. As soon as a program is announced it should become active. The sooner definite results can be shown, the more enthusiasm will be generated, and the faster the program will move forward.

A number of fast possibilities

The number of items that may be done to show immediate results is limited somewhat by the type of plant, but mostly by the imagination and ability of the parties carrying out the program. The following list shows a number of possibilities that may be considered.

• Shut off lights when not needed, such as during lunch and at the end of a shift. Clean dirty fixtures to assure greater efficiency of lighting. Both of these are obvious, but somehow they do not get done.

• Remove lights in areas where they are not needed, such as some stock rooms, aisles, etc. It is amazing to add up the number of lights in a plant that are not needed. It is not unusual to find sections of a plant fully lighted even if all operation has been temporarily suspended.

• Reduce parking lot lighting to minimum, but maintain security and safety.

• Reduce all decorative and advertising lighting. There may be some disagreement as to what is decorative and what is advertising lighting and the value of it, but it should be possible to resolve such questions.

• Shut off all equipment when not in use, such as during relief periods, lunch, and the end of shifts. Perhaps nothing will yield such quick results as assigning someone to walk through the plant at the end of shifts and weekends and turn off the lights, fans, and unnecessary heating equipment. This is even more effective when the offending department is charged a given amount for that man's time since he did what should have been their responsibility.

How many times have you walked into a plant and the upper windows and the monitors are wide open to let the heat out and the unit heaters are running? This is a common situation. Control the operation of the heaters to reduce the heat input to a satisfactory level.

I know of one case where a plant was down over a holiday weekend. The plant used purchased steam, and an engineer kept a record of it. For the same number of working hours as in that long weekend, less than 50 percent more steam was used than during the weekend when the plant was down. It was the Fourth of July weekend so no heat was required for the building. In other words, a high percentage of the total steam paid for was wasted because of a lack of insulation and leaks in the piping and controls.

• Shut off paint booth supply and exhaust fans when not in use. In many cases, even those that are actually used intermittently will be turned on in the morning and run all day.

• Shut off parts washers when not in use; i.e., pumps, steam to tanks and exhaust fans. Keep covers on the tanks closed, and insulate the tanks.

• Shut off drying and curing ovens when not in use. Do not start too early before shift time.

• Shut off interplant, outside truck engines when not in use.

• Analyze interplant truck runs, and consolidate loads and eliminate trips.

• Shut off fork truck engines when not in use. This is particularly valuable if they have combustion engines since this will reduce carbon monoxide levels and lower the amount of ventilation air required.

• Keep heat and smoke relief vents closed in winter, and check to see if the dampers are closed tightly.

• Fix all compressed air leaks. One plant I know claimed to need a new compressor immediately, but they started a conservation program. They since have increased the floor area of the plant without adding any compressors. In fact, two of the original compressors are now used as standby units. Before the program, all of the compressors could not keep up with the demand for air.

• Analyze all air-using equipment to determine the lowest pressure at which the air compressors may be operated.

• Reduce water temperature in rest rooms. This can also be done at parts washing, plating, etc. Do not attempt this with manual control. It will not work as everyone will want to increase the temperature again, not because it is needed but because of habit.

• Insulate all bare steam pipes and condensate return lines. This is particularly true if processed steam is used in an air conditioned area. It will reduce steam usage and chilled water requirements.

HUMANS ARE SUCH STRANGE CREATURES.... THEY MAKE CEILINGS SO WARM AND THEN THEY NEVER WALK ON THEM!

• Fix all faulty equipment: temperature controllers, leaky valves, float valves, steam traps, etc. Most automatic controls have a 10 year life expectancy at the most, yet many people keep them 15 or 20 years if they work or not. They should be replaced where required.

• Reduce heat in areas that are overheated. Do not open windows in winter to cool an area. Fix broken windows, etc. Being too warm is a prime employee complaint. You can save fuel and improve working conditions with better temperature control.

• Check the preventive maintenance program on all air filters and heat exchangers to assure maximum efficiency.

• Readjust outside air dampers to minimize outside air usage. This would probably take place during downtime. Very few plants have more makeup air than necessary.

• Check all process exhaust systems to see that only the proper amount of air is being exhausted. Excessive exhaust is very common. In the past, most problems in this area were handled by installing another exhaust fan, a wasteful solution. I know of a plant that evaluated its practices and reduced exhaust by 35 percent. Many exhaust systems must be tested to comply with the OSHA codes; some tests are directly specified while others must be tested when installed and then repeated periodically. For example, plating exhaust systems must be tested at installation and every 90 days thereafter.

• Reduce heating in unoccupied areas.

• Shroud openings of furnaces, ovens, paint booths, and washers so that the minimum amount of exhaust air will be required. During the design stage, an arbitrary figure, say 10,000 cfm, is selected as adequate to handle requirements; so the system was installed to handle 15,000 cfm, and since no one has checked it since the time of installation, it may be exhausting 20,000 cfm.

• Use cold water detergent in washers wherever possible. Using hot water is a habit rather than a necessity for many operations.

• Where possible, combine operations and reduce the number of washers. This is also a good procedure to follow for some heat treat furnaces. Plants, particularly smaller ones, often have operations that are used spasmodically. These parts could perhaps be sent out for heat treating more economically.

• Plan weekend work so that the whole plant can be shut down on given weekends.

• Reschedule operations wherever possible to second and third shift to get them off the 10 AM to 2 PM peak electric demand period. In many cases, another method of production can be used, or the job can be done equally well on the second shift when electricity is in less demand. If the energy crisis deepens, many operations will be put on the second or third shifts out of necessity.

All of these ideas can be instituted without studies and should be considered as part of a short-term program.

Long-range possibilities

Although the immediate savings program is important, the long-range programs must not be ignored. The following list only touches on the unlimited possibilities that exist to reduce the use of energy in industry.

• Study the plant heating and air conditioning systems to determine if they are correctly designed; many are not. Equipment selection is an important aspect. To digress momentarily, there is never a heating problem in an industrial plant (or anywhere else, including a house) because man is a warm blooded animal and must lose heat to survive, and lose heat at a controlled rate to be comfortable. Thus, a plant has a year around cooling problem, and it follows naturally that equipment should be selected that can also be used year around. When someone complains of being cold, it is because he is being cooled too rapidly, not because he is not being heated. This sounds paradoxical, but remembering it can make a big improvement in providing a better plant environment. An intermittent unit heater certainly does not fit this rule because it emits a blast of hot air on a man that wants to be cooled while another man in close proximity is losing heat too rapidly.

Eliminate stratification of air

• Eliminate stratification of air in the plant in winter. This will cool the ceiling and warm the floor. It is a common sight in plants to see a crane operator flip off all the heaters because he is too hot while the workers on the floor are yelling about being too cold. This can be solved by simply mixing the air. This can be done very easily in an older plant by placing vertical risers or ductwork about one diameter from the floor, and running it upward beside a column to about the level of the crane if there is one. Some kind of fan—propeller, vane axial, or centrifugal—can then pull the air off the floor and discharge it at the upper end of the duct or riser; so its velocity will carry it to the upper part of the plant. A minimum of three air changes per hour will virtually equalize floor to ceiling temperatures.

I know of one plant 113 ft high that uses this system to eliminate stratification. On a zero day, the plant has been able to operate with 8 deg difference between the floor and ceiling, which is much better than many motels. As an experiment, the system was shut off, and a 50 deg temperature difference developed in a few minutes.

This is a simple way of mixing air. There are many other ways. Replacing unit heaters with properly designed and installed units with suitable ducts and controls is a preferable way, but money may not be available. If the situation cannot wait and unit heaters are already installed, it is suggested that the fans be operated continuously, and the heat input be varied to suit space conditions.

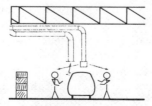

Use spot heating or cooling when people are located far apart.

Reduce quantity of air exhausted

• Reduce the quantity of air exhausted from the building. Use local exhaust not general. Local exhaust uses less air and gives better contaminant control. For example, general exhaust in a welding area where workers are using wire welding guns requires about 2000 cfm for dilution. This does not provide good control for an individual welder. A locally controlled welding gun uses about 60 cfm and provides the welder with excellent control.

Another good example of the superiority of local control and the interaction of codes is a large paint booth job which was used to paint locomotives. The code calls for 100 cfm per sq ft. We designed it for 50 cfm, but we also went to government authorities and got assurance that it would be tested only when painting was going on. A locomotive takes up over half the capacity of the booth, and the code requirements were met. If this had been sent to the authorities without an explanation, it would have been rejected. Codes are meant as guides. I would like to stress that I am not suggesting that the quality of air for people be reduced.

• Where possible, use exhaust air to warm incoming air by mixing the air streams. Too often energy is used to bring in and temper outside air when clean air is exhausted only because it is too warm.

• If exhaust air is contaminated, evaluate air cleaning devices to determine if the air could be cleaned and recirculated. An industrial hygienist does not like this, but there are contaminants, such as cast iron grindings, that are adaptable to this practice. Other contaminants will be able to be handled by expending about $3000 on some form of control device, probably an optical device or a secondary system. The rate of return is not 10 or 5 years, but perhaps weeks.

• Study the possibility of utilizing the energy in contaminated air in production operations before it is exhausted. For example, one plant had a combustion process where it was necessary, of course, to exhaust the products of combustion. Next to the process was a washer and dryer using steam to dry metal parts. Now they are dried by the heat from combustion, which is then exhausted. In another plant, the products of combustion are used in a vestibule to preheat parts before they are heat treated. In the future, more heat will be taken from the exhaust stream by using an air-to-water heat exchanger. This will add heat to the water used to cool the bearings in the furnace. The heat in the water will be given up to heat incoming air by use of a coil, while cooling the water so it can be returned to cool the bearings. This constitutes a complete cycle.

• If all other procedures are not applicable, determine which is the most suitable heat transfer device to incorporate into the exhaust system. Complete packages with heat transfer devices are now available.

Is boiler needed in summer?

• Determine if the boiler plant is really needed in summer. Could small boilers and water heaters be used and the powerhouse be shut down for the summer? One plant uses its system for heating the plant only. The boilers are shut down in summer. No insulation is used on the piping since hot water is in the system only when heat is needed. Flash boilers are used to provide steam for process.

• Study all steam piping, traps, and controls to see if they could be improved.

• Study to determine if pressure blowers could replace some use of compressed air. Do not allow compressed air to be used to cool personnel.

• Use an automatic control to regulate the volume of water or air used. Do not leave this up to the individual worker; he will use too much. The valve is a simple control device to assure constant water volume, regardless of water pressures.

• Do not operate the cooling tower in winter—use heat in the water to temper building or incoming air. I was ridiculed for suggesting this 20 years ago, but a new plant in Manhattan, which handles 40 percent of the meat used in that area, now does this, and it has saved the price of a boiler and even more on an annual basis. Cooling towers are most wasteful in cold climates. It seems foolish to take warm water outside to be cooled while using extra pumps and damaging the tower, and then, add new energy to heat other water so as to warm incoming air.

• Study all solid waste to determine if it can be recycled, burned, or composted. Disney World has installed a compost system that has been used in the Bahamas for eight years. Any substance can be thrown to the grinders from rubber tires to boards. In about two weeks, an enriched soil results that can be used or sold.

• Salvage all oil used in a plant. It can either be reused by refining it, or it can be burned in the boiler. One plant now sells oil to the city to be burned in its power plant. Previously, they paid to have it hauled away. However, it is more economical to recycle or refine it than to sell it.

• Use low volume, high velocity exhaust systems where possible. These include ventilated welding guns, hoods for portable grinding equipment, local traveling hoods for molten metal pouring, as well as a unique method of controlling mist from oil mist lubricators.

• Use push-pull ventilation on open tanks; it will save 50 percent or more of the air required. It is also acceptable to OSHA.

• Insulate hot water and bond rite tanks. Use insulated storage for chilled water to reduce the size of the compressor. Storage systems for hot liquids should also be considered.

Turn off air conditioning at night

• Shut off air conditioning compressors at night and use ventilating fans to maintain the space temperature. The result of such a study was reported to ASHRAE in 1955. It was tested in a bank in Texas. The refrigeration compressors were shut off when employees left the bank, and a large attic fan was operated all night. The next morning, temperatures in the building were the same as when the bank closed.

• Operate the office on two or three shifts, or nights only when more power is available.

• Use spot heating or cooling of people when they are located far apart. Each should have control of the air direction and velocity over them. There are devices on the market that allow for choice of direction, velocity, and volume.

• Use evaporative coolers in place of refrigeration for man cooling. Many engineers do not believe evaporative coolers can be used, yet they are the oldest form of cooling known to man and were used before Christ was born. In the last few years, much information, including weather data, has become available. It is simple, and it costs pennies rather than dollars.

Some claim that mechanical refrigeration is cheaper than evaporative cooling, however, one should be skeptical of such statements. Many of them were made to justify mechanical refrigeration and were not always based on facts. It may indeed be necessary to go backward in order to continue moving forward. All ideas should be evaluated individually and not by just taking somebody's word for it.

Roof cooling is effective

Many of the items listed have been discussed at many meetings, and some of the information has been published previously in *Heating/Piping/Air Conditioning*, and other technical publications. For example, the results of research on using water for roof cooling was reported to ASHRAE in 1966 by John I. Yellott. Not only will roof cooling save energy, but it has been proven

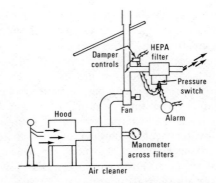

If exhaust air is contaminated, evaluate air cleaning devices to determine if the air can be cleaned and recirculated.

that if the roof temperature varies more than 100°F in a 24 hr period, the roof will be shocked and its life shortened. In the sun, the typical black roof will go up to about 187°F. On a clear night, its temperature will drop below ambient temperature. Spraying the roof intermittently so that water evaporates rather than runs off will maintain it at ambient temperature. Waste water can be used.

Energy recovery systems are also worth exploring as explained in Chapters 24, 25, 26, 27, and 28.

In summary, energy conservation can and should be an important part of operating an industrial plant. When a strong, positive program is carried out, the use of energy can be reduced without adversely affecting the employee's environment or production schedules.

56

Saving air and energy in industrial plants

Ventilation overkill is not the only way to safely control industrial process contaminants and nuisance byproducts. Here is a discussion of alternative methods that may increase capture efficiency and save money.

GEORGE M. HAMA,
Associate Professor, Wayne State
University, Detroit, Michigan

Air conservation involves the control and efficient application of ventilation air so that a plant ventilation system handles only the necessary air quantity and does not exhaust or supply large, needless, and wasteful volumes. With this goal in mind, each industrial process, operation, and ventilation system should be re-examined to determine if modification, or reduction would increase ventilation efficiency without losing control of adverse contaminants or causing unsatisfactory hot and humid working conditions.

In the present energy crisis, plant engineering personnel should study and consider applicable air conservation methods. If fuel and power usage is curtailed further, they may be faced with the unhappy decision either to shut down hazardous operations that produce dangerous contaminants, or to continue to operate them with the risk of adverse exposure to workers unprotected by adequate ventilation. In the furor and panic of unexpected energy restrictions, plant management must never yield to the temptation of operating inadequately controlled processes at the expense of workers' health.

Conservation parameters

Air conservation involves two phases of energy requirements: 1) fuel for heating makeup air to replace air exhausted or discharged from a building, 2) electric power for fan operation and, in some cases, air cleaning and conditioning.

The necessity of replacing exhausted air with tempered makeup air is no longer questioned. Fuel requirements vary directly with outside air replacement rates and can be calculated from the following equation:

$$F = Q \times 1.08\Delta t / E$$

where

F = fuel requirements, Btuh
Q = air flow, cfm
Δt = temperature difference between the room discharge air and outside air, °F
E = efficiency of heating device, stated as a decimal.

Thus, fuel cost for heating 10,000 cfm of air is twice that for heating 5000 cfm. Obviously, fuel savings vary directly with quantity of makeup air supply.

Fan power costs do not always vary directly with air volume ventilation rates. Brake horsepower requirements can be calculated from the following equation:

$$bhp = (Q \times TP)/E_m$$

where

bhp = brake horsepower requirements for a fan

TP = total pressure that a fan is operating against, in. WG.

E_m = mechanical efficiency of a fan at the total pressure and volume flow rate it is operating at, stated as a decimal.

A fan may be required to operate at a total pressure as low as a fraction of an inch water gauge for general ventilation and up to 15 to 20 in. WG or more in a local exhaust system with an air cleaning device attached.

Pressure losses can be reduced in a local exhaust system with power saving by maintaining minimum conveying velocities in ducts and by using low loss fittings, long radius elbows, tapered hood entries, and small angle branches to all of the main entries.

Because a local exhaust system handles a comparatively smaller air volume than a general ventilation system, we cannot assume that it consumes less energy. This may or may not be true. A local exhaust system works against a higher pressure. For example, a local system handling 2000 cfm at 15 in. WG static pressure may require more power than a general ventilation fan exhausting 20,000 cfm.

Recommendations

It is the intent of this chapter to give practical recommendations that have proven effective in conserving air and energy. These are listed and detailed in the following.

• *Reduce exhaust air volumes in existing systems*—Local exhaust systems, which include paint spray booths, plating and pickling tank slot hoods, grinding and buffing hoods, and other process hoods, are often over-ventilated. Only sufficient air quantities

to prevent contaminated air from escaping are needed. A hood can be tested with smoke, injected with a smoke tube, and/or air sampling to see if capture will still be effective at a lower air flow rate (see Fig. 56-1). Air measurements should be made on a system governed by code requirements to determine if ventilation rates are greatly in excess of those specified. Acceptable measurements can be made by drilling a duct and traversing it with a pitot tube and inclined liquid manometer. Measurements of a hood face velocity made with a portable direct reading air meter are less satisfactory.

Some plants have been designed with ventilation rates based on good practice requirements, generally by a certain air flow rate in cubic feet per minute per square foot. In

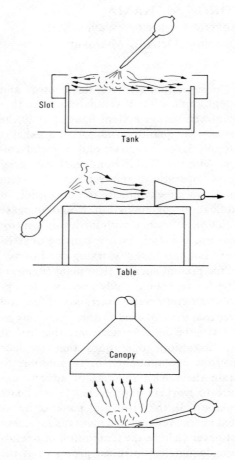

Fig. 56-1. Smoke tubes can be used to test capture efficiency of ventilation systems.

Table 56-1.   Ventilation requirements for gasoline and propane fueled lift trucks.

BASIC DESIGN VENTILATION RATES
- 5000 cfm per propane fueled lift truck
- 8000 cfm per gasoline fueled lift truck

CONDITIONS UNDER WHICH BASIC RATES APPLY
- A regular maintenance program incorporating final engine tuning through carbon monoxide analysis of exhaust gases must be provided. Carbon monoxide concentration of exhaust gases should be limited to 1 percent for propane fueled trucks, 2 percent for gasoline fueled trucks.
- Actual operating time of lift trucks must be 50 percent or less of total exposure time.
- A reasonably good distribution of air flow must be provided.
- The volume of space must amount to 150,000 cu ft per lift truck or more.
- Lift trucks must not have engines larger than 60 hp.

CORRECTIONS FOR CONDITIONS OTHER THAN THOSE ABOVE
- No regular maintenance program—multiply the basic design ventilation rate by three.
- Operating time greater than 50 percent—multiply the basic design ventilation rate by the actual operating time in percent divided by 50.
- Poor distribution of air flow—lift truck operation not recommended.
- Volume of space less than 150,000 cu ft per lift truck—multiply the basic design ventilation rate by a suitable factor based on the following: 1.5 times design rate for 75,000 cu ft; 2.0 times design rate for 30,000 cu ft. Lift truck operation in spaces of less than 25,000 cu ft is not recommended.
- Lift truck engine horsepower greater than 60—multiply the basic design ventilation rate by actual horsepower divided by 60.

plants with efficient local exhaust hoods, these rates may not be necessary to maintain acceptable conditions. If conditions appear satisfactory, air samples and temperature and humidity measurements should be made to confirm this. Then, the ventilation rate should be reduced to what might be considered adequate still and duplicate tests made. This can be best illustrated by an actual example.[1]

In an automotive engine assembly plant with 1,200,000 sq ft of floor area, a ventilation rate of 1,000,000 cfm was provided. Tests were made to measure concentrations of oil mist and total particulates. Under these conditions, tests showed 0.17 milligrams per cu meter for oil mist and 0.37 milligrams per cu meter for total particulates. All general ventilation was then turned off, leaving only local exhaust system ventilation and a corresponding necessary makeup air quantity. This reduced total air supply to 650,000 cfm. Air contaminant tests were repeated with the following results: oil mist, 0.18 milligrams per cu meter; total particulates, 0.34 milligrams per cu meter. Temperature and humidity levels remained satisfactory. A sizable saving, 350,000 cfm, was obtained without adverse exposure to workers.

- *Change of process or operation*—In some cases, certain operations or processes requiring large exhaust air volumes can be substituted by different processes, methods, or equipment needing less ventilation and, in some cases, none. Two examples are substituting electric lift trucks for gasoline or propane fueled ones and using airless spray painting rather than the conventional compressed air method.

Ventilation rates for gasoline and propane fueled lift trucks are listed in Table 56-1.[2] They become very high when a number of trucks are in use. An electric lift truck needs no ventilation for contaminants during operation, but a small quantity is necessary at the battery charging station. If five 60 hp gasoline fueled lift trucks are replaced by five electric ones, approximately 40,000 cfm will be conserved.

Certain paint spraying operations can be adequately performed with airless spray painting. Table 56-2 lists air flow rates for both airless and compressed air methods.[3] The rate can be cut 40 percent with airless paint spraying. From a medium size booth exhausting 3000 cfm, a savings of 1000 cfm can be realized.

- *Switch from general to local exhaust ventilation where possible*—In general, most

[1]Superscript numbers refer to references at end of chapter.

Table 56-2. Ventilation requirements for paint spray booths.

Painting facility	Ventilation requirements, cfm per sq ft of booth face	
	Compressed air spray	Airless spray
Small booth, less than 4 sq ft	200	125
Medium booth, operator outside	150	100
Large booth, operator inside	100	60

Table 56-3. Minimum air volumes and hose sizes for low volume, high velocity system.

Hand tool	Flow, cfm	Hose ID, in.
Disk sanders:		
3 in. diameter, 10,000 rpm	60	1
7 in. diameter, 5000 rpm	120	1¼
9 in. diameter, 2500 rpm	175	1½
Vibratory pad sander, 4 by 9 ft	100	1¼
Router:		
⅛ in., 50,000 rpm	80	1
¼ in., 30,000 rpm	90	1¼
1 in. cutter, 20,000 rpm	100	1¼
Belt sander, 3 in., 4000 fpm	70	1
Pneumatic chisel	60	1
Radial wheel grinder	70	1
Surface die grinder ¼ in.	60	1
Cone wheel grinder	90	1¼
Cup stone grinder, 4 in.	100	1¼
Cup type brush, 6 in.	150	1½
Radial wire brush, 6 in.	90	1¼
Hand wire brush, 3 by 7 ft	60	1
Rip out knife	175	1½
Rip out cast cutter	150	1½
Saber saw	120	1½
Swing frame grinder, 2 by 18 ft	380	2½
Saw abrasive, 3 in.	100	1½

Based on low production requirements with ideal pick-up hoods and branch static pressure at 6 in. Hg. (Courtesy, Reference 5.)

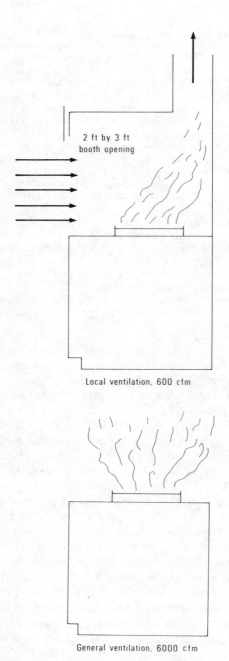

Local ventilation, 600 cfm

General ventilation, 6000 cfm

Fig. 56-2. General and local ventilation arrangements for a process using two pints trichlorethylene per hour. For general ventilation in this case, the K-factor is 4.

contaminant producing operations dependent on dilution by general ventilation can be controlled by local exhaust at a much lower rate. For example, a table cementing operation using two pints of trichlorethylene per hour requires about 6000 cfm general ventilation for satisfactory dilution. This is based on the quantity of air needed to dilute the vapor given off to a threshold limit value of 100 ppm; an average mixing and dilution factor of 4 to 1 is used.[4] A small booth with a 2 by 3 ft opening requires only 600 cfm. Local exhaust with lower air volumes will usually control contaminants to a lower exposure level than will general ventilation.

• *Use low volume, high velocity hoods on applicable operations*—A low volume, high velocity system is a unique application of exhaust air. It utilizes a small volume of air at a relatively high velocity to control dust from portable hand tools and machining operations.[5] A close fitting, custom made hood exhausts air directly at the point of generation. This method incorporates an efficient recirculation dust collector and will save relatively large air volumes over other methods.

Dust is conveyed at high velocities through a small diameter hose. Available sizes and exhaust volumes are listed in Table 56-3. A multi-stage centrifugal turbine capable of producing a static pressure of 130 in. wg provides

exhaust. Figure 56-3 shows a disk grinding operation before and after the system is turned on. Figure 56-4 is a diagram of a commercially available disk grinder.[5] Other applications include rock drilling, highly toxic machining operations and arc welding.

In most cases, a low volume, high velocity system uses only a fraction of the air flow required in other methods. For example, a booth enclosure used to ventilate an auto body disk sanding operation requires from 10,000 to 15,000 cfm, but two low volume, high velocity systems connected to two 9 in. diameter disk sanders use 175 cfm each, or 350 cfm in total.

At least two of these systems are currently available for arc welding.[6] Figure 56-5 shows one for shielded arc welding. The exhaust hood located just above the top of the inert

Fig. 56-3. Photos demonstrate the effects of a low volume, high velocity ventilation system. Photo on left shows disk grinding with no ventilation; photo on right is the same operation after the system has been turned on. (Courtesy Reference 5.)

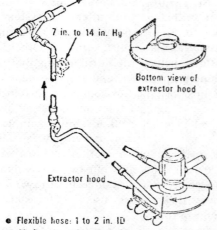

7 in. to 14 in. Hg

Bottom view of
extractor hood

Extractor hood

- Flexible hose: 1 to 2 in. ID
- Air flow capacity: 10 to 30 cfm per in. diameter
- Branch static pressure: 7 to 14 in. Hg
- Slot velocity: 24,000 to 39,000 fpm
- Extension hose: up to 8 ft long (may be extended
  up to 50 ft by using larger sizes between tool
  hose and tubing system)
- Sanding disk sizes: 2 to 9 in. diameter
- Peripheral speed: 4500 to 14,000 linear fpm

Fig. 56-4. Diagram and specifications for an extractor hood for a disk sander. (Courtesy Reference 5.)

gas discharge nozzle does not interfere with normal shielding. The system includes an exhauster that discharges approximately 60 cfm at 60 in. WG static pressure. Air sampling tests of a ventilated gun operation show 1.4 milli-

grams per cu meter, and for the same process without it, 7.2 milligrams per cu meter. Since conventional methods may require up to 2000 cfm per welder, a considerable saving can be achieved.

• *Use push-pull ventilation where it can be properly applied*—When an open exhaust hood located some distance away from its opening is used to capture contaminants, a blowing jet or slot supply air flow can assist capture at a significantly lower flow rate. An important design consideration is the proper selection of supply rate so that primary and secondary air entrainment is minimal. Also, the receiving exhaust hood must be large enough to handle the diverging supply jet.[7]

Figure 56-6 illustrates a push-pull system on a large sulfuric acid pickling tank. In this instance, room air is the source of supply, but it is possible to use outside air.

Three methods of acid tank ventilation, air flow rates, and text data are compared in Table 56-4.

Because an exhaust hood located some distance from its opening is ineffective, slot exhaust alone will not give good control on a large tank. Air sampling data were not available for this method. As shown in Table 56-4, the push-pull method was effective and economical.

On smaller tanks, push-pull methods with

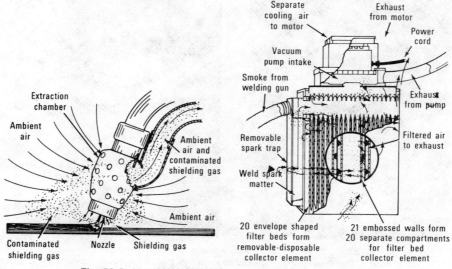

Fig. 56-5. Ventilating welding gun. (Courtesy Reference 6.)

Fig. 56-6. Push-pull ventilation on a sulfuric acid steel pickling tank.

Table 56-4. Comparison of three methods for acid tank ventilation.

Method	Exhaust rate, cfm	Air Concentration, H₂ SO₄ milligrams per cu m
General exhaust	200,000	2.0
Slot exhaust	66,000	NA*
Push-pull	35,000	0.18

*Not available

air jets as a supply or push have been used. With this arrangement, exhaust air flow is approximately 50 percent less than conventional exhaust hoods.

Only operations where no, or only momentary, obstruction exists can successfully use push-pull.

• *Use complete enclosure or a hood with a restricted opening*—Enclosed hoods or those with restricted openings have been employed to control highly toxic contaminants, such as radioactive material, beryllium, and pathogenic organisms. A glove box is an example of an enclosed hood (Fig. 56-7). The hood is totally enclosed, and operations are conducted with long sleeve gloves tightly fitted into port holes. Materials and equipment are inserted through an air lock arrangement. Very little exhaust ventilation is needed, and air flows as low as 20 cfm are sufficient.

The principle of complete enclosure has also been employed for highly toxic opera-

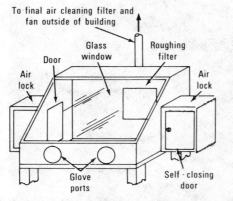

Fig. 56-7. Totally enclosed glove box hood. (Courtesy Reference 4.)

To final air cleaning filter and fan outside of building

Glass window

Door

Roughing filter

Air lock

Air lock

Glove ports

Self-closing door

tions, such as machining, milling, cutting, grinding, and drilling. These enclosures are made of transparent plastic materials to allow observation. Materials are set up, introduced, and removed through hinged or sliding doors that are closed while the operation is performed.

This same arrangement has saved large exhaust air volumes in arc welding operations on small parts.[8] Openings are just large enough to insert, manipulate, and remove materials. Only an amount of air sufficient to provide an indraft into the openings is provided.

• *Recirculate air from makeup air heaters*—It is possible with exchanger type makeup air units to use the same heater for recirculation heating a plant. This is simply done by closing the outside air intake and providing a return air duct to the heater from the building. Since temperatures may be very low in winter, fuel needs for a recirculating air heater will be considerably less than those for heating outside air. The change to recirculated air should be made only when an equivalent quantity of exhaust ventilation is turned off. This may occur when a plant is shut down completely or partially, such as off shifts, weekends, or at night.

Plant air should never be recirculated through the burner of a direct fired makeup air heater. An efficiently engineered direct fired burner, at best, gives off a small quantity of carbon monoxide, usually around 5 ppm or less. With each pass through a burner, more is added. Concentrations up to 400 ppm have been measured in air recirculated past a burner which is totally unacceptable.

It is possible, however, to set up an arrangement where a direct fired makeup air heater operates on 50 percent recirculated air without passing through the burner (Fig. 56-8).[4] Outside air supply is cut to 50 percent, and only this air passes the burner. The other 50 percent is room air. It recirculates to the burner's discharge side and mixes with the heated outside air. Louvers are installed in the burner profile to reduce this area so that the same profile velocity will be maintained with 50 percent of the previous air quantity. No room air recirculates by the burner.

• *Use air cleaning devices*—It has been the policy of official industrial health agencies not to recommend recirculation of exhaust air if the process contaminant might have definite effects on the health of workers. For some contaminants regarded as nuisances, recirculation has been accepted. For these, it has been specified that the effluent from a collector must not exceed 20 percent of the threshold limit of the contaminant.[9]

Because of economic factors and the energy crisis, it is important to examine the feasibility of recirculation more closely.

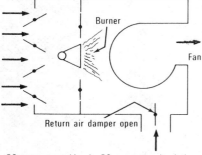

Outside air damper, 100 percent outside air
Profile damper 100 percent open

Burner

Fan

Return air damper closed

No recirculation, 100 percent outside air

Outside air damper, 50 percent outside air
Profile damper closed

Burner

Fan

Return air damper open

50 percent outside air, 50 percent recirculation

Fig. 56-8. Direct fire makeup air heater arrangement for safe recirculation without carbon monoxide buildup.

With carefully designed fail-safe protection, recirculation through an air cleaner can be employed. Factors to be considered are the efficiency of the cleaner and protection provisions in the event of equipment failure or lack of maintenance. More specifically, the following factors should be considered:[4]

1. The needed efficiency of the air cleaner depends on the threshold limit value of the material handled and the distribution and quantity of ventilating air present in the work area.

2. The system should be safeguarded by an efficient secondary air cleaning system or a reliable air monitoring device. The monitoring device shall be fail-safe with respect to failure of power supply or results of poor maintenance.

3. A warning signal shall be provided indicating the need of attention because of above limit concentrations detected by the monitor, or by increased resistance in the secondary air cleaning device as the result of failure of the primary air cleaner to collect the contaminant.

4. The monitor shall actuate a damper for bypassing recirculated air directly to outdoors through a discharge stack when failure of primary air cleaner is indicated.

5. The work area in which the ventilated process is located should be provided with general ventilation air so there is a continuous dilution of recirculated products.

6. Recirculated air from an air cleaning device shall be discharged away from workers' breathing zones and at a point where it mixes with general ventilation air for optimum dilution before it reaches a worker's breathing level.

Figure 56-9 illustrates a dust operation controlled by local exhaust. A fabric filter collector is backed by a HEPA filter. If excessive dust escapes the fabric filter, it is caught by the HEPA filter. Eventually, resistance builds to a point where a pressure switch triggers an alarm and changes the damper system to discharge to the atmosphere through a stack. The HEPA filter safeguards workers from contaminants es-

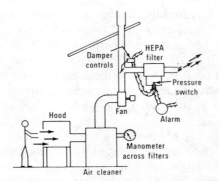

56-9. Arrangement for safe recirculation with an air cleaning device.

caping a faulty primary collector. However, because of its limited loading characteristics, it will plug, and resultant high resistance will reduce air flow. A bypass arrangement allows temporary operation without reduced flow or high contaminant exposure.

Discharge contaminant concentration should be based on factors governing the final desired satisfactory exposure level in a worker's breathing zone. This concentration may be considerably less than the threshold limit value. An example of this is arc welding fumes in which the threshold limit value is 5 milligrams per cu meter. At and around this level, there is a definite visible fume haze that concerns a welder. However, air sampling data indicate that in a ventilated welding operation with a concentration of 0.3 milligrams per cu meter or less there is very little visible fume and no worker complaints.

Some factors that need to be considered in determining a safe air cleaning equipment discharge concentration are:

1. Volume rate of general ventilation air available for mixing and diluting with discharge air from an air cleaner.

2. Location of general ventilation system discharge grilles relative to the discharge from an air cleaner—this is considered in estimating effectiveness of mixing and dilution factors.

3. Consideration as to what might be a satisfactory controlled breathing zone level. In most cases, this will be less than the threshold limit value. One way of considering it might be the worker's breathing zone level with satisfactory exhaust ventilation where the contaminated air is discharged to the atmosphere.

References

1. Robinson, K. E., General Motors Corp., private communication.
2. Hama, G. M. and Butler, K. E., "Ventilation Requirements for Lift Truck Operation," *Heating/ Piping/Air Conditioning*, January 1970, p. 123.
3. Hama, G. M. and Bonkowski, K. J., "Ventilation Requirements for Airless Spray Painting," *Heating/ Piping/Air Conditioning*, October 1970, p.80.
4. *Industrial Ventilation*, 13th ed., American Conference Governmental Industrial Hygienists, 1974.
5. Bulletin D-68, Hoffman Div., of Clarkson Ind., Inc., 103 Fourth Ave., New York, N.Y.
6. Form 188, Dover Corp., Bernard Div., Beecher, Ill.
7. Hama, G. M., "Jet Replaces Vertical Duct in Push-Pull Ventilation of Pickling Tanks," *Heating/ Piping/Air Conditioning*, January 1973, p. 121.
8. Barrett, J., Michigan Department of Health, private communication.
9. *Industrial Ventilation*, 12th ed., American Conference of Governmental Industrial Hygienists, 1973.

57

Nomograph estimates air infiltration due to stack effect

F. CAPLAN, PE,
Oakland, California

Air movement into and out of a building is caused by pressure differences created by fans, wind, and/or differences in air density. The latter factor is called the chimney or stack effect and, generally, is due to a higher temperature inside a building than outside.

Natural ventilation (wind and/or stack effect) is frequently utilized in industrial buildings, garages, farm buildings, residences, and certain types of public buildings.

The following formula is recommended for estimating air flow due to stack effect.*

$$Q = 9.4A \sqrt{h\Delta t}$$

where

Q = air flow, cfm
A = free (open) area of air inlets or outlets (assumed equal), sq ft
h = height from inlets to outlets, ft
Δt = average indoor air temperature minus outdoor air temperature, °F

---

*ASHRAE Handbook of Fundamentals, American Society of Heating, Refrigerating and Air-Conditioning Engineers, Inc., New York, N.Y., 1972, p. 344.

The constant of proportionality, 9.4, includes a value of 65 percent effectiveness of openings for average building construction. For very tight construction, the constant should be reduced to 7.2.

It is assumed in the formula that there is no significant resistance to air flow inside the building.

The nomograph quickly solves the equation (see Fig. 57-1). An example problem is presented below to illustrate its use.

Example

What is the estimated infiltration rate if the average air temperature inside a building is 112°F, and both the air inlets and outlets have free openings of 19 sq ft? There is a 70 ft height difference between these openings. The outdoor temperature is 70°F.

Solution: Align 70 on the h scale with 42 on the Δt scale, and mark the point of intersection on the pivot line. Extend a line from this intersection to 19 on the A scale, and read the answer as 9700 cfm at the point where this line intersects the Q scale.

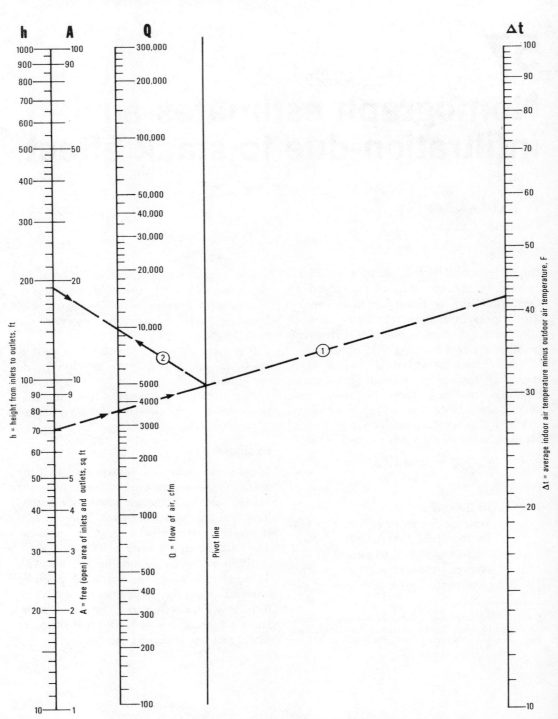

Fig. 57-1. Nomograph solves equation to estimate air infiltration due to stack effect. Broken lines indicate solution to example problem presented in the text.

58

Nomograph estimates air infiltration, heat removal

F. CAPLAN, PE,
Oakland, California

The accompanying nomograph provides a method for estimating air infiltration due to wind and the amount of heat removed by any quantity of heated air (see Fig. 58-1). It supplements the previous nomograph in Chapter 57 on air infiltration into buildings due to temperature differences (stack effect).

The following equation can be used to calculate the amount of infiltration caused by wind:*

$$Q = EAV \qquad (1)$$

where

Q = air flow, cfm
E = effectiveness of openings (0.50 to 0.60 for perpendicular winds and 0.25 to 0.35 for diagonal winds)
A = free (open) area of inlet openings, sq ft (do not consider more than two building sides)
V = average wind velocity, fpm

Actual flow rates depend on the location of inlets and outlets and therefore may be greater or less than the values given in the formula. Inlets should directly face the prevailing wind, and outlets should be placed at one of the following building locations: 1) the side directly opposite the prevailing wind, 2) the side adjacent to windward face where low pressure areas occur, 3)

*ASHRAE Handbook of Fundamentals, American Society of Heating, Refrigerating and Air-Conditioning Engineers, Inc., New York, N.Y., 1972, p. 344.

in a monitor on the side opposite the wind, and 4) roof ventilators or stacks.

Wind velocity data for locations in the United States and Canada may be obtained from Table I, Chapter 33, of the reference* for Equation 1.

The amount of heat removed by air heated to a desired temperature difference is:

$$H = Q \times C_p \times \rho \times 60 \times \Delta t \qquad (2)$$

where

H = heat removed, Btuh
C_p = specific heat of air at constant pressure, Btu per lb
ρ = density of air, lb per cu ft
Δt = temperature difference between indoor and outdoor air, °F

For standard air, ρ is equal to 0.075 and C_p is equal to 0.24. Equation 2 then becomes:

$$H = 1.08 \times Q \times \Delta t \qquad (2a)$$

The nomograph provides for a quick solution to Equations 1 and 2a.

Example 1

How much air will enter a building through 100 sq ft of openings if the average wind velocity is 2.5 mph, and the wind is blowing perpendicular to the walls?

Solution: Draw a line from 2.5 on the V scale to 0.55 on the E scale and read the answer as 121 cfm per sq ft or a total air flow of 12,100 cfm at the point of intersection on the Q/A scale.

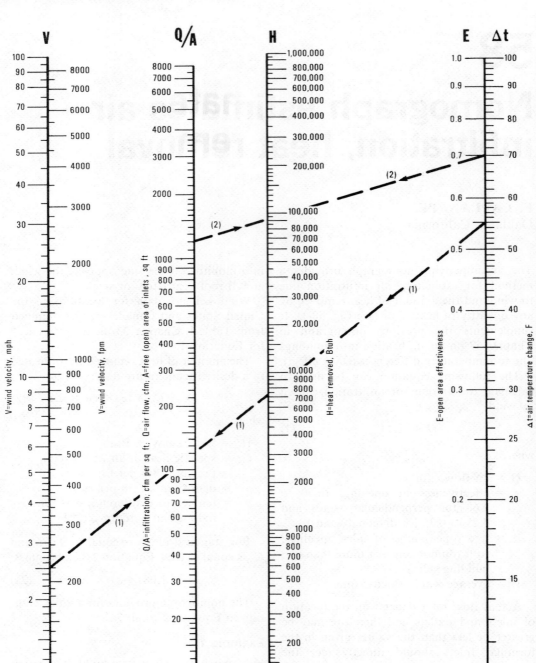

Fig. 58-1. Nomograph estimates air infiltration due to wind and the amount of heat removed by any quantity of heated air. Numbers on broken lines refer to example problems in text. Range of nomograph may be increased by simultaneously multiplying by 10, 100, etc., the Q/A and H scales.

Example 2

How much heat is removed by heating 12,100 cfm of air 70° F?

Solution: Extend a line from 12,100 on the Q/A scale to 70 on the Δt scale and read the answer as 910,000 Btuh at the point of intersection on the H scale. (Both the Q/A and H scales may be simultaneously multiplied by 10, 100, etc.)

59

Industrial ventilation techniques for energy conservation

This discussion of basic techniques also includes criteria for economic justification.

DOUGLAS L. GUNNELL, PE,
Senior Engineer, R. J. Reynolds
Tobacco Co., Winston-Salem,
North Carolina

The critical energy situation and the necessity for environmental control demand that an industrial ventilation design engineer be committed to what is possibly the greatest challenge he has ever encountered—the design of an efficient exhaust ventilation system coupled with a suitable makeup air system that will result in a safe and comfortable controlled environment while keeping energy consumption to a minimum.

The energy saving techniques covered in this chapter will help in meeting this challenge. Six major areas of possible energy saving potential will be explored. Some of the design parameters and suggested considerations are basic knowledge of the experienced design engineer; however, the reiteration of these is intentional to reinforce good design practice from an energy conservation vantage point.

Implementation of many of the suggested applications depends directly on economic

This chapter is based on a paper presented by Mr. Gunnell at the North Carolina State University Seventeenth Annual Industrial Ventilation Conference.

justification since each ventilation system is custom designed with different variables affecting the required results.

Exhaust system components are dealt with first.

Hood and duct design

The required exhaust air volume should be minimized by enclosing the operation as completely as possible with a hood. If enclosure is impractical, locate it as close as possible to the contaminant source. Flanges should be employed where feasible to reduce required air volume.

- *Velocity (transport and vapor systems)*—Since transport systems exhaust particulate matter, minimum duct transport velocity must be maintained to prevent the particulates from settling. The recommended transport velocity should be adhered to as closely as possible to minimize pressure loss.

Since vapor systems do not convey particulates, the recommended duct velocity is lower.

- *Elbows*—A centerline radius of two diameters is recommended. In a recently designed exhaust system, the application of elbows with centerline radii of two duct diameters versus one and one-half diameters was economically evaluated. System air volume was 12,800 scfm, and velocity ranged from 3400 to 4100 fpm. Duct size ranged from 4 to 24 in. in diameter, and the duct run with the greatest pressure loss contained seven 90 deg elbows and three 60 deg elbows. The employment of two centerline radius elbows in lieu of one and one-half yielded a static pressure reduction of 2 in. WG. The required fan horsepower was reduced approximately 8 hp. Based on an electric power cost of $0.02 per kWh and 2000 operating hours per year, the annual operating cost savings will be approximately $240. The cost of the motor and drive was reduced by $440.

• *Branch entries*—An entry angle of 30 deg into a transition, the minimum length of which is two diameters of the smaller duct, is recommended.

• *Enlargements and contractions*—A minimum of 5 in. in length for each inch change in diameter is considered good design.

• *Fan inlet*—Poor inlet conditions are one of the most common causes of reduced fan performance. A straight inlet of approximately six duct diameters in length is preferred, but if space limitations prohibit this and an elbow inlet is necessary, provide an inlet box with splitters or egg crate straighteners to minimize air spin or uneven loading of the fan wheel. Inlet boxes should not be used when the air is dust-laden.

• *Fan discharge*—To obtain a "no pressure loss" condition at the fan outlet, the discharge should be straight for a length of several duct diameters and of the same area as the fan outlet. A conical shaped weather cap is not recommended since it tends to drive effluents down and also results in a pressure loss. For weather protection, the double vertical discharge stackhead (with no pressure loss) is recommended. An evase discharge can be employed to reduce the air discharge velocity efficiently, and a portion of the available velocity pressure will be regained.

Duct system layout

The system should be designed with offsets and elbows at a minimum.

• *Fan location*—The fan should be located downstream from the collector to minimize erosion and abrasion of the fan. Consider locating the fan in the approximate center of the system. This will reflect-economy since less material is required for ductwork and fan pressure requirements are lowered. The location of the fan in this position may present balancing problems in a system designed to achieve the desired air flow without the use of dampers or blast gates.

• *Diversified system operation*—A thorough study of the requirements of an exhaust system for a particular application may yield a system with operational diversity and its associated savings. For example, consider the following hypothetical system. It has eight particulate exhaust points with varying air volumes. Only three of the eight are required to be operative at any given time. This system is feasible if it complies with the governing codes (regarding the contaminant emitted), and if its duct system provides the minimum required transport velocity in the branches and main with any three hoods operating in combination. The system would realize savings in the initial cost of equipment (fan, collector, and makeup air unit) and operating energy. An economic evaluation of the entire system would be necessary because the duct system would be more complex than that of a conventional exhaust system and could require an automated damper-interlock system with monitoring.

Exhaust system equipment

• *Fan*—Fan selection, a complex procedure, is often made without a proper evaluation. When determining the proper type or design of fan for an application, consider the following: the material, if any, that will be handled through the fan (fibrous or heavy dust loads); the temperature and corrosiveness of the gas to be conveyed; and the flexibility offered by belt driven units as opposed to direct drive.

Generally, the most efficient fan (that is, the size fan that will satisfy the capacity and pressure requirements with minimum power requirements) should be chosen.

Final selection should be based on a thorough economic evaluation of first cost, operating cost, maintenance costs, and projected system modifications.

• *Air cleaning devices*—Select the most economical collector based on pressure drop, first cost, and maintenance to obtain the degree of collection required. The degree of collection necessary generally depends on compliance with the applicable air pollution regulations; however, there are instances where the salvage value of a contaminant is appreciable, and the degree of collection must be determined by economic evaluation.

When collected material must be disposed, equipment must be carefully selected to eliminate the possible creation of a costly secondary disposal problem.

Makeup air system

A mechanical makeup air system is preferred and in most instances is necessary to replace the exhausted air volume and maintain environmental control. A mechanical system assures that the exhaust system will operate at design volume, and it will usually conserve fuel required to provide environmental control. In this chapter, only makeup air systems capable of cooling with untreated air and heating are considered, but some makeup air systems are required to heat, cool, humidify, and dehumidify.

• *Heat source*—The heat source is the primary concern in determining the type of makeup air equipment to apply. Electricity, gas, oil, hot water, and steam are heat sources. The selection of one must be based on economics and availability—both present and projected. This is a hazardous business since both conditions depend largely on the President's energy and economic proposals and how they are implemented by Congress.

Other factors can also affect heat source selection. For example, although North Carolina is more heavily dependent on coal than natural gas, the large consumption of natural gas by the state's various industries (over 50 percent) makes the state very vulnerable to shortages of natural gas supply.

• *Air handling equipment*—A complete economic evaluation, including energy conservation factors, should be the basis for determining what type of air handling equipment (indirect or direct fired) to use. Major considerations are: first cost; applicable heat sources; required support equipment and specialties, such as boiler, piping, traps, flues, etc.; efficiency; pressure drop; possible application limitations imposed by state and/or municiple regulations; required controls; and the required maintenance.

• *Air distribution system*—Ductwork should be sized and grilles selected in accordance with good design practice. The minimization of air distribution system resistance is a primary objective.

Heat recovery (exhaust air)

There is a great potential for saving energy in the area of heat recovery. In this chapter, the source of heat available for recovery is limited to exhaust air; however, a multitude of other sources exist that are beyond the scope of this chapter. For further details regarding heat recovery applications, see Chapters 24, 25, 26, 27, and 28.

• *Rotary heat wheel*—This device is a revolving cylinder packed with coarsely knitted mesh or a desiccant impregnated material (see Fig. 59-1). Heat transfer occurs as the wheel rotates between the exhaust and supply air streams. The air streams must be counterflow to achieve maximum efficiency, which ranges between 60 to 80 percent at equal mass flow.

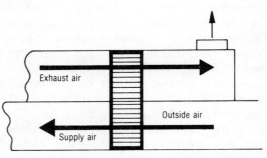

Fig. 59-1. Rotary regenerative air-to-air heat exchanger.

Parallel air flow reduces exchanger efficiency approximately 50 percent.

Cross contamination, which can occur, can be reduced by purging from the supply to the exhaust.

Economic justification is based on the following: some equipment that would normally be required can be omitted. The accrued energy savings can be compared to the cost of the recovery unit (including installation, additional ductwork required, additional horsepower to overcome pressure drop of the unit, and the cost of operating it).

• *Heat pipe*—The basic principles of the heat pipe were developed and patented some 30 years ago. In 1963, NASA applied it when faced with the need of a superconductor of heat (see Fig. 59-2).

The recovery unit has no moving parts. It appears to be a dehumidification coil with a partition separating the face into two sections. The unit is constructed of an array of extended surface tubes sealed at both ends. These tubes are the actual heat pipes. Each heat pipe consists of a tube, a wick, and a working fluid.

The heat pipe is a completely reversible isothermal device. Air streams must be counterflow for maximum efficiency. Typical efficiency varies from approximately 45 percent (4 row coil at 700 fpm face velocity) to approximately 70 percent (8 row coil at 300 fpm face velocity). Cross contamination is not a problem since the air streams are isolated from each other.

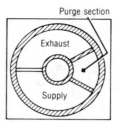
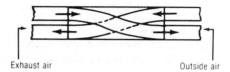

Fig. 59-3. Static heat exchanger.

Conditions similar to those for rotary equipment govern the economic feasibility of applying these units with the exception that no electric power is required to operate the unit.

• *Static heat exchanger*—The recovery unit has no moving parts. The unit resembles an open-ended steel box with a rectangular cross section that is compartmented into a multiplicity of narrow passages in a cellular format. Every other passage carries exhaust air alternating with those carrying supply air. The flow is counterflow, and efficiency varies with the number of passages. A reasonable attainable efficiency is 60 to 70 percent. Heat is transferred by conduction through the passage walls. Cross contamination is not a problem due to the heat exchanger design (see Fig. 59-3).

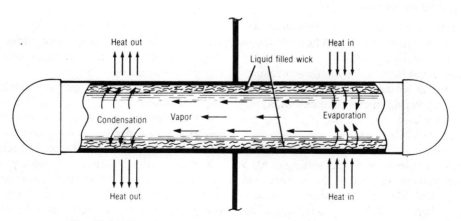

Fig. 59-2. Heat pipe, nonregenerative air-to-air heat exchanger.

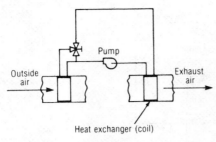

Fig. 59-4. Runaround system.

Criteria for economic justification are similar to those for the application of the heat pipe unit.

• *Runaround systems*—Unlike the three heat exchangers discussed previously, the runaround system can be applied to supply and exhaust ducts that are not located adjacent to each other. The two heat exchangers are connected by a piping system through which a solution of ethylene glycol and water is circulated by a pump. The system can be designed to obtain an efficiency of 60 to 70 percent. Cross contamination is not a problem since the two air streams do not come in contact with each other (see Fig. 59-4).

The inherent adaptability of the runaround system to the application of heat exchange between remotely located supply and exhaust systems (which cannot be practically located or relocated adjacent to each other) should weigh heavily in the justification evaluation, the elements of which are similar to those of the rotary heat wheel.

Testing and balancing

The importance of testing and balancing cannot be overstressed and should be included in the system specifications.

• *Exhaust system*—The air flow measurements should be made to ascertain that the system is operating in compliance with applicable codes and regulations regarding required contaminant control.

Test data will provide the necessary information required for proper setting of blast gates for systems designed by the balance *with* blast gate method or possible required modifications to ductwork and/or hoods for systems designed by the balance *without* blast gate

method. Test data will also often reveal that excessive air volume required to comply with the above established criterion is being exhausted. This will allow a reduction in fan speed that will result in energy savings.

• *Makeup air system*—Air flow measurements should be made to verify that minimum air volume is being delivered to satisfy the exhaust system, maintain environmental control, and provide required area pressure differential (relative to adjoining areas). Test data may indicate that the system is overdesigned to meet the requirements stated above, and a reduction in fan speed will yield energy savings.

Energy saving opportunities

An engineer has many opportunities to apply measures that will conserve energy; however, each measure requires careful economic evaluation for specific applications.

The following should be considered in establishing the operating parameters of a system.

• Heating and cooling level: Reduce the heating level and increase the cooling level when a building is occupied. Shut down the cooling system and further reduce the heating level when the building is not in use.

• Return air: Operate the makeup air unit on return air (if allowable) when the exhaust system is not operating.

• Steam tracing: Turn off steam tracing during mild weather.

The following should be considered in the formulation of a routine maintenance program.

• Fans: Check and maintain the following: bearings for overheating, belt drives for proper tension, fan wheels for accumulations and proper rotation.

• Motors: Maintain proper lubrication and cleanliness.

• Combustion: Perform combustion analysis and adjust burners. Incomplete combustion wastes significant amounts of fuel.

• Steam traps: Repair or replace faulty steam traps. Initial inspections commonly reveal that as high as 7 percent of the traps in a system are leaking. A reduction to 1 percent is reasonably attainable.

- Insulation: Repair faulty insulation.
- Louvers and dampers: Repair faulty louvers and dampers to provide free operation and positive shutoff.
- Filters: Clean or replace air filters regularly to minimize pressure drop.
- Heat exchangers: Keep heat exchanger surfaces clean for maximum heat transfer efficiency.
- Controls: Maintain controls, keeping them in proper adjustment.

A conscientious and imaginative engineer will constantly be disclosing, evaluating, and implementing innovations resulting in energy conservation. This should be accepted as another opportunity for an engineer to comply with his obligation to serve humanity by making the best use of the earth's precious wealth.

Bibliography

1. *Industrial Ventilation: A Manual of Recommended Practices*, 13th ed., American Conference of Governmental Industrial Hygienists, Lansing, Mich., 1974.

2. *Energy Conservation Program Guide for Industry and Commerce*, NBS Handbook, 115, Institute of Applied Technology, National Bureau of Standards, Washington, D.C.

SECTION VIII

Judging the results of an energy conserva-tion program

The results of any energy conservation program can only be judged by the savings that are achieved. The 7 chapters in this section describe several examples and success-ful application methods as follows:

- Case histories of an industrial operation and of an office building.
- Application of value engineering in a school complex.
- Efficient systems for after hours office cooling.
- The impact of ASHRAE Standard 90-75.
- Enthalpy control system principles.
- Chiller plant operations with series and parallel flow.
- Two-phase booster ejector system for air conditioning.

60

Judging the results by checking utility bills before and after implementing an energy conservation program is not enough to prove success

Performance can be evaluated only by establishing an energy inventory by energy use functions. Data from industrial operations and office buildings show how savings can be achieved.

CHARLES J. R. McCLURE, PE,
Consulting Engineer, Charles J. R. McClure & Associates, Inc., St. Louis, Missouri

The vast majority of data published on energy conservation deals with techniques and mechanics of saving energy. Ingenious methods of modifying adjustments, reducing heating and cooling loads, recovering energy normally dispelled to the atmosphere, and other application improvements are described. Often, these descriptions include predictions of energy savings that will be realized, and it is easy to compile a list of conservation measures that, if accumulated, are alleged to save more than 100 percent of the energy presently used.

As long as these application efforts have minor impacts on the quality of performance and have minor capital requirements, management decisions are not concerned. However, any program that may adversely affect the environmental conditions for which customers pay, that may vary the requirements established by the quality control or production management departments of an industrial firm, or that require substantial capital expenditures, will surely require a management program that defines goals, measures results, assigns responsibilities, and audits performance.

This is based on a paper presented by Mr. McClure at a special series of two-day seminars sponsored and developed by *Heating/Piping/Air Conditioning* on How to Save Energy in Existing Buildings.

In all but the simplest projects, the elementary procedure of analyzing fuel and electrical billings before and after implementation of an energy conservation program will be inadequate for determining the effects of various efforts. Energy use may change because of weather deviations from year to year, or changes in hours of occupancy or production programs, or improvements in performance quality, or from one modification rather than another introduced at the same time.

In implementing an energy conservation program it is not uncommon that energy consumption may not change as had been expected. One reason for this occurrence may be that system performance has greatly improved, and the occupants are experiencing comfort previously denied them because the designer's intent had not been realized or was lost over the years due to a variety of factors. Inadequate maintenance and deterioration of apparatus and controls are often found to have reduced performance levels. An increase in comfort may be achieved with an increased use of energy. This is a distinct possibility whenever a comprehensive conservation program is developed. In any case, however, a well conceived and implemented program should assure adequate maintenance of comfort levels at minimum energy consumption levels.

If management intends to monitor the performance of a conservation program, the groundwork for this must be laid during the development phase of the project. Organized effort is required to develop, implement, and monitor a comprehensive energy use control program.

Prepare an energy inventory

An energy inventory must be prepared early in the program developmental phase. This inventory must define both the various energy using functions and the amount of energy consumed by each. Since there are usually records of energy quantities used in the past, the inventory may be accomplished by ascribing a functional use to each kilowatt-hour and Btu that is purchased. This requires consultation with the operating staff, observation of the operation, inventory of energy using apparatus, and providing a lot of imagination to apportion the monthly kilowatt-hours from the client's previous electric and fuel bills—for one or more years—among the various uses that have been identified. A rational program is dependent on identifying the energy consuming functions and the historical level of energy use of these functions.

The different functions that must be identified include:
- *Electrical:* illumination, air handling, pumping, refrigeration, space heating and ventilating, and domestic hot water heating.
- *Fuel:* space heating and ventilating, domestic hot water heating, heat operated refrigeration, and process loads.

Determine electrical loads

Apportioning electrical uses is generally easier than allocating fuel use. Lighting kilowatt-hours per month can be determined by counting the number and size of lamps and ballasts, describing a pattern of use, and accumulating a monthly total hours of use discounted to reflect diversity.

The same method can be used with air handling equipment from motor nameplate data with verification of approximate percent of full load by measuring amperage. However, more accuracy is obtainable with a portable instrument designed to measure current and so indicate the actual power used by the motor. Multiplying the power by the hours of use per month results in an estimate of kilowatt-hours used for air handling. This same procedure can be used for pumping and other motor driven equipment loads.

Determining the consumption of refrigeration equipment may be more difficult. Process cooling tends to be a more uniform load than air conditioning, but it can also peak during hot weather. Air conditioning consumption is a complex function of occupancy, solar effects, weather, internal loads, and the interaction of the temperature control system.

Analyze fuel bills

When analyzing fuel bills, often the only apparent conclusion is that seasonal weather changes caused the monthly consumption to vary. Looking at the summer months, primarily July and August when fuel consumption normally is lowest and little or no space heating is required, it may be possible to let the fuel consumption for these two months represent the process loads and extrapolate the quantities to obtain an estimate of those loads other than space heating and ventilating over the remainder of the year. Degree-days may be used to apportion the remainder of the fuel used, which is assumed to be for space heating and ventilating. This may be sufficiently accurate for preliminary analysis but a word of caution is in order.

Many heating and air conditioning systems use reheat as a means of temperature control and use a considerable amount of heat energy in doing so. Therefore, a false readout of the amount of energy used by process functions can be obtained unless all system characteristics are taken into account in the analysis.

Additionally, certain types of process loads may follow a seasonal pattern. Ventilation rate may not be consistent throughout all weather incidents, and this can cause considerable deviation from extrapolated data.

It would be desirable to meter all fuel used by various types of equipment. Usually, however, all that is available is the consumption figures from a single gas meter or the records of gallons of oil and propane or tons of coal received. Sometimes more than one fuel is used; this may simplify matters if the fuels are used for distinctly separate functions. In some cases, it is possible to submeter specific functions—i.e., direct fired process equipment—if additional data are required.

Delineating energy used for heating is desirable so that the amount of energy used to offset conduction heat losses, heating ventilation air, and energy losses in the combustion process can be examined. Control burden is also an important factor to ascertain, such as the effects of reheat, since most central HVAC systems are to some degree wasteful of heat. Computer programs specifically designed to simulate the operation of the different HVAC systems are a valuable source of data for apportioning to the various functions the recorded fuel energy used. These programs can model an existing facility's energy use, including refrigeration and all electrical loads for statistically accurate weather incidence. Such programs can then be used with confidence to determine what savings are likely to be realized by various systems and/or operating modifications.

After the existing energy consumption is allocated among the various functions, it is possible to establish areas with the most promise of reducing consumption. For instance, if the illumination function is a major user of electricity, then major effort is warranted to revise the lighting system. Similarly, if electricity used by the air handling equipment is a major item, a determined analysis is indicated.

How much can be saved?

The tremendous variety of energy conserving application options can be reviewed. How much energy—and money—will possible alterations save? What will it cost to implement? These questions boil down to dollar economics. (On a long range basis, we must condition our thinking in terms of energy economics as well as dollar economics if we are to successfully deal with the energy problem.)

Once an energy reduction goal for each function is established, an implementation plan must be developed, complete with costs and scheduling. The final step is assigning the responsibility for seeing that the plan is carried out. At this time, it can be said that a comprehensive energy conservation program has been established.

Conservative managers will require a feedback system to be sure that results are commensurate with the investment in time and money put into the plan. Results can only be qualified and quantified by monitoring. The most common monitoring method is a performance report, such as a daily operating log. The operators should enter the time that machinery was started and stopped, performance quality, and energy meter readings.

Evaluate performance

Performance evaluation is an important corollary. Has quality—measured by comfort or by product criteria—increased or decreased? As previously mentioned, it is common to discover performance defects that the owner would like corrected. Correcting these defects often results in increased energy consumption by one or more functions; so the overall program may result in greater energy use. Identifying the probable energy use impact of improved performance should be performed before changes are made so as to not detract from other energy conservation efforts.

The reporting function, once a monitoring function is operating, is the feedback that enables management to determine the relative success of the program. Reporting must initiate at the operating level with the mechanic—the person who actually starts and stops equipment, makes system adjustments, and does the other tasks that will conserve energy.

Getting people involved and letting them see for themselves that their efforts are making a contribution to conserving Btu's and kilowatt-hours is the best way to assure the success of a program.

Good relationships with the operating personnel are important. Operators often feel that they have been getting along quite well without "help." Then, someone not responsible for running the machinery comes along with a lot of ideas on how to change the status quo in the name of energy conservation, often with the effect of operating much closer to the margin of poor performance. Some technique must be devised whereby the operator can translate his actions by means of a simple multiplication or addition into Btu's or kilowatt-hours saved, or he will lose track of them and interest in the program.

It is necessary to acknowledge the results of the operator's interest and efforts. He is the person who makes it happen. At the management level, the results of monitoring are tabulated. Action usually is taken on these tabulations only if the results are not up to expectations. Then, the operator hears from the boss if things are not going well.

Monitoring by metering

The ideal situation for monitoring results would be to use meters to determine the consumption of each energy use function. Me-

tered data provide a visible and fast means of showing the operator the results of his efforts as well as quickly showing management how well the program is working. Electric submeters are a useful tool. If the feeder that serves the lighting circuits for a department can be isolated and a meter installed, it can easily be determined whether the kilowatt consumption increases or decreases with respect to the hours the department is in use.

In many cases, it is not practical to do extensive submetering. However, elapsed time meters often can be used. These inexpensive devices provide a good means of determining the hours of operation of fans, pumps, refrigeration machines, etc.

If daily operating logs are kept, the elapsed time meter also verifies whether or not the motor was operated when it should have been.

Installing fuel meters is another direct approach where various energy functions are served by separate fuel runs.

As a last resort, Btu meters can be used. These devices are expensive to install and require a different order of magnitude in service and maintenance than the metering mentioned above. The effort involved in assuring reliability probably is not warranted for monitoring most energy conservation programs, with the possible exception of installations having skilled instrument service technicians available.

In summation, the use of metering devices for monitoring results should be considered whenever it is practical to do so.

Case histories show results

When reduction in energy use has been achieved, the results can be converted into dollars to determine whether the program was economically justified. Two examples will show what can be accomplished.

Industrial operations offer many opportunities for energy conservation. This example deals with a metal fabricating facility. Considerable gas welding and cutting operations are performed, and several large heat treating furnaces are in operation. Natural gas supplied through a single meter is the primary fuel, and propane is used as a standby fuel.

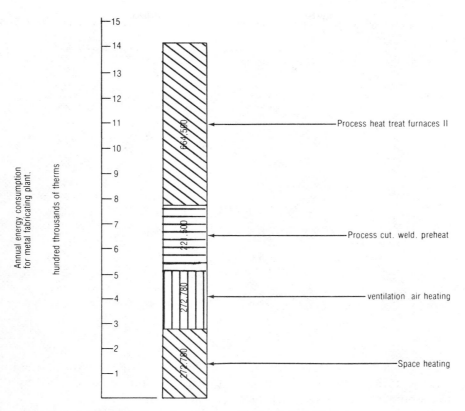

Fig. 60-1. Annual energy use for metal fabricating plant.

Examination of monthly fuel bills for two years showed a repetitive pattern. By relating monthly consumption with degree-days for the month, an estimate of the process load was obtained. The process load was assumed to be relatively constant year around with only minor variations, caused by changing the temperature of air for process combustion. The balance of the fuel represented the energy used for space heating and ventilating. The ventilation component of space heating load was derived from measurements of the makeup and exhaust air quantities and a determination of the daily operating cycle. These data were fed into a computerized energy analysis program, which utilized local weather data to determine the amount of fuel required to heat the ventilation air. This calculation indicated ventilation air heating approximately equaled the space heating load (Fig. 60-1).

Inspection of the fuel using devices inventory showed that the largest fuel use function was the heat treating furnaces, and discussions with the operators led to the conclusion that these four furnaces consumed nearly half of all fuel burned. It was obvious that this was the function to study first.

After studying the implications of the fuel uses inventory, the owner agreed to invest $3000 for meter installations so that the fuel use of one heat treating furnace could be measured.

The furnace selected measures: length, 85 ft; height, 16 ft, and width, 18 ft. The product is normally heated up to 1400°F; product loads during the data collection period ranged from a low of 64,000 lb to a high of 411,000 lb. The fuel consumption did not vary proportionally with the load (Table 60-1).

Calculations were made to estimate where the heat output of the fuel went: heat loss

Table 60-1. Natural gas usage in therms and heat distribution in percent for Heats No. 5-74-40 through 46.

Heat No. and product weight	Heat loss conduction	Heat storage walls	Functional use product	Flue loss	Input fuel total
No. 40 155,000 lb	115 3.9%	335 11.4%	215 7.3%	2265 77.4%	2930
No. 41 68,000 lb	84 4.2%	287 14.2%	81 4.0%	1568 77.6%	2020
No. 42 190,000 lb	99 4.0%	287 11.6%	226 9.1%	1868 75.3%	2480
No. 43 244,000 lb	96 4.5%	292 13.8%	295 13.9%	1437 67.8%	2120
No. 44 411,000 lb	131 4.4%	287 9.6%	488 16.3%	2084 69.7%	2990
No. 45 395,000 lb	115 4.2%	287 10.6%	469 17.3%	1839 67.9%	2710
No. 46 64,400 lb	98 4.5%	287 13.1%	77 3.5%	1728 78.9%	2190
Average 218,200 lb	Average 105 4.2%	Average 294 12%	Average 264 10.2%	Average 1827 73.5%	Average 2490

(conduction) through the furnace wall, heat storage in the furnace structure, heat absorbed by the product, and the amount of heat lost up the flue. These values are tabulated in Table 60-1.

Flue loss was 73.5 percent!

The calculations are relatively simple to perform. The heat loss depends on the area, overall heat transfer coefficient, and temperature difference. The heat storage is a function of the furnace mass, which basically is a brick structure; the specific heat of the furnace materials; and the average temperature that the structure assumes during operation. The heat consumed by the metal is obtained by multiplying the weight of product by its specific heat and the required temperature rise. The flue loss was determined as an average value of 73.5 percent! It was obtained by subtracting the sum of the above computations from the total fuel used. Only 10.2 percent, on an average, of the fuel actually heated the product. The balance went into heating the furnace structure or was vented up the flue to the atmosphere.

An obvious way to conserve heat is to add insulation. Yet, Table 60-1 shows that in this case, the amount of energy that would be affected by adding insulation is only 4.2 percent of the total. The main effort had to be directed at the 73.5 percent that was going up the stack.

Openings in the furnace were partially sealed so that it acted more as an oven rather than a once-through furnace, and the combustion air handling system was modified. With the fuel meter, it was possible to show the operator that if he closed up all unnecessary openings and turned off some of the burners when smaller loads were heated, fuel consumption could be reduced by 35 percent (Table 60-2).

Converting the dial readings on the gas meter to dollars quickly convinced the operator that he was performing a beneficial

---

The energy inventory may be accomplished by ascribing a functional use to each kilowatt-hour and Btu that is purchased.

This requires consultation with the operating staff, observation of the operation, inventory of energy using apparatus, and a lot of imagination.

Table 60-2. Fuel consumption for heat treat oven No. 1 before and after furnace and operating modifications.

Oven No. 1	Before, therms	After, therms	Reduction, %
Small loads, lb (less than 100,000)	2020*	1330	37
Medium loads, lb (100,000 to 200,000)	2190	1450	
	2450	1380	41
Large loads, lb (greater than 200,000)	2120	1670	
	2990	1650	
	2710		36
Estimated annual total	250,000	162,500	35
Cost (1974 prices), $	50,000	32,500	17,500
Duration of batch process, hr	30	24	20

*Each number represents a separate test run.

act that was worth the effort. The operator was sold on the viability of the concept of reducing fuel consumption by manipulating the furnace in a controlled manner. Product temperatures were still the governing criteria, and the operator now had an added complexity concerning the process of combustion.

Many similar conservation opportunities exist in industrial plants. Furnaces generally have been built over the past 25 years to utilize natural gas since it was considered abundant and in many areas available at low cost. When gas was available for $0.045 a therm, management was not overly concerned with the cost. Now that fuel costs are rising and curtailments are a common practice, owners are concerned with cutting fuel use for two reasons: to hold costs down and to assure adequate fuel supplies to continue profitable operation. The industrial firm cited in this example is experiencing reduction in natural gas supply; thus, it is vitally interested in getting the maximum benefit from available fuel.

Study office buildings

Office buildings require quite a different approach. This example involves a high quality corporate office building located in Minneapolis. It is a contemporary structure, circa 1970, and it encompasses approximately 500,000 sq ft of floor area. The skin is reflective glass with insulated metal panels. Heating and cooling energy is supplied from a central hot water-chilled water plant that serves many other buildings since this is part of a manufacturing and research complex. Heating and air conditioning distribution within the building is through a high velocity variable air volume reheat system. Initial lighting levels were 150 fc.

Operating costs were high and rising as electrical costs increased. Building management identified lighting level as a problem. The illumination requirements of the building were restudied by a management task group. As a result, the owner removed approximately 14,400 of the 40 watt fluorescent lamps and associated ballasts. Some lenses were changed, and some fixtures were relocated. The new illumination level is 80 fc.

The lighting system revisions made a significant reduction in electrical consumption. But the VAV reheat system was now grossly oversized because it had been designed on the basis of a lighting load nearly double that which now existed, and the system could no longer provide comfort conditions throughout the building.

The building has four fan systems. Originally, they were sized to supply 140,000 cfm each. Supply air temperatures were determined by the perimeter loads. Control in the interior spaces was obtained by reducing air volume. The VAV boxes have a limited range of adjustment, and the performance of the inlet damper control on the fans was not entirely satisfactory.

The overall result was poor system performance in that interior zones were too cold when exterior zones required low supply air temperatures.

This job was approached on a rigorous basis since complete as-built installation drawings, apparatus data, and testing and balancing reports from a qualified testing agency were available. As the building is existent, it was possible to determine the actual applicable insulation values and infiltration allowances.

The loads for the original project design were recalculated using the refinements of certain knowledge of the completed structure that permitted more precision to the calculation procedure. The building operation was then modeled by means of computer simulation for three modes of operation: 1) as the building had been originally designed, to check the simulation against the actual performance records; 2) as the building now operated with the reduced lighting load and original HVAC design; 3) as the building would operate if modifications were made to the HVAC distribution systems to determine the probable energy consumptions.

The original design required 1800 tons of refrigeration capacity (the calculation indicated that this was somewhat oversized). The amended design load was slightly under 1200 tons. This decrease is a result primarily of the 70 fc reduction in lighting load and the reduction in fan horsepower from 865 to slightly less than 300 hp as a result of the small air volume required with the reduced cooling load.

A significant reduction in electrical consumption in addition to what had been achieved by reducing lighting levels and fan horsepower was obtained by reducing fan operating hours. As an example of the kind of energy consumption impact that may result from a performance difficulty, this project had a history of considerable amount of difficulty with the V-belt fan drives. The belts had a tendency to fly off the pulleys during startup. This is not an unusual situation with high velocity, high pressure systems employing such drives. Starting loads are high, and all components—motor, drive, fan and motor starter—

must be properly selected and matched during the design stage if trouble is to be avoided.

In this installation, downtime due to this problem became a significant factor. The operator solved the situation by running the fans continuously. The belts stayed on and complaints of no cooling were silenced. This solution, however, added approximately $24,000 in electricity costs for the two years prior to modifications.

Reducing the air quantity to accommodate the smaller cooling load required extensive fan system changes. Air volumes from the fan rooms were reduced from 144,000 cfm to 85,000 cfm and less. Multiple fans had been operating in parallel in each fan system. The excess fans were disconnected, and a single fan in each room is now maintaining the required air flow with the new loads. Results of this program are tabulated in Table 60-3.

The owner has been extremely pleased with the results of these efforts. This experience initiated an ongoing corporate energy conservation program. A program of operations management was initiated to assure optimum efficiency, and procedures were introduced to monitor all energy uses in the various buildings at the headquarters, including the number of machines being used, hours of operation, ventilation rates, etc. The results have been dramatic. During the first six months of 1974, $317,000 in electricity and fuel was saved. This was doubly beneficial because the site requirements for fuel oil are high, and oil was hard to get at that time.

Table 60-3. Reduction in energy usage at 500,000 sq ft office building.

Energy use	1973	1974	Percent change
Heating, therms	405,350	412,550	+1.8
Cooling, ton-hr	4,217,890	1,606,975	−61.9
Electricity, kWh	13,940,279	8,574,996	−38.5

Revisions:
- Reduced lighting load from 150 to 80 fc.
- Reduced fan horsepower from 865 to 300 bhp.
- Reduced refrigeration load.
- Reduced hours of operation.

Program managed by an engineer

This conservation program is managed by a competent mechanical engineer. The cost for the first six months of the program is estimated to have been between $50,000 and $60,000 in administrative and engineering efforts to set up the program on a continuing basis.

This success story is an illustration of what can be done with a positive program when someone decides that it is important to save energy. The steps are:

- Establish a program designed to produce results.
- Show management the benefits of investing money to implement the program.
- Set up a program to monitor the performance of the personnel that must make it work, so that it does work.

In summary, the essence of an energy conservation program is to establish an energy inventory by energy use functions. Use this inventory as a guide in setting conservation goals by concentrating first on the functions that appear to have fat in them. Determine the changes that will achieve these reductions, and monitor the results. Remember, one paramount consideration is to make the man who does the work aware of the success of his efforts so that he retains his interest in an important task. Our national goal of energy independence is now understood to require a permanent change in our attitudes about the waste of energy. This learning process must be supported by direct and simple evidence of the results of energy conservation efforts.

61
Energy conservation through value engineering

Valuable construction dollars were saved and reinvested in energy saving systems.

HOWARD McKEW,
HVAC Project Captain, Anderson-Nichols and Co., Inc., Boston, Massachusetts

We know that these are difficult times when we hear words like cutback, energy conservation, inflation, etc., many times a day. But a recent project in Marlborough, Massachusetts, demonstrated that these problems can be handled. The City wanted to design and construct a new high school that would actually cut costs and conserve energy, while simultaneously provide what they desired for the community within the limits of the financial assistance offered by the state. These plans have become reality, and the City has a beautifully structured high school that boasts such facilities as vocational shops, open classrooms, standard classrooms, a field house, an assembly area, plus systems efficient enough

to allow expansion or further additions, as explained in this chapter.

The City selected Earl R. Flansburgh and Associates, Inc. (ERF&A), Cambridge, Massachusetts, to design the structure and its environments to house these facilities. The task of engineering the heating, ventilating, and air conditioning systems, as well as the sanitary, fire protection, electrical, and structural systems for this 300,000 sq ft (28 km²) complex within the financial limits allowed was achieved through value engineering by Anderson-Nichols and Co., Inc., Boston, Massachusetts. ANCO projected a HVAC construction cost of $1,890,000 during the schematic design phase to provide the re-

A typical module sheet metal layout.

quired 600 boiler hp (5.9 kW) heating capacity and 750 ton (2.6 MW) air conditioning capacity.

To make year around utilization of the facility possible, the HVAC system included a primary forced hot water system with a compensated perimeter hot water system, five shield fabricated 40,000 cfm (19 m³/s) HVAC units, 10 factory fabricated heating and ventilating units, two underground wood dust collector systems, an underground metal collector system, and underground carbon monoxide system, numerous fume exhaust and paint spray exhaust systems, energy recovery, and a completely automatic occupied/unoccupied time clock system. When all the HVAC filed subcontractors' bids were in, the awarded bid went to M. J. Flaherty, Newton, Massachusetts, for $1,891,000, just $1000 above the projected cost.

The architectural/engineering phase began with numerous proposed design schemes. These were evaluated from architectural, structural, mechanical, and electrical viewpoints to achieve the most economical building concept on a first cost/operating cost basis to meet present and future needs of the City of Marlborough.

From a heating, ventilating, and air conditioning standpoint, ANCO recommended a three story square structure with a first floor boiler plant and a centrally located mechanical penthouse. This was the foundation for the final scheme selected by ERF&A.

Hot water heating design

The heating system was designed as an outdoor, automatically controlled, cascading compensated hot water heating system. As the outdoor temperature drops below 65°F (18°C), a hot water circulating pump starts, and the water temperature rises to a maximum of 190°F (88°C). As the outdoor temperature reverses upward, the hot water heating temperature lowers until the outdoor temperature reaches 65°F again. At this point, the hot water circulating pump shuts down. The fluctuating heating media temperature in a cascading pattern with the outside temperature avoids maintaining maximum heating water temperature unnecessarily during mild seasonal heating periods, a common phenomenon of the New England area.

Use high velocity air system

The air conditioning and ventilation distribution system was designed as a high velocity, VAV sheet metal system with low velocity reduction boxes and lineal air diffusers incorporated in the light fixtures and a return air ceiling plenum. Air supply would vary depending on student population, student activity, or building heat gain during the air conditioning months. Use of this method rather than a constant volume air system reduced the total calculated air heated or cooled by approximately 17 percent.

The heating, ventilating, and air conditioning system chosen was just the beginning. ANCO strongly believed that with the proper building configuration and some serious thought from a contractor's point of view, dollars commonly wasted on engineering overdesign could be saved and spent on additional energy saving components and systems. These would not only pay for themselves over a short period of years in fuel expense savings but also conserve wasted energy.

Benefits of standardization

The first task toward saving valuable construction dollars was an examination of the engineering layout of the sheet metal system. Of the $1,890,000 estimated HVAC construction cost, $850,000 was for the sheet metal installation. ANCO decided that by the standardization of two high velocity, flat oval sheet metal duct main sizes and of two high velocity, round sheet metal duct main sizes, the air system layout would be simplified. In turn, this simplified the sheet metal contractor's construction bid take-off, inventory, fabrication, and installation. The arrangement at the low velocity side of each variable volume controller was also standardized (Fig. 61-1), which simplified another large area of the air system's fabrication and installation.

During construction, ANCO crosschecked

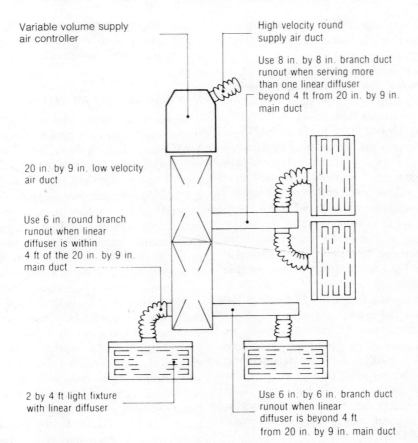

Variable volume supply air controller

High velocity round supply air duct

Use 8 in. by 8 in. branch duct runout when serving more than one linear diffuser beyond 4 ft from 20 in. by 9 in. main duct

20 in. by 9 in. low velocity air duct

Use 6 in. round branch runout when linear diffuser is within 4 ft of the 20 in. by 9 in. main duct

2 by 4 ft light fixture with linear diffuser

Use 6 in. by 6 in. branch duct runout when linear diffuser is beyond 4 ft from 20 in. by 9 in. main duct

Fig. 61-1. How the low velocity side of each variable volume controller was standardized.

its sheet metal decision with William Call, vice president of Worcester Air Conditioning Co. (WACC), the sheet metal subcontractor to M. J. Flaherty. The estimate was confirmed, and the simplified sheet metal system reduced WACC's bid from approximately $931,500 to $810,000. This represented an overall 15 percent savings in the contractor's bid by reducing the shop labor cost by 15 percent and the field labor cost by 10 percent. The standardized process also minimized space conflicts and added greatly to WACC's ability to plan and coordinate the ductwork installation with the other trades.

The second priority in reducing HVAC estimated cost was to simplify the air conditioning equipment layout. All air handling units were standardized. This simplification in turn reduced the pipe fitter and sheet metal construction bid takeoff and installation time.

The long-range benefit to the City of Marlborough was that standardization of much of the sophisticated HVAC system automatically made it less complicated for the operating personnel to understand, and this would lead to more effective preventive maintenance.

With ERF&A developing a basically square three story structure, the HVAC standardized equipment was laid out so that the field house with its support facilities, the assembly area, and the cafeteria with its kitchen, each had a separate air system. This left the administration area and a multitude of assorted learning centers and associated support facilities to be served by five central HVAC high velocity air handling units and associated return and exhaust air fans. These units were identical in physical size, arrangement, and capacities. The second and third floor were similar: each had two central air systems. That portion of the first floor not used for field house, assembly area, or cafeteria required a central air handling system identical to the four units serving the second and third floors.

In addition to standardizing the five central air handling systems, all the shop areas and field house heating and ventilating units were laid out in a similar manner. This arrangement benefited the owner by allowing the option of using the field house, shop areas, or half of any floor during a normally unoccupied period without operating the entire HVAC system for that floor (Fig. 61-2).

The estimate for standardizing equipment was confirmed with the HVAC contractor and his sheet metal subcontractor at the construction site, Gene Towers, M. J. Flaherty's project manager, indicated ANCO's method simplified the cost estimate take-off, making the range of filed sub-bids of all nine competing HVAC contractors within 15 percent of the awarded low bid. WACC followed up this confirmation by stating that the five 40,000 cfm field fabricated HVAC units casing cost alone was reduced by 12 percent, and field installation labor was reduced by 8 percent.

In addition to the initial financial construction savings, the simplification and standardization subsequently reduced the owner's maintenance stock inventory. Only parts to one type of central air handling unit must be kept on hand instead of parts for five different types that would have served the administration and learning areas. The same also applies to the shop and field house heating and ventilating units.

To further reduce cost, the bidding subcontractors had the option of using glass fiber ductwork instead of insulated sheet metal for certain systems. WACC indicated that its bid price was 1.7 percent higher to supply the glass fiber ductwork in lieu of sheet metal. However, this was more than offset by the reduction in the insulation subcontractor's price, since this ductwork did not require insulation. The difference brought a net reduction in cost of 10 percent to the total filed HVAC subcontractor's bid price.

Reinvesting the savings

As ANCO was reducing the estimated cost to furnish and install the proposed heating, ventilating, and air conditioning system, ways of reinvesting the estimated dollar savings in energy conserving equipment and systems were being considered.

ANCO's first energy priority was to provide energy recovery wheels. To comply with the Massachusetts state code for school facilities,

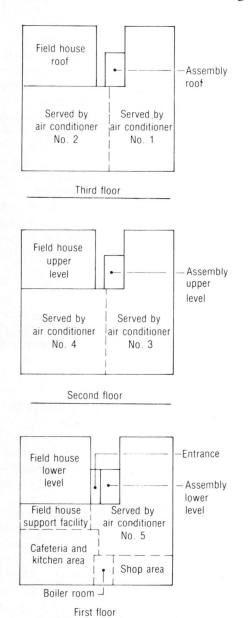

Fig. 61-2. Floor plan and layout for the high school.

restrictive outside air requirements must be met. By using rotating air-to-air recovery wheels, the school air handling system provided the outdoor ventilation while saving energy required to heat or cool this air. Five energy recovery wheels were incorporated into the HVAC total design. The combined cost to furnish, install, and automatically control them was approximately $40,000. Each wheel will save approximately 13 boiler hp (1.3 kW) during the heating season, which represents a fuel expense reduction of $13,300 per year.

The five wheels will save approximately 113 tons (4 kW) of air conditioning during the cooling season. This reduced tonnage converted to fuel cost savings is $5450. The combined energy reduction represents approximately $18,740 annual fuel cost saving. Based on the furnished and installed cost of $40,000, the energy recovery wheels will pay for themselves in just over two years.

In addition to the annual fuel savings, the City has the heating, ventilating, and air conditioning capacity to add substantially to the 300,000 sq ft facility at a future time.

Reduce kitchen exhaust

The second energy priority was to provide an energy saving kitchen exhaust hood and makeup air system. The common problem in most commercial and institutional kitchens is two-fold. First, the kitchen requires adequate smoke, heat, and fume exhaust, and secondly, as a result, an equal amount of makeup air is required. Both are usually met by introducing 100 percent outside air into the kitchen as makeup air. This air must be heated in the winter, but it is not cooled in the summer. It enters the kitchen as warm and humid as the outdoor air may be and, as expected, adds to the already uncomfortably warm kitchen.

ANCO considered this problem, and engineered the kitchen layout to utilize dining room and cafeteria conditioned air as allowed by the state code that would otherwise be exhausted and introduced this conditioned air into the kitchen through transfer grilles.

In addition, ANCO based its design on a kitchen hood that provides the remaining necessary makeup air from a 100 percent outside air heating and ventilating unit. This method provides kitchen makeup ventilation air through the hood's continuous perimeter lineal diffuser. An adjustable supply air thermostat, mounted at the hood, provides an automatic temperature controlled air curtain down-blow effect.

The kitchen hood air curtain system achieves two things simultaneously. First, it prevents cooking equipment heat from radiating out into the kitchen, and secondly, the heat from the exhaust hood enclosure warms the supply air leaving the hood's perimeter diffusers (Fig. 61-3). The adjustable thermostat allows the supply air to be set for colder than room temperature during the heating season, and thus provides spot cooling at the cooking area.

ANCO estimated this air curtain method, in addition to providing comfortable kitchen working conditions, will save approximately $1150 in fuel costs to operate the kitchen ventilation system during the regular school period. Use of the kitchen during the heating season for after-school functions will bring additional fuel savings to the City.

Use sequenced controls

With the bulk of the avoidable energy waste feasibly within the scope of the engineer's design, the sequencing of the automatic temperature control system became very important. ANCO specified a combination of commercial and also, seldom used in school facilities, industrial temperature controls.

This mixture provided the school with economical commercial control devices, where the engineers thought applicable, plus a durable, long life cycle, accurate industrial control devices on all major pieces of mechanical equipment.

A time clock control of all the air systems, the chilled water plant, and the hot water plant was provided to avoid unnecessary operation when the building is unoccupied and at the same time, reduce the school building heating temperature during the heating season when the building is vacant. Each system has a 6 hr manual bypass control to over-

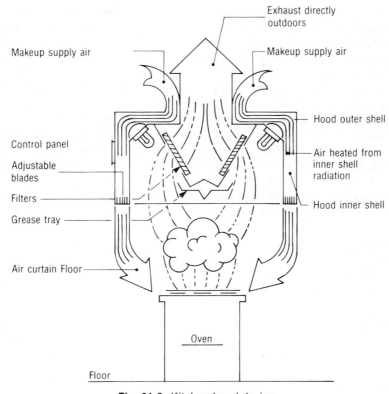

Fig. 61-3. Kitchen hood design.

ride its respective time clock for off-hour use of any of the 15 building zones. Each central air system was provided with a warm-up cycle to allow the respective unit to recirculate 100 percent return air within the building until the building is brought up to occupied room thermostat settings. At that time, the respective air systems would automatically allow its associated outside air damper to open to provide adequate ventilation.

Automatic sequencing of the two hot water boilers on lead-lag control was provided to avoid operating both boilers at one time unnecessarily. Cycling of the four 20 hp (15 kW) cooling tower fan motors to maintain condenser water temperature was provided to avoid operating all 80 hp at one time unnecessarily.

Economizer enthalpy control was also provided in each of the central air conditioning units to take advantage of the outdoor cool air energy to maintain air conditioning during the spring and fall seasons in lieu of operating the chilled water plant.

An indoor cooling tower sump was provided for condenser water availability to serve the chilled water system during unexpected winter warm spells. This sump also provides a reservoir of chemically treated water for year after year use, thereby saving the cost of discharging this water each year.

Summary of savings

Aside from the fact that the City has a very efficient, completely automatic controlled heating, ventilating, and air conditioning plant servicing its new school facility, the City will save annually a minimum $18,740 from the heat wheel operation and $1150 from the kitchen hood operation. The $19,890 total of these two items alone will offset the cost to hire a chief engineer to operate and maintain not only the HVAC systems but also the electrical, plumbing, fire protection, and general maintenance needs of the school. These financial savings estimated are approximate values, but the fuel saving annually will be more than these numbers, because the building is adaptable to a multitude of educational, recreational, and functional uses, and will naturally be used beyond the normal school hours as most schools are.

Knowing that the school facility is an efficient, energy conserving, useful structure, the City will be able to take advantage of utilizing this building in lieu of other local facilities for public functions. With the energy saved, another very important advantage of this facility is, as previously mentioned, the school heating and air conditioning plant has the additional capacity to serve up to 40,000

sq ft of future building expansion. An addition or additions could be for school and/or public use, such as more classrooms, vocational shop area, or athletic facilities. It is also more than adequate for an olympic size swimming pool with support facilities, a branch library, or a theatrical installation.

While the expansion potential is varied, the recovery equipment that has been provided for $1000 above the building construction cost guidelines will be operated daily to conserve energy and taxpayers' money. These are things not every city can claim about its high school.

62

Efficient systems for after hours office cooling

When tenants burn the midnight oil, there is no need to burn excess energy.

ALBERT I. CHO, PE,
Director of Mechanical Engineering,
Skidmore, Owings & Merrill,

ASHRAF ALI,
Project Mechanical Engineer, Skidmore,
Owings & Merrill, Chicago, Illinois

Many high rise office buildings utilize large fan and heating and cooling systems as designed for optimum building operation. Each fan system serves several floors or a section of several floors. Generally, these mechanical systems are shut off weekends and in the evening when the building is not in use. However, quite often, some tenants require after hours cooling. This is usually difficult and sometimes expensive to provide, especially if the particular tenant occupies a small area in the bulding. Recently, an analysis was made for various means of providing after hours cooling. This chapter summarizes some efficient methods for providing this service.

• *Operating the main air conditioning system*—For short, infrequent after hours use of office areas, many tenants utilize the building's main air conditioning system. Building management usually has an arrangement with the tenants for this service; some charge an additional $30 to 60 per hour. This is advantageous if the hours of cooling required are short and infrequent,

but, clearly, the operation of a large system to serve only a few hundred square feet of tenant space is costly and wasteful since to do so requires the operation of a large compressor, pumps, and fans.

• *Installation of independent air system*— An independent air system can be installed in the main mechanical room to serve tenants after hours (Fig. 62-1). The system would utilize chilled water from the central refrigeration system. Such a system is more adaptable to tenants with large areas and who require long hours of use. The longer periods of use justify the cost of the installation.

• *Free standing, water cooled air conditioning package units installed in tenant areas*—This is probably the most frequently applied method of providing after hours cooling for particular tenants. As shown in Fig. 62-2, a local package type air conditioning unit from 3 to 20 tons (10,550.4 to 70,336 W) can be installed in a small mechanical room in the tenant area or in the office space. The unit utilizes city water for

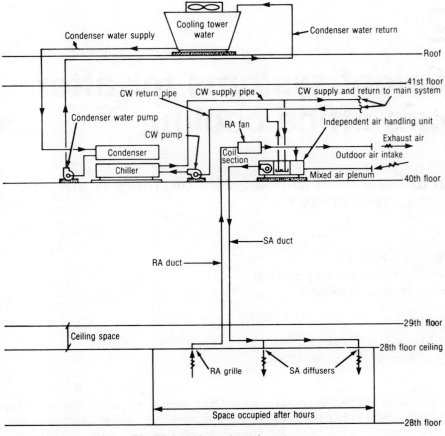

Fig. 62-1. Independent air system.

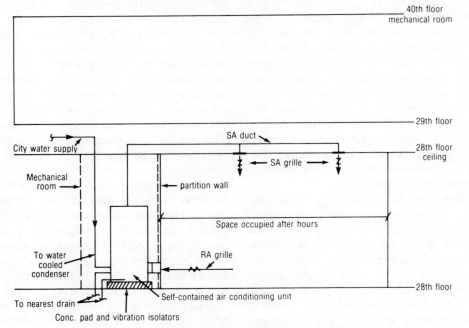

Fig. 62-2. Free standing air conditioning packaged unit.

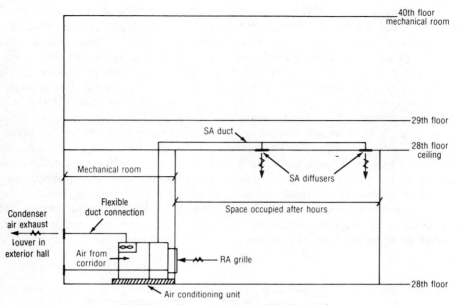

Fig. 62-3. Air cooled air conditioning unit.

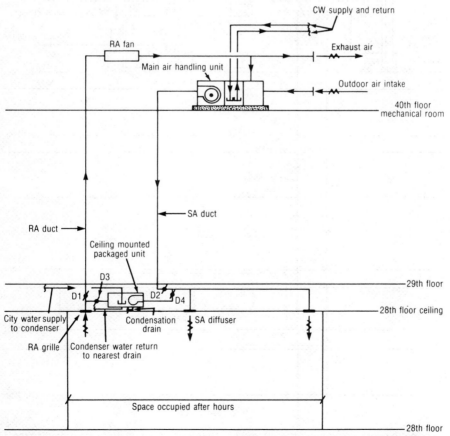

Fig. 62-4. Ceiling mounted packaged unit with duct bypass. Dampers D1 and D2 are normally open and Dampers D3 and D4 are normally closed for normal building operation. The damper positions are reversed for after hours operation.

condenser cooling and is served by electricity from the tenant circuit. The unit is self-contained except for ductwork and water piping. Two problems associated with this arrangement are the installation of a water drain line and the additional ductwork if required. For a relatively small area, the unit can be furnished with a supply and a return air grille. A drain pipe from the unit can be routed in the ceiling of the floor below or removed by a small condensate pump.

• *Air cooled air conditioning unit*—This type of system is limited to small applications—usually around 1 ton (3516.8 W) capacity. It consists of an air cooled air conditioning unit drawing air from the occupied room or corridor at about 80°F (26.7°C) to cool its condenser. The hot discharge air is normally ducted to the outdoors through a flexible connection as shown in Fig. 62-3. Condensate is usually drained to the condenser and evaporated. For a larger application, such as 2 to 3 tons (7033.6 to 10,550.4 W), the unit is installed in a separate mechanical room with condensate drains.

• *Ceiling mounted packaged units with duct bypass*—Ceiling mounted packaged air conditioning units can provide air conditioning for tenants who require 1 to 5 tons (3516.8 to 17,584 W) of cooling. The system consists of city water cooled, horizontal packaged units connected in a T-

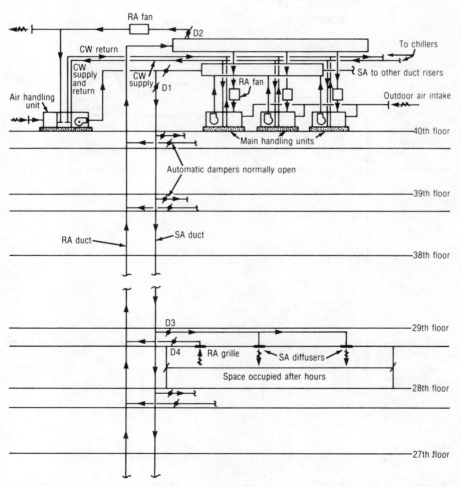

Fig. 62-5. Loop ductwork system. In order to provide after hours cooling in the occupied space, Dampers D1, D2, D3 and D4 must be fully open and all other automatic dampers fully closed.

Table 62-1. Each system compared on the basis of cost.

System	Power demand KW	Annual after hours operating cost, 20 hr per week when peak billing demand is set:		Initial cost, $	Amortized annual first and operating cost (10 yr equipment life expectancy, 10 percent interest)	
		During regular hours incremental cooling cost is 3c/KWH[1]	During regular hours incremental cooling cost is 3.7c/KWH[1]		Demand set, daytime	Demand set, after hours
Main air conditioning system (10,000 cfm fan system)	500	10,400	—	400,000	10,400.00[4]	—
Independent air system (3 tons[2])	33	686.40	—	12,000[3]	2,639.40	—
Free standing water cooled air conditioning packaged units in tenant area (3 tons)[6]	4	124.8 +60.00[5] 184.80	153.92 +60.00[5] 213.92	6,000	1,166.30	1,190.42
Air cooled air conditioning unit (3 tons)[6]	6	187.20	230.88	6,000	1,163.70	1,207.38
Ceiling mounted packaged units (3 tons)[6]	4	124.8 +60.00[5] 184.80	153.92 +60.00[5] 213.92	5,000	998.55	1,027.67
Utilization of loop ductwork system (3 tons[2])	33	686.40	—	7,000[3]	1,825.65	—

[1] 2c per KWH for central systems: main air conditioning system, independent air system, and loop ductwork system.
[2] These systems are generally larger than 3 tons, but for comparison purposes all the units are given the same capacity.
[3] Cost does not include refrigeration machine or pumps.
[4] Does not include amortization of central HVAC system as it will be amortized regardless of whether or not it is used after hours.
[5] Water cost based on 50c per 1000 gal.
[6] Electric utility costs are based on individual tenant metering for the free standing water cooled units in tenant areas, air cooled air conditioning unit, and ceiling mounted packaged unit systems.

assembly to the regular daytime operating ductwork as shown in Fig. 62-4. These systems require no outside windows, and prevent wastage of cooling power due to an automatic thermostat. Pneumatically operated dampers can be provided to isolate sections of the ductwork for after hours operation of the package unit. No additional architectural modifications are necessary. The water supply and drain lines can be routed in tenant areas under construction. The package unit is quiet, efficient in energy consumption, and relatively low in cost.

• *Loop ductwork system*—For high rise buildings anticipating more than the usual amount of after hours cooling, a loop system can be planned. A loop system consists of one or several large systems and one small

air conditioning system that feeds into a loop header, which in turn feeds into other risers of the supply and return air system (see Fig. 62-5). Automatic control dampers are provided at each floor take-off. During the after hours operation, all unused portions of the building would be isolated. A small system can be kept in operation to serve the occupied areas and thus decrease operating costs.

Since energy conservation and life cycle cost effectiveness are essential in choosing the type of air conditioning system for after hours cooling, each individual case must be analyzed. A comparative cost estimate is given in Table 62-1. In general, ceiling mounted package units with duct bypass are found to be an economical approach.

63

The impact of ASHRAE Standard 90-75 on high rise office building energy and economics

Three designs are imposed on the same basic building plan to demonstrate the feasibility of basing a design on Standard 90-75.

N. R. PATTERSON, PE,
Manager, Applications Engineering,
The Trane Co., La Crosse, Wisconsin,

JAMES B. ALWIN, PE,
Sales Engineer, The Trane Co.,
Westchester, Illinois

This chapter deals with a very timely subject in the area of energy conservation— the effects of ASHRAE Standard 90-75, Energy Conservation in New Building Design, which is an official, approved ASHRAE standard. ASHRAE Standard 90-75 will be a fact of life for designers of buildings and systems in the years to come.

The energy consumptions, HVAC system costs, and owning and operating costs for three designs that revolve around the same basic building plan are analyzed in this chapter. By comparison of the three, the reader can examine the possibilities for saving energy by following Standard 90-75.

The three designs may be briefly described as follows:

• *Design No. 1* is based on practices commonly followed in the Chicago area in the last few years. It is a typical design *prevalent before Standard 90-75.*

Two alternate designs, which are both based on Standard 90-75, demonstrate what can be accomplished by following the standard. These are:

• *Design No. 2* conforms to the basic parameters outlined in Sections 4 through 9 of the standard. These sections are a reference for a designer who wishes to depart from any of the standard's basic criteria. Design No. 2 is based on the *criteria established in Sections 4 through 9.*

• *Design No. 3* incorporates the *design freedom permitted under Section 10* of the standard. Section 10 allows a building and system designer to "do his own thing." However, he must show that his method does not consume more energy than if it had been designed based on the prescribed method in Sections 4 through 9. This requirement indicates that a designer who wishes to take advantage of this freedom must perform a comparative energy analysis to prove that his design conforms to the requirements of Section 10.

It must be emphasized that the changes shown in Design No. 3 are not the only possibilities, nor should they be construed as final optimizations. They are simply examples of what may be done with a design based on Section 10. Also, the energy consumptions and costs—energy, investment, and life-cycle—are based on conditions prevailing in the Chicago area. Factors such as local codes, regional design practices, and climate will have a definite influence on the impact that Standard 90-75 will have on energy conservation. Nevertheless, from the results of this particular analysis, it can be seen that the impact will be considerable in many instances.

The basic building plan

Figure 63-1 is an artist's rendering of the building the three designs revolve around. It is a 20 story building with 12 ft floor-to-floor height. The gross floor area is 380,920 sq ft, and the major orientations are north and south. The net air conditioning area is 325,000 sq ft. Overall dimensions are 107 by 178 ft. A 2796 sq ft section in the center core of the building is not air conditioned.

In preparing the input to perform a computerized energy analysis, the building was divided into 10 zones. These are thermal zones that react thermally with time in a similar manner. They are not individual thermostatically controlled zones.

By making this division, the total building can be served by 10 thermal zones—two of which are interior zones and eight of which are perimeter zones with 15 ft setback. There are two interior zones because the top floor of the building would have roof affected loads, while interior floors 1 through 19 would not; therefore, these floors must be served by separate interior zone inputs.

Figure 63-2 shows the general area criteria that must be input to the computer program to establish the various zone loads. Zones 2, 7, 8, 9, and 10 have a percent roof square footage indicated. These are all top floor zones that have roof affected loads.

The computer input also included the orientation of the various zones and their gross wall square footage. The last column in Fig. 63-2 is blank because the amount of glass used is varied in each of the three designs. These data will be supplied later when the criteria for each design are examined.

In addition to the building skin areas for

How ASHRAE Standard 90-75 was developed

ASHRAE Standard 90-75 was developed by ASHRAE at the request of the National Conference of States Building Codes and Standards (NCSBCS). The National Conference of States requested the National Bureau of Standards to prepare a document giving the design and evaluation criteria for energy conservation in new buildings. A draft document dated February 27, 1974 was then delivered to ASHRAE. Earlier ASHRAE had agreed to take over the project since it was felt that the Society was the proper vehicle to complete the work done by NBS and sponsor the document as a consensus standard reflecting the thinking of the total industry. Experts from various disciplines were called together by ASHRAE into committees to write the various sections of this standard. The standard underwent two national reviews for consensus comments, and on August 8, 1975, it was accepted as an official ASHRAE Standard.

To obtain a copy of Standard 90-75

A copy of ASHRAE Standard 90-75, Energy Conservation in New Building Design, can be obtained by writing to: American Society of Heating, Refrigerating and Air-Conditioning Engineers, 345 E. 47th St., New York, NY 10017. Price: $10 for members and $20 for nonmembers. Add 50 cents for postage.

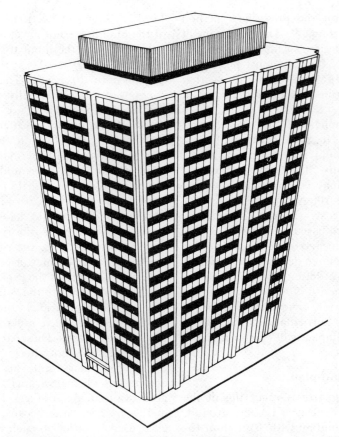

Fig. 63-1. An artist's rendering of the basic building used for the comparison of the three designs.

each zone, the following must also be input: the occupancy density in square foot per person; the miscellaneous loads that would exist in each zone; and the utilization of the building, which includes hourly scheduling of lights, people, base utilities, and miscellaneous internal loads. In this analysis, elevators and parking lot lighting are included in the base utilities. These, of course, do not contribute to the air conditioning load, but they do consume energy and would affect the energy demand rates as well as total energy cost to the building.

All of these parameters remained constant throughout the comparative energy analysis of the three alternate designs.

The building skin construction and the HVAC system *do not* remain constant among the three designs. The three designs are compared in detail in the tables. Table

63-1 summarizes architectural aspects, and design data for each of the three systems are shown in Tables 63-2 and 63-3.

Design details for three designs

Design No. 1, the pre-90-75 design, has a perimeter induction and interior terminal reheat system. It is typical of the type of system used in past years for high glass content buildings in the Chicago area. Such a system is excellent from the standpoint of ability to control humidity and maintain space temperature. However, as will be shown, it can use excessive amounts of energy.

Several economic inputs must be made that include such items as mortgage life, cost of money, maintenance cost, insurance costs, property taxes, depreciation, and other economic factors. In addition,

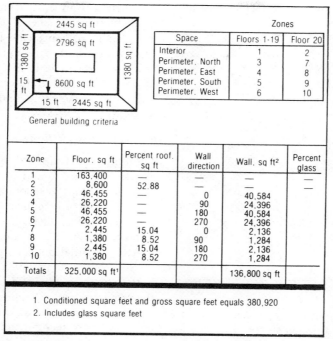

Fig. 63-2. General criteria for the basic building plan.

The table portion of the figure reads:

	Zones		
Space	Floors 1-19	Floor 20	
Interior	1	2	
Perimeter, North	3	7	
Perimeter, East	4	8	
Perimeter, South	5	9	
Perimeter, West	6	10	

Zone	Floor, sq ft	Percent roof, sq ft	Wall direction	Wall, sq ft²	Percent glass
1	163,400	—	—	—	—
2	8,600	52.88	—	—	—
3	46,455	—	0	40,584	
4	26,220	—	90	24,396	
5	46,455	—	180	40,584	
6	26,220	—	270	24,396	
7	2,445	15.04	0	2,136	
8	1,380	8.52	90	1,284	
9	2,445	15.04	180	2,136	
10	1,380	8.52	270	1,284	
Totals	325,000 sq ft¹			136,800 sq ft	

General building criteria

1 Conditioned square feet and gross square feet equals 380,920
2. Includes glass square feet

Table 63-1. Building skin construction.

Design No. 1:
Pre ASHRAE Standard 90-75

Roof: ⅜ in. builtup roofing; 2 in. concrete slab; corrugated metal; metal lath; and ¾ in. plaster; U value is 0.10.

Walls: spandrel; metal or glass skin; 1 in. fiber insulation board; and inside finish; U value is 0.25.

Glass: Single plate bronze; U value is 1.0; shading coefficient is 0.52; glass area accounts for 80 percent of total wall area.

Design No. 2:
Based on Sections 4-9 of ASHRAE Standard 90-75

Roof: The roof is the same as in Design No. 1, except insulation is increased from 2 to 4 in. The U value is 0.68.

Walls: 1 in. expanded polyurethane insulation is substituted for the 1 in. fiber board used in Design No. 1. The U value is 0.15.

Glass: double glazing with air space; bronze coating on outside glass; clear glass inside; U value is 0.54; shading coefficient is 0.37; glass area equals 44 percent of total wall area.

Design No. 3:
Based on Section 10 of ASHRAE Standard 90-75

All factors are the same as in Design No. 2, except the glass area equals 54 percent of the total wall area rather than 44 percent.

Table 63-2. Basic HVAC load data for each of the three designs.

	Standard 90-75 designs		
Factor	Design No. 1: Pre 90-75	Design No. 2: Sections 4-9	Design No. 3: Section 10
Roof	0.10	0.068	0.068
Curtain wall	0.25	0.15	0.15
Glazing:			
U value	1.0	0.54	0.54
Shading coefficient	0.52	0.37	0.37
% glass area	80	44	54
Overall wall U value	0.85	0.32¹	0.36
Outdoor air, cfm per sq ft	0.20²	0.075	0.075
Outdoor design conditions, F			
summer	93DB, 77WB³	90 DB, 75 WB⁴	90 DB, 75 WB⁴
winter	−4 DB	zero DB	zero DB
Indoor design conditions, F			
summer	75 DB, 50% RH	78 DB, 60% RH	78 DB, 60% RH
winter	75 DB	72 DB	72 DB
Lighting, watts per sq ft	5.0⁵	2.5⁶	2.5⁶
Occupancy, sq ft per person	100	100	100
Design cooling load, tons	1585	707.3	725.6
Design heat loss, Btuh	14,886,000	5,227,500	5,611,700

1. Combined U value of glass and curtain wall. For a 6155 degree-day heating season, this value must be ≤0.324 per Section 4 of Standard 90-75.
2. Ventilation code, City of Chicago.
3. 1 percent occurrence, summer; 99 percent occurrence, winter.
4. 2½ percent occurrence, summer; 97½ percent occurrence, winter.
5. General illumination.
6. Reduced general illumination plus task lighting.

input to the computer analysis must include the heating, ventilating, and air conditioning system first cost, on either a dollar per ton or dollar per square foot of conditioned space basis. Design No. 1 has a dollar per square foot input for the heating, ventilat-

Table 63-3. Mechanical system data for each of the three designs.

Factor	Design No. 1: Pre 90-75	Standard 90-75 designs Design No. 2: Sections 4-9	Design No. 3: Section 10
Heating fuel	gas	gas	gas
Cooling energy	electricity	electricity	electricity
Boilers:			
Number	2	2	2
Standard chillers:			
Number	2	2	—
Heat recovery chillers:			
Number	—	—	2
HVAC distribution:			
perimeter	2 pipe induction	hot water radiation	2 pipe fan-coil
Fan static, in. WG:			
supply	8	—	—
return	1	—	—
Interior	terminal reheat	VAV	VAV
Fan static, in. WG:			
supply	6	6.5	6.25
return	1	1.0	1.0

ing, and air conditioning system first cost of $7.70 per conditioned sq ft.

Curtain wall cost is included in the input since the effects of various curtain wall constructions are examined in all three designs.

As curtain wall costs change, the penalty or deduction, whichever the case may be, is added to or subtracted from the heating, ventilating, and air conditioning system first cost when the total economic analysis of the effects of architectural changes on the buildings' owning and operating costs is made.

Designs show effects of 90-75

The effects of Standard 90-75 are reflected by Designs No. 2 and 3. Section 9 of the standard strongly recommends the use of task lighting to reduce energy consumption. Therefore, task lighting is employed in both Designs No. 2 and 3.

Also, ventilation air quantities are reduced from 0.20 to 0.075 cfm per sq ft in both Standard 90-75 designs. This meets the requirements of Section 5.

Since Section 5, which deals with systems, strongly recommends the use of variable air volume, a high velocity shutoff variable air volume system with enthalpy actuated economizer in the interior of a

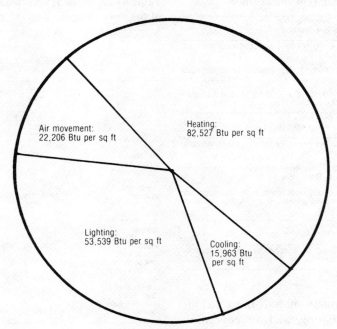

Total energy consumption: 174,235 Btu per sq ft per yr

Fig. 63-3. Energy consumption for the building based on Design No. 1, which is a typical design for high rise office buildings with a high glass content used before ASHRAE Standard 90-75 was adopted in Btu per conditioned sq ft per yr.

building was selected for Designs No. 2 and 3. In Design No. 2, this system is combined with a hydronic radiation skin heating system, and in Design No. 3, it is combined with a two-pipe perimeter fan-coil system.

The equipment for Designs No. 2 and 3 includes centrifugal chillers to produce chilled water, gas fired boilers and air foil variable air volume fans with variable inlet vanes.

Water chillers with double bundle condensers are included in Design No. 3 for heat recovery.

The heating, ventilating, and air conditioning system first cost for Design No. 2 is $4.43 per conditioned sq ft.

A cost penalty was imposed on Design No. 2 because of a change in curtain wall and glass content on the building skin. Glass content was reduced to 44 percent and the glass type changed; hence, curtain wall cost increased from $2.65 (Design No. 1) per conditioned sq ft to $5.62 per conditioned sq ft.

The difference in curtain wall construction costs added to the HVAC system first cost resulted in a system first cost of $7.40 per air conditioned sq ft for Design No. 2 when adjusted for architectural cost increases. Thus, Design No. 2, based on Sections 4 through 9, has a combined skin and system cost 30 cents per sq ft less expensive than Design No. 1, the design based on common practices before Standard 90-75.

Design No. 3 based upon Section 10

Design No. 3, which is based on the freedom allowed in Section 10, was selected to show how the architect and mechanical engineer can maintain design freedom by making architectural and mechanical system and equipment trade-offs without exceeding the energy requirement of Sections 4 through 9. In this design, the glass content is arbitrarily increased from 44 to 54 percent; the same roof, wall construction, and type of glass are maintained.

In accordance with Section 10, the same outdoor and indoor design conditions, lighting intensities, and ventilation requirements are maintained. However, the heating, ventilating, and air conditioning system changes slightly in that the high velocity shutoff variable air volume system is used on the interior with a load shedding economizer to save additional energy. The load shedding economizer merely trims the cooling output from the chiller to match the heating requirements of the building to the double bundle heating capability. Any additional cooling load is handled by the outdoor air economizer to reduce the electrical energy use of the chiller.

The perimeter system in Design No. 3 utilizes a fan-coil heating and cooling system that reacts to the perimeter space net load.

No. 3 utilizes a centrifugal chiller with heat recovery characteristics. This is the primary reason for employing a fan-coil system that can take advantage of lower temperature levels on water from the double bundled condenser on the centrifugal chiller.

The design utilizes the same type of gas fired hot water boiler and fans as used on the interior variable air volume system. Fan static pressure is reduced to 6.25 in. WG on the supply fan because the system does not extend all the way to the building skin. The fans for the fan-coil system, of course, are conventional forward curved fans operating with a free discharge typical of fan-coil equipment.

The last items of input for Design No. 3, since the type of system and glass content

Table 63-4. **Building energy and economic summary.**

	Standard 90-75 designs		
	Design No. 1: Pre 90-75	Design No. 2: Sections 4-9	Design No. 3: Section 10
Building energy consumption, Btu per sq ft per yr	174,235	73,933	72,480
Building first year energy cost, $	260,856	132,477	140,133
Building annual owning and operating cost, $	919,612	648,482	626,851
HVAC system first cost, including architectural cost adjustments	2,502,498	2,405,798	2,182,798

have been changed, are the HVAC system first cost and curtain wall cost effect. The system first cost for No. 3 is $4.76 per conditioned sq ft in lieu of $4.43 for No. 2.

Increasing the glass concentration from 44 to 54 percent results in a curtain wall cost of only $4.60 per conditioned sq ft for No. 3 compared to $5.62 for No. 2. The net effect is a cost reduction of the combined HVAC system and curtain wall cost of 69 cents per conditioned sq ft compared with Design No. 2 and 99 cents compared with Design No. 1. It is important to note while the HVAC system first cost for Design No. 3 increased, the total cost actually decreased because of the combined system and architectural change.

How the designs stack up

Design No. 1: pre-Standard 90-75

The energy analysis for this building indicates that total energy consumption over one year is equal to 174,235 Btu per conditioned sq ft per year. It is interesting to note where the action is as far as energy consumption in this building. The large segments of energy use fall in the area of lighting, base utilities, and heating. This is quite typical of office buildings in colder climates.

The segment representing air movement, which basically covers fan energy, is greater than that of the cooling equipment. The 15,963 Btu per sq ft cooling figure represents the energy of the centrifugal, pumps associated with the centrifugal chiller, the cooling towers, cooling tower fans, and even the energy requirements of the centrifugal chiller control panel.

Looking at these same numbers in the form of a bar chart (Fig. 63-4), it is evident that 47.4 percent of the total energy consumed in this building is used in heating the structure; 30.7 percent is for lighting and base utilities; 12.7 percent is for air movement; and 9.2 percent is consumed by cooling.

Taking this one step further, examine the energy costs in dollars (Fig. 63-5) to determine how total consumption breaks down. In Design No. 1, 15.4 percent of the *energy dollars* are used for heating, which is significantly less than the 47.4 percent figure for energy consumed. This difference in percent use versus percent of total energy cost is related to the utility costs for different forms of energy input into the program. In this design, natural gas was used for heating.

Note the high percentage of energy cost associated with lighting and base utility

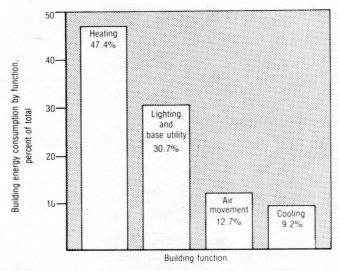

Fig. 63-4. Energy consumption by function for building Design No. 1.

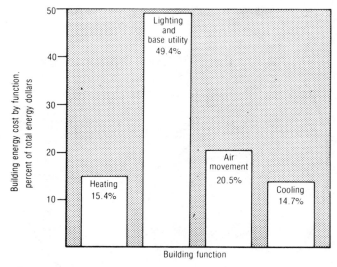

Fig. 63-5. Building energy cost by function for Design No. 1.

costs in Fig. 63-5—49.4 percent of the total energy cost for the building.

In the Design No. 1 building, 20.5 percent of the energy cost is used to transport air, and 14.7 percent of the energy cost is used for cooling.

The total annual energy cost for the first year of the operation of this building is $260,856, based on prevailing Chicago electrical and gas energy rates. For the 20th year, this cost has inflated to $789,233, based on a 6 percent annual energy inflation rate, which incidentally is quite conservative compared to some energy inflation rates experienced in other parts of the country.

To assess the energy costs in subsequent years, the magic number to use is 72. For example, a 6 percent inflation rate indicates that the cost of energy will essentially double in 12 years. This is obtained by dividing 72 by 6.

In the Design No. 1 building, the total annual owning and operating cost, including debt service, property taxes, insurance, etc., would be $919,612.

Design No. 2: Sections 4–9 in 90-75

Let us now look at the Design No. 2 building and review the factors that were changed to comply with Standard 90-75.

• A high velocity variable air volume system is used with an enthalpy actuated economizer and a hydronic radiation skin heating system.

• Glass type and content were changed to 44 percent glass with a glass U value of 0.54 and a glass shading coefficient of 0.37. The new wall construction resulted in a U value of 0.15 and roof construction yields a U value of 0.068 in compliance with the standard.

• Outside air quantities decreased from 0.20 to 0.075 cfm per sq ft.

• The use of task lighting reduced lighting to 2.5 watts per sq ft.

• Design values representing $2\frac{1}{2}$ to $97\frac{1}{2}$ percent occurrence for O'Hare Field in Chicago on outdoor summer and winter conditions were used. Indoor design and summer and winter conditions were changed to reflect Standard 90-75.

• The design cooling load is now 707.31 tons, and design heating load is 5227.5 MBtuh. These values are considerably under those for Design No. 1 because of the changes in building construction, system design, outside air quantities, and lighting intensities.

Figure 63-6 illustrates where the energy action is in the No. 2 building. As in the No. 1 building, lighting and heating still occupy the key energy consuming areas. In

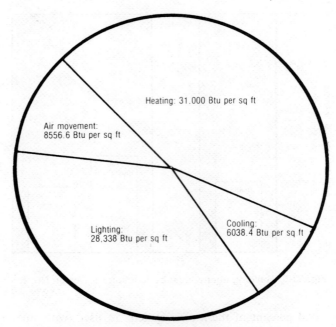

Total energy consumption: 73.933 Btu per sq ft per yr

Fig. 63-6. Energy consumption for the building based on Design No. 2, which is based on criteria in Sections 4 through 9 of ASHRAE Standard 90-75, in Btu per conditioned sq ft per year.

this case, heating at 31,000 Btu per sq ft overshadows the lighting, which is 28,338 Btu per sq ft. Air movement consumed 8556.6 Btu per sq ft, and cooling required 6038.4 Btu per sq ft. Total energy consumption for this building has dropped from 174,235 to 73,933 Btu per sq ft per year. This represents an energy reduction of 57 percent from Design No. 1, which, again, was based on practices common before Standard 90-75.

Figure 63-7 breaks down the energy used within building No. 2. Heating requires 41.9 percent of the total energy; lighting and base

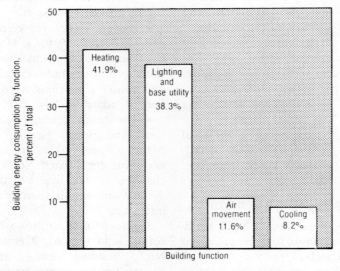

Fig. 63-7. Building energy consumption by function for Design No. 2.

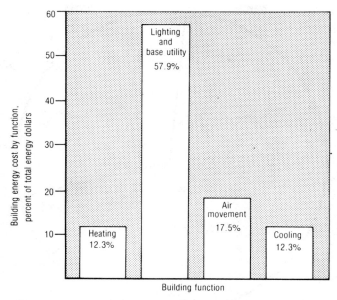

Fig. 63-8. Building energy cost by function for Design No. 2.

utilities consume 38.3 percent; air movement used 11.6 percent; and cooling represents only 8.2 percent of the total energy utilized in this building.

Figure 63-8 breaks down the *energy dollars* for this building. Heating is now 12.3 percent of the total energy dollars, and lighting and base utilities represent 57.9 percent of the energy dollars. Air movement is 17.5 percent, and cooling is 12.3 percent.

The annual energy expenditure for Design No. 2 for the first year is $132,477, compared with $260,856 for the No. 1 building. This represents an energy cost reduction per year of 49.3 percent over No. 1.

The energy cost in the 20th year would be $400,816—again, based on 6 percent annual energy inflation.

The average annual owning and operating costs for this building would be $648,482, compared to $919,612 for Design No. 1. This is a reduction in owning and operating costs of 29.5 percent or $271,130 per year—a substantial amount of money!

Design No. 3: Section 10 in 90-75

In this building, glass content increased from 44 to 54 percent of the total wall area, but the

type of glass remained the same. This reduced the cost of the curtain wall construction substantially for the owner. The change in glass content, however, violates the requirements of Sections 4 through 9 of 90-75. Therefore, trade-offs in the building system's equipment design must be made that will prove that the energy consumed by this design is equal to or less than that consumed by No. 2, which was based on Sections 4 through 9.

The trade-offs are among the architectural, the system, and equipment designs. In other words, a more efficient HVAC and equipment combination is used to offset the higher glass content of the building.

Note that the design cooling load with the increased glass content is now 725.6 tons compared to 707.31 tons for Design No. 2. The design heating load is 5611.7 MBtuh.

However, system Design No. 3 uses heat recovery centrifugal chillers. The perimeter system is fan-coil rather than No. 2's radiation skin system, and the interior system is a high velocity shutoff variable air volume with the load shedding economizer discussed previously.

Figure 63-9 shows the energy consumption breakdown for this system. Only 72,480

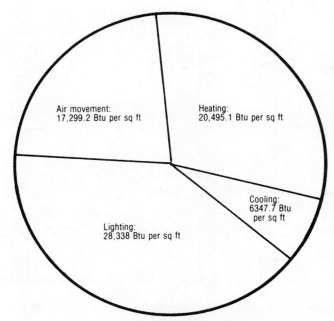

Total energy consumption: 72,480 Btu per sq ft per yr

Fig. 63-9. Building energy consumption for the building based on Design No. 3, which is based on Section 10 of ASHRAE Standard 90-75 and allows for design innovations providing that total energy consumption does not exceed the amount specified in Sections 4 through 9, in Btu per conditioned sq ft per year.

Btu per sq ft per year is consumed. This is under the No. 2 design, which consumed 73,933 Btu per sq ft per year.

The energy use pattern shown in Fig. 63-10 still indicates that heating and light-ing are the key energy users. However, the heating requirement, because of heat recovery capability, has dropped from 31,000 to 20,495 Btu per sq ft per year.

All of this was achieved in spite of the

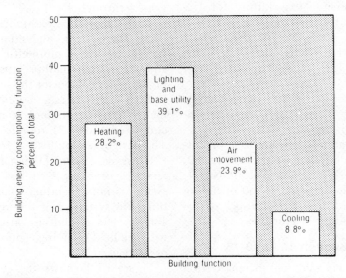

Fig. 63-10. Building energy consumption by function for Design No. 3.

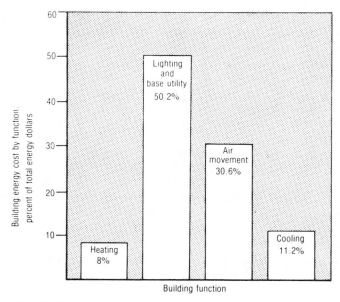

Fig. 63-11. Building energy cost by function for Design No. 3.

increase in glass content on the skin of the building.

The energy profile in Fig. 63-10 shows that 28.2 percent is for heating; 39.1 percent is in lighting and base utilities; 23.9 percent is in air movement; and 8.8 percent is in cooling. The breakdown of *energy dollars* in Fig. 63-11 shows that only 8 percent of the energy dollars are devoted to heating compared to 12.3 percent on Design No. 2.

Lighting and base utilities consume 50.2 percent of the energy dollars; air movement takes 30.6 percent; and cooling requires only 11.2 percent.

But, all of this is not gravy. Since Design No. 3, even though it consumes less Btus per square foot, has redistributed the energy usage from gas to electricity and, in doing so, has increased the annual energy cost from $132,477 for the first year reflected by Design No. 2 to $140,133 per year. This is a 5.76 percent increase in *energy dollars*, not energy use.

Because the cost of the curtain wall was substantially reduced, annual owning and operating costs, which include debt service, dropped from $648,482 to $626,851 for Design No. 3. This is a decrease of 3.4 percent in annual owning and operating costs.

Conclusions offer suggestions

The analysis has shown that Design No. 3, which took advantage of the freedom allowed by Section 10 to be innovative, reduced energy below that achieved by basing a design on Sections 4 through 9. Even though energy costs increased by approximately $7500 per year, annual owning and operating costs were reduced by 3.4 percent, primarily due to the substantial reduction in curtain wall cost with the increased glass percentage. Note that the change in first cost by utilizing Section 10 offers a reduction to the owner of $223,000, or approximately 9.3 percent from the design based on Sections 4 through 9.

The following are the conclusions that are apparent from this analysis:

• ASHRAE Standard 90-75 if properly applied will reduce building energy consumption in office buildings.

• In northern climates, the largest potential for energy conservation is usually in lighting and heating.

• Because of this fact, heat recovery equipment and systems should be considered where the building design and utilization offer coincident cooling and heating loads. In the past, heat recovery chillers have been associated with buildings having high internal heat loads, such as lighting levels at 5 watts per square foot or more. However, from this analysis, it is seen that reduced transmission and ventilation heat loads can result in an economically justifiable heat recovery installation. Trade-offs are involved here. The curtain wall cost in Design No. 3 (versus Design No. 2) was lowered more than the increase in the HVAC system cost. If this had not occurred, the heat recovery design may not have proven to be the most economical. However, it still consumed less energy.

• Standard 90-75 allows flexibility for architectural, mechanical system, and equip-ment trade-offs to reduce energy consumption in buildings.

• Standard 90-75 requires early involvement of all members of the building and system design team to be able to affect the optimum design in the best interests of an owner.

• The last and most striking conclusion of this analysis is the fact that energy conservation makes good economic sense.

Finally, there is very little in this world today that is more expensive to own and operate than a building with a cheap, inefficient heating, ventilating, and air conditioning system—and it is not going to get any cheaper. Therefore, let us use ASHRAE Standard 90-75 to make more meaningful building energy decisions.

64

Enthalpy control systems: increased energy conservation

Data collected in three California cities yield evidence that controlling outside and return air dampers from an enthalpy standpoint offers tangible energy savings over an economizer control system.

DR. GIDEON SHAVIT,
Chairman of Advanced Engineering,
Honeywell Commercial Div.,
Arlington Heights, Illinois

More and more, energy conservation is becoming an essential consideration of good building management. In the past, energy for heating and cooling was in seemingly abundant supply. However, the current energy crisis dictates changes in present and future approaches to building operation and management, especially since significant amounts of energy can be conserved by optimum use of air conditioning controls installed in new buildings or added to existing ones.

One example of energy conservation by optimizing conditioning control is the use of sophisticated control of outside air and return air dampers, based on the enthalpy content of the two air sources. Systems designed to control these dampers are a common feature in most air handling systems in commercial buildings. Control systems operate economically if they select thermodynamic properties of the air mixture that reduce cooling and heating loads in the respective operating modes.

Economizer systems widely used

Several systems are available to control outside air and return air dampers. The most commonly used one is the economizer system. In this system, a mixed air temperature controller operates the dampers so that a minimum of outside air enters the fan system when the outside air dry bulb temperature is below, say, 45°F. As the outside air dry bulb temperature rises, the controller gradually opens the damper to take in more outside air.

Maximum outside air is brought into the fan system when the mixed air temperature reaches, say, 55°F. If the outside air temperature continues to rise, the controller closes the damper to minimum air intake position when the outside air dry bulb temperature reaches a designated cutoff point, usually between 65°F and design return air temperature. Figure 64-1 is an operational graph of a typical economizer system showing schedul-

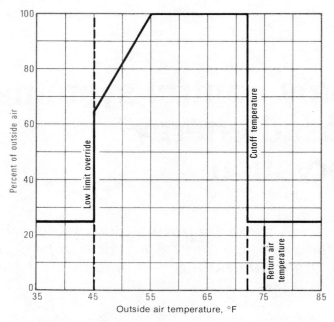

Fig. 64-1. Functional description of an economizer system.

ing of the air supply into the fan system as a function of outside air temperature.

Techniques employed to control dampers vary, but the two most common are:

• A mixed air controller regulates the dampers based on mixed air temperature prior to undergoing any conditioning.

• Dampers and cooling coil valve are sequenced so the cooling coil discharge controller regulates the dampers before the cooling coil valve opens.

Economizer controls sensible heat

The main advantage of an economizer system is that it provides the most economical damper control from a *sensible* energy consideration. However, *latent* heat can impose a significant load on a cooling coil and therefore should be a factor in a control system. Additionally, the designated cutoff temperature is a very important economizer system parameter. Since it is a function of local weather conditions, it is not always at the optimum setting.

Optimum damper control should provide air handling systems with air requiring the least energy to condition to maintain comfort in the controlled space. To do this, the com-

posite energy (sensible and latent) of the air at a given thermodynamic state should be measured. A thermodynamic state of air is defined by any two of the following measurable variables:

• dry bulb temperature
• relative humidity
• dew point temperature
• wet bulb temperature

Measurement of the composite energy of air at a given thermodynamic state is called enthalpy (or total heat or heat content). The amount of energy required to change an air mass at a given thermodynamic state to another state is the enthalpy difference. This energy can be sensible heat only, latent heat only, or any combination of the two. The term generally used is *specific enthalpy*, expressed in Btu per pound dry air.

A generalized implementation of an enthalpy control system is based on the algorithm shown in Fig. 64-2.

Return air enthalpy is not a constraint. It varies as a function of the humidity ratio in the space since the dry bulb temperature is controlled by a thermostat. Therefore, an enthalpy comparison is a dynamic measure of existing conditions in a space versus outside

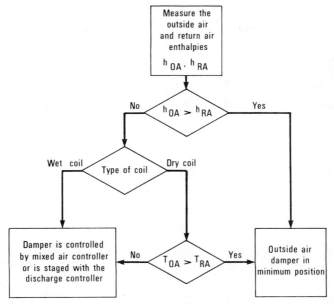

Fig. 64-2. Algorithm on which the generalized implementation of an enthalpy control system is based.

air conditions and minimizes energy consumption under all weather conditions.

Deriving enthalpy control logic

Air entering a cooling coil is a mixture of two air streams: outside and return. Each has a given thermodynamic state at a given time and therefore a corresponding specific enthalpy. An air mixture's thermodynamic state is calculated from the following relations.

Mass conservation is computed:

$$m_{RA} + m_{OA} = m_{MA} \qquad (1)$$

where

m_{RA} = mass flow rate of return air, lb per hr

m_{OA} = mass flow rate of outside air, lb per hr

m_{MA} = mass flow rate of mixed air, lb per hr

Energy conservation is a combination of sensible and latent energy. Sensible energy is computed by the equation:

$$m_{RA}C_{p,a}T_{RA} + m_{OA}C_{p,a}T_{OA}$$
$$= m_{MA}C_{p,a}T_{MA} \qquad (2)$$

where

$C_{p,a}$ = the specific heat of air, Btu per lb dry air per °F

T_{RA} = dry bulb temperature of return air, °F

T_{OA} = dry bulb temperature of outside air, °F

T_{MA} = dry bulb temperature of mixed air, °F,

and latent energy is computed by the equation:

$$m_{RA}(1061 + C_{p,w}T_{RA})W_{RA} +$$
$$m_{OA}(1061 + C_{p,w}T_{OA})W_{OA} =$$
$$m_{MA}(1061 + C_{p,w}T_{MA})W_{MA} \qquad (3)$$

where

$C_{p,w}$ = specific heat of water vapor, Btu per lb per °F

W_{RA} = humidity ratio of return air

W_{OA} = humidity ratio of outside air

W_{MA} = humidity ratio of mixed air.

Therefore, by combining Equations (2) and (3), the total amount of energy conserved is obtained from:

$$m_{RA}h_{RA} + m_{OA}h_{OA} = m_{MA}h_{MA} \qquad (4)$$

where

h_{RA} = specific enthalpy of return air, Btu per lb dry air

h_{OA} = specific enthalpy of outside air, Btu per lb dry air

h_{MA} = specific enthalpy of mixed air, Btu per lb dry air.

While the daily variation in return air enthalpy is relatively small, outside air enthalpy varies greatly. Equation (4) indicates that the enthalpy of an air mixture is primarily a function of outside air enthalpy and hence the percentage of outside air used.

The minimum percentage of outside air required for ventilation constitutes another

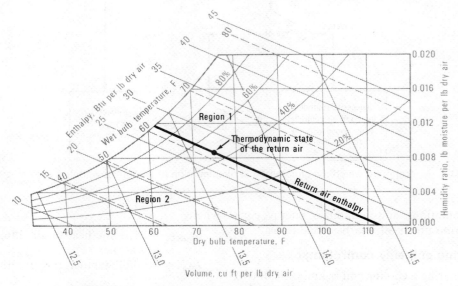

Fig. 64-3. Major subdivisions of the psychrometric chart.

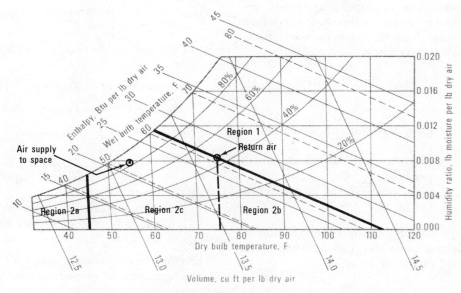

Fig. 64-4. Secondary subdivisions of the psychrometric chart.

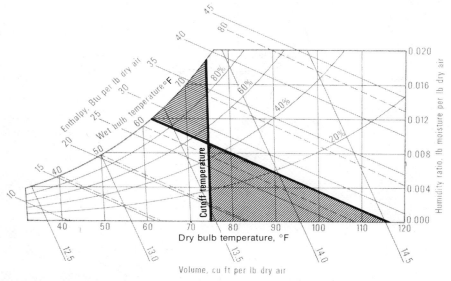

Fig. 64-5. Regions of energy savings for wet (lower area) and dry coil(upper area) systems when cutoff temperature and return air temperature are the same.

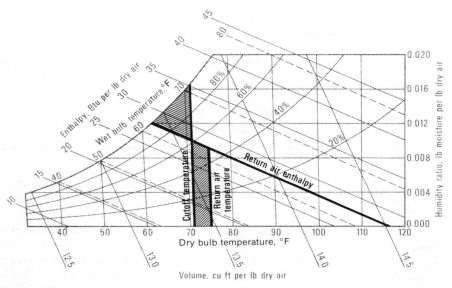

Fig. 64-6. Regions of energy savings for wet (lower area) and dry coil (upper area) systems when the cutoff temperature is 71°F.

constraint. By introducing Equation (1) into Equation (4) and using the definition $\alpha = m_{OA}/m_{MA}$ we obtain:

$$h_{MA} = h_{OA} + (1 - \alpha)h_{RA} \qquad (5)$$

$$\alpha_{min} \leqslant \alpha \leqslant 1$$

where

α = the fraction of outside air
α_{min} = the fraction of outside air required for ventilation.

Return air enthalpy divides the psychrometric chart in Fig. 64-3 into two significant

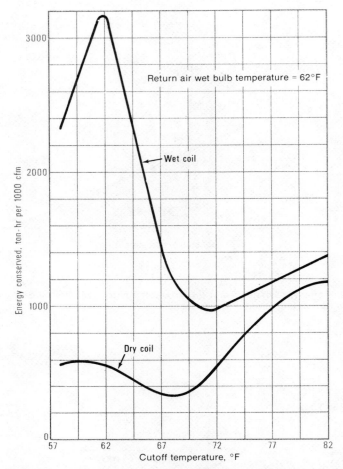

Fig. 64-7. Annual energy conserved in Los Angeles by an enthalpy system over an economizer system.

regions:

- Region 1, area where $h_{OA} > h_{RA}$
- Region 2, area where $h_{OA} < h_{RA}$

In Region 1 enthalpy of the mixture is always greater than enthalpy of the return air, therefore maximum return air should be used. In Region 2 the mixture enthalpy is always lower than the return air enthalpy.

However, it is not necessarily true that all areas of Region 2 should use maximum outside air because it is further divided into subregions (Fig. 64-4) due to the following constraints:

- The outside air damper is at minimum intake position when outside air dry bulb temperature is below a given value, say, 45° F (Region 2a). This constraint is applicable in areas where a preheat coil is used.

- If a wet coil (air passes through a water spray prior to being cooled) is used, maximum outside air is used in all of Region 2 except portion 2a. If a dry coil (humidification is downstream of the cooling coil) is used, outside air dry bulb temperature is greater than return air dry bulb temperature (Region 2b); maximum outside air is used in Region 2c.

Damper control in Region 2c is the same as in an economizer cycle. In Regions 2b and 2c minimum energy is required to condition the air to the desired level.

Damper control with enthalpy override is independent of seasonal changeover and provides optimum year around operating conditions within the prescribed constraints. It should be noted that these constraints are for example only. Individual buildings have

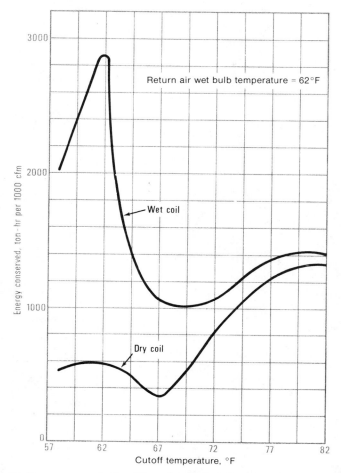

Fig. 64-8. Annual energy conserved in San Diego by an enthalpy system over an economizer system.

unique constraints relevant to their fan systems and geographic locations.

Enthalpy override vs. economizer

Before discussing control system performance, it is necessary to identify regions on the psychrometric chart where enthalpy override control is superior to economizer control. The upper black area in Fig. 64-5 represents improved system operation (energy savings) achieved by an enthalpy override system when a dry coil is used. The lower black area represents regions of enthalpy override superiority when a wet coil is used. In both cases, cutoff and return air temperatures are equal.

Figure 64-6 demonstrates that energy savings are somewhat different when the designated cutoff temperature is lower than the return air temperature. Using a dry coil, the saving (upper black area) results from a changeover from 100 percent outside air (economizer) to minimum outside air (enthalpy override control), or from a changeover to minimum outside air by the economizer system while the enthalpy override controller maintains 100 percent outside air (lower black area). In a wet coil system, energy savings are represented by the two shaded areas referred to above plus the triangular area bounded by the return air temperature and enthalpy lines.

To demonstrate the potential energy savings of an enthalpy override control system over a stand-alone economizer system, it is necessary to analyze weather data of the geographic location in question. A computer pro-

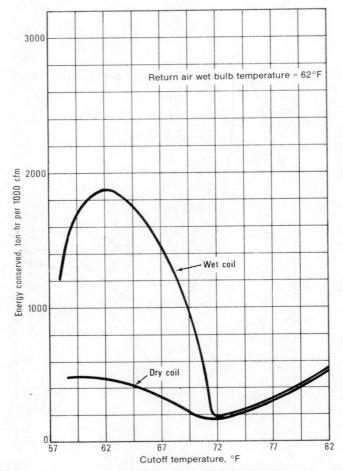

Fig. 64-9. Annual energy conserved in San Francisco by an enthalpy system over an economizer system.

gram analyzed enthalpy override control efficiency versus economizer control efficiency in over 20 U.S. cities based on government weather data.* Three California cities were selected as examples. San Francisco, Los Angeles, and San Diego were computer analyzed based on their weather data. The data collected demonstrate the important effects of geographic location—even within a few hundred miles—on system performance.

Certain assumptions were made about the typical system analyzed.
- 11 hr run time per day
- 7 days run time a week

*Engineering weather data, Departments of Air Force, Army, and Navy, June 15, 1967.

- no allowance for minimum outside air ventilation

The following results may be adjusted for another fan system having different parameters than those assumed above. For example, if a fan system runs 14 hr a day, 6 days a week, and uses 20 percent minimum outside air for ventilation, the results presented here should be multiplied by $[(6 \times 14)/(7 \times 11)] \times (1 - 0.2) = 0.87$.

Three city results presented

Results for the three cities illustrated in Figs. 64-7, 64-8, and 64-9 point up the energy that can be saved by an enthalpy override control system over an economizer system.

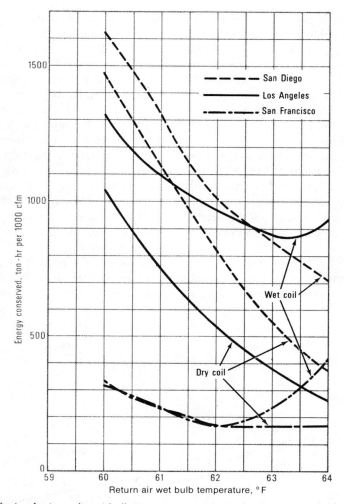

Fig. 64-10. The effects of return air wet bulb temperature on annual energy conservation in all three cities.

Savings are expressed in ton-hr per 1000 cfm as a function of the cutoff temperature and return wet bulb temperature of 62°F. It is assumed that the return air temperature is 75°F whenever the cutoff temperature is below 75°F. When the cutoff temperature is above 75°F, cutoff and return air temperatures are the same.

Figures 64-7, 64-8 and 64-9 show than an enthalpy override control system is superior to an economizer system in all three cities. For example, in a fan system using a dry coil in Los Angeles (Fig. 64-7), annual energy savings with an enthalpy system using a 75°F cutoff temperature are 820 ton-hr per 1000 cfm.

(The figures also show that in all three cities a wet coil is always less efficient than a dry coil.)

The analysis also reveals that the cutoff temperature is a very important factor in an economizer system operation. For example, the optimum setting in San Francisco is 72°F; in Los Angeles, 68.5°F; in San Diego, 67°F. (An enthalpy override system has the advantage that it is insensitive to a cutoff temperature because cutoff temperature is replaced by enthalpy consideration.) The figures show that energy conservation is very much a function of local weather conditions.

Results graphed in Figs. 64-7, 64-8, and 64-9 are for a fixed return wet bulb temperature of 62°F. Savings shown in Figs. 64-6 and 64-7 are not only a function of cutoff temperature but also of return air enthalpy (or wet bulb temperature). Return air enthalpy varies and is primarily a function of the humidity ratio in the space. Thus the region of savings on the psychrometric chart will vary accordingly. For the three California cities, energy conserved in ton-hr per 1000 cfm as a function of return air wet bulb temperature can be plotted (Fig. 64-10). In this case, too, energy savings are a function of local weather data. Savings generally increase as wet bulb temperature decreases, except for the wet coil operations in Los Angeles and San Francisco.

Decrease outside air to save

To look at it another way, the computer study shows (Fig. 64-11) that energy savings increase as the percentage of outside air decreases—assuming that return air is maintained at 75°F DB, 62°F WB, and air leaving the coil is at 55°F and saturated. In Los Angeles, for example, where pollution is a problem, a minimum percentage of outside air is desirable. Thus damper control with enthalpy override coupled with a decrease in the intake of outside air will increase the percent of conserved energy and decrease the intake of pollutants. A fan system with a minimum outside air intake of 15 percent and damper control with enthalpy override will conserve 4 percent of the total cooling load compared to an economizer cycle with optimum cutoff point in San Francisco, 5.7 percent in Los Angeles, and 6.1 percent in San Diego.

To emphasize the sensitivity of the optimum cutoff point in the economizer system,

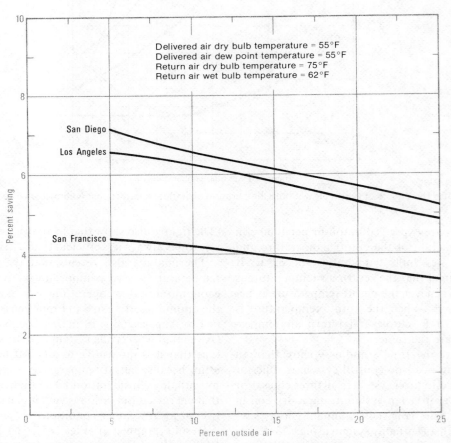

Fig. 64-11. The percentage of cooling load energy saved by enthalpy consideration over the most optimum economizer system versus the percent of outside air.

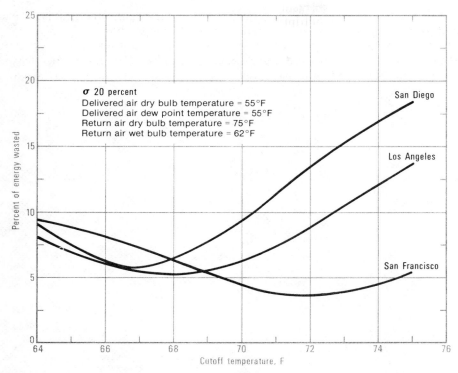

Fig. 64-12. The effects of deviating from optimum cutoff temperature on energy waste by an economizer system with respect to enthalpy consideration.

the percentages of annual cooling load wasted against the cutoff temperature in the three cities are plotted in Fig. 64-12. The graph shows that inaccurately setting the cutoff point doubles or triples energy waste, compared to damper control with enthalpy override.

While this has always been very appealing on paper, the economics of enthalpy control have been, until recently, defeating. Systems to continuously measure both latent and sensible air energy were cumbersome and costly and really not very good. However, recent developments in electronics have brought not only commercial grade controls with excellent temperature and humidity accuracies (temperatures within 1°F and humidities within 1 percent RH) but also have enabled electronically combining the two to get the measurement of the composite energy of air on a continuous basis. The result is enthalpy override control that is economically feasible, even for small, single-fan systems. Since such systems offer tangible savings of

scarce energy as well as money, they will become more important as the fuel crisis becomes more acute.

Conclusions and summary

To summarize the major points of comparison between enthalpy override and economizer damper control, the following points merit emphasis:

• Enthalpy override control provides a dynamic and optimum method of damper control.

• Damper control with enthalpy override conserves energy above and beyond the current most commonly used damper control system, the economizer.

• The efficiency of an economizer system is extremely dependent on an accurate cutoff temperature setting. An economizer system at a given geographical location has an optimum cutoff temperature setting.

• Savings with enthalpy override control are a function of local weather conditions.

• Damper control with enthalpy override

conserved 4 to 6 percent more cooling energy compared to an energy economizer system in the study in California.

- Inaccuracies in setting an economizer cutoff temperature can double or triple energy waste.

- Enthalpy damper control is independent of cutoff temperature.
- The development of low cost electronic systems makes enthalpy control feasible for even small air conditioning systems in many buildings.

65

How to achieve energy savings in chiller plant operations with series and parallel flow through air conditioning chillers

Flexibility in design and control should be provided to assure optimization of operation.

PAUL K. SCHOENBERGER, PE,
Energy Management Consultant,
Columbus, Ohio

Central chilled water plants must operate efficiently to provide various chilled water loads under various ambient conditions. To obtain maximum plant efficiency under all situations, it is imperative that flexibility in system components and operational controls is included within the system. For any particular plant output and ambient condition, minimum energy input to the plant is the primary objective.

Actual plant operations can vary considerably from the theoretical analysis; however, the theoretical performances of system components can be used as a valuable guide toward establishing maximum efficiencies during operations. When maximum flexibility in component design and operational control is provided, actual operations can be

Proceedings of the Central Chilled Water Conference, 1976, published by the Ray W. Herrick Laboratories, Purdue University, Copyright © Purdue Research Foundation, 1976. Reprinted by permission.

varied until minimum energy usage is achieved.

Maintenance must be scheduled

When equipment has been in operation for long periods of time, efficiencies are likely to drop due to wear or fouling in the condenser or chilled water circulating systems. Loss of system efficiency will indicate when maintenance or replacement of equipment is imperative; however, preventive maintenance should be performed at regular intervals in accordance with all equipment manufacturers' recommendations.

System controls must operate within control ranges to prevent short cycling of equipment and unstable system operations. This is a primary reason that actual plant operations can vary considerably from the theoretical methods. Sensitivity of all control systems must be adjusted to provide stable operations.

Analyze energy use for two chillers

Energy consumption of centrifugal water chillers is determined by the evaporation and condensation temperatures of the refrigerant. When the temperature differential between these two temperatures is the closest, the coefficient of performance of the system will be the highest, and minimum energy input is required per ton of output.

The temperature at which the refrigerant condenses approaches the temperature of the condenser water used for heat rejection. The temperature of the water leaving the cooling tower approaches the wet bulb temperature of the ambient air passing through the tower. It has been common engineering practice to size cooling towers for full loads with 95° F water entering the tower and 85° F water leaving the tower at the ARI* design wet bulb of the ambient air. In most instances, ambient wet bulb temperatures may only reach the design level for a few days each year and operation of the tower to provide cooler water for condensing is possible the majority of the time. When prolonged periods of operation at high ambient wet bulb temperatures is imperative, oversizing of the tower can decrease the approach between water outlet temperature and the ambient air wet bulb temperature. It can be more advantageous to oversize the tower and use less energy for additional tower fans than to operate at higher condensing temperatures and increase the energy input to the chiller. For each particular system, various operating characteristics such as evaporator feeding, system depression, etc. will establish a minimum condensing temperature that should be used as a lower limit for the condenser water supply to the chiller.

Size chillers for leaving temperatures

The temperature at which the refrigerant evaporates is determined by the temperature of the chilled water that the chiller is producing. It has been common practice to size and control chillers for 44 or 45° F water leaving the chillers. For air conditioning loads using coils and air handling units, the temperature of the chilled water necessary for maintaining comfort conditions is directly related to the ambient air wet bulb temperature. When wet bulb temperatures are low, there may be no need to maintain low chilled water temperatures since dehumidification of the air passing through the coils may not be necessary. Chilled water supply temperatures as high as 50 to 55° F can be adequate to handle sensible air conditioning loads. Water supply temperatures as low as 44 or 45° F may only be necessary a small portion of the time. Elevating the chilled water temperature to decrease the differential for the evaporating and condensing temperatures will decrease the energy input per ton of refrigeration.

Energy consumption of centrifugal water chillers is affected by the rate of heat transfer of the evaporator and condenser shells. For a given heat exchanger surface, the heat transfer rate in these shells is determined primarily by the water velocity and temperature range of the water. For specific condenser and chilled water temperatures, energy can be saved by oversizing the condenser and evaporator surfaces. An analysis of the life-cycle costs can assist in optimum selection of the evaporator and condenser surfaces as well as water flow rates, velocities, and resulting water pressure drops for efficient selection of pumps.

Examine condenser, evaporator tubes

The condition of the condenser water tubes should be examined periodically to prevent tube fouling. Frequent cleaning of these tubes can produce substantial energy savings. A complete condenser water treatment system for control of suspended solids, algae and pH should be provided. The evaporator tubes should also be checked periodically and cleaned when necessary. A shot type chemical feeder can be provided for the closed chilled water system to add any necessary water treatment for prevention of corrosion.

Free cooling through migration of the refrigerant in the chiller without operating the compressor should be considered if wintertime operation of the plant is necessary. Free cooling operates on the simple principal that the refrigerant migrates to the point in the sys-

*Air-Conditioning and Refrigeration Institute.

tem where the temperature is the lowest. This feature requires additional refrigerant piping and controls; however, when condenser water is available at temperatures below the chilled water temperature, this mode of operation should be considered.

Determine loads by computerization

Computerized programs that include yearly weather tapes for hourly changes in weather for an area can be of valuable assistance in optimizing equipment selections. Full maximum capacity of the chiller plant may only be required for very short periods. The system load factor (ratio of the average load on the system over the maximum output) may only be 50 percent of the maximum output. The computerized programs can assist in establishing the system load factor by modeling loads in relation to actual weather conditions.

Maximum efficiency of the compressors should be sized closer to the average operating point than to the full load point. The power requirement of a chiller is a non-linear function when plotted against the percent of total capacity at which the machine is running because the surface area in the evaporator and condenser is greater at light loads than at heavy loads. For any given machine, an optimum efficiency (kW/Ton) occurs at a specific capacity level that is usually between 60 and 80 percent of the maximum rated capacity. Because of these variations in efficiencies, there are capacity levels at which it is more economical to operate two machines than to operate one.

Select best temperature differential

It has been common practice to size chilled water plants with 10°F temperature differentials between the water supply and return when systems were close-coupled and with 15 to 20°F differentials when long runs of piping were involved for multi-building installations.

The use of temperature differentials larger than 10°F can be advantageous for all systems and should be given serious consideration. Increasing the temperature differential from 10 to 15-20°F reduces the water distribution system pipe sizes and pumping requirements.

Use of the larger temperature differentials through cooling coils in terminal air handling systems requires increased depth in these coils and adds air flow resistance to these systems. Increased air resistances requires additional fan energy. Most existing constant volume air handling systems have cooling coils that were sized using 10°F temperature differentials. Air flow delivery rates of these systems can often be reduced by partially converting their operation to variable volume of supply air along with eliminating excess lighting and reducing supply air delivery rates to offset actual loads. When energy conserving modifications have been instituted for the supply air systems, existing cooling coils in these systems may be quite adequate for use with 15 to 20°F water differentials. A life-cycle cost analysis should be made to assist in optimizing the design water temperature differential for the system.

Series or parallel flow?

There are presently two methods available for obtaining 15 to 20°F water temperature differentials through two centrifugal water chillers. The chillers can have series or parallel water flow through their evaporators. When series water flow is used, single pass evaporators are normally used to hold water pressure drops to the minimum because the entire system flow rate must pass through each chiller. When parallel water flow is used, multi-pass evaporators are necessary to obtain the large chilled water temperature differential between the maximum supply and return temperatures. With parallel flow, each chiller handles one-half of the total water flow rate of the system and must cool this water quantity through the entire differential. With series flow, the lead chiller may cool the water from 60 to 51°F while the lag machine cools from 51 to 44°F. The upstream chiller operates at increased efficiency with series flow because the supply and return water temperatures are higher than those for the downstream machine.

A disadvantage of the parallel mode of operation of the two chillers is that low chilled water temperatures cannot be obtained with one chiller in operation and water passing through the inactive unit. The water passing through the inactive unit must be mixed with the discharge of the active unit and the temperature of this mixture may be higher than desirable. Lowering the discharge of the active unit and the temperature of this mixture may be higher than desirable. Lowering the discharge temperature of the active machine to improve this situation will decrease the efficiency of the machine and waste energy. This situation can occur when the load factor is as high as 40 or 50 percent, and operation with one chiller is desirable. When operation with both chillers is desirable, parallel mode of operation will reduce circulating pump heads.

Consider types of drives

The types of drives used for the chillers should also be given serious consideration. Open electric motors offer some advantages over hermetic motors. Operating voltages can also significantly affect the efficiency of operations. Centrifugal chillers are suitable for various drives such as efficient diesel or turbine engines. In isolated cases these special drives may be justified.

Study heat rejection sources

Sources of heat rejection from equipment within the system may be usable to reduce energy consumption of other systems. A condenser water heat exchanger for use to preheat the cold water make-up to the hot water heating system should be given serious consideration. This is particularly applicable for hospitals or related types of facilities. In many instances, it will be practical to provide additional storage for the domestic hot water system to utilize the daytime peak system condensing temperatures. It should be noted that the use of the condenser water to preheat domestic hot water will lower the temperature of the condensing system and increase the efficiencies of the chillers. A close approach between the condenser water temperature and the preheated hot water should be possible, and in most instances the hot water can be preheated from about 55 or 60°F to above 90°F.

Apply variable flow pumping

Variable flow pumping can be combined with series and parallel flow through two chillers to provide minimum energy usage at partial system loads. System load factors between 40 and 60 percent are not uncommon, and adjusting water flow rates to actual loads can reduce energy consumption significantly. Variable speed pump drives with fluid couplings or solid state pump controllers should be given serious consideration. In systems using two chillers, selection of three parallel pumps can be a reasonable solution for efficient operations. The pumps should be sized to provide full system flow when all three pumps are in operation and parallel flow through the chillers is occurring. When two pumps are operating, circulation should be through two chillers in series. One pump should be used whenever one chiller is in operation. With each pump sized for $\frac{1}{3}$ of the maximum flow rate, water velocities and operating heads will drop when one or two pumps are in operation. One pump should supply more than $\frac{1}{3}$ of the maximum flow rate; however, two pumps may only supply approximately $\frac{2}{3}$ of the maximum flow rate when series flow through the chillers is used in this mode. With system load factors of 40 percent and below, the pumps should operate near maximum efficiency when only one pump is operating. With system load factors between 40 and 80 percent, the pumps should operate near maximum efficiency when two pumps are in operation. Since the majority of the water chiller plants will have system load factors between 40 and 80 percent, selection of the pumps for maximum efficiencies with two of the three pumps running will be the normal situation. A typical centrifugal pump curve is shown in Fig.

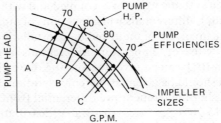

Fig. 65-1. Typical centrifugal pump curves.

65-1, and the following discussion of the pump selection will refer to it.

Pump selection is critical

Selection of each pump to operate at point B with two pumps running is ideal. With one pump running, flow would increase to point C at a reduced head. With three pumps running, flow would decrease to point A at an increased operating head. It should be noted that flat curve pumps should not be used for this application since flow rates can be run off the end of the pump curve with one or two pumps running at a considerable drop in system head. Pumps must be non-overloading and should operate efficiently within the parameters of their intended use. It is also imperative that one pump supplies adequate water through one chiller to maintain the minimum water velocity through the evaporator that is recommended by the manufacturer.

Variable flow pumping of the condenser water system does not offer the same energy savings as for the chilled water system. Ambient air wet bulb temperatures may rarely reach design conditions, and design flow rates for the condensing system will produce water at reduced temperatures. Operation of the chiller with reduced condenser water temperatures will increase chiller efficiency. The savings in energy input to the chiller may offset the pump and tower fan input, and variable flow pumping may be of little or no value. It is imperative that provisions be made in the condenser water system to prevent water temperatures from exceeding a lower limit. Tower fans should be turned off and water flow should bypass the tower to prevent water temperatures from becoming too low.

Central control center saves energy

Control of central chilled water plants for minimum energy usage must involve changes in operations with changes in operating loads and ambient conditions. Minimum energy usage will be difficult to achieve without use of a programmable central control center. Proper monitoring and control of system components is imperative if minimum

energy usage is to be achieved. Energy input to the plant (electrical current and voltage) should be continuously monitored along with energy output (chilled water temperatures and flow rates). The enthalpy of the ambient air should also be monitored and can be used to establish chilled water temperatures necessary for latent loads. Chilled water temperatures should be maintained at the highest levels necessary to handle system loads. Automatic sequencing of chillers, pumps, tower fans, and water flows for equipment bypasses can be programmed to obtain minimum energy input at various plant loads and ambient conditions. Automatic flow control valves and equipment bypass valves must be provided for automatic changes in system operations.

Figure 65-2 shows the necessary system components that should be included for automatic control of the plant to assure maximum efficiency. Switchover points can be established under actual operations and adjustments can be made to offset variances due to equipment wear or tube fouling. Drops in system efficiencies will become apparent and corrective action can be instituted when continuous recording of the energy input and output is provided.

Chiller sequencing control valves can be used to permit operation of either chiller with one pump in operation, operation of both chillers in series with either unit in the lead position with two pumps in operation, or operation of both chillers in parallel when three pumps are in operation. Remote control of the chiller capacities should also be provided at the control center to regulate the percentages of the load that the lead and lag machine will handle.

Consider using rotary screw chillers

The use of two rotary screw chillers should be given serious consideration. Rotary screw chillers are designed for use with high temperature refrigerants and do not require troublesome purge systems associated with the sub-atmospheric refrigerants used in centrifugal chillers. Rotary screw chillers also operate near maximum efficiencies over a wider range of capacities (usually between

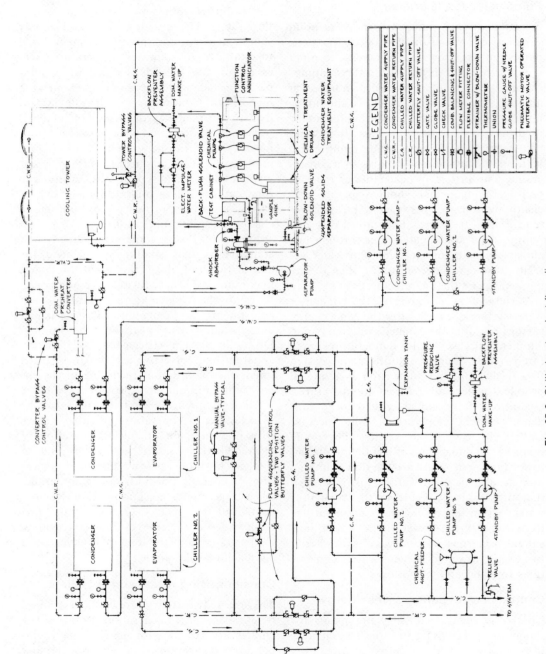

Fig. 65-2. Chilled water plant flow diagram.

40 and 80 percent) and are less sensitive to condenser water temperatures. Screw chillers use less energy input per ton of refrigeration when low chilled water outlet temperatures are necessary. The other items discussed for centrifugal chillers are also applicable for rotary screw chillers, and complete life-cycle screw chillers, and complete life-cycle cost comparisons can permit selection of the most energy efficient type of chillers for any particular application.

References

1. Holbay, Nick, "Energy Conservation Techniques for Centrifugal Chillers," *Heating/Piping/Air Conditioning*, May 1976.
2. Various chiller manufacturers' catalog data on capacities and inputs of chillers.

66

Two phase booster ejector for air conditioning and refrigeration cycles

Principles and examples show how elusive energy lost during throttling process is recovered.

ANCEL L. LEE, II,
Chief Product Engineer, Air Handling
and Terminal Units, York Div. of Borg-
Warner Corp., York, Pennsylvania

In a conventional refrigeration cycle, an expansion or throttling valve controls the flow of refrigerant from the condenser to the evaporator. It also reduces the refrigerant pressure from the high condenser level to the relatively low evaporator pressure. Available energy of the refrigerant decreases as the pressure loss occurs in the throttling process.

Engineers have been tantalized for many years by various schemes to recover this lost energy. Experimental expansion engines and turbines have not been technically or economically successful. Now, a two phase ejector has been applied to a refrigeration cycle to recover part of the available energy normally lost in the expansion device. The ejector is geometrically simple and contains no moving parts or valves, which normally act as parasitic devices in many of the previous schemes.

Conventional cycle operation

Figure 66-1 is a schematic of the components of a conventional cooling system. Figure 66-2 is a plot of the cycle on a pressure-enthalpy diagram for a system utilizing R-22. Numerically designated points are congruent in both figures. Tracing refrigerant around the cycle, it enters the evaporator or cooler at Point 4. Heat is absorbed as the refrigerant travels through the evaporator to Point 1 in an essentially constant pressure process. At Point 4, the refrigerant has experienced the throttling process through the expansion valve, and the proportion of gas in the two phase mixture at the entrance to the evaporator is approximately 31 percent by weight. Refrigerant flow through the evaporator is controlled such that the leaving gas is usually superheated from 3 to 15°F.

Vaporized refrigerant enters the suction side of the mechanical refrigeration compressor at Point 1 on Fig. 66-1 and is compressed to Point 2, where the pressure is approximately equal to the pressure of the refrigerant in the condenser. At the higher pressure and corresponding temperature, heat absorbed in the evaporator can be rejected to the prevailing heat sink, which may

be either ambient air or available water. Refrigerant at Point 2 enters the condenser where heat is rejected from the refrigerant between Points 2 and 3. Desuperheating occurs in the first part of the condenser, after which refrigerant is condensed and in some cases subcooled.

Liquid refrigerant leaves the condenser at Point 3 and passes through the expansion valve before entering the evaporator at Point 4. The throttling process between Points 3 and 4 is adiabatic, and therefore, the enthalpy of the refrigerant entering the evaporator is the same as that of the liquid refrigerant leaving the condenser. Even though the enthalpy per pound of refrigerant is identical at Points 3 and 4, a loss of energy occurs internally during the throttling process. The energy change is consumed in the vaporization of a portion of the liquid refrigerant

as the pressure is reduced between Points 3 and 4 and results in an increase in entropy.

The loss of energy occurring in this process can be depicted on the diagram in Fig. 66-2. The perfect or isentropic expansion process would occur along the dotted line between Points 3 and 4. Because the actual process takes place along the solid line between Points 3 and 4, the loss of energy can be shown as equal to Q_E

$$Q_E = h4 - h4'$$

where

Q_E = energy loss, Btu per lb

$h4$ = actual enthalpy at end of throttling process, Btu per lb

$h4'$ = enthalpy at end of isentropic (ideal) expansion process, Btu per lb

The work of compression in this cycle is Q_C in Fig. 66-2.

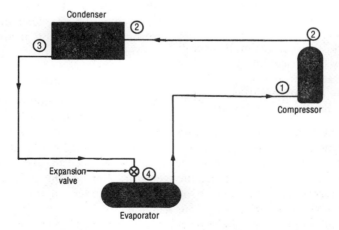

Fig. 66-1. Conventional refrigeration system.

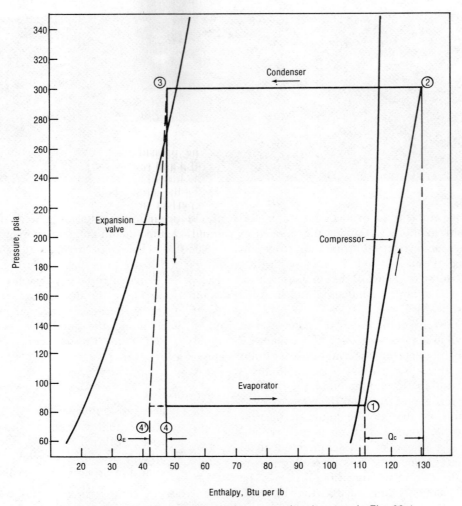

Fig. 66-2. Pressure-enthalpy diagram for conventional system in Fig. 66-1.

$$Q_C = h2 - h1$$

where

Q_C = the work of compression, Btu per lb
$h1$ = enthalpy of refrigerant gas entering compressor, Btu per lb
$h2$ = enthalpy of refrigerant gas at compressor discharge, Btu per lb

In an air conditioning system containing a reciprocating type compressor, Q_C is equal to about 19.7 Btu per lb when utilizing an air cooled condenser. The loss through the expansion device (Q_E) is 3 Btu per lb or about 15 percent of the work of compression.

A number of different devices have been examined throughout the years for the pos-

sible recovery of the normally lost energy represented by Q_E. One that has attracted considerable attention at my company is the multiple phase ejector.

A cycle with a booster

A schematic diagram showing the arrangement of components when a two phase jet booster is utilized in a refrigeration cycle has been depicted in Fig. 66-3. The expansion valve is replaced by the motive nozzle in the ejector. Refrigerant flow exiting from the ejector includes liquid refrigerant to be recirculated through the evaporator as well as gas that must be directed to the compressor. To make a proper division of these two flow

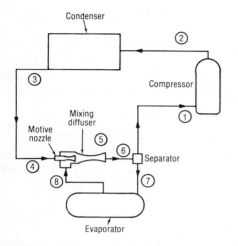

Fig. 66-3. Refrigeration system with a booster ejector.

regimes, a phase separator is installed downstream from the ejector diffuser.

A trace of the booster refrigeration cycle has been made on the pressure-enthalpy diagram in Fig. 66-4. The designated points correspond to those shown on Fig. 66-3.

To become better acquainted with the booster cycle, let us imagine we are taking a trip through the system along with a few particles of refrigerant. Leaving the condenser at Point 4 in a slightly subcooled state, refrigerant passes through the driving steam nozzle and exits at Point 5. The mixture is expanded to a pressure slightly lower than the evaporator pressure in order to induce vapor from the evaporator into the ejector and establish a predetermined pumped

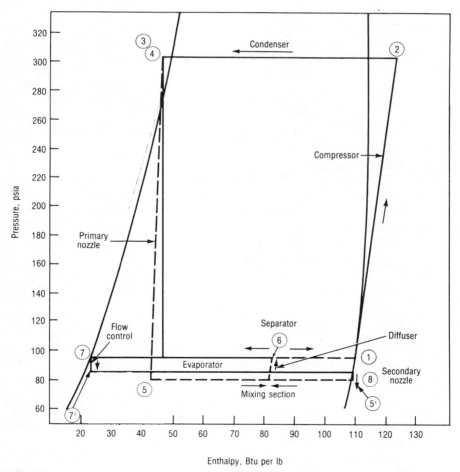

Fig. 66-4. Pressure-enthalpy diagram for a refrigeration cycle with booster (Fig. 66-3).

flow velocity. The refrigerant vapor leaving the evaporator at Point 8 passes through the pumped flow nozzle from Point 8 to Point 5 on Fig. 66-4. The driving steam particles at the exit of the nozzle begin to mix with the pumped stream gas particles, which have been boosted to a certain pre-designed velocity. Mixing of the two streams then occurs between two moving streams rather than between a driving stream of high velocity and a pumped stream with essentially no velocity. This is an additional parameter available to a designer working with a two phase ejector.

Energy from the driving stream nozzle is then utilized to pump the mixed stream of refrigerant to a pressure, at Point 6, that is higher than the evaporator pressure. At this point, the flow is separated into two characteristic streams in the phase separator. The vapor particles journey along the dotted line to the compressor suction port, Point 1, and the liquid particles, being in the saturated condition at Point 7, are directed into the evaporator cooler. The refrigerant particles are essentially saturated liquid upon entrance to the evaporator at Point 7. Refrigerant travels through the evaporator section and absorbs heat from the medium being cooled between Point 7 and Point 8.

Looking back at the vapor particles of refrigerant that left the phase separator at Point 1, this vapor is compressed from Point 1 to the approximate condensing pressure at Point 2. These refrigerant particles are then desuperheated, condensed, and somewhat subcooled between Points 2 and 4.

Immediately, one can visualize two factors that appear to improve the overall performance of the cycle. In the first place, refrigerant entered the evaporator in the booster cycle at essentially saturated liquid conditions; whereas in the conventional cycle, the refrigerant entered the evaporator at a quality of approximately 31 percent. Figure 66-5 illustrates the fact that each pound of refrigerant circulated through the evaporator is capable of absorbing an additional quantity of heat in the booster cycle. This is represented by the difference in enthalpy between Points 10 and 7.

The second immediately discernible fact about the cycle is that the compressor suction is at a pressure somewhat higher than that of the evaporator suction. Therefore, the pressure rise required of the compressor in the booster cycle is smaller than that in the conventional cycle.

Effects of the booster ejector

The booster cycle has been superimposed on the conventional cycle in Fig. 66-5. The booster has the following effects: 1. the compressor suction pressure is boosted from Point 8 to Point 1, and 2. the refrigeration effect of each pound fed to the evaporator is increased.

Boosting suction pressure reduces the work required to compress each pound of refrigerant.

The change in cooling effect per pound of refrigerant is illustrated in Fig. 66-5. Refrigerant from an expansion valve in a conventional cycle enters the cooler at Point 10, and refrigerant from the booster ejector enters the cooler at Point 7. This is an increase of approximately 40 percent. The net result is an improvement in the coefficient of performance, or efficiency of the cycle.

Analysis of components

A booster refrigeration cycle consists of a conventional system basically changed to a two stage compression cycle with intercooler.

A capacity versus evaporating temperature curve is shown in Fig. 66-6. Line A-B is a plot of the conventional compressor capacity curve at a given condensing temperature. Capacity for the combined booster ejector and compressor is equivalent to Curve C-D. Thus, the total capacity at Point 3 is the sum of the compressor capacity and the booster capacity at evaporator temperature Point 3.

Evaporator performance in the form of a plot of capacity versus evaporating temperature provides an almost straight line (Line E-F). Point 1 is the balance point for the compressor and evaporator. Point 2 is the balance point for the compound machine (compressor plus ejector). The booster compressor system then provides added capacity for a given system, but the cycle balances

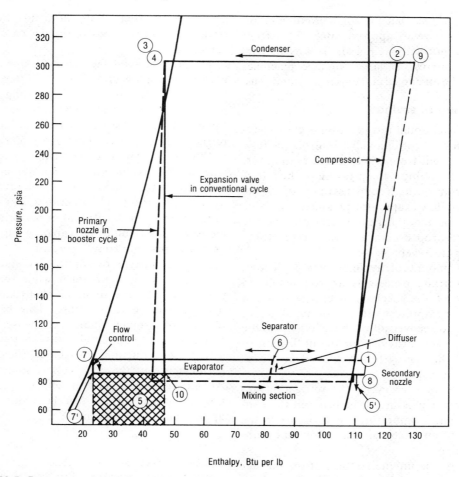

Fig. 66-5. Pressure-enthalpy diagram for a refrigeration cycle with booster and a conventional cycle.

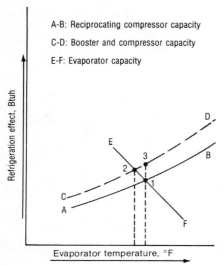

Fig. 66-6. Refrigeration effect for combinations of components in booster system.

at a somewhat lower evaporating temperature. To obtain the full potential of the cycle, with the evaporating temperature equivalent to that of the standard system selected at Point 1, a larger evaporator should be installed to produce an even higher capacity at balance Point 3.

Reducing condenser size

The booster unit can also reduce the size of the refrigerant condenser. The rated capacity of a system can be maintained at a higher overall pressure ratio. This provides a higher than normal condensing temperature and therefore, a greater temperature differential between the condensing refrigerant and the heat sink. When the heat sink for the condenser is ambient air, a larger temperature differential is realized between the refrigerant and the air so that the condenser can be reduced in size to approximately 72 percent of that normally required. This is of particular interest in large tonnage machines where the air cooled condenser is voluminous.

Summary of savings

In a conventional refrigeration cycle, the expansion or throttling valve controls the flow of refrigerant from condenser to evaporator. This valve reduces the pressure of the refrigerant from the high condenser pressure to the relatively low evaporator pressure. Available energy of the refrigerant decreases as the pressure drop occurs because the throttling process is irreversible.

The amount of available energy lost in the throttling process is approximately 15 percent of the mechanical energy required by the compressor. It is not possible, of course, to recover all of this energy, but the booster cycle provides for the utilization of part of the energy totally wasted in a conventional system.

In the booster cycle, the refrigerant experiences a pressure drop between the condenser and evaporator, but it occurs in the nozzle of an ejector instead of in an expansion valve.

One of the unique features of the booster ejector is that the motive or driving stream nozzle operates with a mixture of gas and liquid refrigerant. Much work has been done with ejectors operating with either a liquid or a gas driving fluid. Some new and different techniques are required in the design of the subject booster.

The greatest available energy for operation of the ejector occurs when the differential between condensing and evaporating temperatures is large. The refrigeration cycle for liquid chilling service with an air cooled refrigerant condenser functions under such a differential. An air cooled chiller package with a 37 percent efficient booster ejector can increase the standard cycle cooling capacity by 12 percent. The operating efficiency of the cycle or the coefficient of performance increases as a result of the capacity improvement by approximately 7 percent.

In a second type of application, standard cycle capacity is maintained while the system is made to operate at a higher than normal condensing temperature. This permits the use of a much smaller condenser coil. It provides both an initial cost reduction and a unit size reduction, but the coefficient of performance of the cycle suffers somewhat. At an achievable ejector efficiency level, the condenser surface can be reduced by approximately 28 percent.

References

1. Blum, Harold A. and Olcott, Thomas M., "Thermodynamic Investigation of a Refrigerant Expansion Engine," *ASHRAE Journal*, Aug. 1961.
2. Sears, Francis W., *Thermodynamics: The Kinetic Theory of Gases and Statistical Mechanics*, Addison Wesley Publishing Co., 1955.

SECTION IX

Future Opportunities and Limitations

In any discussion of ways to conserve energy, it is most important to consider the future opportunities that may be available and their limitations. The 12 chapters in this section outline the following opportunities and their limitations:

- Solar energy and its control.
- The heat pump.
- Solar heat pumps.
- Geothermal energy.
- Total energy systems.
- Government regulations and activities.

67

How new technology may save energy in existing buildings

Retrofit of heat pump cycles—both conventional and solar assisted—to existing buildings should be considered when major building changes are under consideration.

FRANK H. BRIDGERS,
Bridgers and Paxton Consulting
Engineers, Inc., Albuquerque,
New Mexico

It must be realized that the energy problem is a people problem. It is people who waste energy; it is people who design energy wasting systems; it is people who operate systems in a wasteful manner; and it is people who fail to do the things they know how to do and use the tools that are available to save energy.

The solution to the world's current energy shortage will embody many things. New technology will provide part of the solution. Examples include large scale coal gasification, recovery of liquid fuels from oil shale, the fast breeder reactor, and fusion power. These are in various stages of development, but it will be some years before any of these "solutions" have a sizable impact upon our economy. Also, these technologies are outside the immediate sphere of the HVAC designer.

Making better use of existing technology and applying it in new ways will stretch the supply of existing fossil fuel and uranium reserves. Decreasing the drain on consumable

energy reserves by both existing and new buildings will contribute to the solution. This is an area where people involved in the design, construction, and operation of heating and air conditioning systems can directly apply their talents to help solve this serious problem.

Solar energy usable today

Much research is currently underway on solar energy. Much remains to be learned. Yet, solar energy technology can be directly applied today by the design mechanical engineer in the area of conserving energy in new and existing structures. It is a usable tool to conserve energy now.

All environmental control processes depend ultimately upon a heat source—typically, fuel burned either on the building premises or remotely at a power plant.

Present solar energy and HVAC technologies lend themselves very well to using the sun as a heat source, which would conserve fossil fuel and uranium. Solar energy is utilized through standard refrigeration system components and proven HVAC systems. The basic water-to-water heat reclaim cycle is

This chapter is based on a paper presented by Mr. Bridgers at a special series of two-day seminars sponsored and developed by *Heating/Piping/Air Conditioning* on How to Save Energy in Existing Buildings.

employed. In addition to the normal internal and external heat gains utilized—lights, machinery, people, solar gain through windows, etc.—solar energy is used to heat stored water as an additional heat source for the heat pump, resulting in a solar assisted heat pump system.

Wide spread solar energy application is currently hampered by high initial cost. A developer borrowing money at high interest rates cannot recover this cost in an acceptable time period. Solar energy is more readily applicable to federal and state projects where monies are appropriated for specific projects. There may be a benefit to a corporation to utilize solar energy in its facilities since it is a visible way (solar collectors cannot be hidden) to demonstrate commitment to conservation.

Solar heat pump retrofit process

Solar energy lends itself readily to retrofit existing heat pump systems. Also, it is sometimes feasible to convert existing air conditioning systems into heat pump systems either with or without solar energy. The retrofit of the heat pump and/or solar energy to an existing building will be discussed to provide insight into how this can be accomplished.

Retrofitting a heat pump cycle to an existing structure generally is economical only if either a major building renovation or addition is contemplated. Both usually involve consideration of changes to the existing HVAC system. Often it is feasible to make extensive HVAC system changes that improve performance, increase comfort, and reduce energy consumption.

A heat pump retrofit is most economical if existing refrigeration machinery is usable in the new system. The economics of the heat pump may justify replacement of the existing plant however.

The performance of an existing compressor under new operating conditions must be determined. If higher condenser water temperatures are required for space heating, can it perform adequately? A reciprocating compressor often will handle an increase in thermal lift adequately. A centrifugal machine originally selected for straight cooling service may be incapable of maintaining stable operation at a higher lift and the reduced space cooling loads associated with winter operation. Unstable operation (surging) is to be expressly avoided to prevent damage to the machine and associated piping.

If the compressor is suitable for the demands of heat pump duty and will provide sufficient heat at a satisfactory water temperature under the new operating conditions, it must be properly interfaced with the heating distribution system.

● *Heating and condenser water circuits*— Typical heat pump water circuits are shown in Figs. 67-1 and 67-2. The system with a double bundle chiller is probably more familiar to readers living in areas having design wet bulb temperatures higher than generally experienced in the southwest. Replacing a single bundle condenser on an existing machine with a double bundle unit is a very expensive operation and might be impractical due to space considerations.

An alternate approach is installation of piping and valves to isolate a single bundle condenser from the condenser water circuit and the building heating lines. Upon changeover from summer to winter operation, the chiller must be drained and flushed before the heating circuit is tied into the condenser. Flexibility is limited because condenser heat is unavailable for reheat during the cooling season. Also, the cooling tower would not be available during the winter to reject excess heat, and this could cause system control problems.

Figure 67-2 shows a heat pump system utilizing chillers with single tube bundle condensers. Cooling tower heat exchangers isolate the condenser from the tower. This provides the same flexibility as double bundle condenser refrigeration machinery.

Often, existing systems have oversized components. Calculated design loads on existing buildings may be greater than those actually experienced, because in the past designers did not have sophisticated computerized load and energy analyses programs. As a result, it is not unusual to find upon analyz-

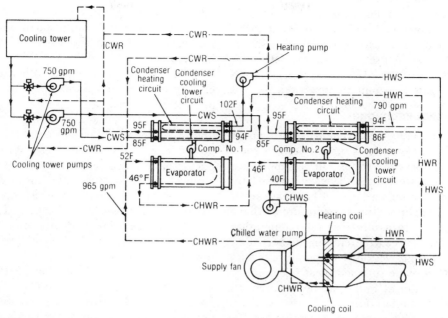

Fig. 67-1. Heat pump system used in a telephone exchange building. The heat pump system, which utilizes double bundle condensers, replaced the original chiller plant when the building was enlarged.

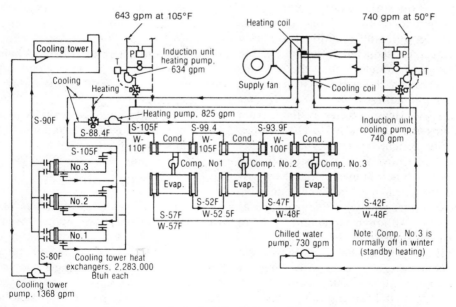

Fig. 67-2. Heat pump system using single bundle chillers. Cooling tower heat exchangers isolate condensers from the tower, permitting condenser water to be used for summer reheat and winter heating. Existing air conditioning systems may be converted (where feasible) into a heat pump by the installation of a cooling tower heat exchanger and suitable heating system modifications.

ing existing buildings with modern tools that such components as chillers, cooling towers, boilers, etc., are oversized—sometimes significantly. This in itself can cause inefficient, wasteful operation. However, this situation can facilitate heat pump conversion.

Add a heat exchanger

The addition of a heat exchanger in a condenser water circuit affects system performance. Lower temperature water will be supplied to the cooling tower due to heat transfer laws and the economic aspects of heat exchanger design. The condensing temperature will have to be raised to assure that the temperature difference between the entering tower water and the design wet bulb temperature is adequate to provide sufficient heat rejection, or a combination of both may be required. In any event, chiller output will decrease. In some cases, excess capacity may provide sufficient margin to permit satisfactory conversion to a heat pump. In other instances, it may be feasible to add additional cooling tower capacity.

It is important to note that if the condensing temperature is increased under summer cooling conditions, the added operating cost for cooling under this new operating condition along with any additional auxiliary power, must be deducted from the heating cost savings when determining the feasibility of converting a chiller to a heat pump.

• *Accommodating lower heating water temperatures*—Heating water available from heat pumps often is in the 105 to 110° F range. Specifically designed machines can provide up to 130° F. Conversion of existing chillers to heat pump duty is limited by the characteristics of the given machine, and it may be necessary to work with water temperatures as low as 95° F. Although temperatures up to 110° F will be feasible in specific instances.

Central air systems where one or two row heating coils have been used with water (or steam) supplied from a boiler around 200° F require additional heat transfer surface to utilize lower temperature heating water. This normally is not a problem because of the following:

• The design heating load is smaller than

the design cooling load, which often is the case in commercial and institutional buildings. Heating can often be done with supply air temperatures as low as 95° F.

• The sensible cooling load rather than the heating load determines the supply air volume in most situations.

• *Air side considerations*—Depending upon the type of air system used, fan horsepower may or may not increase. With single duct systems, variable air volume, or reheat, it must be increased (assuming the original air flow was correct) to overcome resistance from the additional rows of heating coils. Dual air stream systems—double duct or multizone—often do not require additional fan horsepower. Design fan horsepower is based upon the air stream with the greater resistance—generally the cooling side. Cooling coils typically are of six row construction; while heating coils often have one or two rows and, consequently, a lower pressure drop.

Perimeter fan-coil units often have three and four row coils, and it is not unusual for them to operate at lower water temperatures even when higher temperature water or steam is available at the boiler. Therefore, the heating water temperature available from a retrofit heat pump may be adequate.

Control of the fan-coil system should be investigated to determine whether changes in operation should be made to assure satisfactory performance with lower supply water temperatures.

Existing boilers may be available to supply heat for peak loads if the heat pump does not have capacity to meet design loads.

Retrofit example in library

It was possible to take advantage of extra capacity in an existing cooling tower on a job where we converted an existing chiller plant into a heat pump. The two chillers are older models, which are single stage centrifugal units. It was not possible to raise the leaving condenser water temperature significantly above the 95° F design value. Our investigation showed the cooling tower had sufficient excess capacity to install a cooling tower heat exchanger to isolate the tower lines from the

Table 67-1.   The existing building and new addition of Harold B. Lee Library Building at Brigham Young University compared.

Factor	Existing	Addition
Size, sq ft	195,000	225,000
Total cooling, tons	430	445
Internal load, tons	166	233
Heating load, Btuh	4,700,000	3,364,000*

*224 tons of heat rejection.

condenser without adversely affecting chiller output.

The building is a university library. Space heating for the original portion of the structure is supplied from a coal fired, high temperature hot water (HTHW) central plant serving most of the campus. The original library has two centrifugal chillers to provide air conditioning. A central chilled water plant was built after the original library building.

Converting the existing chillers to heat pumps was contemplated when the addition to the library was planned. The large internal loads in both the old and new buildings (see Table 67-1) indicated that the heat pump concept should be investigated. Table 67-2 shows the energy use and energy costs if the central HTHW and chiller plants had been used to serve the new addition compared with the combined central plant and heat pump plan selected. The combined plan yields a 26 percent fuel savings and a 21 percent energy cost savings. Translated into dollars, this is equal to $3242 annually at current costs. An investment of approximately $25,000 was made for the cooling tower heat exchanger and associated pumps, piping, etc., to permit the existing chillers to provide winter heating.

Heating-cooling distribution in the original

building is through a single duct, terminal reheat system. A double duct system was selected for the addition. Operation is as follows:

• During summer operation, air conditioning in the original building is provided by the heat pumps with reheat energy furnished by the central plant.

The addition is cooled by chilled water from the central plant, while reheat energy is supplied by the heat pumps. The 95° F condenser water and a four row heating coil in the hot deck of the double duct system provide 85° F hot deck air, which is sufficient for summer temperature control.

• During winter operation, space heating in the original building is provided by a central HTHW heating system.

During the heating season, the heat pumps do the required internal space cooling for both the old and new areas. This heat is rejected to the four row heating coil in the double duct system serving the addition. The heat pumps provide a major portion of the heating for the addition on a seasonal basis and approximately two-thirds on a design load basis. Hot deck design supply air temperature is 100° F, and the mixed air temperature is 55° F. During colder weather when hot deck temperatures higher than 95° F are required, a two row booster coil—downstream of the four row coil supplied by the heat pumps—fed from the campus HTHW system supplies added heat as required.

Solar assisted heat pump in hospital

A solar assisted heat pump is planned for a 210 bed addition to Shiprock Hospital in New Mexico. A central steam plant currently serves all existing facilities, which includes

Table 67-2.   Comparison of central plant heating and cooling with central plant and heat pump at Harold B. Lee Library addition.

Facility	Million Btu used per year		Heating cost per year, $	
	Central plant	Heat pump and central	Central plant	Heat pump and central
Existing building	13,900.9	12,145.9	10,787	9425
New addition	5,748.4	2,200.5	4,460	2580
Total	19,649.3	14,445.4	15,447	12,005

Saving: 26 percent in raw energy and 21 percent in energy cost.

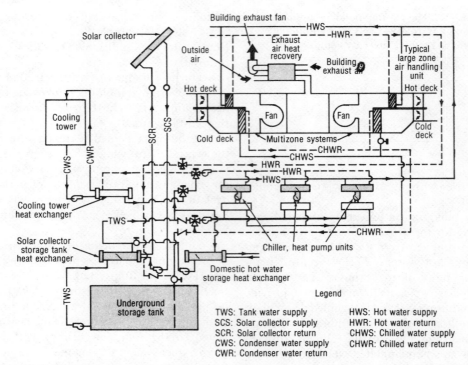

Fig. 67-3. Solar assisted heat pump system for Shiprock Hospital in New Mexico. Standard HVAC components are utilized in conjunction with a solar-to-water collector and storage system.

schools and dormitories as well as the present hospital.

Figure 67-3 shows a schematic diagram of the system planned for the addition. It is designed around the basic water-to-water, internal source heat reclaim cycle.

Storage tanks have sometimes been used with these cycles to capture and store building heat during periods of surplus for use during colder weather and/or during unoccupied times. On this job, the 250,000 gal. storage tank is connected to a solar collector system— total collector area is 30,000 sq ft—and provides an additional heat source to meet the large heating loads associated with this hospital.

During the summer months, the solar storage tank will provide a portion of the domestic hot water load through a heat exchanger. During the winter, it will provide a heat source for the heat pump for morning pick up when the fan systems from areas normally not in use at night go over to the daytime operating mode. Also, it will be a heat source during the night cycle when internal loads are low.

The storage tank will also furnish partial solar heating of the original hospital building. Plans call for converting most of this space into an out patient clinic. The remodeling plans include installation of a perimeter fan-coil system that can be supplied by the heat pump system and supplemented by the existing central heating plant. During milder weather when stored heat is in excess of the requirements of the new building, the fan-coil distribution system in the clinic will be served from the heat pump. This, in turn, will reduce the load on the fuel fired boiler plant.

During severe weather or during extended periods of cloudy weather, heat can be supplied to the storage tank through the domestic hot water heat exchanger at nighttime only to assure adequate heat capacity for the heat pumps. Normally, the heat pump system can reclaim sufficient building heat to meet all daytime heating loads with the exception of morning pickup.

Figures 67-4, 67-5, and 67-6 depict the energy flow for the three normal operating modes: cooling cycle (above 75° F), winter

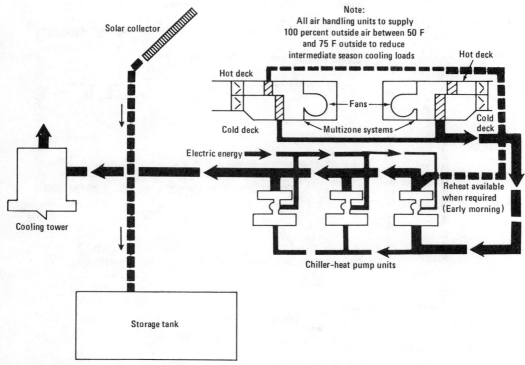

Fig. 67-4. Energy flow diagram—electricity, reclaimed heat, and solar heat—for the cooling cycle (above 75° F outdoor temperature). See note on drawing relative to the intermediate cooling system.

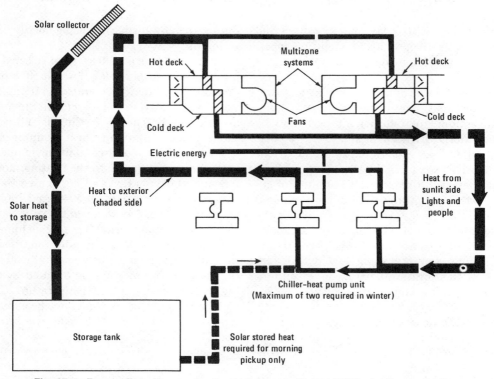

Fig. 67-5. Energy flow diagram for the winter cycle (below 50° F), daytime operation.

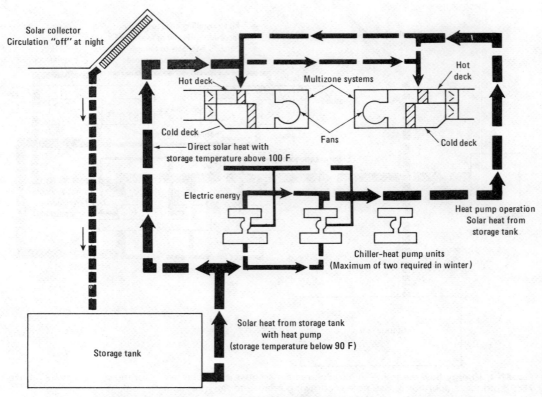

Fig. 67-6. Energy flow diagram for the winter cycle (below 50° F), nighttime operation.

cycle (below 50° F), daytime operation, and winter cycle (below 50° F) nighttime operation. The inside design temperature is 75° F.

Above a 75° F outdoor ambient temperature, the air system operates with the minimum amount of outdoor air permitted by code criteria. Between 50 and 75° F, 100 percent outdoor air is delivered by the supply air systems. This reduces the load on the chillers and conserves a considerable amount of energy since hospital air conditioning is required around the clock. (Enthalpy control of the outdoor air volume will not be used because it is not cost effective due to the inside design conditions and prevailing low outside dry bulb at this location.) Below 50° F outdoor temperatures, a combination of outdoor air and mechanical cooling is employed to handle the interior space cooling loads. Only sufficient mechanical cooling is used to meet the required heating load at a given weather con-

dition, thereby further reducing energy consumption.

Three centrifugal chillers were selected to meet the 400 ton design load. Above 50° F outdoor ambient, they will supply 44° F chilled water with a leaving condenser water temperature of 95° F. The 95° F water is sufficient to meet air conditioning reheat requirements.

Below a 50° F outdoor ambient temperature, the system goes into the heating mode. Leaving chilled water and condenser water temperatures are raised to 50° F and 100° F, respectively. Chilled water at 50° F is sufficient during this period because dehumidification is not required. In addition, raising the chilled water temperature partially offsets the increased power demand dictated by the higher condensing water temperature. The chillers are sequenced so that only the number required to maintain 110° F water will be on-line.

When the chillers extract heat from the

The world's first solar heated commercial building at Albuquerque, New Mexico. This building housed the offices of Bridgers and Paxton Consulting Engineers, Inc. It was equipped with a solar assisted heat pump for six years. When the building was enlarged in 1962, the solar collector system was retired because of the fuel and construction economics that prevailed at the time. All other original system components have been in continuous operation. In 1974, Dr. Stanley F. Gilman of Pennsylvania State University applied for and received a National Science Foundation research grant to restore the solar heating system and provide extensive instrumentation for demonstration and research programs to establish a design procedure for solar assisted heat pump systems that could be put into wide use throughout the country.

Table 67-3. Theoretical and actual performance of refrigeration machines (cooling only) and heat pumps.

- Theoretical performance as a refrigeration machine:

  COP – Ts/Tc-Ts

  where

  COP = Coefficient of performance
  Ts = Evaporator temperature, absolute
  Tc = Condensing temperature, absolute
  COP = (460 + 35 F)/(460 + 105 F)
       – (460 + 35 F) = 7.07

- As a heat pump:

  COP = Tc/Tc-Ts = 460 + 120 F/
        (460 + 120 F)
        – (460 + 45 F) = 7.73

- Manufacturer's data as a refrigeration machine:

  COP = 3,402,000 Btuh/276.3 bhp x 2545 Btuh per hp-hr/0.90 motor efficiency = 4.35

- As a heat pump:

  COP = 4,554,000 Btuh/321.9 bhp x 2545/0.90 efficiency = 5.0

Table 67-4. Net cost of solar energy installation at Shiprock Hospital.

Factor	$
Solar collectors, 30,000 sq ft at $15 per sq ft	450,000
Structure to support collectors at $2 per sq ft	60,000
Storage tank (concrete), 250,000 gal	50,000
Collector-storage tank heat exchanger	39,000
Storage tank—domestic HW heat exchanger	20,000
Cooling tower heat exchanger	26,000
Additional automatic temperature controls	25,000
Additional piping	45,000
Construction cost total	715,000
Additional architect-engineer fee	45,000
Total	760,000
Less cost of 300 HP boiler installed	– 53,000
Net initial cost premium	707,000
Years to recover additional cost =	707,000
20 year average energy cost savings	55,664
	12.7 yr

storage tank, tank water is blended with return chiller water if the temperature of the stored water is above 60° F (too high an entering water temperature could damage the refrigeration machine). Stored water is used directly by the chiller when its temperature is between 50 and 60° F. If the temperature of the water in the tank falls below 50° F, supplemental heat is supplied from the central heating plant by means of the domestic hot water heat exchanger at night only when the domestic hot water demand is low.

The use of the solar assisted heat pump at Shiprock Hospital will reduce fossil fuel consumption by approximatey 62 percent. Fossil fuel consumption for purchased electricity, gas used by the central plant, and/or fuel for a separate boiler in the new addition would be around 19.54 billion Btu per year. Annual fossil fuel consumption with the solar assisted heat pump including the fuel burned to generate electricity to power the heat pump system will be approximately 7.33 billion Btu. This savings, 12.21 billion Btu, is equivalent to nearly 500 tons of coal.

Consider all factors first

The use of the heat pump principle to recover heat is a proven technology. Anytime an existing building is extensively altered, the heat pump should be considered. Solar heating via heat pump systems has been proven from an engineering feasibility and an energy conservation standpoint. Solar retrofit of existing heat pump systems is also technically feasible.

Solar systems are not always feasible from an economic standpoint. Each possible application requires detailed studies. Economic factors include the added investment cost of the solar system, estimates of future cost escalation of fossil fuel, geographical location, and how the project will be funded. Another factor that can apply is that when a company faces curtailment of its fuel supplies, energy conservation in some form may mean the difference between continued operation of the facility or shutting it down.

Retrofit of heat pump systems with or without solar energy is not offered as a panacea. It is an additional energy conservation tool to be applied when it offers the optimum solution to a problem.

68

The prospects for solar energy since 1967

In 1966 and 1967, John I. Yellott authored a four-part series on solar energy. Much has happened since then. Here is a broad overview of developments in this expanding field.

JOHN I. YELLOTT, PE,
John Yellott Engineering Associates,
Inc., and Visiting Professor in Architecture, Arizona State University, Tempe,
Arizona

Nearly a decade ago, a series of comprehensive articles[4] was published in *Heating/Piping/Air Conditioning* as a state of the art report on the use and control of solar energy. At that time, there was very little interest in the use of solar energy; most of the emphasis was placed on sun control because air conditioning was becoming prominent. Those responsible for specifying the size and capacity of air conditioning apparatus were much concerned about solar heat gain through windows.

Fortunately, by 1967 ASHRAE had developed a rational procedure for estimating solar heat gain through fenestration, and data had been accumulated for clear day solar heat gain factors through vertical windows facing north, northeast, east, southeast, south, southwest, west, and northwest. These tables, which were produced from earlier information published by ASHRAE and from work

---

Superscript numerals refer to references at the end of the chapter.

done at the University of Minnesota, have proven to be of such accuracy that they continue to be used. A variation based on the same equations is now available to those who wish to study the utilization of solar energy; tables of the insolation falling on south facing surfaces tilted at various angles to the horizontal are now available (see Reference 1).

While most of the solar energy research and testing has been concerned with residential structures, these data and developments are very important to the HVAC engineer. What happens in residential solar applications will indicate future trends in solar technology applied to large commercial and industrial buildings.

Areas of interest

Ten years ago, there was much interest in solar distillation of sea water or other brackish supplies; today, this interest has dwindled to the Greek Islands in the Mediterranean, to Australia, and to a few locations in southern California. It is quite likely that the types

487

of solar stills shown in the 1966 and 1967 *HPAC* articles[4] will be built in much larger numbers in the United States as supplies of good, low cost drinking water become exhausted, and new developments spring up in areas where there is plenty of water, but which is nonpotable in its natural form.

During the past decade, there was sporadic interest in solar house heating and more recently, in solar cooling. A major report, written in 1972 for the National Science Foundation and the National Aeronautics and Space Administration, gave convincing evidence that solar energy could indeed play a significant part in helping the United States to balance its energy budget.

The National Science Foundation began a modest program of financing solar energy investigations by authorizing three industry-university teams to study the state of the art of solar energy, the degree of acceptance that it would probably obtain in various geographical locations throughout the country, and every other aspect of the prospective use of solar energy that the respondents to this particular RFP (request for proposal) could envision.

The three winning industry-university teams in this competition were General Electric Co. in cooperation with the University of Pennsylvania; TRW Systems, Inc. with Arizona State University; and Westinghouse with Carnegie-Mellon University and Colorado State University. Their reports were submitted in June of 1974 at a workshop held by the National Science Foundation, and there was considerable agreement among the three companies as to the length of time that it would take to get widespread use of solar energy. The crystal balls into which each company had looked were rudely smashed in October of 1973 when the Arabs suddenly realized that they held the key to the lock that could readily be applied to the world's liquid fuel supplies. There was very little indication in the three final reports from these companies that an irreversible change had taken place in the world's energy situation. The four-fold rise in the price of oil had made a quantum change in the price of energy, and it will never be the same again.

The National Science Foundation responded by setting forth on a broad program of research and development to establish the basis for a solar energy heating and air conditioning industry, but it was cut short in the midst of its operations by the passage of a new piece of legislation, entitled Energy Research and Development Administration Act of 1974. This bill established a completely new organization, which was, in fact, very largely a reorganization of the former Atomic Energy Commission, which in one of the first clauses of the bill was "deactivated." The new bill established an administrator; a deputy administrator; and seven subordinate administrators, each of whom will be responsible for one phase of the energy research program. Four of the seven will devote their activities to atomic energy, and of the approximately 900 employees of ERDA, well over 800 are former employees of the Atomic Energy Commission.

The budget that has been proposed to establish ERDA is 90 percent directed toward atomic energy and only 10 percent toward alternate sources of energy, of which solar radiation is certainly the most promising.

Solar breakthroughs

Since the publication[4] of the 1966–67 series of articles in *Heating / Piping / Air Conditioning*, there have been several important breakthroughs in the field of solar heating and cooling. First of these was the successful demonstration in 1967–68 by Harold Hay of Los Angeles and the writer at the Yellott Solar Energy Laboratory in Phoenix of the Hay Skytherm system. This system uses the absorption of solar heat by roof ponds in the winter to heat a structure and to store the heat for use at night by rolling insulating panels over the heat absorbing water surfaces. In summer, cooling is accomplished with the same water surfaces by rolling the insulation away during the night hours, thus permitting heat to be dissipated to the sky by convection (under suitable weather conditions), by radiation, and particularly by evaporation.

The first full-scale house of this type (Fig. 68-1) has undergone its first year of testing at Atascadero, California, under the auspices of

Fig. 68-1. Skytherm residence built at Atascadero, California, for its inventor, Harold Hay of Los Angeles. Architecture by Professor Kenneth Haggard, California State Ploytechnic University of San Luis Obispo.

California State Polytechnic University, San Luis Obispo. Harold Hay, the inventor, put up the construction cost of the demonstration house, and HUD invested approximately an equal amount in the cost of monitoring the house for the first year. The results are available in a final report of the year's investigation, and it was clearly shown that the structure could be kept within the comfort zone without the use of any auxiliary energy what-

soever. This confirmed the earlier findings of the tests conducted in Phoenix in 1967–68.

Since 1968, dozens of solar heated and a few solar cooled houses have been built. The crosshatches on Fig. 68-2 show the locations of these structures. Some are still in operation, and some have been taken out of service for one reason or another. Figure 68-2 also shows isopleths of the average daily horizontal insolation given the units used by the U.S.

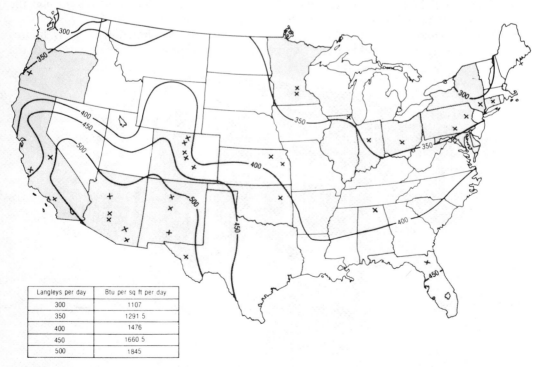

Langleys per day	Btu per sq ft per day
300	1107
350	1291 5
400	1476
450	1660 5
500	1845

Fig. 68-2. Locations of solar buildings in the United States. Isopleths show annual average daily horizontal insolation in Langleys. One Langley per day equals 3.69 Btu per sq ft per day.

Weather Bureau, which are Langleys per day. One Langley equals one calorie per square centimeter, and one Langley per day equals 3.69 Btu per square foot per day. The distribution of solar houses pays very little attention to the prevalence of sunshine. For instance, New England has a very large number, relatively, of solar heated structures primarily because the necessary funds and degree of curiosity have been available. New Mexico is another center of solar energy activity, as is Colorado, chiefly due to the high intensity of solar energy available in those states and the shortage of natural gas that has already developed there.

The details of the structures indicated on Fig. 68-2 can be found in a publication commissioned by the American Institute of Architects Research Corporation, entitled *Solar-Oriented Architecture*, compiled by the Architectural Foundation of Arizona State University, Tempe, Arizona (see Reference 2).

Solar Structures prove feasibility

The solar houses developed during the last 20 years can be divided into four categories. The simplest type allows solar energy to enter through very large south facing windows (Fig. 68-3), which are generally double glazed. The windows are provided with insulating drapes, blinds, or shutters that can be used to convert the normally low resistance glass window into a high resistance wall for retention of heat at night. The David Wright house, Santa Fe, New Mexico, is an excellent example of this type of structure. Its large overhang protects the south wall against the heat of the summer sun, and its floor, which is 2 ft of adobe blocks plus a layer of face brick, constitutes the major heat storage.

The walls of the building are adobe brick of double thickness with 4 in. of insulation between the two courses of brick. The interior design of the hoouse is quite open; so radiant heat from the floor can circulate throughout the house during the evenings when there is no sun shining. Auxiliary heat is provided by a wood-burning fireplace, as is customary in that part of New Mexico. This house is similar in principle to the Keck houses, which were tested near Chicago in the 1930s and again at Purdue University in that same era, but they differ in principle by adding the all-important insulating blinds that can be lowered at night to minimize heat loss through the south wall.

The second classification of solar houses developed during the last decade uses the

Fig. 68-3. The David Wright House, Santa Fe, New Mexico, employing large south facing glass, internally shaded at night to minimize heat loss and using the floor of the building as the primary heat storage means.

Fig. 68-4. Trombe-Michel house, developed for use in the south of France, has a massive concrete south facing wall used as the heat absorber, insulated by double glazing with convection space provided between the concrete and the glazing. The vertical south wall provides both heat storage and heating of the building at night.

fabric of the building both as a heat collection and heat storage means. This is exemplified by the structures originated by Dr. Felix Trombe and his collaborating architect, M. Michel, in France (Fig. 68-4). In these structures, a very massive south facing concrete wall covered by a single or double glazing provides a means of absorbing and storing solar heat during the day and also of heating air by convection by permitting air flow through the space between the dark painted concrete and the glass. Openings at the bottom and top of the concrete wall permit flow of air from the floor of the building up to the ceiling and back again by natural convection. The relatively slow movement of the wave of heat, which is absorbed during the daytime, through the concrete wall at night brings heat indoors when it is most needed; and then, as the heat wave recedes outward during the early morning, the wall is again cool and ready to receive the heat that is freely donated by the sun. A number of houses of this construction have been built adjacent to the great French solar furnace at Odeillo, and they appear to be working very satisfactorily in that relatively mild climate.

Using the same principle, but in a com-

pletely different manner, is the Hay Skytherm house shown in Fig. 68-1. Here, the heat storage and collection functions are performed by water bags that constitute the roof of the building. The ceiling and the roof are one single long panel of steel, which is supported appropriately by masonry walls and steel beams. Panels of insulation are so arranged that they can be rolled back and forth across the water bags to shut off the sun's rays during summer days and to admit them during the winter. On summer nights, the bags are exposed to the sky; so they can lose heat by radiation and evaporation and also by convection when the air temperature is suitable. This system was thoroughly proven in Phoenix in 1966-67, and it has completed several years of successful operation with no auxiliary energy whatsoever in the moderately demanding climate of Atascadero, California.

Use separate solar collector

The third system, which is by far the most widely used, employs a separate solar heat collector, generally in the form of flat plates covered with glass or other suitable transparent material to trap the sun's rays, with an

Fig. 68-5. The Thomason residence in Washington, D.C., uses the system shown in Fig. 68-6 to provide heating by relatively low temperature sun-warmed water. This house now has more than 12 years of experience to demonstrate its feasibility.

indoor solar heat storage means, which may be a large tank of water, a bin of rocks, or a combination of both. This system was demonstrated late in the 1950s by Dr. Harry E. Thomason in a series of homes built near Washington, D.C. (Fig. 68-5), and these now have a dozen years of experience behind them. His most recent structure applies the results of the use of his years of experience, and it employs the service hot water heater of the house to provide auxiliary heat when needed. Storage of heat is provided by a 1600 gal. tank of water surrounded by 60 tons of rocks. The principle on which the house and the collector operates was shown clearly in the articles[4] published in *Heating/Piping/Air Conditioning* in 1967 (Fig. 68-6).

Among the improvements incorporated into his latest structure, are the elimination of dampers and the use of a very small compression-refrigeration air conditioning system that can be used at night during the off-peak periods to chill the water in the storage tank, and thus store up "coolth" for use during the following afternoon.

The Solaris system pioneered by Dr. Thomason is now being employed successfully as far north as Minnesota. Since it is a fail-safe system, it is relatively free of the danger of freezing that besets most of the

other systems employing water as a heat transfer means.

Another strikingly handsome house using separate collectors and heat storage means is Phoenix of Colorado Springs (Fig. 68-7), which was built in 1973 when the City of Colorado Springs became aware of the fact that its supply of natural gas was critically short. This structure uses two banks of flat plate collectors to heat an antifreeze solution, which transfers the collected heat to a 16,000 gal. buried storage tank. This tank provides the hot water used to heat the building under normal circumstances, and it will store enough heat for about five days with virtually no sunshine under the relatively severe winter weather of Colorado Springs.

Use heat pump as auxiliary

As an auxiliary energy source and to provide the fan and filtering services necessary for a warm air system, a heat pump is used. This also provides the cooling needed occasionally in Colorado Springs. The domestic hot water is preheated by passing through a heat exchanger in the storage tank. This building, erected as a result of civic action by the City, is now being thoroughly monitored thanks to a grant from the National Science Foundation. Its first year of operation has indicated

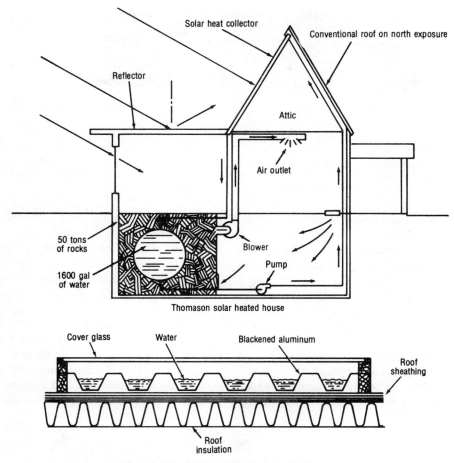

Thomason solar heated house

Cross section of roof mounted solar water heater

Fig. 68-6. Operating principle of the Thomason Solaris solar heating system in which heat is collected at moderate temperatures by a steeply tilted south facing collector and stored in a 1600 gal. water tank; 60 tons of rocks pick up heat from the warm water in the tank. The rocks become both additional storage and the heat transfer surface to warm the air going to the house.

Fig. 68-7. Phoenix of Colorado Springs is a solar heated and heat pump cooled house erected in Colorado Springs, Colorado, in 1973. Flat plate collectors provide the heat required in winter, and air circulation is provided by the indoor fan of the heat pump, which serves as the auxiliary heat source and as the summer cooling means.

Fig. 68-8. The Löf house, erected in Denver in 1959. Two banks of overlapping glass collectors are used to warm air, which in turn heats gravel in two storage cylinders. Heating at night and during cloudy weather is accomplished by the stored sun-collected heat in the rock filled cylinders.

that it is able to carry most of the heating load by its solar collectors alone.

A much older structure of the same general type is that built by George Löf in Denver in 1959 (Fig. 68-8). Two banks of glass type collectors are used to heat air, and the air in turn heats cylinders filled with gravel, which provide the all-important heat storage. Approximately 40 percent of the energy required to heat the Löf home during the past 12 years has come from its solar system.

A quite different system is used in the residence of the Paul Davis family in Albuquer-

que. A bank of air heaters is mounted on a natural slope below the patio, which faces south. A storage bed of rock is mounted under the patio, and natural circulation causes the air heated by rising through the solar collectors to fall again through the cool rock bed and to be reheated by passing again through the collectors. This is the first known application of the thermosyphon circulation system for an air-rock bed combination, and the house itself is heated at night by air circulated through the rock bed by fans. Once again, a fireplace is used as the auxiliary

Fig. 68-9. Paul Davis residence in Albuquerque has air heaters providing hot air to charge by thermosyphon action rock bed storage enclosed beneath the patio. The house is warmed at night by heat stored in the rock bed, and a fireplace provides the necessary auxiliary heat. Cooling in summer is adequately provided by ventilation through open windows.

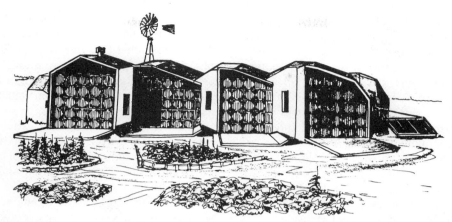

Fig. 68-10. The Baer house near Albuquerque uses water barrels mounted horizontally behind insulating panels, which can be lowered during the daytime to admit solar radiation and raised again at night to conserve the collected solar radiation.

source of heat, and a windmill produces a portion of the electricity. (See Fig. 68-9.)

The unique house built near Albuquerque by Steve Baer (Fig. 68-10), uses a version of the Trombe south wall system with the substitution of barrels of water mounted horizontally behind glass for the concrete walls of the Odeillo buildings. The walls have insulating panels in front of them that can be lowered during the daytime to allow the sun's rays to warm the blackened ends of the barrels, and at night, these walls are raised again to retain the heat within the building.

The most sophisticated of all the solar heated houses is Solar One, erected by the Institute of Energy Conversion at the University of Delaware, at Newark, Delaware, under the direction of Dr. Karl Boer and Dr. Maria Telkes. Much of the heating is done by vertical south facing windows, which are shown as bay windows in Fig. 68-11, and by using roof mounted photovoltaic cells to serve the double purpose of heat absorbers and electrical generators. These cells must be kept cool since they are of the cadmium sulfide variety and deteriorate in temperatures much above 160°F.

During the winter, the heat collected by the cells will be stored in heat-of-fusion materials developed by Dr. Telkes. The electrical

Fig. 68-11. Solar One, the solar-electric house erected at Newark, Delaware, by the Institute of Energy Conversion, University of Delaware. Solar heat is provided during the winter by vertical south facing panels in bay windows and by solar-electric panels on the roof, tilted at 60 deg to the horizontal. Cadmium sulfide cells produce both electricity and heat when irradiated, and since they must be kept relatively cool, the heat is stored in heat of fusion devices located in the basement. The electricity generated during the day will be stored in 120 volt DC storage batteries for use at night to run a heat pump that will cool the building and have its excess cooling power stored in another heat-of-fusion storage bin. This is the only Type IV house in the shape of things to come!

energy from the roof will be stored in 120 volt lead-acid storage batteries, and this energy will provide most of the electrical needs of the building, including the power required under favorable circumstances to operate the heat pump used to keep the building cool in the summer. The solar system will be activated primarily during the winter, and the hot air accumulated during the summer will simply be vented to keep the cadmium sulfide cells cool. The objective of this house is to cooperate with the utility system by pumping energy back into the system during the afternoon peak periods of the summer.

Solar One is the first house in the world to take the giant step forward of generating electricity on the rooftop, while, at the same time, heat is being collected. This is likely to be done in the future by one of a number of ways in which low cost photovoltaic cells will undoubtedly play a very important part.

Concentrating collectors capable of producing sufficiently high temperatures to operate small Rankine cycle vapor engines will also play a part when collectors able to produce high temperatures at high efficiencies become available.

During the balance of 1975, the newly formed Energy Research and Development Administration was extremely busy in getting itself organized and under way, but its major charge from Congress was to implement the Solar Heating and Cooling Act of 1974, which was passed overwhelmingly by both houses of Congress and signed by President Ford on September 3, 1974. This called for the erection of a very large number of demonstration structures in all parts of the United States to demonstrate the feasibility of space heating and water heating by solar energy. After the first three years of operating this program, it will be broadened to include solar cooling, as well, since it is expected that this technology will be perfected to the commercial stage by that time.

The day is coming

In summary, there have been only a few advances in the technology of solar energy since the publication of the 1966–67 four-part series[4] of articles in *Heating/Piping/Air Conditioning*, but thanks to the unexpected action of the Arabs in 1973, who multiplied by at least four the price of conventional fossil fuels, solar energy is now becoming a very attractive alternative. This is particularly the case in those parts of the country, now increasing very rapidly in number, where there will no longer be the possibility of making new connections for natural gas heating of houses, apartments, and other residential and commercial buildings.

The heating and ventilating engineer will be driven to the use of solar energy before the end of the 1970s by the simple force of economics. His only alternative will be kilowatt-hours, and this he cannot afford unless he employs a heat pump, and here again he will find that solar energy is an invaluable adjunct, as was demonstrated very clearly well over ten years ago by the successful operation of the Bridgers & Paxton building at Albuquerque.

According to the new legislation, ERDA will have responsibility for the implementation of the Solar Heating and Cooling Act, with Housing and Urban Development, National Aeronautics and Space Administration, and the National Bureau of Standards all playing important parts. The role of the National Science Foundation will return to that of basic and fundamental research, exploring advanced concepts in using the nondepletable sources of energy with which our earth is supplied. It will be a very wise engineer who keeps a careful eye on the developments in the fields of solar and geothermal energy because these are certain to come into much wider use as the price and availability of conventional fuels become higher and smaller, respectively.

For technical details on the availability and use of solar energy for space heating and cooling the reader is referred to Chapter 59 of the *1974 ASHRAE Handbook of Applications*.[1] This chapter is also available through the Superintendent of Documents at the price of 70 cents. Another extremely valuable source of data on solar energy and other nondepletable sources is to be found in *Energy Primer* (see Reference 3).

The field of solar space heating and cooling is about to explode with the powder keg set up by congressional action and the fuse ignited by the impending increase in cost and reduction in availability of natural gas. The farsighted HVAC engineer will start his program of self-education in this field immediately by reading everything that he can find on this fascinating subject.

References

1. *ASHRAE Handbook of Applications*, Chapter 59, 1974 ed., American Society of Heating, Refrigerating and Air-Conditioning Engineers, New York, N.Y.

2. *Solar-Oriented Architecture*, American Institute of Architects Research Foundation, Washington, D.C., 1975.

3. *Energy Primer*, Portola Institute, 558 Santa Cruz Ave., Menlo Park, CA 94025, at $4.50 per copy, prepaid.

4. John I. Yellott, Solar Energy: Its Use and Control, *Heating/Piping/Air Conditioning*, Sept., 1966, p. 101; Oct., 1966, p. 135; March, 1967, p. 114; April, 1967, p. 105.

Illustration credits

Illustrations of solar houses prepared by the College of Architecture, Arizona State University for the AIA Research Corp., under contract with the National Bureau of Standards and the U.S. Dept. of Housing and Urban Development.

69

Solar energy overview

A comprehensive look at solar developments during 1976.

JOHN I. YELLOTT, PE,
John Yellott Engineering Associates,
Inc., and Visiting Professor in
Architecture, Arizona State University,
Tempe, Arizona

"It is particularly ironic that, in the year of its Bicentennial, America must confess that in energy—the very thing that makes greatness possible—it was *not* independent." This statement is confirmed by the fact that we are now importing more than 50 percent of the petroleum that is currently the most important component of our energy budget. An even more dramatic summary of the situation was made before the Atomic Energy Forum on March 20, 1976, by Dr. Richard Roberts of the Energy Research and Development Administration (ERDA): "We stand on the edge of an energy famine, of a dark age, of approaching chaos."

Despite the fact that only in the area of natural gas are we experiencing real shortages now, we are rapidly approaching the point where *all* fuels, even including coal, will be in short supply. The reasons for this situation are many and far too complex to analyze here, but we are unquestionably on the downhill portion of the fluid fuel availability curve. Coal mines are not being developed at a sufficiently rapid pace to meet the ever-increasing demand for energy, and the atomic age is encountering difficulties that were not foreseen by the scientists who pushed nuclear technology from the theoretical to the practical stage in less than a single generation.

The only completely renewable source of energy that remains is solar radiation. The purpose of this chapter is to summarize what is happening today in the private and public sectors of our economy to put this endless resource to work in those areas for which it is particularly well adapted. The generation of heat at moderate temperatures for space and service water heating is the application that offers greatest promise of success in both the technological and economic fields.

In the public sector, the federal government, through ERDA and a multitude of other agencies, and many of the states, is beginning to mobilize resources for the purpose of implementing, after 25 years of neglect, the recommendations made by the Paley Committee in 1952. This report stated that, even then, the technology of solar energy utilization had advanced to the point where it was ready for use where sunshine was abundant and conventional fuels were expensive. Two decades later, another blue-ribbon committee of solar energy specialists echoed the same recommendations and concluded that: "Solar energy is received in sufficient quantity to make a major contribution to future U.S. heat and power quirements (Ref. 1)."

In the private sector, the rapidly increasing cost and diminishing availability of the fluid fuels have tremendously increased interest in the use of solar energy. It is apparent that the demand for solar heating and cooling systems for use *now* is far exceeding the infant solar energy industry's ability to produce and install foolproof systems at competitive prices. The areas of most intense interest are, listed in order of the temperatures that are required to make them work: swimming pool heating, space heating, absorption refrigeration, and power generation on both small and large scales. Let's take a look at each of these in turn.

Low temperature water heating

One of the oldest applications of heliothermal technology is the heating of water. Swimming pools offer an excellent opportunity to use solar energy; because pool water is usually reasonably close in temperature to the ambient air, and hence the efficiency of collection, even with an unglazed collector, can be very high. Starting in California and rapidly moving eastward, the use of simple black plastic collectors is expanding with amazing rapidity, particularly in those areas where the use of gas fired heaters is being curtailed both by high costs and by restrictive regulations.

Since pool water is usually acid, to prevent the growth of algae, plastic piping and extruded plastic collectors are being used increasingly. Their cost is relatively low, installation is simple, and they can generally be piped directly into the pool filter circulating system. Figure 69-1 shows a schematic diagram of a typical pool heating system, with the collectors mounted on a patio roof or some other suitable south facing surface.

High temperature water heating

Very little work has been done in the United States on the subject of heating water to high temperatures by the use of flat plate collectors. The Australians, however, have investigated this application in great detail, and they find that, by putting three kinds of flat plate collectors in series, they can attain temperatures up to 212° F or even higher if pressure vessels are used for thermal storage. The lowest temperature level is attained with unglazed collectors, the intermediate level is reached with single glazed collectors, and the top temperature requires double glazed collectors with selective surfaces.

Industrial hot water is needed for literally hundreds of applications, and many of these have been fueled with natural gas in the past. Now, most of these users find themselves on interruptible contracts, with oil or electricity as the only available alternative sources of energy. Oil is very likely to be rationed when the Oil Exporting Countries decide to double the price of their only exportable commodity, and the stand-by oil tanks are very probably going to be empty assets within a few years.

Figure 69-2 shows the Australian solution to this problem with unglazed, single glazed, and double glazed collectors connected in series. In the U.S., ERDA has

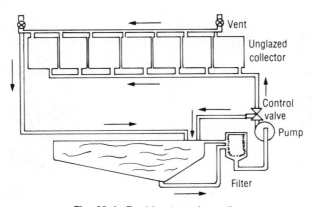

Fig. 69-1. Pool heater schematic.

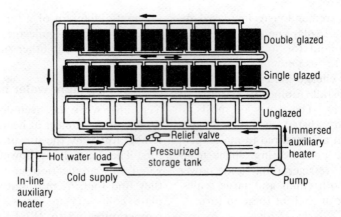

Fig. 69-2. High temperature water heater for industrial applications.

undertaken a major program to discover applications for which solar energy can be an economical solution. Where electricity is the only alternative fuel, solar energy for the base load and electricity for the auxiliary energy source will make a good combination.

Passive systems

The term "passive" is now generally applied to heating systems in which the transfer fluid (air or water) is moved by gravity alone, and neither pumps nor fans are required. The first solar heating system to be patented in the United States (E. L. Morse, Salem, Massachusetts, 1882, Fig. 69-3) used a heavy south facing slate covered brick wall to absorb solar radiation in winter. The wall was single glazed with ordinary window glass, spaced about 4 in. in front of the slate. Four dampers were used to permit the following modes of operation:

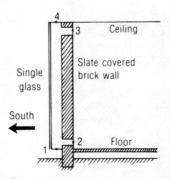

Fig. 69-3. Passive solar room heater invented by E. L. Morse in 1882.

• Dampers 1 and 4 closed, 2 and 3 open. Room air will flow upward through the air space whenever the sun is shining, due to the chimney effect caused by the heated slate, and re-enter the room near the ceiling. After the sun has ceased to shine on the wall, the dampers were closed, and radiant heat from the warm inner surface of the wall provided evening warmth to the room.

• Dampers 2 and 4 closed, 1 and 3 open. Outdoor air is drawn into the space between the glass and the slate, heated, and admitted to the room near the ceiling. Venting is accomplished through an open window (not shown) on the north side of the room.

• Dampers 1 and 3 closed, 2 and 4 open. Room air is drawn out through the wall opening near the floor and exhausted outdoors. Cool outdoor air is drawn in through the north side window.

This system has been revived by Dr. Felix Trombe, director of the great French solar energy laboratory in the Pyrenees Mountains near the Spanish border, and architect J. Michel. Starting with a single small prototype in 1967, they have applied this system to a number of residences in France, including at least one relatively large multifamily apartment complex. During the past year, a number of similar structures have been built in the United States by individuals who have read of Trombe's work.

Morse's invention has been virtually forgotten, but copies of his patent are still available from the United States Patent Office, and an account of his work is to be found in Ref. 2. As is the case with solar devices today, the published account apparently exaggerated the scope of his achievement, because he actually heated one room only and not his entire house. However, the validity of the concept was demonstrated nearly a century ago, proving once again that "there is nothing new under the sun." There is one exception to this maxim, but most of the solar-thermal ideas being heralded today as brilliantly new were tried many, many years ago.

New Mexico is the center of interest in passive systems in the United States, primarily because its high altitude produces cold winters and cool summers, with brilliant and intense sunshine. Heating is required throughout most of the year because of the characteristic 40°F diurnal swing in outdoor air temperature. Summer cooling, when it is needed, may be obtained simply by opening windows or by storing the night's "coolth" in rock beds.

The Santa Fe and Taos areas contain many examples of passive solar heating and natural cooling, as exemplified by the David Wright house, shown in Fig. 68-3 in Chapter 68. His residence, now produced in various forms in a number of homes near Santa Fe, uses a very large double glazed south facing wall to admit solar heat during winter days. The incoming radiant energy is converted into heat when it strikes the adobe walls and floor of the structure, and the fabric of the house provides the necessary storage. Loss of heat at night is minimized by using retractable panels of fireproofed foam insulation, which can be folded, accordion-wise, and drawn up into a pocket near the ceiling when they are not needed. Temperature control is also provided by simply adjusting the extent to which the panels are lowered when more sunshine is available than is needed.

Another New Mexico invention is the beadwall. It uses airborne styrofoam pellets to fill the space between the glass panes in a double glazed wall. During the daylight hours, when solar heat is needed to warm the interior of a building equipped with this ingenious system, the beads are withdrawn and stored in metal cylinders. When the sun no longer shines on the glazed surface, a small blower is used to return the beads to the interglass volume; and in as little time as 10 seconds, a 4 ft by 8 ft window is transformed from a low resistance to a high resistance but still translucent element of the building. Beadwalls, invented by David Harrison and developed by Steve Baer (Ref. 4), are now in use in many buildings in the Rocky Mountain states.

Largest system in Colorado

The largest single installation is in the south wall of the airport at Aspen, Colorado, located nearly 8000 ft above sea level and subject to extremely cold winters with heavy snowfalls. The snow is an asset of major proportions, since the reflectivity for solar radiation of newly fallen clean snow is very high, approaching 80 percent. South wall radiation in winter approaches 275 Btuh per sq ft because of the low solar elevation and the favorable incident angle between the sun's rays and the glazing. A layer of fresh snow on the ground on the south side of a vertical window can easily add another 200 Btuh per sq ft.

Many of the New Mexico solar homes use air systems to eliminate the freezing problem that must be solved if a liquid is used as the heat transfer fluid. Rock beds are generally used for heat storage with air circulation provided either by gravity alone (see Ref. 3, Chapter 68) or by fans. Adobe is the preferred building material throughout the northern half of New Mexico. It can generally be made from the earth that is excavated to provide space for the rock bed storage bins and to lower the floor level below grade for the purpose of insulating at least part of the structure against the extreme cold experienced in Rocky Mountain winters. Used by the Indians for a thousand years before the white man came to the southwest, adobe has good insulating qualities and adequate bearing strength for low buildings.

In the warmer, low elevation portions of

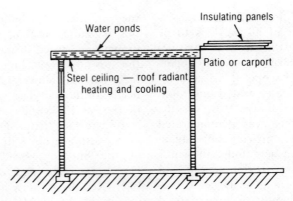

Fig. 69-4. Schematic cross section of the Hay Skytherm system showing thermoponds on roof and movable insulating panels stacked over unenclosed area of house.

the southwestern states, and even over into southern California, the Skytherm system invented by Harold Hay of Los Angeles (Ref. 3, Chapter 68) has proven its ability to provide both winter heating and summer cooling without the use of any auxiliary energy. Skytherm uses metal roof panels to serve as the ceiling of the house and to support plastic bags filled with water to a depth of about 8 in. Movable urethane insulating panels are mounted above the "thermoponds," and a combination of tracks and rollers permits the panels to be moved to cover or uncover the roof ponds. First treated rigorously in Phoenix for 18 months in 1967–68, the system proved to be capable of providing adequate heat during all but a few winter hours without any auxiliary energy supply other than the body heat of the occupants and a small amount of lighting used for a few hours at night.

System tested by HUD team

During 1974 and 1975, the first full-scale Skytherm residence at Atascadero, California, was tested with meticulous care by a HUD supported team of architects and engineers from California State Polytechnic University at San Louis Obispo, and it was found to remain in the comfort zone, despite the very wide range of temperatures (105°F in summer down to 20°F in winter) that are experienced in California's coastal range. Summer cooling is accomplished by rolling back the insulating panels at night

and allowing the heat gained during the day to be dissipated to the sky by radiation, convection, and evaporation.

Early in the Phoenix testing program, it was found that enclosing the water in the thermoponds was necessary to prevent undesirable cooling by evaporation during the fall, winter, and spring months when heating was the primary requirement. However, during periods of high nighttime dew point temperature, it is necessary to call upon the very potent cooling power of evaporation (Ref. 5). This is accomplished by flooding the enclosed ponds with an inch or two of water and allowing some of it to evaporate at night, thus cooling the enclosed water. This in turn provides radiant cooling for the building. Forced air movement under the insulating panels during the day was also found to be effective although it involved the use of a small quantity of electric power, which nocturnal cooling did not need.

It is interesting to note that in the Solar Heating and Cooling Demonstration Act of 1975, which established the program that ERDA is implementing, solar cooling specifically includes nocturnal cooling by radiation, convection, and evaporation. The Skytherm system was the first to be approved by the Department of Housing and Urban Development (HUD) for solar heating and cooling. The house at Atascadero, built at Mr. Hay's personal expense, has been recognized by the Bicentennial Commission as one of the 200 concepts brought to reality since July 4, 1776, that possesses potential for maintaining and improving the American way of life during the coming "third century" of our nation's existence. A complete report on the evaluation of the Atascadero house, carried out with a modest grant from HUD, is now available (Ref. 6).

Active solar heating and cooling

The distinction between passive and active solar heating and cooling systems is made by determining the amount of nonrenewable energy that the system requires to make it function. A south facing glazed area using the beadwall system is classified as passive, since the amount of energy used by its miniature blowers during their two brief periods of daily operation is negligibly small, as is the energy used at Atascadero by the small geared-down motor that opens and closes the insulating panels in response to commands from a thermostat. A system that uses a pump or a fan to circulate its working fluid is classified as active, because the energy used by the circulating system is not negligible.

Active systems have proliferated to a very large extent during the past year, primarily as a result of the rapidly rising cost of electricity and the decreasing availability of natural gas. LPG and light oil are both difficult to obtain and far more expensive than they were before the Arab oil embargo, and most builders of new structures of all kinds are having to turn to electricity to provide all of their energy needs. At 5 cents per kWh, which is the current residential price in Arizona, and 10 cents per kWh, which is the going rate in New York City and northern New Jersey, solar energy is already competitive for space heating and domestic hot water, and so a discussion of recent developments in this field is in order.

The impact of ERDA

The major new entrant into the solar heating and cooling scene is the Energy Research and Development Administration (ERDA), which was created by Congress late in 1974 to draw together all of the nation's energy related research activities as explained in Chapter 68. More than 94 percent of ERDA's budget for fiscal 1977 will be spent on nuclear projects, including weapons, power systems, and the highly controversial breeder reactor. The remaining 6 percent will be spent on all other sources of energy, including solar, wind, geothermal, ocean temperature gradients, and any other new projects that ERDA considers worthy of serious attention, including improved methods of using fossil fuels in general and coal in particular.

A major project that ERDA had to finalize before the end of 1976 was the locations and activation of the National Solar Energy Research Institute, which is specifically required by the legislation that created ERDA. One of the stipulations issued by ERDA was that the winning team must be ready to begin work by January 1, 1977.

The original intention of Congress, in including the requirement for the establishment of NSERI in the ERDA legislation, was that a national center of excellence in solar energy research and development would be created. ERDA has greatly limited the scope and importance of NSERI, however, and has stated that no outdoor experimentation would be done for the first three years of the Institute's existence. By the end of 1979, ERDA will evaluate NSERI's accomplishments and then decide whether to expand the Institute into something that may resemble the original Congressional intent, or to abolish it. The Request for Proposals stated that the winner of the competition must be able to guarantee the availability of an outdoor site covering at least 300 acres in close proximity to the office space that will be NSERI's initial home. The original concept of NSERI as spelled out in the ERDA Request for Proposals calls primarily for a management team that can supervise ERDA's far-flung research activities. No in-house research is contemplated for the first three years of NSERI's existence.

The Solar Heating and Cooling Demonstration Act of 1975 called on ERDA to establish a nationwide program to demonstrate the feasibility of solar heating and the provision of service hot water, and subsequently to include solar cooling in the program. ERDA has delegated to HUD responsibility for the residential demonstration program, and HUD has awarded contracts for the application of solar heating equipment to approximately 100 residences to be located in a somewhat smaller number of carefully chosen cities throughout the nation. This is a far cry from the 4000 demonstration residences that were envisioned by the Congressmen who pushed the enabling legislation through the House and Senate.

The nonresidential portion of the demonstration program is being handled by the National Aeronautics and Space Administration (NASA) under a subcontract from ERDA. Thirty-three buildings, ranging from small office buildings to very large secondary school and university complexes, have been selected by ERDA from among several hundred proposals, and work on the detailed designing or actual construction of these systems is currently under way. Eight of the awards went to heating and cooling applications; one (in the Virgin Islands) is for cooling alone, since no heating is needed in that location; and the remainder are for heating and, in most cases, provision of service hot water as well.

The private sector

At least five hundred buildings are in various stages of design or construction throughout the U.S. in the private sector, with solar energy selected primarily on economic grounds. The virtual elimination of new natural gas connections and the rapid escalation in electric rates are responsible for this sudden rise in real, as opposed to emotional, interest in solar energy utilization. Since the solar energy industry is what the economists would call "marginal," in that there is no requirement for large amounts of capital or equipment to enter it, hundreds of small companies

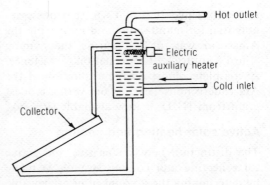

Fig. 69-5. Side view of a typical thermosyphon hot water heater with electrical insertion auxiliary heater.

across the nation are getting into the picture along with a few industrial giants.

Solar water heating coming

Solar water heating for domestic purposes is rapidly emerging as a widely used application of solar technology. The thermosyphon heater, Fig. 69-5, is making a comeback to the prominence that it enjoyed half a century ago. Requiring no moving parts and suitable for use at regular city pressure, it can produce most of the domestic hot water requirements for homes in the southern half of the U.S. With double glazing and added insulation, it can provide up to half of the needs of homes as far north as New England.

A problem that must be faced is freezing, which will occur if long periods of subzero weather are encountered. If an antifreeze solution is employed to overcome this problem, a heat exchanger must be used to separate the circulating fluid from the potable water supply. In some areas, the separation must be accomplished by the use of two metal barriers to eliminate any possibility of contamination. One way to accomplish this is to wrap a coil of copper tubing tightly around the outside of a conventional metal hot water tank and then to put a thick layer of insulation around the entire assembly. This is being done successfully in both Florida and California, and since the idea is not patentable, it will probably be more widely used in the future.

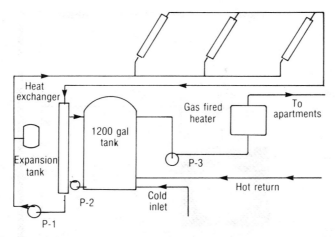

Fig. 69-6. Forced circulation hot water heating system as used in project SAGE.

Utility companies show interest

A number of large utility companies, both gas and electric, are taking serious looks at solar energy as a means of supplementing their diminishing reserves. The largest such operation is Project SAGE (Solar Assisted Gas Energy), a joint venture of the Southern California Gas Co. and ERDA, with the Jet Propulsion Laboratory (JPL) of the California Institute of Technology handling the technical end of the project. SAGE uses a forced circulation system, as shown in Fig. 69-6, to transfer heat gathered by south facing roof-mounted collectors to a storage tank from which the hot water requirements of a large number of apartments are met.

Economic as well as technical problems are being studied to see what is the most effective manner for a gas utility to get into the solar energy business. Should the collectors be leased or sold to the apartment owners, or should the utility provide hot water and charge for it on a Btu delivered basis? Who should be responsible for maintaining the equipment? How long will today's collectors, pumps, and storage tanks last? These are among the questions that SAGE seeks to answer.

The Solaris system

Marked improvements have been made during the past year in the Solaris system,

invented 20 years ago by Dr. Harry Thomason of Washington, D.C. Figure 69-7 shows the arrangement of his Solaris No. 6, erected near Washington during 1975 and now about to undergo a year of careful testing under the auspices of George Washington University, financed by ERDA. The latest version of Solaris, now represented by some 25 residences ranging in location from North Carolina to New England, is shown by Fig. 69-7. The "trickle collector" described in detail in Ref. 3, Chapter 68, is mounted on a steeply tilted south facing roof, where its corrugated aluminum collector surface is well insulated by the 6 in. thick glass fiber batts mounted between the rafters. Water runs down through the corrugations in a thin film from a perforated copper pipe along the roof line to a gutter at the bottom of the collector. Gravity then takes the water, heated to about 100 to 110°F, to the storage tank. When the pump stops, either at the command of the collector thermostat or because of a power outage, the collector runs dry. The freezing problem is thus solved without the need for antifreeze fluids.

Approximately 60 tons of rocks, poured into the bin around the water tank, provide both additional heat storage and a very large expanse of heat transfer surface by which the rocks can transfer heat to the air that is circulated throughout the house. Relatively low temperatures can be used

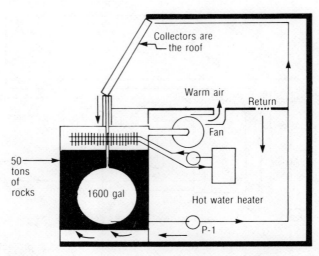

Fig. 69-7. Schematic diagram of Dr. Harry Thomason's Solaris system.

successfully to heat the house, and this enables the simple and inexpensive Solaris collectors to function with good efficiency. Stand-by heating is provided by a 200 ft long finned tube heat exchanger that receives hot water when necessary from the oil fired domestic water heater.

Summer cooling also provided

Summer cooling can be accomplished by using a small compression refrigeration system, running only at night, to chill the water in the tank. The rocks are cooled in turn, and when the house air is circulated through the rock bed, it is both cooled and dehumidified.

Solaris is by far the least expensive of the active heating and cooling systems that are now available. The collectors cost about $4 per sq ft, and the lightweight storage tanks cost about $700. Controls are extremely simple, consisting primarily of one collector mounted thermostat to start and stop the pump, and a room thermostat to control the blower's operation. The first of the Thomason houses was built in 1959, and it, along with succeeding editions, is still working satisfactorily.

Complex systems also perform

More complex systems, using closed collectors with the water or water plus anti-freeze circulating through tubes or integral water passages, are now coming into wide use where heat is the only requirement. Figure 69-8 shows such a system schematically. Collectors, mounted in an unshaded south facing location, are provided with water from a well insulated storage tank. The circulating pump (P-1) is controlled by a differential thermostat, which has one temperature sensor mounted on the collector surface and the other located near the bottom of the tank. When the sun is shining, the collector surface will be heated to the

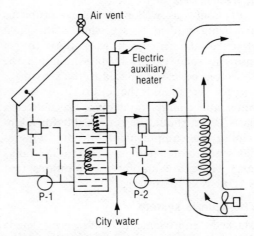

Fig. 69-8. Solar space and hot water heating system with electric auxiliary heaters.

point where it becomes warmer than the water, and P-1 will begin to operate. When the sun is no longer capable of heating the water, the thermostat will shut down the pump.

One demonstrated method of preventing freezing with this system is to allow an air vent at the top of the collector to open when the pump stops, thus returning the water to the storage tank. Another widely used system employs antifreeze plus water in the circulating system, with a heat exchanger, as shown in Fig. 69-6, to transfer the heat to potable water in the storage tank.

In most parts of the country, refrigeration is needed to provide cooling, particularly for large buildings with high internal loads. Dehumidification also generally requires the use of chillers, which can reduce the temperature of the recirculated air below its dew point. Heat from highly efficient solar collectors can be used to "fire" (a term the writer deplores!) lithium bromide water absorption chillers as long as the delivered water temperature remains above 190°F. Figure 69-9 shows the system that is now in wide use in many installations. Three pumps and two valves are required, with suitable controls to turn pumps on

and off, to start and stop the fan, and to position the control valves. The collectors used in these systems must have the ability to maintain high efficiency at temperature differences between ambient air and collector water exceeding 100°F.

When heating is needed, P-2 goes to work and valves V-1 and V-2 are positioned to direct the sun warmed water through the heating coil, which is mounted in the air duct. Moderate temperatures, below 140° F, are needed for this purpose, but since ambient temperatures on cold winter days will generally drop below 40°F, the collectors must still be able to gather solar heat efficiently with a 200°F difference between the water and the air. P-1 operates whenever the collectors are hot enough to add heat to the water stored in the tank.

When cooling is needed, valves V-1 and V-2 are repositioned to direct hot water from the storage tank into the absorption chiller, which in turn provides cold water to the cooling coil, requiring a fourth pump for this purpose. A cooling tower is also needed, with P-3 to circulate water from the chiller to the tower. Air cooling is not adequate for today's lithium bromide water chillers, particularly when they are oper-

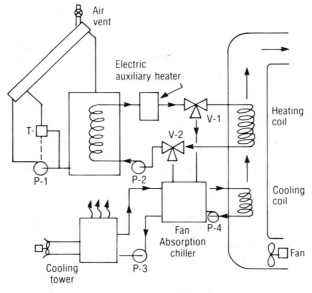

Fig. 69-9. Solar heating and cooling system.

ating at the very bottom of their temperature range near 190°F. Derating of the chiller also occurs when its operating temperature falls below 220 to 200°F, so collectors capable of operating efficiently at temperatures up to 250°F are needed to make solar cooling attractive from an economic point of view.

A solar collector that can maintain high efficiency at elevated temperatures has been developed and demonstrated during 1975. It uses an extruded acrylic linear Fresnel lens to concentrate a 1 ft wide beam of direct solar radiation onto a 2 in. wide copper receiver tube. The acrylic material used in the lens has been shown to have a life in the New Mexico desert of at least 20 years.

Use concentrating collectors

Concentrating collectors characteristically must be able to follow the sun as it makes its daily journey across the sky. This can be accomplished in several ways but perhaps the simplest is to use a polar mount, tilted at the angle of the local latitude and directed toward the North Star, to support the collector. Rotation at the rate of 15 deg per hr will then keep the concentrator trained on the sun.

Another method of regulating the collector's position is to use two small silicon photovoltaic cells, mounted on a wedge that is located on the center line of the concentrator. When the solar rays are striking both cells with equal intensity, the circuit is balanced, and the collector remains stationary. As soon as the sun moves (actually it is the earth that moves and not the sun) so that one cell receives more radiation than the other, the circuit becomes unbalanced, and a small geared-down motor moves the positioning device to make the collector face the sun again and thus return the control circuit to a balanced condition. At the end of the day, a limit switch is tripped, and this reverses the positioning motor and returns the concentrator to the sunrise position.

A sun following collector has a major advantage in summer over a fixed flat plate collector, because the sun follower can begin to collect energy soon after the sun rises in the northeast quadrant of the sky. The fixed collector must wait until the sun moves into the southeast quadrant before it can receive enough radiation to begin working. The concentrating collector can use only the direct rays from the sun, while the flat plate unit can capture both the direct and diffuse radiation.

The photos show a battery of three concentrating collectors, which are controlled by a single solar cell combination, mounted on the center collector. A single actuator, enclosed in the box at the top of the array, positions the three units simultaneously; the same actuator can operate as many as 30 units by means of a cable-pulley arrangement.

When the linear Fresnel lens is combined with a tubular receiver coated with a selective surface and mounted within an evacuated glass tube to eliminate convection heat losses, temperatures in the 500°F range can be attained with good efficiency. This combination, now under development, offers promise of providing a working fluid, probably "freon" vapor, which can run turbines or refrigeration compressors with very good cycle efficiency. The absorption chiller has a Coefficient of Performance below 1.0, while compression refrigeration systems, particularly when they can operate at variable speed to accommodate load changes, can reach COPs in the range of 4 to 5. The concentrating collector, despite its inability to use a diffuse radiation in its present stage of development, offers an immediate answer to the problem of making absorption function at their rated output. In the future, concentrators may well make possible the provision of both heating and cooling, with electric power as the third contribution of the system.

Growing use of solar energy

Solar energy is now beginning to be used on a scale far larger than has ever been the case in the U.S. and in other industrialized nations as well. Heating of space and hot water with simple flat plate collectors is the most widely used application, but the development of simple passive space heating systems is making solar energy a valuable asset for builders and homeowners in the cold Rocky Mountain part of the United States. Process hot water at moderate temperatures for washing and other essential applications is now being studied by industries that are confronted with exorbitantly high energy costs. Absorption refrigeration can be done with flat plate collectors of advanced design under favorable conditions; concentrating collectors, which can attain and maintain temperatures above 250°F can make a major improvement in the performance of lithium bromide water chillers. For year around applications, the solar assisted heat pump, first developed in Japan 30 years ago, can make good use of low temperature flat plate collectors.

Electricity is certain to be the back-up energy source for most solar energy installations, since natural gas is rarely available for new installations, and both light oil and LPG are expensive and uncertain in availability. Solar radiation, despite its relatively low intensity and its climatic uncertainty, can be used and is being used now. Its use will increase rapidly in the future, as the effect of fuel shortages is felt with rising intensity.

References

1. *Solar Energy as a National Energy Resource*, National Science Foundation, Washington, D.C., 1972, p. 5.
2. *Scientific American*, May 13, 1882, account of E. L. Morse's sun heated house in Salem, Mass.
3. Yellott, J. I., "The prospects for solar energy since 1967," Chapter 68.
4. "Proceedings of the Conference on Passive Heating Systems for Solar Heating and Cooling," held at Albuquerque, N.M., May 18–20, 1976; available from: Solar Energy Section, Los Alamos Scientific Laboratory, Los Alamos, NM 87545.
5. Hay, Harold and Yellott, John, "Natural Air Conditioning with Roof Ponds and Movable Insulation"; Yellott, John and Hay, Harold, "Thermal Analysis of a Building with Natural Air Conditioning," ASHRAE Transactions, Vol. 75, Part I, 1969.
6. Niles, Philip; Haggard, Kenneth, and Hay, Harold, "Research Evaluation of a System of Natural Air Conditioning," Document PB 243498, $10, National Technical Information Service (NTIS), Springfield, VA 22154.

70
Solar gain control

A computer simulation study analyzes energy consumption as related to solar gain and finds that the situation can be turned to your advantage with rotatable louvers.

DR. GIDEON SHAVIT,
Chairman of Advanced Engineering,
Honeywell Commercial Div.,
Arlington Heights, Illinois

Solar load is one of the primary factors in a building's energy consumption. Unfortunately, solar load is usually considered only in the initial design stages of a building and then only to size the load requirements of the central plant. So solar load remains pretty much an uncontrollable energy source throughout the life-cycle of a building. An uncontrollable energy source is one that imposes a load on the central plant and thus increases a building's overall energy consumption.

Think positive about solar load

The most common methods to reduce a building's solar load are:
- Minimum window areas
- Internal shading
- Reflective glass
- Orientation
- Aspect ratio
- External shading

All of these methods cut the solar load and thereby reduce the annual energy consumption. But they all treat the solar load as a negative factor, something to be overcome; none try to make maximum utilization of solar energy.

Each of the six techniques has associated advantages and disadvantages. The following discussion details these for each method.

Minimum window glass

Minimizing window areas has the following advantages:
- Reduces the cooling load during the summer.
- Reduces heat gain and loss by conduction and convection throughout the year.

The following are considered disadvantages:
- Increases heating load on sunny winter days.
- Decreases the outdoor diffuse light, thus increasing the required artificial lighting level in perimeter zones.
- Window area is a design parameter and not a control variable.

Internal shading

The advantages of internal shading are as follows:
- Decreased solar load on the perimeter zone.
- Increases comfort by reducing direct solar radiation on occupants.

Disadvantages of internal shading are:
- Solar energy absorbed by the shading

devices and trapped between them and the window is convected and radiated to interior spaces.

• Reduces diffused outdoor light and thus increases the artificial light requirements.

• A design parameter and a manual control variable.

Reflective glass

Advantages of using reflective glass are:

• Reduces the solar energy transmitted into the space during the cooling season.

• Reduces artificial lighting levels in perimeter zones.

Disadvantages of using reflective glass are:

• Reduces the contribution of solar energy in winter.

• Increases cost of glass.

• Requires an intensive analysis to establish the type of glass and window area that will minimize the annual energy consumption.

• A design parameter and not a control variable.

Orientation

The advantage provided by an optimum orientation is:

• Reduces a building's annual energy consumption.

The disadvantages are:

• Requires an intensive analysis for a given structure at a given geographical location.

• A design parameter and not a control variable.

Aspect ratio

The advantage of applying aspect ratio is:

• Reduces annual energy consumption.

The disadvantages are:

• Usually governed by lot size and shape.

• Requires intensive analysis to establish optimum configuration.

• A design parameter and not a control variable.

Fixed external shading

The advantages provided by fixed external shading are:

• Reduces solar load in summer.

• Uses solar energy to help heat perimeter zones in winter.

• Reduces convective gains and losses by reducing air movement near windows.

The disadvantages are:

• Increases the cost of a structure.

• A design parameter and not a control variable.

Building orientation, aspect ratios, and external shading can all help optimize the annual solar load. Therefore, it was decided to make a computer simulation study of how these parameters affect solar energy gain. The simulation capabilities were developed to analyze energy consumption in a building as a function of the structure, external and internal loads, mechanical system, and associated controls.

Select building for study

A building in Pittsburgh (latitude 40 deg) was selected for the simulation study. For clarity, only a module of the external zone, including the building envelope, for each exposure was simulated. To reduce the complexity of the problem, the same window area and type of glass (double strength, single pane with shading coefficient of 0.3) were used throughout the study. The window assumed an area of 49 sq ft (6.125 by 8 ft).

The simulation was performed for two different module configurations: a window without overhang (Fig. 70-1a) and a window with overhang (Fig. 70-1b). Measurements were made at the following angles of orientation for each configuration: 0, 22.5, and 45 deg. An orientation angle is the angle between the north and the perpendicular to the north exposure, clockwise rotation. Thus at a 0 deg orientation angle, the north exposure of a building faces true north.

Figures 70-2, 70-3, and 70-4 show the results for the configuration without overhang (Fig. 70-5, 70-6, and 70-7), show the results for the configuration with overhang (Fig. 70-1b).

Figures 70-2 through 70-7 give the total solar load on the space for January, April, July, and October. Charts were also prepared for the same months at an orientation angle

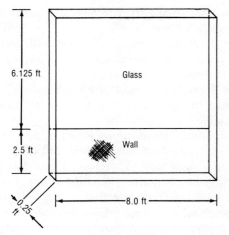

Fig. 70-1a. Window test module without overhang.

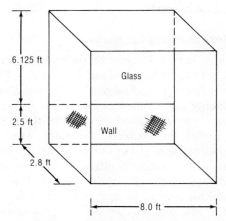

Fig. 70-1b. Window test module with overhang.

of 67.5, but in the interest of brevity, these were not included. Results obtained at this angle can be approximated since 67.5 is inverse to 22.5. The solar load is comprised of the transmitted energy plus the energy absorbed by the glass and convected into the space.

Figures 70-2, 70-3, and 70-4 reveal that there is a preferred orientation and aspect ratio of a window without an overhang at different times during a year. But to establish the optimum, it is necessary to simulate the entire building with the proposed mode of operation and to compare annual energy consumptions for the various combinations. Once the structure is completed, there is no way to control solar energy except with internal shading.

Figures 70-5 through 70-7 indicate a definite decrease in solar load on the external zone in the summer, and a smaller reduction in the winter for windows with overhang. This solar load reduction during winter is not necessarily a positive factor since the sun can help heat the building's exterior.

The effects of aspect ratio can be easily assessed by counting the desired number of modules in each exposure and multiplying by the calculated solar load per module for the given exposure.

The aspect ratio, orientation, and external overhang can be used effectively to optimize the total contribution of the solar load without changing the window area. However, there are three major disadvantages. The building lot does not always give the necessary freedom to optimize the aspect ratio. Optimum orientation and configuration require intensive analysis. Finally, the solar load cannot be utilized by a space on a daily basis as a function of the space energy requirement. Additionally, orientation, aspect gain, and fixed external shading do not provide feasible solutions for existing buildings. Therefore, there is a need to control solar load on the space to overcome these deficiencies.

Turn the situation to your gain

External shading provides a design solution to this problem. Ideally, such external shading devices would increase or decrease the solar load on the space as a function of the zone requirements. A simple solution is external louvers that by rotation change the effective area of the window subjected to direct solar radiation. This change, then, is a function of the control system in the external zone. Figure 70-8 is a simple schematic of how variable external shading might be integrated with control sequencing.

The advantages of movable external shading (rotating louvers) are as follows:

• Optimizes the effect of solar load on buildings every day of the year.

• Reduces convective losses and gains as a function of wind velocity.

• Minimizes obscuration of diffused outdoor light, hence reduce level of artificial

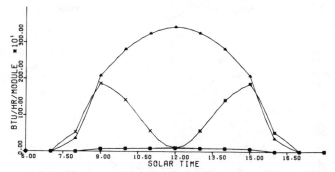

Fig. 70-2a. Solar load measured for January, 0 deg orientation angle, window without overhang.

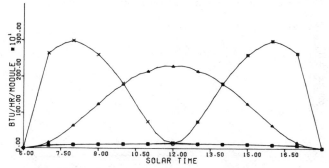

Fig. 70-2b. Solar load measured for April, 0 deg orientation angle, window without overhang.

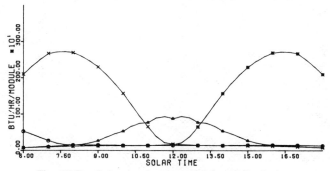

Fig. 70-2c. Solar load measured for July, 0 deg orientation angle, window without overhang.

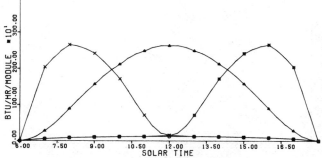

Fig. 70-2d. Solar load measured for October, 0 deg orientation angle, window without overhang.

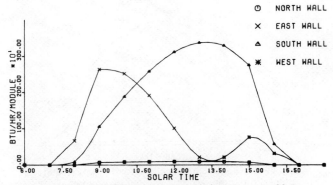

Fig. 70-3a. Solar load measured for January, 22.5 deg orientation angle, window without overhang.

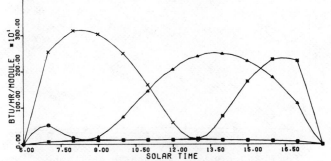

Fig. 70-3b. Solar load measured for April, 22.5 deg orientation angle, window without overhang.

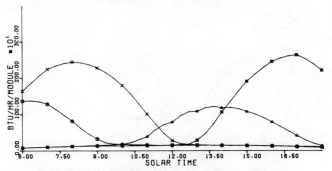

Fig. 70-3c. Solar load measured for July, 22.5 deg orientation angle, window without overhang.

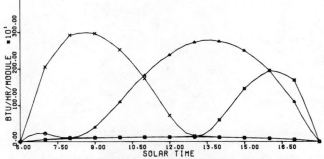

Fig. 70-3d. Solar load measured for October, 22.5 deg orientation angle, window without overhang.

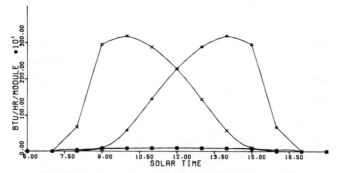

Fig. 70-4a. Solar load measured for January, 45 deg orientation angle, window without overhang.

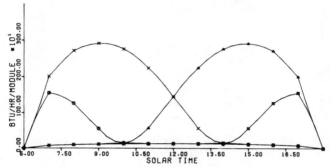

Fig. 70-4b. Solar load measured for April, 45 deg orientation angle, window without overhang.

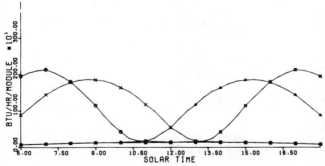

Fig. 70-4c. Solar load measured for July, 45 deg orientation angle, window without overhang.

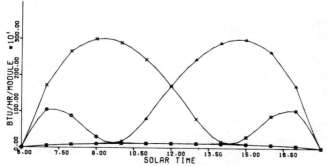

Fig. 70-4d. Solar load measured for October, 45 deg orientation angle, window without overhang.

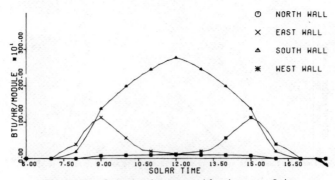

Fig. 70-5a. Solar load measured for January, 0 deg orientation angle, window with overhang.

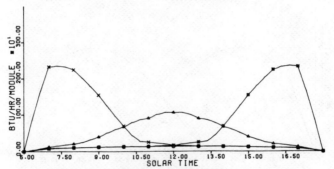

Fig. 70-5b. Solar load measured for April, 0 deg orientation angle, window with overhang.

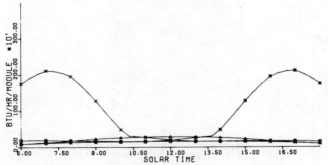

Fig. 70-5c. Solar load measured for July, 0 deg orientation angle, window with overhang.

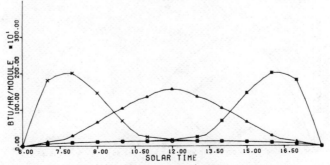

Fig. 70-5d. Solar load measured for October, 0 deg orientation angle, window with overhang.

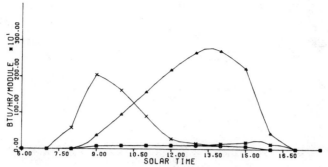

Fig. 70-6a. Solar load measured for January, 22.5 deg orientation angle, window with overhang.

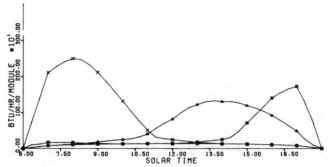

Fig. 70-6b. Solar load measured for April, 22.5 deg orientation angle, window with overhang.

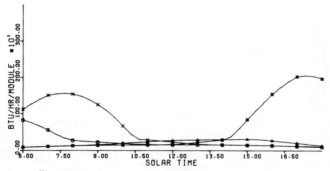

Fig. 70-6c. Solar load measured for July, 22.5 deg orientation angle, window with overhang.

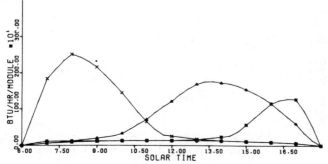

Fig. 70-6d. Solar load measured for October, 22.5 deg orientation angle, window with overhang.

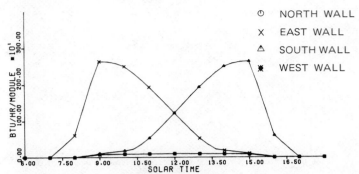

⊙ NORTH WALL
× EAST WALL
△ SOUTH WALL
✱ WEST WALL

Fig. 70-7a. Solar load measured for January, 45 deg orientation angle, window with overhang.

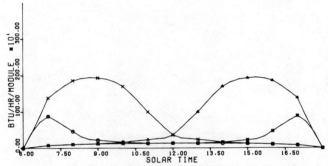

Fig. 70-7b. Solar load measured for April, 45 deg orientation angle, window with overhang.

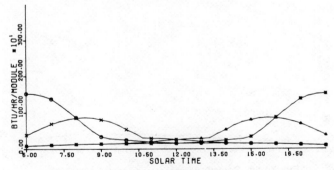

Fig. 70-7c. Solar load measured for July, 45 deg orientation angle, window with overhang.

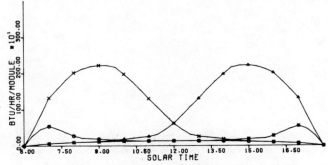

Fig. 70-7d. Solar load measured for October, 45 deg orientation angle, window with overhang.

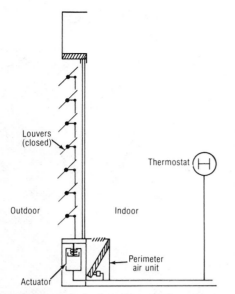

Fig. 70-8. Variable external shading (movable louvers) integrated with control sequencing.

lighting in external zone (compared to internal shading or smaller window area).

• Provides a control variable that can be integrated into the control system without difficulty.

• Offers a system that can be designed into new buildings or retrofitted to existing buildings.

• Does not require intensive analysis.

• Less expensive than fixed external shadings that are an integral part of a building.

• Eliminates internal shading.

• Possibly reduces glass costs by eliminating or minimizing reflecting glass.

• Allows any desired orientation and aspect ratio.

One disadvantage is that it may increase building costs somewhat.

Latitude 40 deg was simulated to provide the total energy required to maintain comfort in the external zone using a module with fixed external shading (Fig. 70-9). The simulation includes the building envelope, outdoor and indoor conditions, the mechanical system, and the associated control. The results for the north exposure represent energy required to overcome conduction and convection losses only. All deviations from this line contribute to the solar load.

All energy consumed by the building (below the heavy line, Fig. 70-9) is a function of direct solar radiation on the window and could be saved if there were any external means to control the solar radiation.

Figure 70-10 represents the results for the same configuration for a mild winter

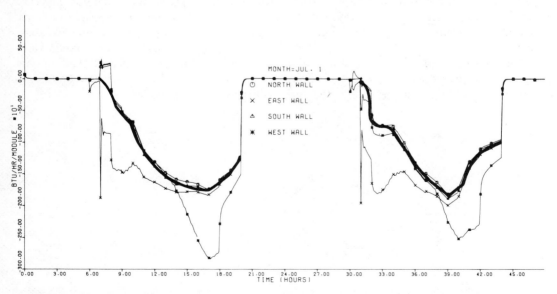

Fig. 70-9. Energy requirements to maintain comfort in the external zone using a fixed external shading for the month of July, 40 deg latitude. The area below the heavy line represents all energy consumed by the building that is a direct function of direct solar radiation on the window and could be saved by an external method of controlling solar radiation.

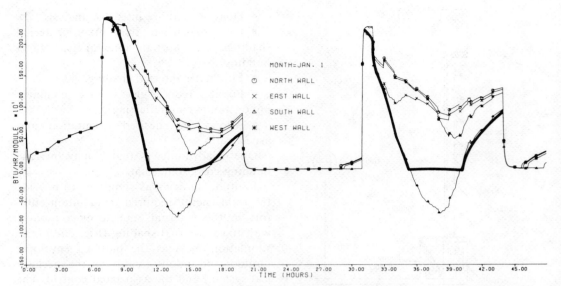

Fig. 70-10. Energy requirements by the external zones on a configuration with overhang, 40 deg latitude, on a mild winter day in January. The area below the heavy line represents the energy savings possible by controlling the solar load in the external zones.

day. All the exposures call for heating, except the southern one needs cooling due to a high solar load on the space. Optimum control of this solar load on the southern exposure will eliminate this unnecessary cooling while maintaining comfort for the same period of time without any heat. Similarly, the heating load on the eastern and western exposures could be compensated by solar load if the shading coefficient of the glass were greater than 0.3.

Orientation, fixed external shading, and aspect ratio can reduce the energy consumption by a building. Unfortunately, the op-timum configuration is not always feasible. Thus, control of the solar load as a function of the energy required by the space provides the optimum solution. A movable external shading (rotating louvers) provides a mechanism to regulate incident solar radiation on the window, meaning an optimum solution to the problem throughout the year. Additionally, this provides all the functions that are obtained by other parameters, such as orientation, aspect ratio, and fixed external shading. Movable external shading can be put on both existing and new buildings, which is a definite advantage.

71
Solar energy in Europe

A. A. FIELD,
London, England

Before the energy crunch, solar collectors were considered something of a curiosity in Europe. The few existing installations were mostly in regions bordering the Mediterranean, and the technology was imported—Australia, South Africa, and Israel.

Now, in spite of very long payback periods and the use of collectors almost exclusively as supplementary base load units, solar collectors are attracting widespread interest throughout Europe. Germany recently launched a research program funded at the rate of 20 million DM per year; and the government backed French research organization CoSTIC opened a teaching and research center to specialize in solar collector design.

In the United Kingdom

The United Kingdom was one of the first European countries to experiment with full scale solar heating, but recently a British branch of the International Solar Energy Society was formed. In 1975, the Copper Development Association (CDA) sponsored a competition for space heating or hot water service by means of solar heat collectors, and there were nearly 100 entries.

Recent observations by Professor Page of Sheffield University showed that the potential for solar energy in the United Kingdom is about half that of Australia and about two-thirds that of the southern United States. Although United Kingdom's sunshine record is not particularly good compared with these countries, there is always a significant gain from diffuse radiation on sunless days; such radiation accounts for roughly one-half of the total annual insolation.

A pioneer in the field in the United Kingdom was Professor H. Heywood, whose work in the late 40s and 50s at Imperial College, London, established most of the basic design parameters for flat plate collectors. Heywood proved, in the face of considerable scepticism at the time, that England's climate did not necessarily exclude solar energy. Although at the time, the economics were completely against it. His original collector was made from two corrugated steel sheets placed together so that the corrugations formed waterways. The construction was similar to present day units with glazed front and insulated back. Average daily recovery varied from 300 to 600 Btu per sq ft—spring to summer—with a maximum of 800 to 1000 Btu per sq ft. Efficiency ranged from 47 to 54 percent.

An ambitious attempt at all-solar heating was made by architect Emslie Morgan in 1961 with his design for a two story school building in the north of England (53 deg latitude). The building has a total volume of 250,000 cu ft, and heat is collected by a 10,000 sq ft double glazed solar wall. Insulation is the equivalent of 5 in. of expanded polystyrene on walls and roof. Natural convection and radiation from screens backing the inner pane heat the 38 ft deep rooms, and temperature swings are reduced by the 9 in. thick masonry inner walls.

Observations over a period of years showed that average monthly mean internal temperature varied from 66°F in winter to 73°F in summer. The relatively small temperature rise in summer, when the solar gain is roughly twice that in winter, results from part of the winter energy input being from lighting and the reduced ventilation rates. In summer, free use is made of windows and reflecting screens turn back solar radiation. The mass of the inner partitions limits daily temperature swing in winter to about 8°F. So far, this remains the only building of its size to be solar heated in the United Kingdom, and its design is unlikely to be repeated. It does not offer sufficient control over internal temperature, and it relies on an undesirably low ventilation rate in winter to reduce the load. A design rethink today would probably use mechanical ventilation with heat recovery and automatic solar gain control.

New city solar development

A number of solar heated houses have recently been completed in the United Kingdom, and many more are projected or under construction. Possibly the most interesting so far—because it could form the prototype for standard construction dwellings—is the government financed project at the new city development of Milton Keynes, north of London. Designed from a feasibility study by S. V. Szokolay, the two story house is intended to provide a major part of the annual energy needs for heating and service hot water from the collector, with the balance coming from a gas fired heater. The single glazed, flat plate unit takes up part of the 30 deg sloping roof and consists of black anodized sheets with integral waterways, ensuring a temperature difference of not more than 3°F between plate and circulating water. The total receiver area of 400 sq ft relates to a total floor area of 970 sq ft. Water storage capacity—a total of 1200 gal. in two tanks, both insulated with 4 in. of glass wool—provides a one day reserve in winter and prevents boiling in summer. Design water conditions are 104 to 77°F with forced convection emitters. Service hot water comes from separate exchangers within the tanks. Initial operating results are expected to show over 50 percent savings in energy.

Aiming at a possible future need for complete decentralization of energy and group services, such as water and sewage, is Cambridge University's design for the so-called *autarkic* (self-sufficient) house (Fig. 71-1). The project was launched in 1972 with a three year grant of $50,000 from the Science Research Council. The theoretical work has been completed, and a prototype building constructed. The *autarkic* house will get heat from a solar collector; power from an aerogenerator, rain water via a purification system and condensed water from a solar still;

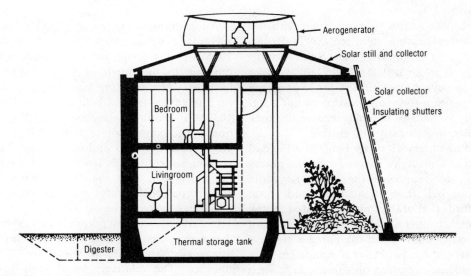

Fig. 71-1. Cambridge University's concept of the *autarkic* (self-sufficient) house, designed to be independent of conventional fuels and piped services. (Courtesy of University of Cambridge Department of Architecture.

and will have its own sewage digestion plant. Low spots of ambient energy will be guarded against by massive water storage.

Solar energy space heating is still largely experimental, and most industry activity is aimed at direct water heating. Several manufacturing firms supply the market in the United Kingdom; most offer a limited module size range from about 10 sq ft upward. The average four person household would need 30 to 50 sq ft of collector surface to supply the normal 40 gal. storage cylinder, and this would save at least one-third of the annual energy requirements, depending on the weather conditions. Most current designs use metallic collectors with bonded or integral pipe grids headered at each end. Single glazing is the rule, and generally the additional cost of double glazing is not justifiable unless high water temperatures (over 140°F) are needed. Efficiencies range from 40 to 50 percent.

Recently on the United Kingdom market is a Japanese designed unit that uses large diameter glass tubes to form both the collector and a mass of water storage. Capacity of the standard 67 by 57 in. module (ten tubes) is 48 gal. Advantages of glass tubes are: they absorb radiation directly, have no corrosion problems, and have an indefinite life. Direct connection with the storage cylinder is therefore possible, but freezing safety devices must be used.

Joint prizewinners of the CDA competition were architects Masini/Franklin who entered a bonded pipe/plate unit with secondary pumped circuit to a heat exchanger in the storage cylinder. They proposed a panel size range of 30 to 100 sq ft, a compromise based on payback period and the need to limit temperature overshooting in summer. A panel 88 sq ft in area is calculated to yield 3350 kWh per year—a savings of 60 percent in energy.

Solar research in France

France's HVAC industry research organization CoSTIC has recently completed the most detailed simulation yet attempted on solar collector performance. The original computer assisted appraisal used insolation and associated climatic data (for example, wind speed and direction, humidity, air temperature) from about 30 meteorological stations throughout the country, and this is now being extended to cover 105 stations. The analysis was made for every hour of the day and covers five consecutive years' records. Interaction between the collector and the building is also considered. Next phase of development will be the establishment of calibration standards in conjunction with the industry's test laboratory CETIAT, which verifies equipment ratings. CoSTIC's new teaching center for the solar engineering profession at Digne with its computer program will be available for promoting economical designs of solar collectors.

Asked in a recent interview how he saw the development of the solar collector market, R. Cadiergues, CoSTIC's director, said that now a reliable design base existed, and the industry could concentrate on equipment manufacturing techniques, better production control, training and qualification of personnel, and clarification of the legal position. The basic starting point for collector design should be the price, which must be reasonable in relation to performance, or the market will not develop at all. The price will then be the major constraint.

First cost comparisons with conventional forms of heating, such as electricity or gas, are unrealistic, since there is the hidden factor of services cost, which is paid for in the price of the fuel. Solar heat service operators could be a logical development in the industry, and is one way in which the installation cost could be kept to a minimum. A central solar collector array could supply heated water to a group of buildings, and the operator would amortize the cost by charging users for the solar "fuel."

Agreed standards of construction and of qualifications of personnel will be essential to protect the consumer. An important legal point will be the question of shading of existing solar collectors by future buildings, invoking the old laws of the right to light.

A number of organizations have been studying the performance of solar houses.

Under a grant from the French government's science research council (CNRS), three houses have been built at Odeillo in the Pyrenees, not far from the world famous 1000 kW solar furnace. Designed by architect Jacques Michel and Felix Trombe, director of the solar furnace research group, the houses use part of the facade as the collecting surface—triple glazing backed by an 18 in. thick concrete wall and separated from the glazing by an air space. Radiation falls on the blackened surface of the wall where the prodigious mass stores the heat, thus leveling out the peaks of solar gain. Air is circulated around the house by natural convection.

An example of a prefabricated solar house, also using the Trombe solar storage wall, is one of 1100 sq ft floor area built at Chauvency Chateau in the Meuse region. A first balance sheet showed 20,000 kWh gain during the year; 12,000 kWh are needed from back-up electric heaters. The construction of the house is particularly interesting since it is built within the government's moderate rental cost limits.

Electricite de France has built solar houses at Le Havre and Aramon that are expected to show only a 10 year payback period on installation cost. The houses have 400 sq ft double glazed roof panels with low temperature water distributed either to fan convectors or to floor coils.

Long-term studies of a group of solar houses with normal occupation have been made by a heating contracting firm, Ets Missenard-Quint. The region in which the houses are built—St. Quentin, north of Paris—is said to have the least number of sunshine hours in France and will therefore be an effective test of performance on diffuse radiation. Three houses have been built, identical in construction and orientation, well insulated, and all mechanically ventilated. The control house is electrically heated only; the second one has a solar collector for service hot water; and the third is all-solar heated by warm air using a broken stone matrix to store heat. The collectors form part of the roof structure and are recessed about 3 ft into the roof slope to give protection from wind

cooling. Low solar gain periods will be backed up by electric heaters. It is hoped the experiment will give a reliable cost balance for the two limits of solar energy recovery—hot water only and heating. Metering is made easy by the use of all-electric back-up appliances, and other household equipment is electric.

Solar developments in Germany

Germany's minister for research Han Matthofer has announced a 110 million DM ($45 million) program for the study of solar energy recovery. A chief priority is the development of collectors for service hot water—oil fired boilers are least efficient in summer when solar potential is greatest. (Solar reception in Germany is about 1000 kWh per yr, which is about half that of Japan and Australia.)

An organization has recently been formed in Munich to study and document the solar heating industry—*The Deutsche Gesellschaft fur Sonne-energie*—although there have been some objections from the German Engineers' Association (VDI), who claim they already cover the field.

Perhaps the most spectacular of the German projects so far is the solar heated swimming pool now under construction in Wiehl. The 30,000 sq ft collector will have an average input to the swimming pool in the May through September season of 4000 kWh per day, with a maximum in summer of 10,000 kWh per day. Total input during the season is calculated to be 650,000 kWh—the equivalent of 23,000 gal. of oil.

A solar house has been built by the Philips Co. in collaboration with the Rhine-Westphalia Electricity Co. (Fig. 71-2). It is basically a standard construction but with greatly increased insulation and reduced window sizes, in which the double panes are filled with Krypton gas to reduce conduction losses still further. Compared to the normal house, it has a heat loss of only one-sixth, and even compared to the so-called fully insulated construction, loss is still only one-third. Yearly energy requirement for conduction losses is 6300 kWh and for ventila-

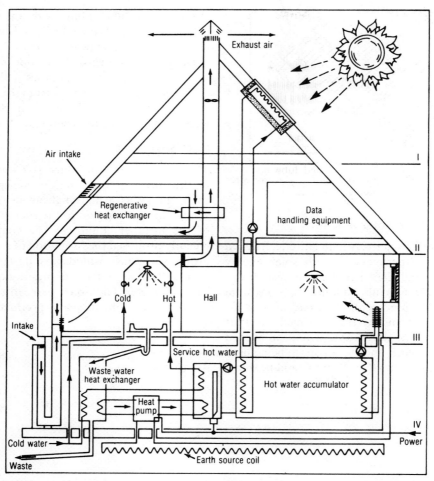

Fig. 71-2. Section through the Philips low energy house in Aachen (West Germany). The roof mounted solar collector is made up of banks of evacuated glass tubes with silvered inner surface concentrating radiation on heating pipes within the tubes.

tion, 2000 kWh—a total of only 8300 kWh for a floor area of 1250 sq ft (compared to about 50,000 kWh for normal construction). Hot water needed for washing and laundry amounts to a further 3980 kWh, but waste water heat recovery reduces this to 980 kWh. An additional 2915 kWh for the electrical equipment (television, cooking, etc.) goes to heat the house. Details of the tube collector in the Phillips house are shown in Fig. 71-3.

The solar collector set into the sloping roof is intended to generate water at temperatures up to 200°F and uses an array of evacuated glass tube 2¾ in. in diameter and 39 in. long with the heating pipes (flow and return) located within the hydro-electric stations. Solar energy for water heating has now come into sharp focus, and in June 1974, the Swiss Solar Energy Society was formed. Its 500 members are architects, consulting engineering firms, and industry.

Under average Swiss climatic conditions, a simple flat plate collector with an area of 85 sq ft will deliver enough domestic hot water for a four person household (45 gal. per day) for seven months of the year and will give a significant boost during the winter. For partial solar heating, a 1100 sq ft collector would be needed, according to figures by the Swiss Solar Energy Society.

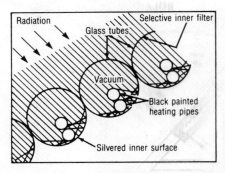

Fig. 71-3. Detail of the evacuated tube collector array in the Philips house.

Scandinavian solar activities

In spite of its relatively high latitude, Sweden has been experimenting with solar energy recovery, and David Sodergren described a scheme for a bank building in Stockholm that would use an all-glass facade with air as the heat transfer/transport medium. Part of the double glazed area would be backed by a third pane, and the air space being used as a duct for return air. Heat would be trans-ferred across the building by means of channels cast into the concrete floor beams, the latter also acting as heat storage.

Denmark has a zero energy house, similar to the Philips' house, that uses a flat plate collector with heat recovery from the exhaust air and waste water. There is an 8000 gal. water storage tank to accumulate heat.

EEC prospects

The EEC (Common Market) Commission in Brussels is taking a serious viewpoint on solar energy as a marginal but nevertheless signficant contribution to the energy balance of the member countries. The commission recently proposed the formation of a research group whose task would be to design a standard solar collector for domestic hot water. The parameters laid down were: efficiency 50 to 70 percent, life 10 years, production cost $1.85 per sq ft.

Figures 71-2 and 71-3 courtesy of the publishers of Heizung Luftung Klimatechnik.

72

How the heat pump conserves energy resources

A review of the heat pump concept; heat sources, including sewage and solar energy; available equipment; and the role this concept can play in conserving energy resources.

T. E. PANNKOKE, PE,
Engineering Editor, Heating/Piping/Air Conditioning, Chicago, Illinois

In this time of energy shortage, it is imperative that any method that promotes energy conservation be examined and used to best advantage. One such method is the heat pump. Proposed by Lord Kelvin in 1852, the first installations were made around 1930. While it has since been proven in numerous installations, the heat pump has never been fully exploited.

Energy may be looked upon as consisting of two forms: capital energy and income energy. Capital energy comes from sources such as fossil fuels, which when used are gone forever. Income energy implies replenishment of the energy by nature as it is used.

Sources of income energy can be air or ground water. Heat from these sources can be transferred to the interior of a building to provide a major portion or all of the required heating energy. These are the conventional air and water source heat pumps successfully applied in residential, commercial, and industrial applications.

The internal source heat pump was developed to reclaim building heat normally exhausted to the atmosphere. In buildings with the proper configuration—distinct interior and perimeter areas—and with sufficient internal heat generation from lights, people, machinery, etc., this heat can be transferred from areas having an excess amount (interior areas) to spaces having a deficiency (perimeter). This is accomplished by mechanically cooling the interior spaces and utilizing the perimeter as a heat sink instead of (or in conjunction with) a cooling tower, air cooled condenser, etc. Diagrams and descriptions of the many different variations of heat pump cycles are found in the references at the conclusion of the chapter.[1,2,3,4]

Definitions of heat pump cycles

Heat pumps are refrigeration systems in which the rejected heat is at least as important as the refrigeration effect.* Heat pumps have been referred to as *reverse cycle refrigeration systems*. Some cycles, notably the air-to-air and water-to-air systems, generally do em-

---

[1]Superscript numerals indicate references at end of the chapter.

*Technically, any refrigeration machine is a heat pump in that it transfers or pumps heat from one location to another. However, with the heat pump, this heat is a useful product.

ploy reversal of refrigerant flow to accomplish cooling or heating as required.

With the various water-to-water heat pump cycles, refrigerant flow usually is fixed, and the water circuits are reversed to accomplish the required heating and cooling.

Heat pump cycles are ideally suited to reclaim and redistribute heat. Therefore, external source systems can be designed to utilize the internal heat loads of a building in addition to the heat extracted from the source, thereby improving the overal seasonal efficiency or *performance factor*.

From the sewer and the sun

Air, water, and internal building heat are the commonly thought of heat sources for heat pump applications. Yet, many other sources can be utilized. Some of these are described in the following discussion.

• *Sewage*—Nuffield College in England has a sewage-to-water diesel engine driven heat pump that was installed around 1961.[5] The 42 hp engine drives two 17.3 bhp compressors and some of the system pumps. The output of the condensers is 330,000 Btuh, and engine heat recovery brings the total available heat to 434,800 Btuh.

During the first ten years of operation, this heating-only system had operated approximately 11,000 hours.

• *Solar energy*—At least as far back as 1948, the heat pump was considered as a possible means to utilize solar energy. From 1950 to 1955, tests were conducted on a heat pump in which solar energy vaporized refrigerant in a solar collector evaporator.

Studies have been done on combined solar-earth heat pumps, and procedures for developing such units have been developed. Calculations show that the system would be more efficient than a ground source alone.[6,7]

In 1958, a solar-to-water heat pump installation was designed and installed in a consulting engineering firm's office in Albuquerque, New Mexico. (See Chapter 67.) This installation embodied solar collectors through which water was circulated. A storage tank carried the building overnight and during cloudy periods.[8]

Currently, there is considerable interest in the utilization of solar energy, and it has been demonstraated that present technology in the form of the heat pump permits its use for selected applications.

• *Thermal pollution*—Surface waters generally are not considered to be good heat sources. As the water temperature drops during the winter, the possibility of freezing and resultant damage to the system exists as heat is extracted from this colder water.

Some bodies of water, however, are affected by thermal pollution and maintain sufficiently high temperatures in these affected areas so that the danger of freezing does not exist. Areas downstream of the condenser water discharge from generating stations and industrial plants are locations where this form of pollution can be found. With the great need for increased generating capacity in the future, this "source" can be expected to become more widely distributed. A form of energy conservation could be implemented by locating building and industrial complexes in the vicinity of power plants where the rejected heat could be utilized either directly or through water source heat pumps.

• *Other sources*—Considerable heat is wasted in industry that could be reclaimed by heat pumps and other heat conservation devices. It is not unusual to see the plumes from cooling towers at industrial plants during cold weather. These plumes indicate heat is being dissipated from a process. At the same time, a boiler or furnaces are supplying heat to the offices and the plant. This heat could be reclaimed to supply at least a portion of the space heating load.

Jacket water heat from air compressors and engines and engine exhaust are other sources. Heat in exhaust systems is also another source. Injection molding operations are another good source. The molds are water cooled, and often this heat is dissipated through a heat rejection device.

In food processing operations, the heat rejected from a freezing operation may be used elsewhere in the manufacture of frozen orange juice. The rejected heat is used in evaporating or concentrating the juice prior to freezing.[9]

Heat pumps will not always prove to be the best answer for heat recovery purposes, but whenever heat is wasted, various forms of recovery should be investigated.

Air or well water as heat source

Much has been written on the two most common external sources, air and well water; so these will be reviewed briefly.

Outdoor air is for all practical purposes an inexhaustible heat source. Single stage compression is effective down to around 20°F outdoor ambient, although some auxiliary heating may be required prior to that. Compound compression—compressors operating in series—permit lower evaporator and/or higher condensing temperatures. Compound heat pump systems have been designed to extract heat from outdoor air at ambient temperatures as low as −20°F, meeting the entire space heating load. Several compound air source heat pump installations were designed and installed during the 1950s. Installations included office buildings, a large department store, and a large precision manufacturing plant and associated offices. This latter project had a design cooling load of 640 tons.[9]

Ground water is an excellent source of heat; temperatures generally are in the 50 to 55°F range. Several factors discourage its use, however. Capital investment tends to be very high. Supply wells have to be drilled, and generally, the used water must be reinjected into the ground by means of a re-entry well; otherwise a prodigious waste of water would occur.

Re-entry wells sometimes pose problems. Pumping water into the ground, however, tends to pack the sand, increasing the resistance. Sometimes resistance reaches a degree that re-entry is difficult to achieve.

A site must be carefully tested to assure that an adequate supply of water exists and that wells will deliver and accept the required water quantities. The services of a geologist are essential in this situation.

In some areas, water shortages exist. Some metropolitan areas are experiencing severe drops in water tables as water is pumped out faster than nature can replenish it.

In spite of the aforementioned problems, a number of successful installations have been made. Like the compound air source heat pumps, the installation peak for water source systems was reached during the 1950s. Successful installations were made for a variety of applications. These include office buildings, schools, and shopping centers. Some of these have utilized engine drive with engine heat recovery, increasing the overall cycle efficiency.

Compressors may be used

Any compressor type suitable for refrigeration applications can be used for heat pump applications. The most commonly used types are reciprocating, centrifugal, and helical rotary (commonly called a screw compressor). Large volume rotary compressors used for low temperature refrigeration work have also been employed. This type has been confined to low stage compression in compound air source heat pump system applications.

A reciprocating machine is most commonly used and is available in capacities up to approximately 250 tons. It has been almost exclusively used in air source heat pumps, and it can be used in any heat pump cycle. Since it is a positive displacement machine, it can maintain a high discharge head and consequently, high condensing temperatures at low loads.

A reciprocating unit may be used in chilled water systems and direct expansion (DX) systems, either air or water condensing. This makes it an ideal unit for any heat pump cycle.

Centrifugal compressors range in size from approximately 100 to 10,000 tons. Due to cost factors, a centrifugal compressor generally is not used in heat pump installations below 150 to 200 tons. It has much lower vibration levels than a reciprocating compressor since it is a rotating machine. Capacity reduction is good with operation to as low as 10 percent of full load.

A centrifugal compressor cannot operate over as great a lift (condensing temperature minus suction temperature) as can positive displacement machinery. Therefore, its application to heat pump systems has generally been confined to the external and internal water source systems where reasonably constant suction temperatures can be maintained.

Virtually all centrifugal machines are employed as water chillers, and most installations are water cooled.

A screw compressor has some of the best features of both the reciprocating and centrifugal machines. It is a positive displacement unit, hence, it can maintain a high head at reduced load. Capacity modulation is good—down to 10 percent of full load. It also has the advantage of being a rotating piece of machinery and therefore runs smoother than a reciprocating unit.

Capacities range from approximately 20 to 1000 tons. It may be used in DX and water chiller applications with either air or water condensing.

System design considerations

Heat pump installations differ from other types of HVAC systems in that the heating media temperatures generally are lower. Air-to-air and water-to-air systems will deliver lower temperature air on heating due to the limits on condensing temperatures. Lower temperature water is used when water is the heating medium for the same reason.

Excellent heating results can be achieved with 100 to 105°F leaving water temperatures available from a water cooled compressor, and temperatures to 130°F can be obtained with the correct machine selection if a higher temperature is needed.

With lower design supply water temperatures, more water must be circulated, and larger piping and greater pump horsepower are required. Also, heat transfer surfaces will be considerably larger than if greater temperature drops were available. Thus, first costs for heat pump installations can be higher than for a fuel fired installation.

The amount of supplemental heat required over a heating season has a definite bearing on whether a heat pump cycle is practical. While a heat pump may have a coefficient of performance around 4.5, supplemental heat supplied by electricity has a COP of 1. If the seasonal requirement for supplemental heat is high in relation to the seasonal output of the heat pump, the performance factor is greatly lowered. Because of such considerations, a thorough analysis of building structure and configuration, heat sources (internal, external, or both), outdoor temperature profiles, occupancy schedules, what methods of supplemental heating should be used, and fuel and power costs, etc., usually is necessary to make a valid owning and operating analysis on which to base a decision to go with the heat pump system.

Other heat recovery devices and systems may be used to good advantage in the overall design, such as heat wheels, run-around cycles, etc. A word of caution is in order. It is unwise to try to salvage every last Btu. This usually will boost the cost and make the system unnecessarily complex. The economic benefit of each option should be determined, and the dubious ones discarded.

Supplemental heat, heat storage

Supplemental heating was mentioned previously. Many heat pump installations require supplemental heating at lower outdoor ambient temperatures. Air source systems lose capacity as the evaporator suction temperature decreases, which results in greater lifts (condensing temperature minus suction temperature). Internal source systems also lose capacity. A greater share of the internal cooling load is met by the outdoor air provided for ventilation and the increased heat loss through the roof as the outdoor temperature falls. Therefore, the need for mechanical cooling decreases.*

*During mild weather, a greater percentage of outdoor air may be used to supplement the mechanical cooling so that the refrigeration machine output is limited to what is required to meet the space heating load. This reduces power consumption of the chiller since there is no excess heat to dispose of through the cooling tower. However, if humidification is provided, increased outdoor air volume will increase the amount of energy required for humidification.

External water source systems can be designed so that the entire space heating load can be met by heat pumps, since ground water temperature is relatively constant year around.

In an all-electric system, supplemental input is provided by resistance heating: baseboard radiation, duct coils, electric boilers, etc. Gas and oil fired equipment can be used; however, with such an arrangement, the owner may not get the most favorable electric or gas rate.

Turning on lights during unoccupied periods to provide heating is sometimes done. In a reclaim system, running this heat through the compressor *does not* yield a COP greater than one. A brief discussion will show why this is true.

During unoccupied periods, lighting needs are greatly reduced, and turning on lights is the same as using resistance heat to supply transfer heat for the heat pump. The lighting is not essential for performance of specific tasks; therefore, all lighting and compressor energy is chargeable to heating and the COP (output/input) becomes unity. The use of lighting as a heat source during off-hour heating can, where codes permit, reduce the amount of auxiliary heating equipment required. However, during this time of energy shortage, the appearance of high illumination in an unoccupied building would have a negative effect on the public.

The internal source system can provide its own heat source for unoccupied periods by transferring excess heat during occupied periods to a storage tank. Water from the tank—stored at 100 to 130°F—can be used directly in the system until it cools down to a point where it no longer can handle the load. (This is determined by outdoor conditions.) At this point, the compressor is run to extract more heat from the water and elevate the temperature to where it satisfies the heating requirement.

Use heat from stored water

An alternate approach is to use the compressor initially to extract the heat from the stored water. From a control standpoint,

this is simpler than the previous method where at some point a changeover from direct use to heat pump operation must be made.

Safeguards must be provided to prevent water at too warm a temperature from entering the chiller and causing serious damage. The water from the storage tank must be blended with water in the distribution system before it is fed to the chiller. This is easily accomplished by means of a three-way valve.

Whether or not the use of a heat storage tank is justified depends upon a number of factors. Some of these are:

- The outdoor heating design temperature.
- The balance temperature. This is the outdoor temperature at which the heat generated in a building equals the transmission and ventilation loads. The balance temperature also determines whether or not the internal source system is practical by itself.
- The occupancy schedule. Will the building follow a normal office building schedule or is there extended occupancy and/or use of the facility weekends? A building with a low balance temperature—for example, 0 to 10°F—and occupied 24 hr per day has little need for stored heat.
- The number of hours during the heating season that excess heat is available. This is a function of the balance temperature, occupancy schedule, and temperature profile.
- The number of hours that stored heat can be used.

All of these factors must be carefully analyzed. The use of a storage tank does increase the complexity and cost of a system.

The size of a heat storage tank is a matter of economics. Ideally, it would be desirable to store all available heat if it could be used. However, the law of diminishing returns applies to this decision. For example, it might be possible to utilize a 20,000 gal. tank, but a 10,000 gal. one would meet the need most of the time. Under this condition, it would be difficult or impossible to justify the expense of the larger tank.

Equipment availability

The system designer has a wide variety of equipment to choose from. Air-to-air packages from about 2 to 20 tons are available (all sizes given are the nominal cooling capacity). Larger air-to-air units from approximately 40 to 120 tons are available as are air-to-water units in this size range.

Water-to-air units of about one to five tons are available. These operate in conjunction with a closed water loop having a boiler and closed circuit evaporative cooler.

Packaged chillers are commonly applied to internal source systems.

There are many successful examples of properly designed and applied packaged equipment performing well. There are, however, applications where built-up cooling systems may be preferable. A compressor operates many more hours per year when used for both heating and cooling duties, and compressor life becomes a more significant factor. This is particularly true of reciprocating machinery. Heavier duty, industrial or semi-industrial type compressors may be used to good advantage.

These heavier duty machines have removable cylinder liners to permit the rebuilding of a unit, more rugged valves, larger oil crank case storage, and superior lubrication systems. Lower speed is the rule above 50 tons capacity; 1200 rpm rather than the 1750 normally used in air conditioning work. Higher compression ratios are available. This is a desirable feature for air source applications due to the decrease in suction temperature and significant loss in heating capacity as the outdoor temperature falls.

Matching the refrigerant and compressor to the operating conditions may yield a slight horsepower savings.

Numerous factors may enter into the decision of whether to use a packaged or built-up unit. A built-up system requires an experienced designer in order that components and accessories are properly selected and matched to assure reliability. Cost nearly always enters into the picture, however, and true cost can be determined only by a complete owning and operating cost analysis.

Engine drives as prime movers

Gas and diesel fueled internal combustion engines have been applied as prime movers for refrigeration, air conditioning, and heat pump applications to a limited degree. Higher equipment and installation costs and maintenance considerations have precluded widespread use.

Nevertheless, an engine is attractive from an overall energy efficiency basis when heat recovery is considered. Up to 70 percent of fuel input energy may be utilized. This is at peak load; seasonal values would be lower. Overall electric efficiency (from coal pile to shaft) is seldom greater than 30 percent.

With heat recovery, higher heating water temperatures are possible than those obtained from a condenser alone.

Where operating cost savings make an engine drive economical, and when *qualified* maintenance people are available, an engine drive installation should receive consideration.

The heat pump's future

Application of heat pump cycles can be expected to find greater acceptance in the future. The experience of one oil embargo has served notice on the public that the imbalance between energy demand and supply is a genuine problem. The possibility exists that the foreign oil producers may again shut off the supply in an effort to achieve their political goals.

Various state bodies have developed energy conservation programs, and it is possible that ASHRAE Standard 90-75, which deals specifically with energy constriction in new buildings, will be adopted in some form.

Reduced electrical consumption—primarily lighting—will reduce the heat generated in buildings. More insulation, increased use of double glazing, and reductions in the amount of outdoor air used for ventilation will reduce the heating and cooling loads.

While it appears that the use of energy for air conditioning reheat will be sharply curtailed, most types of equipment and systems currently on the market will continue to be used; although some equipment may have to be redesigned to meet possible mandatory efficiency levels.

Increasing pressures to reduce energy consumption are bound to occur. This will be from both a political standpoint (energy conservation takes its place right up there next to motherhood and apple pie) and from a cost standpoint. Tremendous amounts of capital will be required both to develop new fossil fuel deposits and to develop new forms of energy.

At the energy price levels that existed prior to 1970, heat pump systems were not always competitive on owning and operating cost bases. Designs utilized only the basic refrigeration cycle due to higher first cost in some cases. At today's energy costs, some of these projects—if designed today—could have gone the heat pump route.

External water source heat pumps can only remain a small factor in the market due to the reasons enumerated previously.

Buildings with sufficient heat generation within interior spaces to meet a major portion of heating season requirements will continue to be prime candidates for the heat pump reclaim cycle.

A market definitely exists for air source equipment for smaller buildings that do not generate sufficient internal heat to make a reclaim system economical by itself due to the ability of the heat pump cycle to both reclaim heat and take heat from an internal source.

Improvement of equipment is an ongoing process. Many suggestions have been made on how to improve heat pump systems further. Some of these are described in the literature.[10]

Engine drives with engine heat recovery are more efficient than electric motors. It would appear that a market could exist for engine drives. However, prime mover manufacturers generally have not been enthusiastic proponents of engine driven air conditioning equipment. Engine driven packages have come and gone from the market place. Built-up installations are few in number. Yet, some packaged and some built-up installations have been notable successes.

Development of another form of heat engine that does not have the inherent complexity disadvantage of reciprocating power and yet is not as expensive as the gas turbine could further enhance the contribution of the heat pump to energy conservation.

Finally, all forms of energy conservation should be used where practical or where they show promise of being improved to make them practical if possible. In this latter category is the total energy concept. It is more complex to design and install. There have been failures as well as successes, but properly applied, it, along with the heat pump, are perhaps the two most efficient energy utilization concepts that are beyond the theoretical or the laboratory stage.

References

1. Ambrose, E. R., *Heat Pumps and Electric Heating*, John Wiley & Sons, Inc., New York, N.Y., 1966.
2. *ASHRAE Guide and Data Book—1972 Equipment*, Chapter 43, "Heat Pumps," American Society of Heating, Refrigerating and Air-Conditioning Engineers, Inc., New York, N.Y.
3. *ASHRAE Handbook and Product Directory—1973 Systems*, Chapter 11, "Applied Heat Pump Systems," American Society of Heating, Refrigerating and Air-Conditioning Engineers, Inc., New York, N.Y.
4. Japhet, R. A., "Considering Heat Reclamation? Here are Your Options," *Heating/Piping/Air Conditioning*, April 1969, p. 89.
5. Kell, J. R., and Martin, P. L., "The Nuffield College Heat Pump," *Journal of the Institution of Heating and Ventilating Engineers*, London, 1963, p. 333.
6. Penrod, E. B., and Prasanna, K. V., "Will Solar Energy be the Heat Source for Tomorrow's Heat Pump?," *Heating/Piping/Air Conditioning*, May 1960.
7. Penrod, E. B., and Prasanna, K. V., "A Procedure for Designing Solar-Earth Heat Pumps," *Heating/Piping/Air Conditioning*, June 1969, p. 97, and November 1969, p. 104.
8. Bridgers, Frank H., "Application and Design of Electric Heat Pump Systems," *Heating/Piping/Air Conditioning*, February 1967, p. 98.
9. Miner, Sydney M., "Air Conditioning with the Air Source Compound Heat Pump," *Proceedings of the American Power Conference*, Vol. XIX, 1967. Illinois Institute of Technology Center, Chicago, Ill.
10. Ambrose, E. R., "The Heat Pump Performance Factor, Possible Improvements," *Heating/Piping/Air Conditioning*, May 1974, p. 77.

The author wishes to express his appreciation to Thomas S. Beers, consulting engineer, Red Bank, N.J., and John Levenhagen, manager, Air Conditioning Dept., Vilter Manufacturing Corp., Milwaukee, Wis., for their assistance in the preparation of this chapter.

73

Solar heat pump integrated heat recovery

Energy conservation for new hospitals.

BOGGARM S. V. SETTY, MS, PE,
formerly with Gershon Meckler
Associates, PC, Washington, D.C.*,
presently Director of Mechanical-
Electrical Engineering, Benham-Blair-
Winesett-Duke, Inc., Architects,
Engineers, Consultants, Falls Church,
Virginia

The operating and maintenance costs of mechanical systems in hospital facilities are expected to increase sharply in the future. Fortunately, many present day engineers and architects realize that energy conservation is essential in all areas. This chapter, however, is focussed on one part of the problem—energy conservation for new hospitals.

Heat is generated from the interior spaces of a hospital, year around. If properly utilized, this heat can meet the energy demand during the heating season. Also, the energy wasted through the exhaust stack of the incinerator can provide part of the requirements for service hot water generation and hot water for reheat coils.

Finding the right system

A comprehensive analysis of various systems was conducted to determine a suitable and economical system for a hospital. The following systems were investigated:

* A constant volume system with reheat coils and heat recovery wheels for interior zones, fan-coil units for exterior zones wherever possible, and a four-pipe system with oil or gas fired boilers for heating.
* A constant air volume system with reheat coils and heat recovery wheels for interior zones, induction units for perimeter zones wherever possible, and oil or gas fired boilers for heating.
* A double duct system with mixing boxes for interior zones, fan-coil units for perimeter zones wherever possible, and oil or gas fired boilers for heating.
* A constant volume system with reheat and heat recovery wheels for interior zones, self-contained heat pumps for perimeter zones, and electric boilers supplemented with solar collectors and incinerator exhaust recovery coil for heating.

The first three systems were analyzed using oil as the prime source, and the fourth was based on electricity as the prime source.

The system evaluation was based on data from a few hospitals that covered building energy cost studies, mechanical space re-

---

*Data for this chapter were prepared while the author was with the firm of Vosbeck, Vosbeck, Kendrick, Redinger, Alexandria, Va.

quirements, ceiling space, structural requirements, electrical distribution, and conditioned air distribution.

Important factors in favor of the use of
electricity are: the impact of the oil shortage, environmental regulations on fuel oil
emissions from stacks, the uncertain availability of fuel oil, increasing demand for
No. 6 fuel oil, and the rate at which the
price of fuel oil is rising compared with the
escalating cost of electricity in the years
ahead. The latter point was determined by a
comparative analysis of oil and electricity
based on the current rates. The study pointed
to the conclusion that the price of oil will
rise sharply, and therefore the operating
cost of using oil and electricity will even out
after five to six years; however, electricity
will become relatively cheaper per million
Btu thereafter.

The comparative system analysis found
that the initial cost of an all-electric heat
pump unitary system with central chillers
was 20 to 30 percent less expensive than the
other systems. This is due to its inherent
characteristics, such as running only two
pipes for both cooling and heating and simple operating controls.

Other favorable points for an all-electric
system are its significant advantages over
other systems. These include: accuracy of
control, low maintenance costs, and the
availability of electricity from a nuclear
power source in the years ahead.

These considerations confirmed the selection of a heat pump energy reclaim system
with an integrated heat recovery loop and
using electricity as the base fuel as the most
suitable and economical for new hospitals.

Electric steam generators would provide
supplemental heat for reheat coils, humidifiers, terminal units, and sterilizers when
heat from other sources is not available. The
following points played a dominant role in
this decision.

• Handling, distribution, and storage of
fuels is eliminated.

• Breeching, breeching insulation, and
the need for a chimney are eliminated, as is
the need to meet strict pollution standards
governing smoke emissions.

• Lower initial cost is incurred.
• It can be installed easily.
• Long pipelines are not needed.

Integrated heat recovery loop

The integrated energy loop absorbs, rejects, and stores heat. Solar heat, incineration waste heat, and hospital interior heat
are injected into the integrated loop. The
terminal units, such as heat pumps, reheat
coils, and air handling unit (AHU) absorb
and reject heat to the loop as required. Even
during the summer, solar, incinerator, or
building heat can be utilized through the
integrated loop as the necessary heat for
the reheat coils.

The heart of the system is the integrated
heat recovery loop with two centrifugal
chillers connected in series, which decreases
the energy consumed per ton of refrigeration. The chillers operate between 92 and
103°F (33 to 39°C) condenser water temperatures. A leaving water temperature of
103°F optimizes the additional heat requirements for reheat coils, as indicated in
Fig. 73-2. Centrifugal chillers provide cooling for the interior zones throughout the
entire year. Heat pumps provide the necessary heating and cooling to perimeter areas
and patient rooms. The reheat system provides the necessary heating for the interior
zones as an artificial load under partial load
conditions that occur due to changes in environmental conditions in the interior areas.

A comprehensive study was made to evaluate the leaving and entering water temperatures of the reheat system. This resulted in a realistic entering water temperature
of 103°F being used to design the reheat
coils for interior zones without sacrificing
either the efficiency of the coils or the temperature of the leaving air. Clearly, the integrated heat recovery loop acts both as a
source for the reheat coils and sink for the
centrifugal chillers during summer, as shown
in Fig. 73-2. The reheat integrated heat recovery loop acts as a source when supplementary heating is required during the
severest periods of the winter by obtaining
heat from the electric boilers. Steam-to-
water converters add heat to the condenser

water reheat coils, and the unused heat returns to the integrated heat recovery loop and raises the temperature of the integrated loop within a short period of time.

Kitchen heat can be used

In most hospitals, the kitchen is operated in the morning, afternoon, and night. Substantial quantities of outside air are introduced into the kitchen and exhausted through the kitchen hood. Especially in winter, the energy used to heat the induced outside air is wasted. A runaround coil system with glycol and water as the energy transfer media can be used to recover the exhaust heat. The reclaim coil can be housed in the kitchen exhaust hood and equipped with spraying nozzles for dispensing hot water every 24 hours to dissolve any grease buildup generated by frying fatty foods. Figure 73-1 illustrates how this system works.

First cost can be recovered in one to two

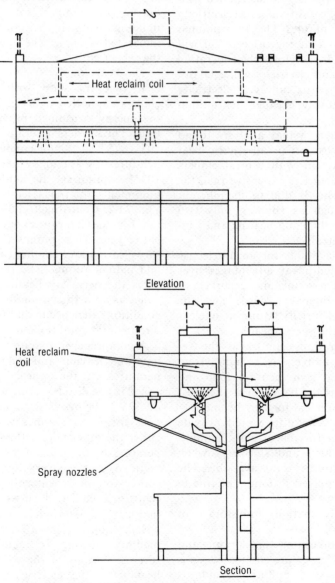

Elevation

Section

Fig. 73-1. Kitchen exhaust heat recovery system.

years. When the kitchen is not in use or during periods of light activity when the exhaust heat is insufficient to heat the outside air, the integrated heat loop acts as a source and provides the necessary heat to a standby coil housed in the outside air handling unit.

Summer and winter operation

• *Reheat coils*—Condenser water at 70° F (22°C) in winter and 103° F in summer from the integrated loop provides the necessary heat for the reheat coils. During either season, initially, condenser water is passed through the solar storage tank and is heated

to 110° F. If the temperature in the tank falls below 70° F, a sensing element in the tank operates a three-way valve and diverts all water to the steam-to-hot water converter. When the solar tank can provide heat in either winter or summer, the sensing element at the discharge point of the solar tank modulates the three-way valve and maintains a discharge temperature of 110° F downstream from the solar tank.

• *Domestic hot water*—Cold water from the city mains is preheated partially in a heat exchanger using condenser water as the heat source by absorbing the heat available from

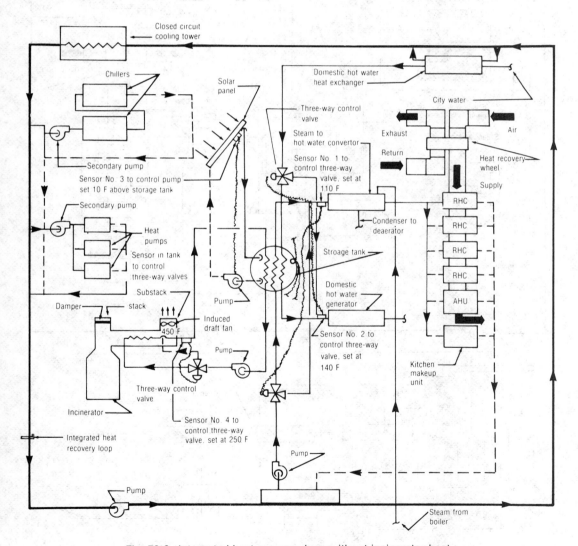

Fig. 73-2. Integrated heat recovery loop without incinerator heat.

the integrated loop during both seasons. The remaining heat is supplemented primarily by the solar tank or the service hot water generator. If the solar tank temperature is lower than the preheated service hot water, the sensing element in the solar tank diverts all the water to the electric service hot water generator. If the temperature in the tank reaches a level sufficient to provide more

heat than is required, the sensing element downstream of the solar tank modulates the three-way valve, as shown in Figs. 73-2 and 73-3 to maintain its setting.

Flat plate solar collectors

Flat plate solar collectors consisting of copper plates with integral tubular passages, inclined at an angle of 38 deg to horizontal,

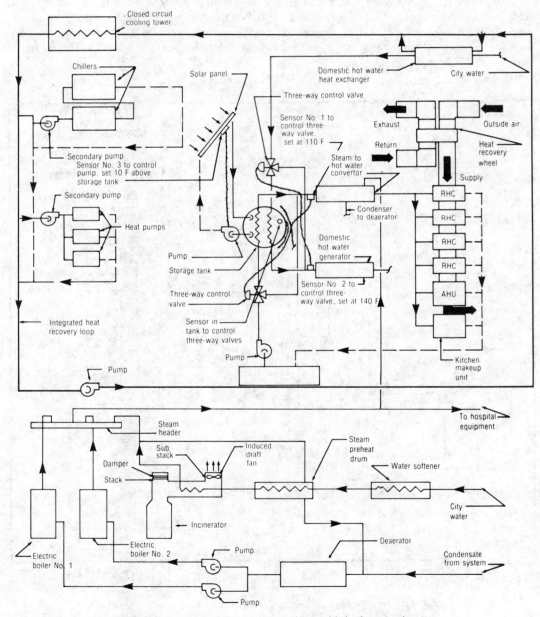

Fig. 73-3. Integrated heat recovery loop with incinerator heat.

with 2 in. insulation and double tempered glass transfer heat to the solar tank throughout the year. The solar water tank stores the available heat and acts as a source to the reheat coils and domestic hot water. The solar tank should be designed to obtain optimum tank temperatures without sacrificing the available solar heat.

Incinerator heat can be used

Pathological waste usually is burned in the hospital incinerator, and the heat generated during the process is discharged through the stack. The gases carrying the heat can be passed through a stainless steel coil housed in the exhaust stack, and the heat can be recovered as steam and used for the hospital's requirements. (See Fig. 73-3.)

City water is preheated in a steam drum to avoid thermal shock conditions to the stainless steel coil housed in the stack. The preheated water is then passed through the stack coil to obtain steam. The electric boiler will supply the steam to meet the hospital's demands when waste heat from the incinerator is not available. A study concluded that a stainless steel coil can pay for itself in about four to five years.

Air side heat recovery

To achieve lower installation cost and provide the lowest operating and maintenance costs, total heat recovery wheels on the air side must be used. All hospitals are generally governed by federal and local building codes. Some codes recommend using 100 percent outside air for sensitive areas, and some recommend 25 percent outside air. The use of 100 percent total heat recovery wheels not only satisfies the stringent codes but also produces savings in first and operating costs of the refrigeration equipment.

The advantages of using total heat recovery wheels are:

• They can be operated at a suitable speed in spring and fall nights to maintain a suitable discharge temperature to meet the space demand without taxing refrigeration equipment by operating the wheels on the economizer cycle.

• No separate exhaust ducts are required for toilets, janitor's closets, and storage areas.

• Heat can be recovered from exhaust air in all seasons.

• In a year around operation of both heating and cooling equipment, the savings in capital outlay and in maintenance cost of the refrigeration equipment will usually pay for the heat recovery equipment during the first year of operation.

A recovery integration system can be adopted for hospitals by using naturally available heat from the hospital. This system can save not only on operating and maintenance costs, but also on the initial cost. This type of system is a real possibility for hospitals, and the feasibility of this system must be analyzed quantitatively and qualitatively in regard to recoverable heat integration and the type of hospital.

74

Solar energy
caps unique system

Total solar heat may not yet be truly economical for school heating, but a little bit can go a long way.

WILLIAM E. FENNER, PE,
Fenner and Proffitt, Inc., Consulting
Engineers, Wilson, North Carolina

The elementary school designed by Griffin-Flynn, Architects, of Goldsboro, N.C., for the Goldsboro Schools, has a combination of energy saving systems. When these systems function as designed, in addition to saving taxpayers on operating cost, the building may serve as a showcase for energy conservation.

The proposed school is of a compact design, with a minimum of outside walls and a relatively small amount of window area. There is a large amount of interior space with no outside walls. This makes the building very easy to heat, and a cooling load exists in the interior areas all year around.

With this type of building, the engineers, Fenner & Proffitt, Inc. began to look for some type of system that could capture the heat produced by the lights and people in the interior areas and use this to heat the exterior rooms. The solution was obvious in that the last similarly styled school built by the Wayne County Schools, Eastern Wayne Junior High School, also designed by Griffin-Flynn and Fenner & Proffitt, Inc., had used a heat recovery system to achieve the desired results. This system used some 40, four ton water-to-air heat pumps. These removed the heat in the interior areas; put

it into a circulating water system; then, this same heat was taken out of the circulating water system and used to heat the exterior rooms. The system was one of the first of its kind installed in a school in North Carolina and has received national publicity praising its effectiveness and operating cost. The fact that Eastern Wayne has used 17.5 kWh per sq ft of area per year, while other air conditioned and electrically heated schools have used 20 to 40 kWh per sq ft per year, is an indication of the energy saving characteristics of the system.

So, in this school, the engineers wanted to use the same system as was used in Eastern Wayne, but they wanted to take it a step further. At times, a system such as used at Eastern Wayne will cool during a mild day. When the units cool, heat is taken out of the rooms and put into the water circulating pipe. If the water temperature gets too hot, a cooling tower will come into action, remove the heat from the water, and reject it into the air. Then at night, or in the early morning, heat will be needed to warm up the school, and the electric heat will have to come on. If the heat that was removed during the previous day could be stored, rather than removed by the cooling tower, it could be

Architect: Griffin-Flynn
Architects Ltd., Goldsboro, N.C.

used the next morning. Similarly, if there were several warm days followed by several cool ones, the heat captured during the warm days could be stored and used during the cool days.

Storage tank advantages

In order to store the heat, the engineers specified a 20,000 gal. underground water tank. This tank is connected to and becomes a part of the water circulating loop. If sufficient heat can be removed from the building and stored in the tank to raise the water temperature from 60° to 90°F, some 5,000,000 Btu will be stored. This is the equivalent of the heat available from 60 gal. of oil.

An additional advantage of the storage tank is that, when sufficient heat cannot be captured from the building and the electric heat must be used, a smaller electric heater can be employed. It can operate during the night (when it will not add to the electric demand of the building), heating the water in the tank, then that heat can be used the next day. In future years, the electric companies may give a reduced rate for off-peak power

(as they already do in some parts of the country), and the storage tank will let the school take advantage of that rate. Similarly, the tank can be used for storage during the cooling season. Heat removed from the building during the day can be stored in the tank and rejected at night when the cooling tower operates most efficiently.

With the school being equipped with a storage tank, and with heat being stored at relatively low temperatures (60 to 90°F), it became apparent that this was a good possibility to use some solar heat.

Solar heat has not yet become truly economical for school heating in all parts of the country. The few installations operating with solar heat have been built at tremendous cost, with government subsidies, as experimental units. However, with the heat storage tank, which meant we could capture solar heat whenever it was available and store it for future use, and the fact that the heat pump units could use water at low temperature as a heat source (solar panels are 70 percent efficient at capturing solar heat when they are operating at 90°F, versus only 30 to 40 percent when operating at

200° F), meant that we could skim the cream off the solar heat without trying to use it for the whole heating job. In addition, for somewhat nominal cost, the school board could do a little research of their own into the economics of using solar heat.

Utilize solar panel

The engineers have incorporated a small solar panel, about 600 sq ft of collector area, into the design. A complete solar heating system would take some 10,000 sq ft of collector panel; so the panel specified will only be 6 percent of the panel required to do the whole job. However, since the panel will be operating at top efficiency all the time sun is available, it should provide much more than 6 percent of the heat required. It could produce 10 to 15 percent of the yearly heat needed.

To further maximize the use of the solar panels, they will be used to heat the domestic hot water when heating the building is no longer required. When they do this, the panels will be operating at temperatures from 150 to 200° F, and efficiency will go down, but they will still produce a lot of hot water in the summertime.

How it all works together

• Heat pump Units A (see Fig. 74-1) are operating in the cooling cycle. They absorb heat from the room air and put it into the water circulating Loop C.

• Heat pump Units B are operating in the heating cycle. They remove heat from the circulating Loop C and put it into the room air.

• At any one time all heat pump units can be operating in the cooling cycle, or all can be operating in the heating cycle, or they can be operating in any combination with some cooling while others are heating.

• All the water in the loop flows through storage Tank D. The tank in effect becomes a part of the loop.

• A Pump E circulates water through the system. This pump is controlled by a clock switch and runs whenever the building is occupied. Electrical interlocks prevent any

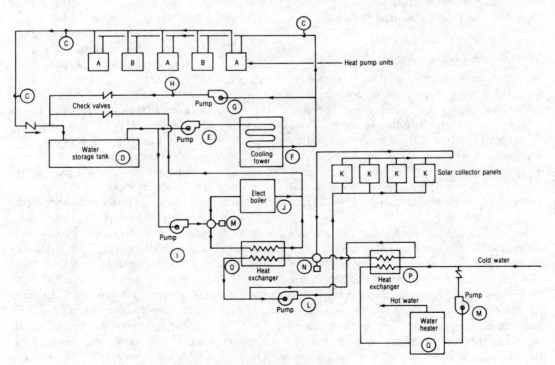

Fig. 74-1. System piping diagram.

of the heat pumps from running unless Pump E is running.

- When many of the heat pump units are running in the cooling cycle, rejecting heat into the loop, the water temperature in the loop will gradually rise. The large volume of water in the storage tank will make the rise very slow.

- When the water temperature in the loop exceeds 90°F, cooling Tower F will come into operation and remove heat from the water, rejecting it into the air. This cooling tower will have two 20 hp fans; so it will be an energy saving to keep this fan off as long as possible. The storage tank does this.

- During hot weather operation, at night, when the cooling tower operates most efficiently and can reject heat at part capacity without the 20 hp fans running, a smaller Pump G will start and circulate water through a bypass Loop H from the storage tank, through the cooling tower, and back to the storage tank. This reduces the temperature of water in the tank at the least possible cost.

- When many of the heat pump units are in the heating cycle, they will remove the heat from the circulating water loop, and the temperature will gradually drop. The amount of water in the storage tank will make this drop occur slowly.

- If the circulating loop water temperature drops below 60°F, Pump I will start and circulate water through an electric Boiler J to put heat into the circulating water loop.

- During the coldest weather, a clock switch will start Pump I and the electric boiler at night and build up the water temperature in the storage tank to 90°F; so the boiler will not have to operate during the day. This will reduce electrical demand, allowing the electric service to be smaller and also reduce the size of the electric boiler. If Carolina Power & Light Co. starts charging the schools a demand charge (and this is very probable since other utilities already do this), a substantial savings will be realized.

- When the sun is shining, the water (actually antifreeze solution) in the solar collector Panels K will get hot. A thermostat sensing sun heat will turn on Pump L.

- If heat is needed in the circulating water loop, a thermostat will Start Pump I and position motor operated three-way Valves M and N so that loop water and solar panel water go through heat Exchanger O. The heat collected by the solar panels will go into the circulating water loop. During a day of maximum sunlight, this can be as much as 900,000 Btu, which is equal to the heat available from 11 gal. of oil.

- If heat is not needed in the circulating water loop, Pump I will stop. The solar loop will continue to build up heat until it reaches 150°F, when Pump L and M will start. Solar loop water will be directed by motor operated three-way Valve N so that it will flow through heat Exchanger P, and the solar heat will be put into the 400 gal. domestic water Heater Q.

75

Solar system employs total heat concept

This system utilizes both sensible and latent heat in the liquid or gaseous state with an overall increase in efficiency and effectiveness.

RICHARD R. STUBBLEFIELD,
Vice President, Broyles and Broyles,
Inc., Fort Worth, Texas

Most solar collector systems in use today function on the basic principle of converting the radiant solar energy into thermal energy— sensible heat. The collector is the heart of the solar energy system. Its cost and efficiency are key elements in the practical application.

Thus far, solar collectors have been primarily utilized to collect solar energy in a sensible heat form and make it available for useful purposes. There have previously been two basic types of collectors: the common flat plate and the focusing collector. Collector efficiencies are closely related to the temperature required or desired and deal totally with sensible heat.

A third dimension in the collection and use of solar energy has evolved. At Broyles & Broyles, Inc., we are in the process of perfecting a patented process and apparatus that should significantly increase collection efficiency and greatly widen the application of solar energy. The basic process involved will utilize the sun's energy in a combination of processes, as shown in Fig. 75-1, to allow for:

- Sensible heat with air only.

- Sensible heating and humidifying with air only.
- Humidifying only.
- Sensible heating with liquid as a result of latent heat conversion plus humidification as required.

Our approach to solar energy collection and conversion will utilize three basic components:

- Solar collector assembly.
- Converter section.
- Heat transfer and storage vessel.

The solar collector assembly is composed of three components:

- A sensible heat section.
- A sensible heat plus latent heat section.
- The liquid solution.

The sensible heating collector section is a basic air heater. The sun's energy is transmitted through a glazed surface and absorbed by a highly conductive surface. The energy conducted through this surface is transferred by forced convection into the air stream.

The sensible and latent heating section is used to elevate the temperature of the solution and evaporate moisture from the solution. The sun's energy is transmitted

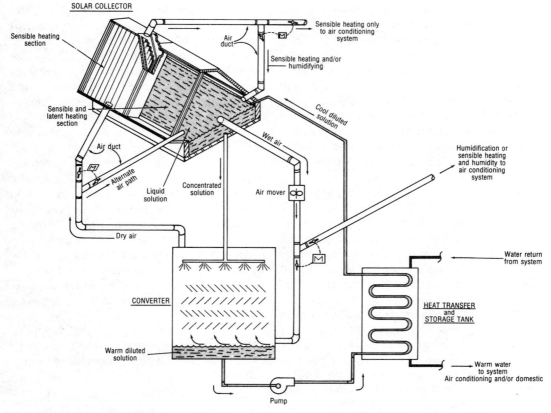

Fig. 75-1. Schematic of solar energy system employing total heat concept.

through a glazed surface and partially absorbed by the absorbtive base of the pool. The infrared energy of the sun will cause a more direct evaporative process without elevation of temperatures. Both collector sections are mounted at the optimum angle for winter heating applications—at the latitude angle plus 15 deg.

The horizontal pan, or pool solution, provides the base of the collector assembly. Between the inclined collector sections and the solution pool, an air chamber is formed. The sun's energy, in combination with a differential in vapor pressures, will cause evaporation of moisture from the solution. This converted latent energy is then transported by means of a ducted air system for conversion to usable sensible heat or for delivery directly to the building air conditioning system for humidification purposes.

General process description

The sun's energy can be used for sensible heat only, or in combination to vaporize moisture within the solar collector section. Since most solar collector systems utilize the same basic method, sensible heat collection, we will deal more specifically with the latent heat extraction and conversion process.

Air flow will be induced through the collector section for absorption of water vapor. The air induced into the solar collector will be at a lower vapor pressure than that of the water, thereby allowing for the absorption process.

For some applications, a salt solution will be utilized. The solution will provide the means for recovery or transfer of the latent heat of vaporization from the air stream. The proper solution will have a vapor pressure greater than the air stream.

The air flow through the solar collector section will be at a low humidity ratio. Water will be vaporized by the sun's energy from the solution, and thus the solution will become more concentrated. This process will have the resulting effect of increasing the vapor pressure of the air stream and decreasing the vapor pressure of the solution in the flow direction toward the exit of the solar collector section.

The moisture-laden air from the solar collector will flow to the collector converter section. The concentrated solution (relatively moisture-free) from the solar collector will flow to the collector converter section. In the collector converter section, the solution will pass over mixing baffles. The moisture-laden air will be induced at the lower part of the collector converter and flow in counter-flow to the solution. Since the solution is concentrated, it will re-absorb the water vapor from the air stream. In this process, the latent heat will be converted into sensible heat. The air will give up its moisture. The solution will become diluted with the water vapor. The temperature of the solution will be increased. The relatively moisture-free air will be returned to the solar collector section to start another cycle. The heated, moisture-laden solution will flow to the heat transfer storage tank. (See Fig. 75-1.)

The heat transfer storage tank is filled with water. Immersed within the water is a heat transfer device. The heated solution is circulated through the heat transfer device where its heat is given up to the water within the heat transfer and storage tank. The cooled solution is then returned to the solar collector to repeat the process. The heated water within the heat transfer and storage tank is available for heating application as required.

Adding moisture costs

Numerous approaches to energy conservation have been developed and implemented within the past few years. One of the significant and widely used conservation methods is the outside air economizer cycle whereby outside air is used for cooling below a predetermined outside temperature. This technique, when applied properly, can and will provide significant savings in energy consumption. However, when the outside air is cold, it is most often lacking in moisture content to provide an acceptable level of relative humidity within the occupied space. When such occurs, moisture is normally added to the HVAC system by utilization of some form of adiabatic heat process or by the injection of steam. For each pound of moisture added with the adiabatic heat process, approximately 1050 Btu of sensible heat must be added to the system. For each pound of moisture added by steam, approximately 1500 Btus are required (assuming a 70 percent plant efficiency).

An air conditioning system supplying 100,000 cfm and operating on an economizer cycle basis could typically operate under the following conditions:

- Supply air: 100,000 cfm at 55°F DB.
- Return air: 55,000 cfm at 75°F DB.
- Outside air: 45,000 cfm at 30°F DB, 40 percent RH (12.5 grains per lb).
- Space design: 72°F DB and 35 percent RH (40.6 grains per lb).

The approximate required moisture addition would be:

$$Q = (45,000/13.34) \times 60$$
$$(40.6 - 12.5)/7000 = 812 \, \text{lb}$$
$$\text{per hr}$$

where:

Q = moisture addition, lb per hr
13.34 = specific volume per pound of dry air
60 = minutes per hour
7000 = grains of moisture per pound

The latent heat involved would be:

$$HL = 812 \times 1050 = 852,600 \, \text{Btuh}$$

From the previous analysis, one can readily see that moisture addition can be an expensive and energy consuming process.

Our solar energy collection system can be designed with the required components to achieve maximum utilization and efficiency for the particular application. We feel that our approach will be a significant contribution to effective utilization of the sun's energy.

76

Geothermal energy rediscovered

A. A. FIELD,
London, England

The existence of vast quantities of heated underground water has been verified in France. A large housing development is drawing some energy from this source, and an intensive study program has been launched by the French government to find the best means of exploitation.

In 1973, the two geothermal power stations at Larderello and Monte Amiata in Italy produced 2400 million kWh. The current total operating power from geothermal sources in the country is 406 MW, but like France, Italy wants to boost this with new drillings. Professor Tongiorgi, head of the International Geothermal Energy Center in Pisa, has announced a new geological survey to identify the major heat producing zones in Italy.[1] Very little is known about the geothermal potential in Italy, and it is generally believed to be very much higher than ever imagined in the past. The uncertainty of the state of knowledge is illustrated by the explosion of a geothermal stream from an area declared sterile by the National Research Council only a few years ago.

France's recovery methods

Figure 76-1 shows the distribution of medium and high temperature deposits revealed by the survey by the Bureau de Recherche Geologique et Miniere. Temperatures of 212°F and above are regarded as power sources

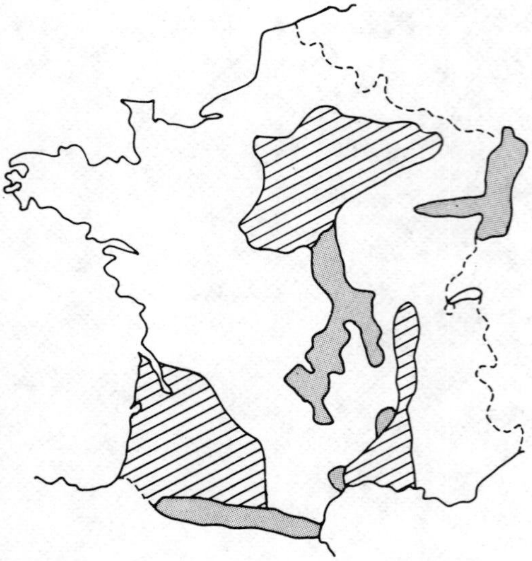

Fig. 76-1. Distribution of geothermal resources in France. Water temperature ranges from 160 to 180°F in light areas. Dark areas indicate water temperatures over 212°F.

for turbines, and temperatures below this range can be used for district heating. In addition to these deposits, there are others that as yet have not been delineated but are generally considered to be equally as large with temperatures ranging from 90 to 160°F. In the upper temperature band, water can be used as the primary medium in a heat exchanger, but in the lower temperature range, more sophisticated recovery techniques are necessary—for example, the heat pump, or a combination of counter flow heat exchanger and heat pump.

---

[1]Superscript numerals refer to references at end of the chapter.

548

Geothermal temperature gradients can vary from 0.5 to 5°F per 100 ft. The steeper gradients are associated with volcanic areas. In general, the gradient in France is around 2°F per 100 ft. This means that usable deposits begin at around 3000 ft.

Depending on its temperature, water can be used to drive a steam turbine (212°F plus), act as a primary medium for a heat exchanger (140 to 200°F), or form the source for a heat pump (90 to 110°F). It generally follows in heating applications that the deeper the bore hole, the less complex the hardware needed to utilize the heat; but low temperature water needs a heat pump to raise it to a realistic temperature.

R. Cadiergues, director of the French HVAC research center of CoSTIC, presented some interesting ideas on the combination of heat exchangers and heat pumps in a paper to the Journal *Promoclim A*.[2] By using a counter flow heat exchanger for the first stage of the transfer, it is possible to keep the exchanger to a reasonable size while leaving a moderate second stage entry temperature. For example, water from a well enters the first stage of a heat exchanger at 158°F and leaves at 122°F, producing a 158°F exit temperature from the secondary (heating installation) side (Fig. 76-2). The water then enters the second stage unit, which in this case is the evaporator or the heat pump, and is cooled to 68°F before finally being reinjected into the underground source.

For an expenditure of 1 kW in the heat pump, the amount released in the two stages (for the sake of continuity, this will be expressed in kilowatt units also), would be: heat exchanger, 2.85 kW; and heat pump, 5.15 kW. This represents a total input of 8 kW for only 1 kW expended at the heat pump. Water pumping costs have been omitted since they are fairly small by comparison.

This arrangement obviously makes the best use of the source, but the initial cost may be difficult to justify since at 158°F, the water could be used directly. Heat pumps alone would be the only solution when the source temperature is 110°F or less. In this temperature range, it would be possible to reach a coefficient of performance of about 4 or 5 with a heat pump, providing the heating system was designed to work at 120°F—for example, embedded floor or ceiling panels or warm air systems.

By studying economic and technical models of geothermal systems, the CoSTIC laboratory concluded that the optimum flow rate from the well could be determined solely in relation to a temperature drop of 45°F and not the temperature level itself. Figure 76-3 is based on the CoSTIC analysis and shows the technical and economic changeover points for the three systems: heat exchanger only, heat exchanger plus heat pumps, and heat pump only. The ratios of energy used in the various stages can be derived from the intersection points on the vertical ordinates from the source temperature. A small proportion is heat from the compressor.

Paris applications and costs

The principal heated water layers in the Paris region are the Albien layer at 86°F and the Dogger layer at 158°F. The first layer has been used for some time by the water authorities since the water is particularly pure. Further exploration, however, is limited by law. Water from this layer is still serving as the source for the heat pump installation at the Radio Paris building, commissioned 12 years ago.

While water from the Albien layer must

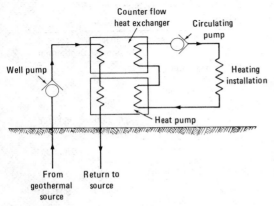

Fig. 76-2. Geothermal water used in a two stage heat recovery technique.

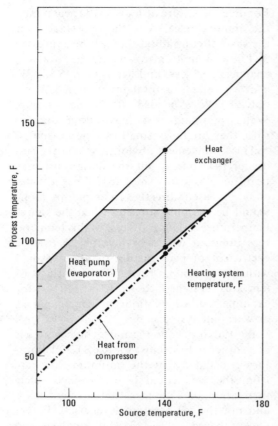

Fig. 76-3. Optimization of heat recovery techniques for geothermal sources.

mer, the heat pump is used as a refrigerator for the air conditioning plant.

A cost comparison at the time showed almost the same figures for the boiler plant and heat pump systems, including sinking the well (see Table 76-1).[3]

A running cost comparison ranks the heat pump installation much lower than the traditional boiler plant, even based on oil costs and electricity rates at the time. A further bonus was that the source water was so pure that it could be reused for water services in the building and thus avoid the cost of purchasing water from the city.

More recently, geothermal energy from the Dogger layer has been used to heat part of a 3000 dwelling development of social housing near Paris.[4] In this development, 2000 of the dwellings, which are mostly apartments, are connected to the system. The structural insulation for the units was made particularly high at the time of construction, which resulted in a load of only 0.05 Btuh per cu ft per deg F. Each apartment has a total floor area of 900 sq ft and a total of 7000 cu ft. For the design conditions in this part of France and for an average internal temperature of 65°F, the heat loss per apartment is 17,000 Btuh.

Two wells were sunk at the site, and each

be treated by a heat pump to make it usable, water from the Dogger layer can go straight to a heat exchanger. However, it is much more aggressive since it contains a high proportion of dissolved salts. The Dogger layer is nevertheless the most extensive, covering an area of 2700 sq miles located between 5200 and 6200 ft below the surface.

The Radio Paris building at the time it was built had a total volume of 16 million cu ft, of which 5.3 million required air conditioning, 9 million required heat only, and the remaining 1.7 million cu ft is comprised of underground areas and other spaces. The heat pump installation has an output of 8 million Btuh into a low temperature water circuit that serves heated acoustic ceilings. A 108°F condenser exit temperature gives a gas condensing temperature of 117°F. During the sum-

Table 76-1.   A comparison of relative first costs between a traditional heating system and the geothermal system installed at the Radio Paris building. The comparison is based on 1957 prices.

Item	Relative first cost, $	
	Traditional	Geothermal
Boiler plant and chimneys	188,000	—
Fuel handling and storage	50,000	—
Refrigeration compressors	250,000	250,000
Evaporative coolers	62,000	—
Well sinking and pumping gear	—	180,000
Heat exchangers	—	100,000
Standby electric boiler (possible failure of source)	—	50,000
Pipework	—	25,000
Total	$550,000	$605,000

is pumped at a rate of 370 gpm. Each well is connected to heat exchangers, individually rated at 10 million Btuh. The geothermal input over the year is just under half the total requirement for heating and hot water services. The remainder is provided by two peak load boilers, each rated 22 million Btuh. The total investment cost was $1.75 million.

Heating was charged at a flat rate of $0.20 per sq ft of floor area, and hot water was billed $0.50 per 100 gal. Consumers paid in four installments. The first three covered 95 percent of the total flat rate charge, and the fourth made up the balance and the cost of hot water service over the year.

Analysis of the load curves presented in Fig. 76-4 shows that in one year, geothermal energy accounts for 60 billion Btu. The cost of heavy oil to generate this quantity of heat would have been $190,000 in May 1974.

The piping hookup for the geothermal system was designed so that the system would always predominate at low load and form a permanent base load at all times. Water was taken from the well at a rate of 370 gpm, yielding a 45°F drop in the heat exchanger. Water enters the secondary side of the ex-

changer at 130°F and rises about 11°F. The oil fired boilers boost the temperature to 185°F into the network. When the heating circuits are closed down in the summer, geothermal heat is connected directly across the coils of the hot water storage cylinder.

The system has been in operation since 1970, and the only major trouble encountered was the failure of the heat exchanger after six months due to corrosion. The battery has since been replaced with one made with titanium, and the failure has not re-occurred.

The future of geothermal energy

R. Cadiergues gave the potential for geothermal energy as 30×10^{12} kWh per year (based on a consideration of radioactive rocks) in cold and temperate countries.[2]

Compared to other hitherto neglected sources of energy, such as wind or solar energy, geothermal energy offers perhaps the greatest reserves, with a steady output unaffected by climatic conditions. A question that is not entirely certain at the moment is the size of the reserves in terms of heat production, and how they will be influenced by reinjection of water.

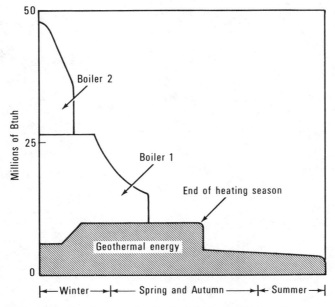

Fig. 76-4. Load curves for the installation at Melun, France.

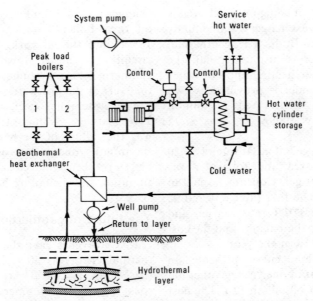

Fig. 76-5. Piping hookup at the Melun geothermal installation.

References

1. Energia Geotermica, Conference Report, *Installatore Italiano*, September 1974.
2. Cadiergues, R., "Energies et Procedes Nouveaux," *Promoclim A*, May 1974.
3. Conturie, L. and Testamale, H., "La Pompe de Chaleur de la Maison de la Radio de Paris," *Industries Thermiques*, April 1957.
4. Guichemerre, A., "Les Eaux Souterraines et le Chauffage des Locaux," *L'Installateur*, September 1974.

Figs. 76-2 and 76-3 are based on diagrams from Reference 2. Figs. 76-4 and 76-5 are based on diagrams from Reference 4.

77

Geothermal energy heats campus

Almost unbelievable savings are provided by the Oregon Institute of Technology's on-campus geothermal heating installation.

JAMES HITT,
Geothermal Coordinator, Physical Plant Superintendent, Oregon Institute of Technology, Klamath Falls, Oregon

Klamath Falls, a city of 16,000 population, is near the center of an elongate, northwest trending structural valley called the Klamath graben. It is a complex structural valley about 50 miles long and 10 miles wide. Upthrown rocks, or horsts, bound the valley on both east and west sides. Upper Klamath Lake, Oregon's largest lake, and numerous other smaller lakes, ponds, and swamps occupy the lowest parts of the valley. Unlike the other large graben valleys of the basin and range of the western United States, which have internal drainage, Upper Klamath Lake is drained by the Klamath River that flows south and then west through the high cascades to the Pacific Ocean.

In the early days of Klamath Falls, several groups of hot springs and boiling mud pots were present in the flats and low, rolling hills where the city is now built. The Indians were the earliest users of the hot springs, and they cooked fish and game and also, because they had great faith in the healing power of the water, bathed and soaked in the overflow pools.

Early settlers used hot springs

The early settlers used them frequently for various things, including the scalding of hogs, and by 1915 there was a greenhouse built over one of the hot springs. By 1928, Klamath Falls boasted a modern hot springs natatorium in which the pool water was completely changed daily. There are more than 360 relatively shallow wells that tap the natural hot water to heat schools, industrial buildings, apartments, and houses.

Complex faulting, most of which occurred in late Pleistocene time, has broken the whole Klamath area into a characteristic pattern of northwest trending fault block ridges and intervening down-dropped graben valleys. Tilted fault blocks are numerous within the Klamath graben.

Geologic evidence explained

Perhaps nowhere else in the United States is the geologic evidence of recent faulting so well displayed. At nine separate locations, the shiny, polished fault surfaces are exposed. All are high angled normal faults and dip from 55 to 70 deg into the valleys. One of the most spectacular faults has been exposed along the east side of Upper Klamath Lake, where removal of talus has uncovered the fault plane for nearly a third of a mile in length and as much as 250 ft in height. Grooves in the slick sided surfaces indicate that the last and perhaps all of the displacement has been vertical. At least 1600 ft of vertical displacement is indicated in the vicinity of Klamath Falls, and movements of this order of magnitude can be estimated at other places where the steep fault escarpments are elevated about this much above the valley floors.

The Klamath Falls geothermal area is near the center of the graben in slightly tilted fault blocks that are elevated a few hundred feet above the valley floor. These tilted blocks are made up of impure diatomite, thin beds of tuffaceous sandstone, clayey tuff, and minor intercalated basalt flows. The tilted blocks have, in turn, been complexly faulted into elongate ridges that generally trend northwest. Because they are so easily eroded, they are now seen as low, rolling hills.

Well logs indicate that the diatomaceous tuffs and layered sediments are at least several hundred feet thick in most places, are impervious to the flow of water, and act as a cap at the surface. Broken lava flows and zones of scoria and cinders are encountered at various depths, and in most cases in the thermal areas, these horizons yield large quantities of live hot water. At least one strong northwest trending fault is present on the east side of the geothermal area, and the rise of hot water from a deeper reservoir could be associated with the brecciated rocks that are common to this type of fault. Although the heat source is not known and no surface rocks of recent age are present, the Pliocene-Pleistocene dikes and sill-like masses that are intercalated in the lacustrine deposits may indicate the presence of a larger intrusive rock mass cooling at not too great a depth.

A long history of hot spring activity is shown by the presence of bleached silicified rocks, deposits of calcite and gypsum, and minor mercury mineralization in a broad halo surrounding the geothermal zone.

The Oregon Institute of Technology campus is located approximately three miles north of the downtown business district of Klamath Falls. Elevations on campus range from 4250 to 4400 ft.

A carefully chosen move

Oregon Institute of Technology was founded in 1947. Dr. Purvine, the founder of the school, concluded that numerous factors present at this first site created heating problems and high costs at the original location.

In 1959, the Oregon State Board of Higher Education was willing to gamble with funds granted to them to establish a new campus on the site that Dr. Purvine had selected, after careful study of the winter conditions (early morning dissipation of ground frost as opposed to earlier afternoon dissipation of frost). Dr. Purvine's knowledge of the area prompted him to determine that the heat that warms the water more than likely rises to the surface through the fault zones. He carefully charted every fault or faults known to exist near to the campus. Based on these findings, he directed a drilling crew to a portion of the campus above what he considered to be an old fault and gave orders to begin drilling. The well was drilled to 1205 ft. They found water, but it was a mere 78°F. He moved the crew to the other side of the fault. Again water was found, but this time it measured 176°F. A total of six wells were drilled, three hot and three cold. The campus today consists of eight buildings and is heated with water from just one of the six geothermal wells. The three cool wells measure temperatures of 65, 78, and 92°F. Information on all wells is listed in Table 77-1.

Well No. 1 is approximately 1205 ft deep, and the water temperature is 78° F. This water is used for irrigation and domestic purposes. It is located at an elevation of 4530 ft above sea level and produces 510 gpm. A typical section of one of the six drilled wells is shown in Fig. 77-1.

Large volumes of hot water

Klamath Hills is a large, isolated fault block within Klamath graben about 10 miles south of Klamath Falls. Large volumes of hot water

Table 77-1.   Summary of data on the six geothermal wells at Oregon Institute of Technology

Data	Well No. 1	Well No. 2	Well No. 3	Well No. 4	Well No. 5	Well No. 6
Depth of well	1205 ft	1288 ft	1150 ft	1224 ft	1716 ft	1800 ft
Static water level	449 ft	332 ft	110 ft	315 ft	358 ft	359 ft
Water level at testing	532 ft	550 ft	355 ft	550 ft	393 ft	540 ft
Temperature of water	78°F	191°F	65°F	92°F	191°F	191°F
Volume of water pumped at testing	510 gpm	107 gpm	175 gpm	400 gpm	442 gpm	250 gpm
Water use	Domestic, irrigation	Domestic heating	Domestic, irrigation	Domestic, irrigation	Domestic heating	Domestic heating
Casing used						
12 in.	—	441 ft 3 in.	—	733 ft	530 ft 3 in.	416 ft 4 in.
10 in.	—	—	—	315 ft 6 in.	814 ft 6 in.	867 ft 6 in.
8 in.	686 ft	803 ft	707 ft 7 in.	207 ft 7 in.	318 ft 6 in.	294 ft 6 in.
6 in.	544 ft 10 in.	515 ft	—	—	648 ft 1 in.	677 ft 8 in.
Cost of well						
Drilling	$10,038.00	$11,082.50	$ 7,418.00	$15,317.75	$20,322.50	$21,466.50
Testing	2,076.00	3,425.00	1,325.00	1,750.00	1,750.00	2,000.00
Casing	4,836.12	7,592.00	3,012.87	5,939.32	9,180.00	8,816.74
Total	$16,950.12	$22,099.50	$11,755.87	$23,007.07	$31,252.50	$32,283.24

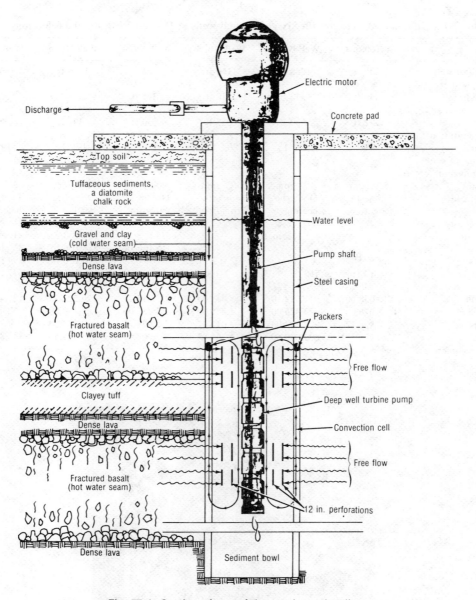

Fig. 77-1. Section of one of the geothermal wells.

(200°F) are found at shallow depths in a narrow zone along the southwest side. The geologic environment of this thermal zone is similar to that of the Klamath Falls area, and a halo or border of silicified lake sediments and tufa are present. High angled normal faults with a general northwest trend are common, and again, the recency of faulting is shown by slick sided fault surfaces. Basalt flows and associated breccias, scoria, and cin-

ders predominate in the Klamath Hills fault block. Drillers' logs from a few water wells indicate several hundred feet of lake sediments and layered tuffs in the lower Klamath Hills. A large area in the southeastern part of the Klamath Hills is underlain by cinders, scoria, and palagonitic tuff breccias, indicating that here a basaltic magma encountered water or saturated sediments near the surface and violent explosive eruptions resulted.

The original hot springs of Klamath Falls have disappeared through lowering of the water table and cultural changes. Of the 360 wells that have been drilled, the depth of the wells range from as little as 100 ft to a maximum of 1800 ft. Most are in the 200 to 350 ft range. The water table generally seems to coincide with the elevation of the Upper Klamath Lake, and static water level depths vary with the topography. Much of the rock penetrated in the geothermal zone is quite "tight" or impermeable, and perched water influences the height of the local water table in many cases. Drillers report drops of water levels amounting to as much as 50 ft from levels initially encountered as drilling proceeds to greater depths.

Temperatures of the wells in our geothermal zone range from 140 to 235° F.

Klamath Falls is 297 miles southeast of Portland and 383 miles north of San Francisco. Major cities within an approximate 1000 mile radius are: San Francisco and Los Angeles, California; Reno and Las Vegas, Nevada; Salt Lake City, Utah; Boise, Idaho; Portland, Salem, and Eugene, Oregon; and Seattle and Spokane, Washington. Because of its geographic location, it is at the center of population of more than 26 million people.

Well No. 2 is the original hot water well. It produces 107 gpm and is available for emergency purposes. It is 4355 ft above sea level in elevation. The temperature is approximately 191° F.

Well No. 3 is 1150 ft deep and was tested at 175 gpm of 65° F water. It was capped as uneconomical for domestic and irrigation uses. It is located at an elevation of 4352 ft.

Well No. 4 is used for domestic and irrigation purposes. It is situated 4402 ft above sea level. It yields 400 gpm of 92° F water. (Thus, wells No. 1 and No. 4 supply the domestic and irrigation needs of the campus.)

Well No. 5 is one of the two hot water wells used for hot water production. No. 5 is situated at 4374 ft above sea level and produces 442 gpm. The water temperature is approximately 191° F.

Well No. 6 is the second of two wells to be used for heating purposes. It produces 250 gpm of about 191° F water. This is the deepest of the wells with a depth of 1800 ft.

Common underground source

By alternate pumping, it was determined that the three hot water wells have a common underground source. At the highest use level, drawdown of the static water level has been experienced at a minimum of 2 ft and a maximum of 7 ft. This seems to relate to the general groundwater levels in the area. The water appears to be a portion of the normal groundwater since it contains no impurities indicating direct contact with molten rock or water from such magmatic sources. A comparison of the corrosive effects of some hot water wells in the Klamath Falls area determined the water there to be carrying chemical elements corrosive to metals. Due to the comparatively pure characteristics of the Oregon Institute of Technology well water, it is forced through the heating system and discharged at about 120° F and flows via the storm sewer into Klamath Lake. All geothermal water waste is allowed to naturally discharge into Klamath Lake. The discharge is controlled by the city and is based upon a ratio formula of cubic feet of waste of geothermal water allowed versus the square feet of floor space heated.

Use closed system

All geothermal hot water is maintained in a closed system, and corrosion is very minimal. Slight, but apparently increasing, subsidence originating from the geothermal wells is producing cracks in excess of 1000 lineal feet across the campus. Serious consideration is being given to geothermal reinjection to stabilize this condition.

During the first few seasons, considerable expense was incurred while operating the hot water wells. Submersible pumps were installed with the bowl located well below static water level. Engineering assumptions had been made that water lubricated bearings would be satisfactory as had been demonstrated in cold water wells. This proved to be an error. The lubricating capability of hot water was so low as to cause frequent bearing failures. The schematic flow diagram of the heating system is shown in Fig. 77-2.

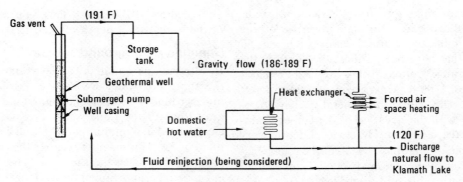

Fig. 77-2. Schematic flow diagram of heating system.

Expansion rate a problem

Another problem was associated with the rate of expansion of the pumping components. As the pumps operated, hot water was forced up through the pump column made of pipe with the drive shaft from a surface mounted electric motor located in its center. Variable rate of expansion of the pipe and the shaft produced a requirement of 5.5 in. expansion allowance, both at the motor mounting and in the pump bowl. Eventually, redesign and reinstallation produced a successfully operating mechanical system with added lubrication.

Cost down from $94,000 to $14,000

Original costs of this early experimentation came to about $90,000 or an average of $10,000 per year for mechanical renovation.

The original campus presented extremely high heating costs. The heating of housing and instructional buildings amounted to an average of $94,000 per year. The geothermal heating system at the new campus is now a valuable operation. The costs range from $12,000 to $14,000 per year to heat over 440,000 sq ft of building space.

The unique factor of the heating system is the use of standard motors and pumping equipment without the need for expensive custom built machinery. Lack of proper cooling of motors, shaft stretching, and corrosion of pipes were the basic obstacles to be overcome. These tasks were accomplished by first elevating the pumping equipment to above ground level. This permitted air cooling of all motors. Secondly, it was determined that the 550 ft shaft would initially expand as much as

3 in. in a 3 min period. By having custom made 5 in. lateral bowls substituted for the normal 1¼ in. lateral bowls, the expansion factor was eliminated. Finally, air venting released corrosive gases that had caused much corrosion of pipes.

Motors now run continuously

In the preliminary installations, the life span of $8000 motors was anywhere from 10 min to several months. Since the problems have been solved, one motor has been running continuously for 24 hr per day for the past five years without any problems. A motor in combination with a clutch type that permits water to be supplied on demand is used in the installation. This system is probably the only one of its type in the United States, drawing water from a 550 ft depth.

The original motors had insufficient ventilation and frequently burned out. In the beginning, these electric motors were set down in a sump, and later on, it was found necessary to raise them above the surface and to build a housing for improved ventilation and to protect the equipment from vandalism. At first, there were some troublesome periods during which a standby boiler was required. This has not been necessary for the past ten years. After numerous alterations, the original equipment was found adequate and serviceable for our particular needs. In a region where moderately severe winters are experienced, Oregon Institute of Technology has benefited greatly from the use of natural hot ground water in heating not only in comfort, but also economically.

Glossary

basalt—A dark gray to black dense to fine grained igneous rock that consists of basic plagioclase, augite, and usually magnetite.

breccia—A rock consisting of sharp fragments embedded in a fine grained matrix (as sand or clay).

diatomite—A light easily crumbled siliceous material derived chiefly from single celled algae remains.

dike—A tabular body of igneous rock that has been injected while molten into a fissure.

escarpment—A long cliff or steep slope separating two comparatively level or more gently sloping surfaces and resulting from erosion or faulting.

fault—A fracture in the earth's crust accompanied by a displacement of one side of the fracture with respect to the other and in a direction parallel to the fracture.

graben—A depressed segment of the earth's crust bounded on at least two sides by faults.

horst—A block of the earth's crust separated by faults from adjacent relatively depressed blocks.

magma—Molten rock within the earth's crust.

Pleistocene—An epoch within the Cenozoic Era of the geological time scale, usually taken to embrace the last two million years.

Pliocene—An epoch within the Cenozoic Era.

scoria—Small fragments of porous volcanic rock.

talus—A slope formed especially by an accumulation of rock debris; a rock debris at the base of a cliff.

tuff—A rock composed of the volcanic detritas usually fused together by heat (tuffaceous, adjective).

tufa—A porous rock formed as a deposit from springs or streams.

78

A critical look at total energy systems and equipment

A detailed review of associated problems and solutions.

MAURICE G. GAMZE, PE
Gamze-Korbkin-Caloger, Inc.,
Consulting Engineers, Chicago, Illinois

The concept of total energy is not new. Applications using prime movers to provide shaft work or drive generators with reclaimed waste heat used for process heat, space heating, etc., have been around for years. This discussion is confined to systems that generate electricity for buildings and furnish a significant portion of the heating, air conditioning, and process requirements from the heat recovered from prime movers. Furthermore, only specific hardware problems within these systems will be discussed.

During the 1960s, the total energy concept got a big boost from the natural gas industry. It offered an alternative to the all-electric concept that was becoming a serious competitive threat. While the energy conservation aspects of total energy were recognized, it did not receive direct emphasis during those days of so-called cheap and unlimited energy. The potential monetary savings inherent with a total energy installation were stressed in sales efforts.

Due to large scale promotional efforts by segments of the gas industry, total energy installations were made in such diverse applications as shopping centers, industrial plants, office buildings, motels, and hospitals. Installed generating capacities ranged from a low of 50 kW to several thousand.

Unfortunately, the existing hardware generally was not up to the demand of the concept in an era when owners and tenants were growing accustomed to increasingly sophisticated systems that performed with a reasonable degree of reliability. Misapplication of the concept compounded the problem. Total energy was sometimes installed where, at best, the economies were marginal. Also, poorly conceived and designed total energy plants were installed. Maintenance was greater than anticipated, and this had a definite bearing on reliability and economics.

Manufacturers should have learned a lot during this period about their equipment due to the rigors of around the clock, seven days a week operation. However, today, when energy conservation is vitally important, the existing hardware—primarily the prime movers and ancillary equipment—still has shortcomings that must be circumvented in the design stage.

In spite of these obstacles, enough well conceived and competently engineered jobs have been installed to demonstrate the potential of total energy for wider use if system technology is improved. Wider application will significantly reduce the energy consumption of buildings.

Total energy technology advances will also

benefit what are called selective energy systems. Here, the total energy plant is an extension of the electric utility system, and optimum efficiency of the plant's electrical output can be matched to the building's thermal demands at all times, with power either flowing from the plant into the utility grid or vice versa as conditions require.

Analysis of the magnitude, duration, and coincidence of electrical and thermal loads, plus selection of prime movers and waste heat recovery methods determine system feasibility and design. Integration of design between the project's electrical and thermal needs and the energy plant provides optimum economic benefit.

Basic system components

The basic components of a total energy plant are:

- Prime movers—normally either an internal combustion engine or a gas turbine.
- Electrical generators.
- Waste heat recovery systems—to capture the heat rejected by the prime movers.
- Waste heat rejection systems—for periods when the building's thermal demand is lower than the heat of the prime movers.
- Prime and supplemental heating apparatus.
- Fuel systems.
- Control systems.
- Connection to building mechanical and electrical services (Fig. 78-1).

The first part of the selection process is to determine the user's requirements and plant capacities for electrical, heating, cooling, and other energy demands and usages, such as: maximum electrical demand, hourly electrical loads, maximum heating and cooling demands, hourly heating and cooling loads, hourly amount of salvageable heat available from the prime movers, and the hourly amount of heat energy (for either heating or cooling) required in excess of available salvageable heat (Fig. 78-2).

If the plant, because of utility rate structure or legal constraints, must provide all of the user's electricity, determination of electrical loads and usage profiles for the particular user is the first step.

Electrical demands are determined by totaling the various individually connected loads as corrected by application of both load demand factors and time of usage. Electricity used by heating and air conditioning systems, lighting, power, and other requirements must be included to determine demand.

The maximum heating and cooling demands are normally established by computing the building thermal loads in accordance with ASHRAE standards. The computed demands are then applied on a 24 hr basis to obtain the total kilowatt per hour consumption as well as profile the electrical load. These data guide selection of the number and size of prime mover modules.

Since combinations of prime movers and prime heating and cooling equipment are infinite, selection is dictated by plant economics, equipment service records, and locally available service capability. The economics are normally evaluated with the aid of data processing techniques that can simulate possible system configurations as related to both first cost and thermal usage. Since economic considerations are iterative, this discussion deals with plant design and equipment selection as related to the not so readily apparent parameters.

State of the art

Presently, the total energy system designer and installer must circumvent equipment problems. (Technical problems associated with total energy plants are listed later.) Therefore, reliable total energy systems are complex, redundant, and costly. Thorough preventive maintenance procedures must be rigorously followed to obtain minimum component failure.

Actual component selection depends on factors, such as service experience and availability, machinery operating limits as tempered by experience, available fuels, and available raw water quantities and conditions. For example, piston engine manufacturers vary the ratings of engine output in accordance with their laboratory and design

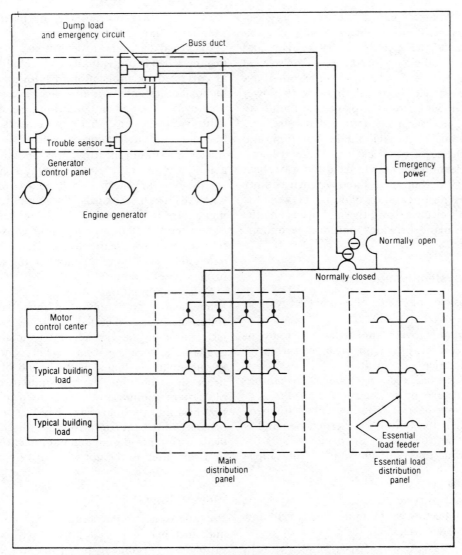

Fig. 78-1. Total energy plant schematic of electrical power system.

parameters. However, in the 1200 rpm range, we have found maximum long-time loading must be limited to between 60 and 80 percent of the so-called continuous rating capability to achieve the specified time between overhauls. From our operating experience, the life of slower speed engines compared to higher speed models does not materially improve until the 3000 to 4000 kW size ranges are used.

A study of 21 engines indicated that prime mover failure rate is a function of the man-

ufacturer's design more than it is of maintenance. The engine studies were all maintained under the same conditions, organization, and personnel (Table 78-1).

In other plants, we have experienced failure of control switchgear relays, turbochargers, turbine thrust bearing assemblies, etc. On one project, four piston engines failed after about 1300 operating hours due to improper gasket materials used by the manufacturer. At another installation, only after 90,000 hours of operation, turbochargers failed approx-

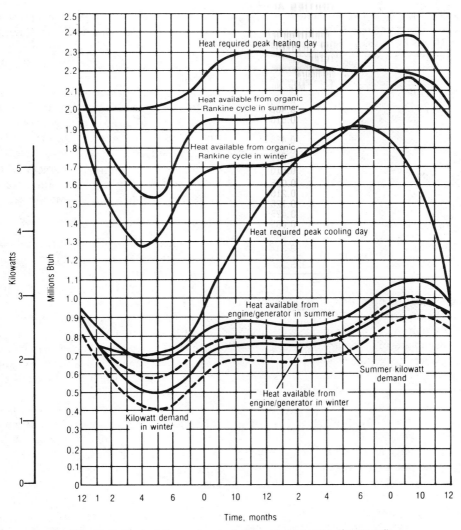

Fig. 78-2. Typical thermal and electrical load profile (for motel).

imately every six months. After $2\frac{1}{2}$ years of investigation, the oil temperature was determined as the source of the problem. Apparently, for this turbocharger design, the shaft lubricant must be heated above the manufacturer's recommended temperature in order to maintain the lubricant film. When the oil was below 170° F, the combination of lubricant pressure and viscosity used in the bearing assembly did not provide sufficient oil flow to lubricate the bearing shaft.

We have experienced prime mover failures due to oil level regulator failures, after cooler leaks, and numerous minor component failures. As stated previously, total energy plant failures have been limited basically to prime mover equipment and ancillary devices.

Engine cooling methods

If piston engines are used, the basic system configuration is determined by the engine cooling media—either steam or hot water.

The advantages of steam are:
- Smaller heat rejection devices.
- Less expensive to control.

Table 78-1. Piston engine maintenance survey. Operating hours and incidents per 10,000 hr by manufacturer.

Component	Incident hours	
	Mfg. No. 1	Mfg. No. 2
Connecting rod	0.275	0.085
Valve	0.092	0.468
Cylinder head	0.276	0.510
Turbocharger	0.138	
Bearing	0.138	0.425
Cylinder liner	0.046	0.128
Manifold	0.046	0.212
Piston	0.046	0.510
Crankshaft	1.057	0.085
Short block		0.210
Governor		0.128
Rod bolts		0.425
Crankshaft		0.085
Piston rings		0.043
		3.314
Total number of engine hours	217,710	325,228

• Less parasitic power required to circulate coolant through the engine jackets.

The disadvantages of steam systems are:

• Oxygen entering through open vents and makeup water to replace vented steam and flashed condensate contaminates the system. This causes scale formation on the inside of engine jackets and reduces heat transfer, which causes both hot spotting and thermal distortion of the jacket.

• Fast acting backpressure valves are required to maintain a constant fluid/steam level above the engine, regardless of the user's steam consumption.

Heat rejection apparatus selections also relate to past history. The basic types of heat rejection apparatus are:

• Cooling towers.
• Evaporative coolers.
• Dry coolers (radiators).
• Indirect water coolers through heat exchangers.
• Direct water cooling.

The disadvantage of cooling towers operating at the normally elevated temperatures associated with engine cooling is that evaporation rates are noticeably greater than when the tower is used for the same Btu rejection as compared to refrigeration condenser water cooling applications. This increased evaporation, atmospheric pollutants, and airborne materials, such as leaves, insects, etc., increase the probability of heat exchanger and engine jacket corrosion and fouling. The cost of water treatment is not as governing a factor as the cost of component or system failure.

Evaporative coolers exhibit the same evaporation, scale formation, and deterioration problems as cooling towers. However, scale formation and heat exchanger fouling, etc., occur basically on the exterior of the heat rejection surface, which allows slightly easier maintenance. Evaporative coolers cost more than cooling towers, but the water treatment costs and problems are about the same.

Three types of dry coolers

Dry coolers are generally of three types. The commonly used cooler normally furnished by engine manufacturers is designed to operate at low pressures with limited heat transfer surface. This requires serious attention to both physical location of the radiator and its operating power demands.

So-called industrial coolers can be made to operate over any pressure condition and with the heat transfer surface correlated to the level of fan power usage.

Approaching the industrial type, is the remodeled refrigerant condenser equipped with a water coil instead of a refrigerant coil and a fan motor suitable for high ambient temperature. It is selected for the same operating characteristics as the industrial unit but generally is not of the same quality. However, in most applications, the differences are not sufficient to warrant using the industrial type since the cooler's operation will be periodic and not continuous.

It is preferable to select a dry cooler equipped with direct drive fans to avoid belt drive failure and with high ambient temperature water-proof motors. Also, dry coolers should be broken into separate circuits and cells so that in case of tube failure or outage, the system can continue to operate.

Water cooling is safest

Water cooling through heat exchangers is the safest method of rejecting heat. Using sufficient water quantities reduces both the level of fouling and scale formation within the tubes. This method is extremely expensive.

Direct water injection should not be used in total energy plants without strict control since entrapped air can cause hot spotting, and cold water can thermally shock the block, possibly destroying the engine.

Cleanliness of the engine cooling water system is essential because the internal cooling passages of engines are not readily accessible for service. Also, engines generally cannot be effectively drained and cleaned internally. Therefore, coolant fluids must be noncorrosive and free from salts, minerals, or chemical additives that deposit on hot engine surfaces or form sludge in areas where flow velocities are relatively low. The installation of coolant piping, heat exchangers, valves, and appurtenances associated with the cooling system must include provisions for internal cleaning before they are placed into service to prevent construction debris from being carried into the engine cooling passages.

Cooling systems should be tight and well maintained to remain free from leaks in order that fresh water makeup be held to a bare minimum. Softened or mineral-free water is excellent for initial filling of a system and for makeup. Therefore, forced circulation hot water systems may require nothing more than the addition of minor corrosion inhibiting additives to assure long, trouble-free service. This is a major asset of hot water heat recovery systems.

Ebullient cooling, where the coolant turns to steam as it passes through the engine, making it in effect a steam generator, creates critical water treatment problems. The concentration of dissolved minerals increases at the point where evaporation occurs. This can be minimized by reducing the addition of minerals into the system by using treated makeup water. It is important that the treatment inhibits corrosion from free oxygen in the makeup water and returned condensate. Deaerators may be required to control the oxygen level. Water level control in multiple ebullient cooled engine installations must account for the possibility of backflow through the steam nozzles of units that are out of operation. Each should be equipped with a steam check valve or a backpressure regulating valve to avoid this problem.

It should be noted that most engines being built are designed for forced circulation cooling and are adapted to ebullient cooling applications.

Control problems

Another troublesome area is controls. There are two basic types of automated control switchgear. The solid-state logic circuit type and the so-called hardware or relay control variety. The only problem we have found in the solid state models is the use of several open wired relays in the control. Replacement of these by the manufacturer with closed plug-in units have rectified the problem of dirt accumulation on contact points.

Problems with the hardware switchgear have been basically caused by the use of the same relay for both 120 volt AC and 24 volt DC service. When used on DC circuits, a dropping voltage resistor is placed in-series with the relay. Since the DC amperage is constant, the resistor acts as a heat generator. Therefore, over a period of time, burning of the bakelite enclosure occurs, causing the relay to malfunction when the charred bakelite coats the contactors. In addition, those relays with multiple contactors or "adders" tend to act up mechanically due to the extra shaft length required to accommodate the added circuits.

We have forced conditioned air into control rooms housing these cabinets, but the longevity of the relays has been increased by only a factor of two. Hence, so-called ghost outages occur about once a year instead of every six months. With solid-state switchgear, this problem is nonexistent.

Required research and development

In addition to the need to upgrade performance and reliability of components, efforts in the followng areas are required:

• Engine jacket configuration should reduce parasitic power requirements and allow internal cleaning by direct access.

• Organic Rankine cycle equipment, should be developed (combination boiler, turbine, heat recovery apparatus) and coupled with a plug-in control system and combination type heat recovery-heat rejection apparatus. The latter should be modular and suitable for remote installation. This cycle offers the best promise of competing on an energy consumption level, installed cost basis, and service cost basis with conventional total energy plants.

- A gas turbine cycle should be developed specifically for total energy application. This would require the following revisions to presently available gas turbines:

  1. A combustion system similar to that of so-called premix burners to allow injection of low pressure gas directly into the compressor air intake, eliminating the need for gas compression equipment.

  2. An integral heat recuperation, heat recovery, and heat rejection unit in response to thermal demand. When electrical demand exceeds thermal demand, this recovered energy should be directed into the air stream between the compressor and combustor, acting as a recuperative system and increasing the cycle's efficiency.

- Modular control systems with "umbilical" type connections.
- Control switchgear for interconnection with utility systems.
- Thermal storage devices and systems, including the use of building structure, fluid storage tanks, and solid storage systems (such as Glaubers salts).
- Electric storage devices including development of high density batteries and high energy rotating wheels.
- Combination thermal-electric storage, including use of stored heat to energize thermoelectric generators.

Conclusion

The problems outlined are many, and the solutions are not easy. But neither are there easy solutions to the problems total energy can help alleviate. The days of cheap and plentiful energy appear to be at an end, at least for the foreseeable future. Total energy has a realistic potential to conserve dwindling fossil fuel reserves. If it can be developed into a successful and viable concept, it will yield the important added benefit of reducing air and thermal pollution as well.

For these reasons the institutional and technical problems associated with the total energy concept should be solved.

Plant problems

Concept and application

- The noise level inside equipment rooms is higher than in conventional central plant systems.
- Total mechanical equipment space is approximately double that required for central plants.
- Further system expansion is often difficult, particularly within confined sites.
- Load factors are poor in certain types of buildings, such as schools, offices, and simple residential buildings.

Prime mover related

- Part load performance of natural gas fueled reciprocating engines and all turbines is poor.
- Engine head gaskets tend to fail at operating temperatures required for total energy operation.
- Engine wet sleeve liner O-ring seals have a high failure rate in certain manufacturers' models.
- Failure of piston engine center main bearings, plugging of oil galleries, and high lubricating oil consumption indicate insufficient attention has been paid to engine lubricating systems.
- Piston engine components, such as lubricating oil level regulators, water pumps, thermostats, ignition systems, vibration mountings, exhaust gas connectors, oil cooler and after cooler heat transfer surface are other areas that have exhibited high problem rates.
- Gas turbines have been subject to: thrust bearing failures, lubricating oil coagulation; nozzle, vane, and combustion chamber cracks; oil and dirt accumulation causing unbalance either in bearing system or turbine blading.
- Turbine exhaust ducts and waste heat

recovery units reduce turbine efficiency due to backpressure sensitivity.

• Turbine and axial flow compressors require periodic cleaning to maintain efficiency.

• Fire resistant lubricating oils often reduce turbine capacity.

• Gas turbines require high pressure fuel source (both oil and natural gas) in order to maintain precise fuel flow control (required for automatic systems).

• Gas compressor used to boost low pressure gas to high pressure cause surges in the outlet pressure, thereby affecting the stability of the turbine performance.

Control

• Control of the user's electrical distribution and equipment required in order to prevent their destruction in case of malfunction is a problem.

• Prime movers (both reciprocating engine and gas turbine) require extensive safety control systems to prevent their destruction in case of malfunction.

• There may be control system failures, particularly with so-called hardware types when switchgear is not properly cooled.

• Control equipment is often poorly manufactured and requires extensive field correction.

Engine cooling system

• Engine water jackets do not provide for sufficient flow in certain areas of block, causing thermal distortion of block.

• High pressure drops at design flow rates through engine jackets, oil coolers and after coolers, represent a high parasitic load on the plant.

• Interior of engine jackets are inaccessible for cleaning.

• Engine jackets are sensitive to hot spotting caused by sludge often encountered in ebullient or steam recovery systems.

Maintenance

• Maintenance costs can be high unless properly monitored.

• Maintenance personnel require extensive training to service both the generation equipment, control equipment, as well as heating and cooling equipment.

Miscellaneous

• The thermal characteristics of available absorption water chillers are poor.

• The actual annual operating efficiencies, of boilers are lower than cataloged efficiencies due to both standby losses and poor part load operation.

• Waste heat cannot be used in mild weather.

• Waste heat from oil coolers cannot be used economically.

• Battery starting equipment is often marginal in life and cranking time. Batteries are generally heat sensitive, causing high deterioration rate when located in warm areas.

• The radiators normally used are subject to both erosion and pressure induced failures.

Energy consumption

Energy consumption of the plant is affected by a wide variety of items such as:

• Fuel to electric efficiency of the prime movers as compared to the particular prime mover and the prime mover load factor.

• The mix of absorption and electric cooling as related to both the plant's recoverable and actually recovered heat, the cooling used by the site, and the percentage of part load vs. part load operating hours.

• The used recovered heat.

• The amount of supplemental heat required for both heating and absorption machine loading.

• The selection and operation of the site should carefully consider both the electrical and thermal energy uses to prevent partial operation of the boilers and absorption machines at low load factors.

79

Government regulations and activities directed toward energy conservation in buildings

A look at what various agencies have done and intend to do for energy conservation.

PAUL R. ACHENBACH,
Chief, Building Environment Div.,
Center for Building Technology,
National Bureau of Standards,
Washington, D.C.

The major near term opportunity for energy conservation in buildings occurs in the existing building inventory. There are approximately 70 million residential buildings and several million commercial and institutional buildings that were constructed before energy conservation became a national need. Despite these impressive numbers, there has been more activity in developing guidelines and criteria for energy conservation in new buildings than in existing ones, perhaps because the technology of retrofitting is less straightforward and the cost effectiveness less certain than for new buildings.

In this chapter, I will concentrate first on identifying the types of documents that

now exist that relate to retrofitting existing buildings and then documents that are under development for both new and existing buildings that may have an impact on regulations, standards, and codes in the future if the energy shortage continues over a period of time. Federal and state government activities related to energy conservation can be classified under several major subheadings—constituting incentives for the building owner or operator to carry out some activity directed toward energy conservation.

- The first classification is *education*. This includes seminars, conferences, publications, etc., that have wide dissemination and can be used as guidelines by building owners and operators.

- The second classification might be called *persuasion*. This includes exhortation by government officials and community leaders. It also includes efforts by salesmen from

This chapter is based on a paper presented by Mr. Achenbach at a special series of two-day seminars on How to Save Energy in Existing Buildings, sponsored and developed by *Heating/Piping/Air Conditioning* magazine.

commercial organizations to influence people to modify existing buildings or to include energy saving components in new buildings.

- Next, there is the broad heading of *regulatory incentives, which includes standards, specifications, regulations, and codes.* This class of incentives ranges from voluntary to mandatory.

- Another classification is *financial incentives.* These include tax breaks, low interest loans, direct subsidy for low income homeowners, and other incentives that involve money and financial support for energy conservation activities.

Publications for existing buildings

There are four documents that have been issued by various departments or agencies of the Federal Government with mandatory provisions for energy conservation in existing buildings. All are related to government buildings or to construction of buildings for the Federal Government. In addition, the Federal Government has enacted four pieces of legislation which include energy conservation requirements for new or existing buildings, whether publicly or privately owned. This latter group is discussed later in this chapter.

The first agency document is the General Services Administration's (GSA) document entitled "Energy Conservation Guidelines for Existing Office Buildings," which was issued February 1975. It was prepared for GSA under the auspices of the American Institute of Architects Research Corp. and is used by the GSA to evaluate office buildings constructed for federal use under GSA supervision. The GSA document contains suggested energy budgets for office buildings as guidelines for designers. The energy is expressed in terms of annual energy use per square foot of floor area at the building boundary, on the one hand; and if desired, a larger number can be used as the energy budget to adjust the energy required for conversion of fuel to electric energy at the power plant.

The GSA document identifies a large number of options for energy conservation

in the choice of the site, design, and operation of buildings; and it is intended for use as a checklist for designers as well as operators of buildings.

The Federal Housing Administration Minimum Property Standards, with which almost everyone is familiar, were originally prepared to apply to new construction. However, there is a companion set of requirements for the rehabilitation of housing. The newest regulations require that for any building for which a loan is desired for rehabilitation, the construction must be upgraded to meet the energy conservation requirements set forth in the companion document.

The Department of Defense (DOD) has some limited and rather simple requirements for energy conservation in buildings constructed for it and for rehabilitation of its existing buildings. One relates to the energy efficiency ratio of air conditioners; this is limited to unitary systems. These must have an energy efficiency ratio of either 8 or $8\frac{1}{2}$ Btu per watt, depending on the operating voltage of the unit. In 1974, when these requirements were first issued, the mandatory efficiency ratios eliminated approximately the lower one-third of the current models of room air conditioners. However, there has been a general improvement and upgrading of energy efficiency ratios for room air conditioners, partly as a result of the voluntary appliance labeling program carried out by the National Bureau of Standards (NBS).

On a more limited basis, NBS has prepared for the Department of Defense a revised federal specification on heat pumps for military family housing that contains requirements for the coefficient of performance and the fraction of the total heating load that must be supplied by the compression system of the heat pump during the heating season. This specification has been in use for about two years and has been employed to purchase several thousand residential heat pump units.

Guidelines for retrofitting

Several documents have been prepared in the last year that contain guidelines for retrofitting housing, commercial buildings, institutional buildings, religious buildings, etc. These, in the form issued, are not mandatory and not directly related to procurement by the agencies of the Federal Government. They constitute informative guidelines to designers, owners, and operators of buildings on how energy might be conserved in existing structures.

First, the Federal Energy Administration (FEA) had, by contract to Dubin-Mindell-Bloome Assoc., prepared a manual in two volumes. They are identified as ECM-1 and ECM-2. Retail, office, and religious structures are emphasized. ECM-1, "Guidelines for Saving Energy in Existing Buildings," is aimed at owners and operators, and ECM-2, "Energy Conservation Opportunities and Guidelines for Existing Buildings," is directed toward engineers, architects, and operators.

ECM-1 is specifically directed toward those modifications that might be made to save energy without the expenditure of appreciable amounts of money. ECM-2, on the other hand, is directed toward conservation options that would involve investment of money and modification of parts of a building or its equipment. These volumes are available from the Federal Energy Administration.

Three other publications have been issued related to energy conservation in housing with specific guidelines that consider the economics of retrofitting and the ability of such modifications to pay back an owner during the operating lifetime of the building. One is a document issued by NBS and identified as Building Science Series 64 entitled "Retrofitting Existing Housing for Energy Conservation: An Economic Analysis." It presents the technical basis for choosing the optimal combination among several retrofit options that will be the best investment of a given amount of money in retrofit activities.

A companion publication, "Making the Most of Your Energy Dollars in Home Heating and Cooling," is a layman's treatment of the economics of retrofitting housing and is written in a form and illustrated in a way

that a homeowner can follow in retrofitting his residence.

Another document, "In the Bank or Up the Chimney," is described as a dollar-and-cents guide to energy saving home improvements. This publication was issued by the Department of Housing and Urban Development (HUD). It is a rather detailed account of how to make modifications to existing housing to save energy. It contains information on the economics of the retrofit of buildings but does not attempt to identify in detail the best combination or the best choice among several retrofit options to optimize the return on the investment, as do the two previous publications.

The National Electric Manufacturers Association and the National Electric Contractors Association have jointly developed a practical handbook on energy conservation and management carrying the short title "Total Energy Management." While endorsed and supported by the FEA, it was produced under private sponsorship. It provides a rather comprehensive set of guidelines for surveying and inspecting an existing building, its mechanical systems and equipment, and for identifying promising opportunities for energy conservation. It also contains a questionnaire that can be used to record the physical and equipment features of a building for an analysis of conservation opportunities. This handbook is intended for use by consulting engineers and architects on behalf of building owners and does provide some limited guidance on how the information can be analyzed to identify the best opportunities in any given building for energy conservation.

Designing new buildings

There are two documents available on the design of energy efficient new buildings. One is "Energy Conservation Design Guidelines for New Office Buildings," a GSA publication. This document will be used by the GSA as a guide and reference document in evaluating design proposals submitted on various contracts for federal buildings. It is not written in mandatory language but has an extensive checklist to determine whether the design team has taken advantage of opportunities for energy conservation in its design.

The second document is ASHRAE Standard 90-75 Energy Conservation in New Building Design by the American Society of Heating, Refrigerating and Air-Conditioning Engineers. The standard is based on an earlier set of criteria produced by the NBS in response to a request from the National Conference of States on Building Codes and Standards (NCSBCS). NCSBCS wanted an interim standard such as ASHRAE Standard 90-75 that might be used for regulatory purposes in the states while awaiting a consensus standard on the subject. The document produced by NBS in February 1974 was transmitted to ASHRAE with the intent that they would process it through the Society's standards generating mechanism and eventually submit it to the American National Standards Institute for promulgation as a national consensus standard. The ASHRAE Standard was issued in October 1975, and has been used for numerous educational seminars for architects, engineers, and regulatory officials, under the sponsorship of ASHRAE. ASHRAE also plans to produce several additional standards for dealing with the retrofit of various types of existing buildings for energy conservation.

Standards for mobile homes

When the NBS criteria for energy conservation were released to NCSBCS in February 1974, the National Fire Protection Association (NFPA) as sponsors of the American National Standard (ANSI A119.1) for Mobile Homes was asked to prepare an addendum on energy conservation to that Standard. A task group of NFPA has prepared design requirements for energy conservation and a manual for calculating heating and cooling loads for mobile homes. In the meantime, the Federal Government enacted legislation in 1974 directing HUD to promulgate standards covering all newly constructed mobile homes. The HUD Mobile Home Construction and Safety Standard includes provisions for energy conservation based largely on the

NFPA work, and became effective in June 1976.

Three energy saving laws

In addition, the Federal Government has enacted three major pieces of legislation related to energy conservation in various types of new and existing buildings. These are:

1. Public Law 93-409, Solar Heating and Cooling Demonstration Act of 1974.
2. Public Law 94-163, Energy Policy and Conservation Act, December 1975.
3. Public Law 94-385, Energy Conservation and Production Act, August 1976.

The Solar Heating and Cooling Demonstration Act provides for an extensive demonstration of solar heating and cooling technology in residential and commercial buildings over a 3 to 5 year period, and authorizes research programs to improve the efficiency of solar cooling processes. It also provides for the drafting of procurement criteria and testing procedures for purchase and evaluation of demonstration systems.

The Energy Policy and Conservation Act of 1975 provides for:

1. A voluntary program of assistance to the States for the implementation of State energy conservation plans.
2. Federal financial assistance contingent on mandatory thermal and lighting efficiency standards and insulation requirements for new and renovated buildings in their State conservation plans.
3. Technical assistance in developing model State laws.
4. A directive to the President to develop a ten-year energy conservation plan including mandatory thermal efficiency standards for all Federal buildings.
5. Labeling most household appliances for energy performance and annual energy cost, and development of test procedures in support of such labeling.

The Energy Conservation and Production Act of 1976 provides for:

1. The development of energy performance standards for residential and commercial buildings in three years, with promulgation six months later, and a final effective date one year after promulgation. Compliance with these standards is mandatory for Federal buildings. For other residential and commercial buildings, activation of the mandatory compliance provisions requires the approval of both houses of Congress.
2. Monitoring the progress of States and local jurisdiction in adopting and enforcing energy conservation standards, with reports every six months to Congress and the President.
3. Development of energy efficiency improvement targets for most of the simpler appliances by November 1976, and for the more complex ones by August 1977.
4. A demonstration program, with some subsidy, to stimulate the use of systems employing renewable-resource energy measures in existing dwelling units.
5. A weatherization program to provide assistance, upon application, to low-income families in adding the most essential energy conservation measures to their homes.

Model energy conservation codes

In the last three years, the States have been active in developing enabling legislation for energy conservation regulations and in formulation of energy conservation criteria for building code purposes. Several of the United States have adopted some type of energy regulations and many others have such regulations under active consideration. Under a contract between ERDA and NCSBCS, the model building code organizations, Building Officials and Code Administrators of America (BOCA), the International Conference of Building Officials (ICBO), and The Southern Building Code Congress (SBCC), are developing a Model Energy Conservation Code based on the technical provisions of ASHRAE Standard 90-75. The steps toward implementation of energy conservation legislation in several States will probably be related to the actions taken by the model code groups because many individual States are using earlier issues of the model codes. Figure 79-1 shows various degrees of authority for regulating building

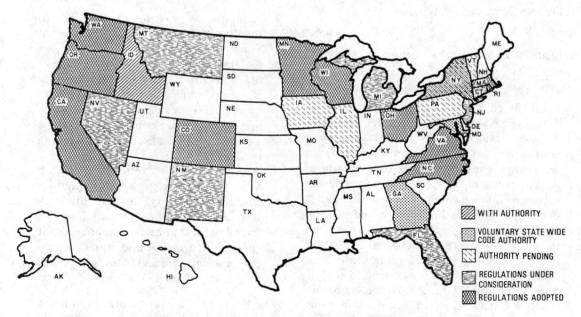

WITH AUTHORITY

VOLUNTARY STATE WIDE CODE AUTHORITY

AUTHORITY PENDING

REGULATIONS UNDER CONSIDERATION

REGULATIONS ADOPTED

Fig. 79-1. Map shows various degrees of authority for regulating building energy conservation for each state as of November 25, 1975.

energy conservation for each state as of November 25, 1975.

A significant related problem is the need for implementation procedures that would make ASHRAE Standard 90-75 effective in actual construction. Aside from the enabling legislation to incorporate the industry standard into state regulations, there is the need to develop means by which designs for specific buildings submitted by builders, architects, engineers, etc., can be evaluated for compliance with the standard. Most states do not have adequate staffs to do this within the governmental structure. There will be a need to train architect-engineering groups and others in the interpretation required to evaluate not only the more or less conventional or component performance designs but innovative designs that are provided for in Standard 90-75. The establishment of curricula for such training courses, the carrying out of training, and the use of such trained staff in the various states to evaluate design proposals will require a major effort, such as the one initiated by NCSBCS. This will undoubtedly require a good deal of attention both from Federal and State Governmental organizations if the standard is to be effective in actual practice. These comments are also applicable to the implementation of performance energy conservation standards to be developed under the Energy Conservation and Production Act, P.L. 94-385.

Retrofitting can save energy

Quite a number of building owners, both private and government, have carried out energy conservation modifications in recent years. These have generally revealed that up to 20 percent of the energy used in existing buildings can be saved by improving and modifying the methods of operation and control of existing equipment. In other words, this amount of savings can be obtained without investment of additional money to accomplish conservation. There is also strong indication that perhaps another 10 to 20 or up to 40 percent of the energy used in conventional construction in the past can be saved by retrofit or modification of the building and its equipment.

NBS is in the process of carrying out one study of this kind on its facilities at Gaithersburg, Maryland. This facility comprises 21 laboratories, most of which require both temperature and humidity control for performing standard test procedures or for calibrating equipment against existing standard environmental conditions. Because of these requirements, relaxation of humidity control can be effected in only certain buildings. Through reductions in lighting levels by approximately 33 percent during the past year, the shutdown of several buildings at night and weekends, and the reduction of winter room temperatures to 72°F, NBS has already accomplished about 11 percent saving in electricity and about 18 percent in heating fuel. Figure 79-2 illustrates the reduction in electrical consumption at the NBS laboratories as a result of energy conservation methods implemented over a three year period.

In addition to this short term conservation program on the Bureau's campus, a longer range study program was initiated to look for additional energy conservation potential. The topics that were considered to merit additional detailed study were: evaluation of window loads, optimization of year around inside temperature, reduction in ventilation rates, evaluation of shutdown of HVAC units, optimization of HVAC control, reduction in the number of exhaust hoods, optimization of chiller operation, and evaluation of the cost effectiveness of exhaust recovery units.

A number of these have already been evaluated using a simulation model and a computer program for predicting reduction in energy use by implementing the various options. It appears that the window loads would be reduced approximately 10 percent if all of the glass area in the various buildings on the campus were fitted with storm windows. However, it appeared that the amount of reduction in energy use would not repay the cost of the retrofitting job in a reasonable time and therefore is not very attractive from an economic viewpoint.

Although the indoor temperature during the winter was reduced to 68°F, in accordance with the GSA recommendation for federal

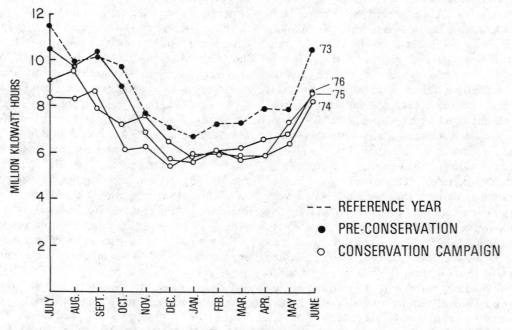

Fig. 79-2. Dashed line represents electrical consumption at the NBS Laboratory complex prior to initiating energy conservation measures which is the reference year, '73. Conservation campaign, shown by solid lines for '74, '75, and '76, indicates how electrical consumption dropped as steps were implemented.

buildings, an analysis indicated that this was not the optimal setting if one must choose a single setting for year around operations. In order to operate at different temperatures winter and summer, approximately 2000 thermostats would have to be reset twice a year—a task requiring approximately 700 man-hours. The computer analysis indicated that a 72° F year around temperature was an optimal setting when both energy use and year around comfort were considered. Figures 79-3 and 79-4 illustrate the reduction of heating fuel use at the NBS laboratories as a result of energy conservation measures being implemented for three years.

Reheat systems provided temperature and humidity control to the Bureau laboratories. This influenced the optimization of the year around temperature because a lower dry bulb setting for winter would have been more economical in the use of energy. But, 68° F would hardly be acceptable from a comfort viewpoint in summer. On the other hand, the higher the dry bulb temperature is in summer, the more reheat energy is required in any space where humidity control is necessary.

Reduce ventilation, lighting

The ventilation rate could be reduced by about 20 percent in the office space that constitutes the peripheral rooms of the various office buildings as a result of the reduction in lighting since the air circulation rate was largely determined by the amount of energy that had to be picked up from lighting fixtures in the summertime. Reductions in these two areas would constitute a significant opportunity for energy savings.

Analysis also indicated that considerable additional energy could be saved if many of the systems could be shut down overnight and on weekends. Again, this would require a

HEATING FUEL USE FROM JULY 1972 TO JUNE 1976 (NBS, GAITHERSBURG)

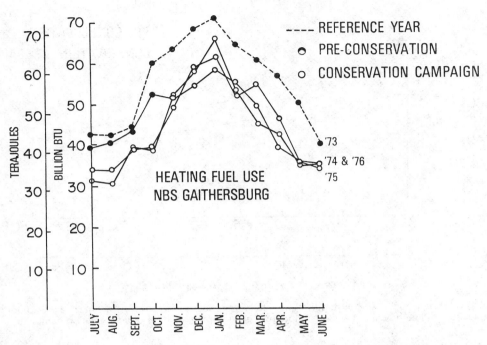

Fig. 79-3. Fuel consumption at the NBS Laboratory complex prior to implementation of energy conservation measures (dashed line) which is the reference year, '73. Conservation campaign, shown by solid lines for '74, '75, and '76, indicates how fuel consumption dropped as conservation steps were implemented.

large amount of operator's time if done manually. It would also be difficult to provide for special schedules of operation when certain laboratory staff wanted to work during the evening or some other schedule other than the 8:30 AM to 5 PM normal working day. Therefore, on a pilot basis, an automatic control system is being installed in one of the major general purpose buildings as a trial to determine what the potential for energy conservation is by automatically programming the various HVAC units for nighttime and weekend shutdown. If the results indicate substantial energy savings, the system will be extended to the entire group of buildings on campus. It is anticipated that this will produce significantly larger energy savings than those already cited.

Many of the laboratories contain exhaust hoods to handle chemical fumes and other undesirable products from various experi-

ments. The hoods are usually located in interior units and are supplied in part by the exhaust air from the offices around the periphery. Therefore, the interior and exterior modules are interdependent, indicating that one cannot arbitrarily shut off some hoods and leave others operating. Another study is being made to determine whether the laboratory activities that must have exhaust hoods can be grouped and served by one or more selected HVAC systems in the attic of a given building. The others could then be shut down overnight without disturbing the overall ventilation pattern of the office and laboratory module.

An analysis was made of the potential benefit of installing heat recovery units between incoming ventilation air and exhaust ventilation air to recover some of the thermal energy from the exhaust air. It was determined, however, that the cost and

HEATING FUEL USE vs OUTDOOR TEMPERATURE
FROM JULY 1971 TO APRIL 1976 (NBS, GAITHERSBURG)

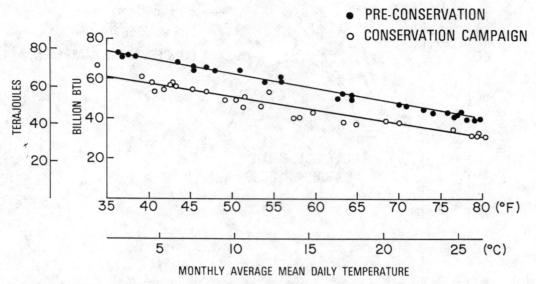

Fig. 79-4. Heating fuel use vs outdoor temperature from July 1971 to April 1976 at the NBS Laboratory complex. Solid line indicates fuel consumption prior to implementation of conservation campaign.

complexity of making these installations in the existing systems were such that the cost would not be recovered by the energy saved, and the idea was abandoned.

Modify refrigeration machines

Air conditioning for the NBS laboratory complex is furnished by four centrifugal compression refrigerating units of 3000 nominal tons each. Laboratory activities require some air conditioning year around, but the load is usually less than the capacity of one 3000 ton unit during several winter months. Several schemes for modifying the refrigerating machines were investigated. It was found that hot gas bypass control modifications, compressor speed reduction, and installing two-speed drives for some machines not only save significant amounts of energy but are also cost-effective.

The anticipated effects of some of these major modifications are summarized in Table 79-1. It indicates that while the adjustment of interior temperature makes a significant difference in heating energy, the reduction

in ventilation air rate by 20 percent would result in a major savings in both winter and summer. The greatest savings can be achieved by shutting off various systems at night and over the weekend. This is being tested by the pilot installation, which became operational in the Fall of 1976.

Other NBS activities for saving

Retrofit activities being carried out at NBS include a demonstration of the energy savings possible in a typical frame house built about 20 years ago by:

• Reducing infiltration through caulking and weather stripping.

• Installing storm windows and adding insulation in the attic.

• Incorporating insulation in the cavity walls.

These studies show that insulating the cavity walls and the additional ceiling insulation reduced the energy required for heating by about 58 percent. These retrofit measures had a very small effect on the summer cooling load, possibly because the daily average in-

Table 79-1.   Calculated effects of various conservation measures on annual heating and electric (chiller) energy consumptions.

	Annual heating energy			Annual chiller's electric energy	
	10^9 Btu	10^6 kWh	% 1972	10^6 kWh	% 1972
Reference year, 1972	626	183	100	35.6	100
Insulating glass	559	164	95	32.0	90
Temperature set to 78° F (interior)	714	209	114	33.3	94
Temperature set to 72° F (interior)	548	160	87	35.2	99
20 percent air reduction	528	155	84	30.5	86
Unoccupied shutdown	215	63	34	14.4	40

door and outdoor temperatures were nearly equal in this climate. The caulking and weather stripping produced no measurable reduction in air leakage, perhaps because the house was relatively tight to begin with.

In addition, the Bureau is collaborating with the New York Board of Education in studying energy use in secondary school buildings. As a background for the design of new buildings, data were collected in a number of existing school buildings to serve as a base for comparison. Data showed that the ventilation rate could be substantially reduced in school buildings without developing noticeable occupant discomfort and without creating respiration CO_2 levels beyond those permitted by safety and health considerations. The survey of existing schools showed that 75 percent of the thermal energy consumed in a typical school went for heating ventilation air and that about 80 percent of the electrical energy was consumed by the lighting system. Therefore, these constitute the most promising areas for energy conservation in new school building design.

Study oil burner efficiencies

The Bureau has conducted under contract a survey of various types of oil burners in the New England area to gather factual information about potential savings that might accrue from regular servicing of the burner and adjustment of the air/fuel ratio for efficient combustion. This was done by surveying several hundred existing installations; collecting information on combustion efficiency before adjustment, carrying out a typical service operation, and immediately taking a new reading of combustion effi-

ciency; and then, several months later, retracing these steps and taking new data on combustion efficiency.

This study has revealed that the deterioration of performance from a combustion efficiency standpoint was not large over a three or four month period and that it did not appear to constitute a major opportunity for energy savings. However, it did show that in many installations, the capacity of the heating plant was 200 to 400 percent in excess of that needed. This oversizing in itself represents a substantial opportunity for conservation. Consideration of what can be done about improving efficiency through better sizing of equipment to load is the subject of the second year's effort in the study. Modifications will include installation of smaller burners in some cases, and smaller boilers in other situations.

NBS has done a study on modular boilers where several boilers of equal capacity are used to make up the total capacity needed to service a load rather than a single large boiler. This concept permits operation of one or more of the small boilers in a sequence to more nearly match the load, under varying weather conditions. The experiment in the laboratory indicated that the modularization concept could save approximately 10 percent of the energy when applied to a typical winter cycle of weather and heating load.

References

"In the Bank or Up the Chimney?" Department of Housing and Urban Development, Superintendent of Documents, U.S. Government Printing Office, Washington, DC 20402. Price: $1.70. Stock No. 023-000-00297-3.

"Making the Most of Your Energy Dollars in Home

Heating and Cooling," National Bureau of Standards, Superintendent of Documents, U.S. Government Printing Office, Washington, DC 20402. Price: 70 cents. No. C13.53:8.

"Retrofitting Existing Housing for Energy Conservation: An Economic Analysis," National Bureau of Standards in cooperation with Federal Energy Administration, Superintendent of Documents, U.S. Government Printing Office, Washington, DC 20402. Price: $1.35. No. C13.29:2/64.

"Energy Conservation Guidelines for Existing Office Buildings," and "Energy Conservation Design Guidelines for New Office Buildings," General Service Administration, Price: $2 each. Copies can be obtained from GSA Business Service Centers in the following cities: Boston, Mass.; New York, N.Y.; Washington, D.C.; Philadelphia, Pa.; Atlanta, Ga.; Chicago, Ill.; Kansas City, Mo;

Fort Worth, Tex.; Denver, Colo.; San Francisco, Calif.; and Seattle, Wash.

"Guidelines for Saving Energy in Existing Buildings," Buildings Owners and Operators Manual ECM 1; and "Energy Conservation Opportunities and Guidelines for Existing Buildings," Engineers, Architects, and Operators Manual ECM 2, available from the Federal Energy Administration.

"Total Energy Management," National Electrical Manufacturers Association and the National Electrical Contractors Association. Price: single copies, free; multiple orders, $1 per book. NEMA, Project Daybreak, 155 E. 44th St., New York, NY 10017.

ASHRAE Standard 90-75, Energy Conservation in New Building Design, ASHRAE, 345 E. 47th St., New York, NY 10017.

SECTION X

Summary

The list of 133 ways to save energy in new and existing buildings as outlined in the summary was compiled from the information presented in the previous 79 chapters. It must be pointed out that not all of the 133 items are applicable to all buildings. Each situation should be examined individually as in every engineering problem before applying the solution.

133 ways to save energy in new and existing buildings

Structure

1. Add additional insulation to roofs, ceilings, or walls where practical.
2. When reroofing, use light colored material to reduce solar gain on air conditioned structures.
3. Ventilate attic spaces.
4. Put solar film on windows to cut cooling loads.
5. Install solar screens on windows to reduce cooling loads.
6. Install double glazing in place of single glazing.
7. Recaulk window and door frames.
8. Reseal curtain walls.
9. Eliminate excessive crackage between double entry doors.
10. Install weather stripping around windows and doors.
11. Repair broken windows.
12. Keep garage and warehouse doors closed as much as possible.

Lighting and power

1. Install higher efficiency lighting systems where possible.
2. Reduce overall illumination levels.
3. Implement a lighting maintenance program to obtain maximum efficiency from existing systems.
4. Use supplemental lighting for specific tasks instead of increasing the overall illumination for a given area.
5. Utilize natural lighting in perimeter office spaces.
6. Utilize multiple switching for selective lighting levels in offices, conference rooms, etc.
7. Reduce lighting in areas not requiring higher levels: stock rooms, corridors, etc.
8. When redecorating, use light colors on ceilings and walls to achieve good illumination levels with less lighting.
9. Reduce decorative and advertising lighting.
10. Use timers or photocells to control outdoor lighting.
11. Reduce parking lot lighting to minimum levels required for safety.
12. Use properly sized motors. Grossly oversized motors operate at a low power factor.
13. Apply power factor correction where applicable.
14. Install demand limiting equipment.

Controls

1. Recalibrate all controls.
2. Lock thermostats to prevent resetting by unauthorized personnel.
3. Check room thermostats for proper location—not on cold walls, in drafts, receiving direct lighting radiation.
4. Install individual room control whenever possible.
5. Install temperature control valves (self-contained) in radiators rather than controlling by use of hand valves.
6. Install enthalpy control to optimize use of outdoor air for building cooling.
7. Install building automation system if feasible.

HVAC and miscellaneous

1. Study system carefully before making changes—some changes may increase energy usage.

2. Retest, balance, and adjust systems.
3. Turn off air conditioning machinery during unoccupied hours.
4. Revise cleanup schedule so lights and system can be turned off earlier.
5. Optimize system startup times.
6. Shut off outdoor air intakes during unoccupied hours.
7. Reduce outdoor air quantity for ventilation.
8. Reduce system air volume.
9. Reduce air duct leakage.
10. Adjust outdoor air dampers for tight closure.
11. Replace dampers with higher quality ones whenever possible.
12. When balancing or rebalancing a system, consider outdoor air leakage when making minimum outdoor air settings.
13. Adjust dampers in mixing boxes and multizone units so that they shut off tight to reduce leakage.
14. Avoid use of preheat coils if possible.
15. Raise the mixed air temperature.
16. Reset hot and cold deck temperatures in direction of reduced heating and cooling.
17. Set reheat schedule as low as can be tolerated.
18. Reset chilled water and heating water temperature in accordance with loads.
19. Use lowest possible radiation temperature possible in perimeter spaces.
20. Do not permit perimeter and interior systems to buck one another.
21. Optimize multiple chiller operation.
22. Run heating and cooling system auxiliaries only when they are required.
23. Whenever possible, only operate return air fans for heating during unoccupied hours.
24. Install auxiliary air risers to reduce fan horsepower.
25. Convert constant volume fan system to Variable Air Volume.
26. Reduce heating in unoccupied areas.
27. Reduce heating in overheated spaces. Do not open the window to cool these areas!
28. Shut off exhaust fans during unoccupied cycles.

29. Check exhaust systems to ensure they are exhausting only the amount of air required.
30. Reduce exhaust air quantities from toilet rooms, laboratories, etc. when feasible.
31. Convert toilet room exhaust fans to operate only when room is occupied.
32. Reduce supply temperature of domestic hot water systems.
33. Use condenser water to preheat domestic hot water.
34. Use condenser water for air conditioning reheat.
35. Retrofit solar collectors to the building to preheat domestic and process water.
36. Install heat recovery devices to reclaim heat from building, kitchen, and process exhaust.
37. Replace forced air heaters with infrared heaters.
38. Replace indirect fired makeup air units with direct fired equipment.
39. Insulate piping and duct work located in unconditioned spaces.
40. Replace worn insulation on boilers, furnaces, pipes, ducts, etc.
41. Reduce hot and chilled water flow rates.
42. Trim pump impellers to match load.
43. Convert three-way valves to two-way operation and install variable speed pumping.
44. Clean strainer screens in pumping systems.
45. Check system vents in hot water and steam systems for proper performance.
46. Check expansion tank size. Undersized tanks can cause excessive water loss.
47. Determine whether the boiler plant can be shut down for the summer and small boilers and water heaters used during this period.
48. Do not waste condensate; return it to the boiler.
49. When high pressure steam is available, consider use of steam turbines for pump and fan drives. Turbines can operate as a pressure relief valve to supply low pressure steam needs.
50. Repair all leaks; water, steam, gas, compressed air, etc.

51. Use proper water treatment to reduce fouling of transfer surfaces in boilers, chillers, heat exchangers.
52. Use no more water treatment chemicals than necessary.
53. Check cooling tower bleed-off periodically to ensure that water and chemicals are not wasted.
54. Maintain cooling towers, evaporative coolers, and air cooled condensers to maintain peak efficiency.
55. Periodically inspect and repair faulty equipment: steam traps, float valves, valves, etc.
56. Implement a filter maintenance program to ensure peak efficiency.
57. Clean and maintain cooking equipment to maintain peak efficiency.

Combustion equipment

1. Check building for negative pressure, which can reduce combustion efficiency.
2. Check flues and chimneys for blockages or improper draft conditions.
3. Clean combustion surfaces.
4. Check and adjust fuel-air ratios.
5. Replace atmospheric burners with powerburners.
6. Install pressure controls on furnaces (industrial).
7. Install automatic air-gas combustion controls.
8. Do not overfire equipment.
9. Repair furnace linings frequently.
10. Reduce production equipment preheat times to minimum required.
11. Reduce production equipment (furnaces, ovens, etc.) temperatures to holding temperatures when production stops for relatively long periods.
12. Shut off drying and curing ovens when not in use; do not start too soon prior to the start of the shift.
13. Seal all cracks in furnaces, ovens, etc.
14. Preheat combustion air with waste heat.

Industrial plants

1. Study plant heating and air conditioning systems to determine if they are of correct design.

2. Reschedule operations whenever possible to second and third shift to get them off of the 10:00 AM to 2:00 PM peak electric demand period.
3. Plan work so that the whole plant can be shut down on given weekends.
4. Shut off machinery when not in use: lunch, etc.
5. Keep covers on tanks and vats closed to reduce evaporation losses.
6. Use push-pull ventilation on open surface tanks; 50 percent or more of the air can be saved.
7. Use immersion heaters whenever possible.
8. Use cold water detergent in washers whenever possible.
9. Combine operations where possible to reduce the number of washers.
10. Shroud openings of furnaces, ovens, paint booths, and washers so that only the minimum amount of exhaust air will be required.
11. Eliminate stratification of air in the plant during the winter thereby warming the floor. This can easily be done with fans blowing down ducts terminating close to the floor.
12. Use spot heating or cooling of people when they are located far apart. Each should have control of the air direction and velocity over them.
13. Use evaporative cooling for human cooling whenever practical.
14. Determine whether pressure blowers could replace some compressed air usage.
15. Do not use compressed air at higher pressures than required.
16. Do not permit compressed air to be used for "people" cooling.
17. Reduce the quantity of exhaust air; use local exhaust, not general.
18. Use low volume, high velocity exhaust systems whenever possible, such as: ventilated welding guns, hoods for portable grinding equipment, local traveling hoods for molten metal pouring.
19. Analyze all solid waste to determine

whether it can be recycled, burned, or composted.

20. Salvage all oil used in the plant. It can either be reused by refining it or burned in the boilers.

21. If exhaust air is contaminated, evaluate air cleaning devices to determine if the air could be cleaned and recycled.

22. Determine feasibility of utilizing energy in production operations before exhausting it.

23. Analyze interplant truck runs. Consolidate loads and eliminate trips.

24. Shut off interplant truck engines when not in use.

25. Shut off fork lift truck engines when not in use.

26. Replace worn out machinery with modern, efficient equipment.

27. Keep heat and smoke relief vents closed during the winter.

28. Consider using waste water for roof sprays during the summer to reduce heat load on the plant.

29. Use automatic regulators to control the volume of water used.

Index